Fundamentals of
Renewable Energy Processes

Fundamentals of
Renewable Energy Processes

Aldo Vieira da Rosa
Stanford University

ELSEVIER
ACADEMIC
PRESS

Amsterdam • Boston • Heidelberg • London • New York • Oxford
Paris • San Diego • San Francisco • Singapore • Sydney • Tokyo

Elsevier Academic Press
30 Corporate Drive, Suite 400, Burlington, MA 01803, USA
525 B Street, Suite 1900, San Diego, California 92101–4495, USA
84 Theobald's Road, London WC1X 8RR, UK

This book is printed on acid-free paper. $\otimes$

Library of Congress Cataloging-in-Publication Data
Application Submitted

British Library Cataloguing in Publication Data
A catalogue record for this book is available from the British Library

ISBN -13: 978-0-12-088510-7
ISBN -10: 0-12-088510-7

For all information on all Elsevier Academic Press publications
visit our Web site at www.books.elsevier.com

Printed in the United States of America
07 08 09 10 9 8 7 6 5 4 3

Table of Contents

	Page Chapter	Book
Foreword		xv
Acknowledgments		xvii

Chapter 1   Generalities

	Chapter	Book
1.1 Units and Constants	1.1	1
1.2 Energy and Utility	1.2	2
1.3 Conservation of Energy	1.3	3
1.4 Planetary Energy Balance	1.4	4
1.5 The Energy Utilization Rate	1.5	5
1.6 The Population Explosion	1.8	8
1.7 The Market Penetration Function	1.9	9
1.8 Planetary Energy Resources	1.13	13
1.9 Energy Utilization	1.16	16
1.10 The Ecology Question	1.19	19
1.10.1 Biological	1.20	20
1.10.2 Mineral	1.20	20
1.10.3 Subterranean	1.21	21
1.10.4 Undersea	1.22	22
1.11 Nuclear Energy	1.22	22
1.11.1 Fission	1.25	25
1.11.2 Fusion	1.27	27
1.11.3 Cold Fusion	1.31	31
1.12 Financing	1.36	36
References	1.39	39
Problems	1.41	41

Part I   Heat Engines

Chapter 2   A Minimum of Thermodynamics and of Kinetic Theory of Gases

	Chapter	Book
2.1 The Motion of Molecules	2.1	53
2.2 Temperature	2.1	53
2.3 The Perfect-Gas Law	2.2	54
2.4 Internal Energy	2.3	55
2.5 Specific Heat at Constant Volume	2.3	55
2.6 The First Law of Thermodynamics	2.4	56
2.7 The Pressure-Volume Work	2.5	57

2.8	Specific Heat at Constant Pressure	2.5	57
2.9	Adiabatic Processes	2.6	58
2.9.1	Abrupt Compression	2.6	58
2.9.2	Gradual Compression	2.10	62
2.9.3	p-V diagrams	2.11	63
2.9.4	Polytropic Law	2.12	64
2.10	Isothermal Processes	2.13	65
2.11	Functions of State	2.15	67
2.12	Enthalpy	2.16	68
2.13	Degrees of Freedom	2.17	69
2.14	Entropy	2.19	71
2.14.1	Changes in Entropy	2.20	72
2.15	Reversibility	2.21	73
2.15.1	Causes of Irreversibility	2.23	75
2.15.1.1	Friction	2.23	75
2.15.1.2	Heat Transfer Across Temperature Differences	2.23	75
2.15.1.3	Unrestrained Compression or Expansion of a Gas	2.24	76
2.16	Negentropy	2.24	76
2.17	How to Plot Statistics	2.25	77
2.18	Maxwellian Distribution	2.26	78
2.19	Fermi-Dirac Distribution	2.29	81
2.20	Boltzmann's Law	2.31	83
	Appendix (Symbology)	2.33	85
	Problems	2.34	86

Chapter 3 Mechanical Heat Engines

3.1	Heats of Combustion	3.1	89
3.2	Carnot Efficiency	3.4	92
3.3	Engine Types	3.5	93
3.4	Efficiency of an Otto Engine	3.9	97
3.5	Gasoline	3.13	101
3.5.1	Heat of Combustion	3.13	101
3.5.2	Antiknock Characteristics	3.13	101
3.6	Knocking	3.13	101
3.7	Hybrid Engines for Automobiles	3.17	105
3.8	The Stirling Engine	3.18	106
3.9	The Implementation of the Stirling Engine	3.23	111
3.10	Cryogenic Engines	3.25	113
	References	3.28	116
	Problems	3.29	117

Chapter 4   Ocean Thermal Energy Converters

4.1	Introduction	4.1	125
4.2	OTEC Configurations	4.2	126
4.3	Turbines	4.4	128
4.4	OTEC Efficiency	4.6	130
4.5	Example of OTEC Design	4.7	131
4.6	Heat Exchangers	4.9	133
4.7	Siting	4.10	134
	References	4.11	135
	Problems	4.12	136

Chapter 5   Thermoelectricity

5.1	Experimental Observations	5.1	139
5.2	Thermoelectric Thermometers	5.6	144
5.3	The Thermoelectric Generator	5.8	146
5.4	Figure of Merit of a Material	5.11	149
5.5	The Wiedemann-Franz-Lorenz Law	5.12	150
5.6	Thermal Conductivity in Solids	5.16	154
5.7	Seebeck Coefficient of Semiconductors	5.17	155
5.8	Performance of Thermoelectric Materials	5.18	156
5.9	Some Applications of Thermoelectric Generators	5.20	158
5.10	Design of a Thermoelectric Generator	5.22	160
5.11	Thermoelectric Refrigerators and Heat Pumps	5.25	163
	5.11.1 Design Using an Existing Thermocouple	5.25	163
	5.11.2 Design Based on Given Semiconductors	5.29	167
5.12	Temperature Dependence	5.32	170
5.13	Battery Architecture	5.33	171
5.14	The Physics of Thermoelectricity	5.33	171
	5.14.1 The Seebeck Effect	5.34	172
	5.14.2 The Peltier Effect	5.37	175
	5.14.3 The Thomson Effect	5.38	176
	5.14.4 Kelvin's Relations	5.39	177
5.15	Direction and Signs	5.43	181
	Appendix	5.45	183
	References	5.46	184
	Problems	5.47	185

Chapter 6   Thermionics

6.1	Introduction	6.1	199
6.2	Thermionic Emission	6.3	201
6.3	Electron Transport	6.6	204
	6.3.1 The Child-Langmuir Law	6.8	206

6.4   Lossless Diodes with Space Charge Neutralization   6.12   210
    6.4.1   Interelectrode Potentials   6.12   210
    6.4.2 V-J Characteristics   6.14   212
    6.4.3   The Open-Circuit Voltage   6.14   212
    6.4.4   Maximum Power Output   6.15   213
6.5   Losses in Vacuum Diodes with No Space Charge   6.16   214
    6.5.1   Efficiency   6.16   214
    6.5.2   Radiation Losses   6.17   215
        6.5.2.1   Radiation of Heat   6.17   215
        6.5.2.2   Efficiency with Radiation Losses Only   6.19   217
    6.5.3   Excess Electron Energy   6.21   219
    6.5.4   Heat Conduction   6.22   220
    6.5.5   Lead Resistance   6.22   220
6.6   Real Vacuum Diodes   6.22   220
6.7   Vapor Diodes   6.23   221
    6.7.1   Cesium Adsorption   6.24   222
    6.7.2   Contact Ionization   6.27   225
    6.7.3   Thermionic Ion Emission   6.28   226
    6.7.4   Space Charge Neutralization Conditions   6.29   237
    6.7.5 V-J Characteristics   6.30   228
6.8   High-Pressure Diodes   6.34   232
    References   6.37   235
    Problems   6.38   236

Chapter 7   AMTEC

7.1   Operating Principle   7.1   241
7.2   Vapor Pressure   7.3   243
7.3   Pressure Drop in the Sodium Vapor Column   7.4   244
7.4   Mean Free Path of Sodium Ions   7.6   246
7.5 V-I Characteristics of an AMTEC   7.7   247
7.6   Efficiency   7.9   249
7.7   Thermodynamics of an AMTEC   7.12   252
    References   7.14   254

Chapter 8   Radio-Noise Generators

    References   8.5   259

Part II   The World of Hydrogen

Chapter 9   Fuel Cells

9.1   Introduction   9.1   263
9.2   Electrochemical Cells   9.2   264

9.3   Fuel Cell Classification   9.6   268
  9.3.1   Temperature of Operation   9.6   268
  9.3.2   State of the Electrolyte   9.7   269
  9.3.3   Type of Fuel   9.7   269
  9.3.4   Chemical Nature of the Electrolyte   9.8   270
9.4   Fuel Cell Reactions   9.8   270
  9.4.1   Alkaline Electrolytes   9.9   271
  9.4.2   Acid Electrolytes   9.9   271
  9.4.3   Molten Carbonate Electrolytes   9.10   272
  9.4.4   Ceramic Electrolytes   9.10   272
  9.4.5   Methanol Fuel Cells   9.10   272
9.5   Typical Fuel Cell Configurations   9.12   274
  9.5.1   Demonstration Fuel Cell (KOH)   9.12   274
  9.5.2   Phosphoric Acid Fuel Cells (PAFC)   9.13   275
    9.5.2.1   A Fuel Cell Battery (Engelhard)   9.13   275
    9.5.2.2   First-Generation Fuel Cell Power Plant   9.14   276
  9.5.3   Molten Carbonate Fuel Cells (MCFC)   9.15   277
  9.5.4   Ceramic Fuel Cells (SOFC)   9.16   278
    9.5.4.1   High Temperature Ceramic Fuel Cells   9.21   283
    9.5.4.2   Low Temperature Ceramic Fuel Cells   9.23   285
  9.5.5   Solid-Polymer Electrolyte Fuel Cells   9.25   287
    9.5.5.1   Cell Construction   9.26   288
    9.5.5.2   Membrane   9.28   290
    9.5.5.3   Catalysts   9.29   291
    9.5.5.4   Water Management   9.30   292
  9.5.6   Direct Methanol Fuel Cells   9.31   293
  9.5.7   Solid Acid Fuel Cells   9.33   295
  9.5.8   Rechargeable Fuel Cells (NiMH)   9.34   296
  9.5.9   Metallic Fuel Cells—Zinc-Air Fuel Cells   9.36   298
9.6   Fuel Cell Applications   9.37   299
  9.6.1   Stationary Power Plants   9.38   300
  9.6.2   Automotive Power Plants   9.38   301
  9.6.3   Other Applications   9.40   302
9.7   The Thermodynamics of Fuel Cells   9.41   303
  9.7.1   Heat of Combustion   9.41   303
  9.7.2   Free Energy   9.43   305
  9.7.3   Efficiency of Reversible Fuel Cells   9.46   308
  9.7.4   Effects of Pressure and Temperature on the
          Enthalpy and Free Energy Changes of a Reaction   9.47   309
    9.7.4.1   Enthalpy Dependence on Temperature   9.47   309
    9.7.4.2   Enthalpy Dependence on Pressure   9.49   311
    9.7.4.3   Free Energy Dependence on Temperature   9.50   312
    9.7.4.4   Free Energy Dependence on Pressure   9.54   316
    9.7.4.5   Voltage Dependence on Temperature   9.56   318
9.8   Performance of Real Fuel Cells   9.57   319

9.8.1   Current Delivered by a Fuel Cell   9.57   319
9.8.2   Efficiency of Practical Fuel Cells   9.58   320
9.8.3   Characteristics of Fuel Cells   9.59   321
    9.8.3.1   Scaling Fuel Cells   9.62   324
9.8.4   More Complete V-I Characteristics
         of Fuel Cells   9.63   325
9.8.5   Heat Dissipation by Fuel Cells   9.71   333
    9.8.5.1   Heat Removal from Fuel Cells   9.73   335
         References   9.74   336
         Problems   9.76   338

Chapter 10   Hydrogen Production

10.1   Generalities   10.1   353
10.2   Chemical Production of Hydrogen   10.3   355
    10.2.1   Historical   10.3   355
    10.2.2   Modern Production   10.4   356
        10.2.2.1   Partial Oxidation   10.4   356
        10.2.2.2   Steam Reforming   10.5   357
        10.2.2.3   Thermal Decomposition   10.5   357
        10.2.2.4   Syngas   10.6   358
        10.2.2.5   Shift Reaction   10.6   358
        10.2.2.6   Methanation   10.7   359
        10.2.2.7   Methanol   10.7   359
        10.2.2.8   Syncrude   10.8   360
    10.2.3   Hydrogen Purification   10.8   360
        10.2.3.1   Desulfurization   10.8   360
        10.2.3.2 CO_2 Removal   10.8   360
        10.2.2.3   CO Removal
                    and Hydrogen Extraction   10.9   361
    10.2.4   Hydrogen Production Plants   10.11   363
        10.2.4.1   Compact Fuel Processors   10.11   363
10.3   Electrolytic Hydrogen   10.16   368
    10.3.1   Introduction   10.16   368
    10.3.2   Electrolyzer Configurations   10.17   369
        10.3.2.1   Liquid Electrolyte Electrolyzers   10.17   369
        10.3.2.2   Solid Polymer Electrolyte Electrolyzers   10.18   370
        10.3.2.3   Ceramic Electrolyte Electrolyzers   10.19   371
    10.3.3   Efficiency of Electrolyzers   10.19   371
    10.3.4   Concentration-Differential Electrolyzers   10.22   374
    10.3.5   Electrolytic Hydrogen Compressors   10.24   376
10.4   Thermolytic Hydrogen   10.25   377
    10.4.1   Direct Dissociation of Water   10.25   377
    10.4.2   Chemical Dissociation of Water   10.31   383
10.5   Photolytic Hydrogen   10.33   385

10.5.1   Generalities   10.33   385
10.5.2   Solar Photolysis   10.34   386
10.6   Photobiologic Hydrogen Production   10.35   387
        References   10.37   389
        Problems   10.38   390

Chapter 11   Hydrogen Storage

11.1   Compressed Gas   11.3   399
11.2   Cryogenic Hydrogen   11.5   401
11.3   Storage of Hydrogen by Adsorption   11.7   403
11.4   Storage of Hydrogen in Chemical Compounds   11.8   404
        11.4.1   Generalities   11.8   404
        11.4.2   Hydrogen Carriers   11.10   406
        11.4.3   Water Plus a Reducing Substance   11.11   407
        11.4.4   Metal Hydrides   11.11   407
                11.4.4.1   Characteristics of Hydride Materials   11.17   413
                11.4.4.2   Thermodynamics of Hydride Systems   11.21   417
11.5   Hydride Hydrogen Compressors   11.25   421
11.6   Hydride Heat Pumps   11.29   425
        References   11.32   428
        Problems   11.33   429

Part III   Energy from the Sun

Chapter 12   Solar Radiation

12.1   The Nature of the Solar Radiation   12.1   445
12.2   Insolation   12.4   448
        12.2.1   Generalities   12.4   448
        12.2.2   Insolation on a Sun-Tracking Surface   12.7   451
        12.2.3   Insolation on a Stationary Surface   12.7   451
        12.2.4   Horizontal Surfaces   12.10   454
12.3   Solar Collectors   12.11   455
        12.3.1   Solar Architechture   12.11   455
                12.3.1.1   Exposure Control   12.11   455
                12.3.1.2   Heat Storage   12.11   455
                12.3.1.3   Circulation   12.12   456
                12.3.1.4   Insulation   12.13   457
        12.3.2   Flat Collectors   12.14   458
        12.3.3   Evacuated Tubes   12.15   459
        12.3.4   Concentrators   12.15   459
                12.3.4.1   Holographuc Plates   12.16   460
                12.3.4.2   Nonimaging Concentrators   12.17   461
12.4   Some Solar Plant Configurations   12.18   462

12.4.1 High Temperature Solar Heat Engine	12.18	462
12.4.2 Solar Chimney	12.20	464
12.4.3 Solar Ponds	12.20	464
Appendix A (The Measurement of Time)	12.22	466
The Duration of an Hour	12.22	466
Time Zones	12.22	466
Time Offset	12.23	467
The Calendar	12.23	467
The Julian Day Number	12.25	469
Appendix B (Orbital Mechanics)	12.26	470
Sidereal versus Solar	12.26	470
Orbital Equation	12.28	472
Relationship Between		
Ecliptic and Equatorial Coordinates	12.32	476
The Equation of Time	12.33	477
Orbital Eccentricity	12.36	480
Orbital Obliquity	12.37	481
References	12.39	483
Problems	12.40	484

Chapter 13   Biomass

13.1 Introduction	13.1	493
13.2 The Composition of Biomass	13.1	493
13.2.1 A Little Bit of Organic Chemistry	13.2	494
13.2.1.1 Hydrocarbons	13.2	494
13.2.1.2 Oxidation Stages of Hydrocarbons	13.3	495
13.2.1.3 Esters	13.4	496
13.2.1.4 Carbohydrates	13.4	496
13.3 Biomass as Fuel	13.6	498
13.3.1 Wood Gasifiers	13.7	499
13.3.2 Ethanol	13.8	500
13.3.2.1 Ethanol Production	13.8	500
13.3.2.2 Fermentation	13.11	503
13.3.3 Dissociated Alcohols	13.13	505
13.3.4 Anaerobic Digestion	13.14	506
13.4 Photosynthesis	13.21	513
References	13.28	520
Problems	13.29	521

Chapter 14   Photovoltaic Converters

14.1 Introduction	14.1	525
14.2 Theoretical Efficiency	14.6	530
14.3 Carrier Multiplication	14.13	537

14.4	Spectrally Selective Beam Splitting	14.14	538
14.4.1	Cascaded Cells	14.15	539
14.4.2	Filterd Cells	14.16	540
14.4.3	Holographic Concentrators	14.17	541
14.5	Thermo-photovoltaic Cells	14.17	541
14.6	The Ideal and the Practical	14.19	543
14.7	The Photodiode	14.20	544
14.8	The Reverse Saturation Current	14.40	564
14.9	Practical Efficiency	14.42	566
14.10	Solar-Power Satellite	14.44	568
14.10.1	Beam from Space	14.46	570
14.10.2	Solar Energy to DC Conversion	14.46	570
14.10.3	Microwave Generation	14.47	571
14.10.4	Radiation System	14.48	572
14.10.5	Receiving Array	14.49	573
14.10.6	Attitude and Orbital Control	14.50	574
14.10.7	Space Transportation and Space Construction	14.50	574
14.10.8	Future of Space Solar Power Projects	14.51	575
	Appendix A	14.52	576
	Appendix B	14.53	577
	References	14.60	584
	Problems	14.61	585

Part IV   Wind and Water

Chapter 15   Wind Energy

15.1	History	15.1	597
15.2	Wind Turbine Configurations	15.4	600
15.2.1	Drag-Type Wind Turbines	15.4	600
15.2.2	Lift-Type Wind Turbines	15.6	602
15.2.3	Magnus Effect Wind Machines	15.7	603
15.2.4	Vortex Wind Machines	15.8	604
15.3	Eolergometry	15.8	604
15.4	Availability of Wind Energy	15.9	605
15.5	Wind Turbine Characteristics	15.10	606
15.6	Principles of Aerodynamics	15.12	608
15.6.1	Flux	15.12	608
15.6.2	Power in the Wind	15.13	609
15.6.3	Dynamic Pressure	15.13	609
15.6.4	Wind Pressure	15.13	609
15.6.5	Available Power	15.14	610
15.6.6	Efficiency of a Wind Turbine	15.15	611
15.7	Airfoils	15.16	612

15.8	Reynolds Number	15.19	615
15.9	Aspect Ratio	15.21	617
15.10	Wind Turbine Analysis	15.23	619
15.11	Aspect Ratio (of a wind turbine)	15.30	626
15.12	Centrifugal Force	15.31	627
15.13	Performance Calculation	15.33	629
15.14	Magnus Effect	15.35	631
	References	15.36	632
	Problems	15.37	633

Chapter 16 Ocean Engines

16.1	Introduction	16.1	651
16.2	Wave Energy	16.1	651
16.2.1	About Ocean Waves	16.1	651
16.2.1.1	The Velocity of Ocean Waves	16.2	652
16.2.1.2	Wave Height	16.3	653
16.2.1.3	Energy and Power	16.4	654
16.2.2	Wave Energy Converters	16.4	654
16.2.2.1	Offshore Wave-Energy Converters	16.5	655
16.2.2.1.1	Heaving Buoy Converters	16.5	655
16.2.2.1.2	Hinged Contour converters	16.6	656
16.2.2.1.3	Overtopping Converters	16.7	657
16.2.2.2	Shoreline Wave Energy Converters	16.8	658
16.2.2.2.1	Tapered Channel System	16.8	658
16.2.2.2.2	Wavegen System (OWC)	16.9	659
16.3	Tidal Energy	16.11	661
16.4	Energy from Currents	16.11	661
16.4.1	Marine Current Turbine System	16.13	663
16.4.1.1	Horizontal Forces	16.13	663
16.4.1.2	Anchoring Systems	16.14	664
16.4.1.3	Corrosion and Biological Fouling	16.14	664
16.4.1.4	Cavitation	16.14	664
16.4.1.5	Large Torque	16.15	665
16.4.1.6	Maintenance	16.15	665
16.4.1.7	Power Transmission	16.16	666
16.4.1.8	Turbine Farms	16.16	666
16.4.1.9	Ecology	16.16	666
16.4.1.10	Modularity	16.16	666
16.5	Salination Energy	16.16	666
16.6	The Osmotic Engine	16.19	669
	References	16.22	672
	Problems	16.24	674

Subject Index   677

Foreword

This book examines the fundamentals of some renewable energy processes. A limited effort is made to describe the "state of the art" of the technologies involved because, owing to the rapidity with which these technologies change, such description would soon become obsolete. Nevertheless, the underlying principles discussed in the book are immutable and are essential for the comprehension of future developments. An attempt is made to present clear physical explanations of the pertinent principles.

The text will not prepare the student for detailed design of any specific device or system. However, it is hoped that it will provide the basic information to permit the understanding of more specialized writings.

The topics were not selected by their practicability or by their future promise. Some topics are discussed solely because they represent good exercises in the application of physical principles, notwithstanding the obvious difficulties in their implementation.

Whenever necessary, rigor is sacrificed in favor of clarity. Although it is assumed that the reader has an adequate background in physics, chemistry, and mathematics (typical of a senior science or engineering student), derivations tend to start from first principles to permit the identification of basic mechanisms. The organization of the book is somewhat arbitrary and certainly not all-encompassing. Processes that can be considered "traditional" are generally ignored. On the other hand, the list of "nontraditional" processes considered is necessarily limited.

Acknowledgments

I wrote this book. Without Aili, I could not.

My thanks to Dr. Edward Beardsworth who, incessantly scanning the literature, alerted me to many new developments.

My gratitude goes also to the hundreds of students who since 1976 have read my notes and corrected many typos and errors.

Chapter 1
Generalities

1.1 Units and Constants

Although many different units are employed in energy work, we shall adopt, whenever possible, the "Système International," SI. This means **joules** and **watts**. If we are talking about large energies, we'll speak of MJ, GJ, TJ, and EJ—that is, 10^6, 10^9, 10^{12}, and 10^{18} joules, respectively.

We cannot entirely resist tradition. Most of the time we will express pressures in **pascals**, but we will occasionally use **atmospheres** because most of the existing data are based on the latter. Sometimes **electron-volts** are more convenient than joules. Also, expressing energy in **barrels of oil** or **kWh** may convey better the idea of cost. On the whole, however, we shall avoid "quads," "BTUs," "calories," and other non-SI units. The reason for this choice is threefold: SI units are easier to use, they have been adopted by most countries, and are frequently better defined.

Consider, for instance, the "calorie," a unit preferred by chemists. Does one mean the "international steam table calorie" (4.18674 J)? Or the "mean calorie" (4.19002 J)? Or the "thermochemical calorie" (4.18400 J)? Or the calorie measured at 15 C (4.18580 J)? Or at 20 C (4.18190 J)?

Americans like to use the BTU, but, again, there are numerous BTUs: "steam table," "mean," "thermochemical," at 39 F, at 60 F. The ratio of the BTU to the calorie of the same species is about 251.956 with some variations in the sixth significant figure. Remember that 1 BTU is roughly equal to 1 kJ, while 1 quad equals roughly 1 EJ. The conversion factors between the different energy and power units are listed in Table 1.2. Some of the fundamental constants used in this book are listed below.

Table 1.1
Fundamental Constants

Quantity	Symbol	Value	Units
Avogadro's number	N_0	6.0221367×10^{26}	per kmole
Boltzmann constant	k	1.380658×10^{-23}	J K^{-1}
Charge of the electron	q	$1.60217733 \times 10^{-19}$	C
Gas Constant	R	8314.510	J kmole^{-1}K^{-1}
Gravitational constant	G	6.67259×10^{-11}	m^{3}s^{-2}kg^{-1}
Planck's constant	h	$6.6260755 \times 10^{-34}$	J s
Permeability of free space	μ_0	$4\pi \times 10^{-7}$	H/m
Permittivity of free space	ϵ_0	$8.854187817 \times 10^{-12}$	F/m
Speed of light	c	2.99792458×10^8	m s^{-1}
Stefan-Boltzmann constant	σ	5.67051×10^{-8}	W K^{-4}m^{-2}

Table 1.2
Conversion Coefficients

To convert from	to	multiply by
Energy		
BARREL OF OIL	GJ	≈ 6
BRITTISH THERMAL UNIT (Int. Steam Table)	joule	1055.04
BRITTISH THERMAL UNIT (mean)	joule	1055.87
BRITTISH THERMAL UNIT (thermochemical)	joule	1054.35
BRITTISH THERMAL UNIT (39 F)	joule	1059.67
BRITTISH THERMAL UNIT (60 F)	joule	1054.68
CALORIE (International Steam Table)	joule	4.18674
CALORIE (mean)	joule	4.19002
CALORIE (thermochemical)	joule	4.1840
CALORIE (15 C)	joule	4.1858
CALORIE (20 C)	joule	4.1819
CUBIC FOOT (Methane, STP)	MJ	≈ 1
ELECTRON VOLT	joule	1.60206×10^{-19}
ERG	joule	1.0×10^{-7}
FOOT LBF	joule	1.3558
FOOT POUNDAL	joule	4.2140×10^{-2}
kWh	joule	3.6×10^{6}
QUAD	BTU	1.0×10^{15}
TON of TNT	joule	4.2×10^{9}
Power		
FOOT LBF/SECOND	watt	1.3558
FOOT LBF/MINUTE	watt	2.2597×10^{-2}
FOOT LBF/HOUR	watt	3.7662×10^{-4}
HORSEPOWER (550 Foot LBF/sec)	watt	745.70
HORSEPOWER (electric)	watt	746
HORSEPOWER (metric)	watt	735
Other		
ATMOSPHERE	pascal	1.0133×10^{5}
DALTON	kg	1.660531×10^{-27}

LBF stands for pounds (force).

1.2 Energy and Utility

In northern California, in a region where forests are abundant, one cord of wood sold in 1990 for about \$110. Although one cord is a stack of 4 by 4 by 8 ft (128 cubic feet), the actual volume of wood is only 90 cubic feet—the rest is empty space between the logs. Thus, one cord contains

2.5 m^3 of wood or about 2200 kg. The heat of combustion of wood varies between 14 and 19 MJ/kg. If one assumes a mean of 16 MJ per kilogram of wood burned, one cord delivers 35 GJ. Therefore, the cost of energy from wood was \$3.2/GJ in northern California.

In 1990, the price of gasoline was still approximately \$1.20 per gallon, the equivalent of \$0.49 per kg. Since the heat of combustion of gasoline is 49 MJ/kg, gasoline energy costs \$10/GJ, or three times the cost from burning wood.

Notwithstanding electricity being inexpensive in California, the domestic consumer paid \$0.04 per kWh or \$11.1/GJ.

From the above, it is clear that when we buy energy, we are willing to pay a premium for energy that is, in a more convenient form—that is, for energy that has a higher **utility**.

Utility is, of course, relative. To stoke a fireplace in a living room, wood has higher utility than gasoline and, to drive a car, gasoline has higher utility than electricity, at least for the time being. For small vehicles, liquid fuels have higher utility than gaseous ones. For fixed installations, the opposite is true.

The relative cost of energy is not determined by utility alone. One barrel contains 159 liters or 127 kg of oil. With a heat of combustion of 47 MJ/kg, this corresponds to 6 GJ of energy. In mid-1990, the price was \$12/barrel or \$2/GJ, somewhat less than the price of wood at that time notwithstanding oil being, in general, more useful. However, oil prices are highly unstable depending on the political circumstances of the world.

Government regulations tend to depress prices below their free market value. During the Carter era, natural gas was sold in interstate commerce at the regulated price of \$1.75 per 1000 cubic feet. This amount of gas corresponds to 1 GJ of energy. Thus, natural gas was cheaper than oil or wood.

1.3   Conservation of Energy

Energy can be utilized but not consumed.[†] It is a law of nature that energy is conserved. Instead of consuming it, we degrade or randomize energy, just as we randomize mineral resources when we process concentrated ores into metal and then discard the final product as we do, for example, with used aluminum cans. All energy we use is degraded into heat and eventually radiated out into space.

[†] It is convenient to distinguish *consumption* from *utilization*. The former implies destruction—when oil is consumed, it disappears being transformed mainly into carbon dioxide and water, yielding heat. On the other hand, energy is never consumed—it is utilized but entirely conserved (only the entropy is increased).

The consumable is not energy; the consumable is the fact that energy has not yet been randomized. The degree of randomization of energy is measured by the entropy of the energy. This is discussed in some detail in Chapter 2.

1.4 Planetary Energy Balance

The relative stability of Earth's temperature suggests a near balance between planetary input and output of energy. The input is almost entirely that of the solar radiation incident on Earth. This amounts to 173,000 TW $(173,000 \times 10^{12}$ W$)$.

Besides solar energy, there is a contribution from tides (3 TW) and from heat sources inside the planet, mostly radioactivity (32 TW).

Some 52,000 TW (30% of the incoming radiation) is reflected back to the interplanetary space: it is the **albedo** of Earth. All the remaining energy is degraded to heat and re-emitted as long-wave infrared radiation. Figure 1.1 shows the different processes that take place in the planetary energy balance mechanism.

The recurrence of ice ages shows that the equilibrium between incoming and outgoing energy is oscillatory in nature. Some fear that the observed secular increase in atmospheric CO_2 might lead to a general heating of the planet resulting in a partial melting of the Antarctic glaciers and consequent flooding of sea level cities. The growth in CO_2 concentration is the result of the combustion of vast amounts of *fossil*[†] fuels and the destruction of forests in which carbon had been locked.

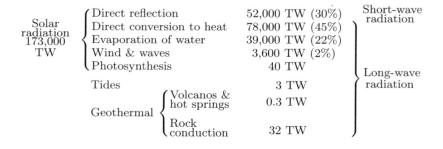

Figure 1.1   Planetary energy balance.

[†] Fuels derived from recent biomass, such as ethanol from sugar cane, do not increase the amount of carbon dioxide in the atmosphere—such fuels only recycle this gas.

1.5   The Energy Utilization Rate

The energy utilization rate throughout the ages can only be estimated in a rough manner. In early times, man was totally nontechnological, not even using fire. He used energy only as food, probably at a rate somewhat below the modern average of 2000 kilocalories per day, equivalent to 100 W. Later, with the discovery of fire and an improved diet involving cooked foods, the energy utilization rate may have risen to some 300 W/capita.

In the primitive agricultural Mesopotamia, around 4000 B.C., energy derived from animals was used for several purposes, especially for transportation and for pumping water in irrigation projects. Solar energy was employed for drying cereals and building materials such as bricks. Per capita energy utilization may have been as high as 800 W.

The idea of harnessing wind, water and fire to produce useful work is ancient. Wind energy has been in use to drive sailboats since at least 3000 B.C. and windmills were described by Hero of Alexandria around 100 A.D. Extensive use of windmills started in Persia around 300 A.D. and, only much later, spread to China and Europe.

Hero described toy steam engines that apparently were built and operated. Vitruvius, the famous Roman architect and author whose book, first published at the time of Hero, is still on sale today, describes waterwheels used to pump water and grind cereals.

In spite of the availability of the technology, the ancients limited themselves to the use of human or animal muscle power. Lionel Casson (1981), a professor of ancient history at New York University, argues that this was due to cultural rather than economic constraints and that only at the beginning of the Middle Ages did the use of other energy sources become "fashionable." Indeed, the second millennium saw an explosion of mechanical devices starting with windmills and waterwheels.

The energy utilization rate in Europe was likely 2000 watts per capita around 1200 A.D. when there was widespread adoption of advanced agriculture, the use of fireplaces to heat homes, the burning of ceramics and bricks, and the use of wind and water. Since the popular acceptance of such activities, energy utilization has increased rapidly.

Figure 1.2 illustrates (a wild estimate) the number of kilowatts utilized per capita as a function of the date. If we believe these data, we may conclude that the annual rate of increase of the per capita energy utilization rate behaved as indicated in Figure 1.3. Although the precision of these results is doubtful, it is almost certain that the general trend is correct— for most of our history the growth of the per capita energy utilization rate was steady and quite modest. However, with the start of the industrial revolution at the beginning of the 19th century, this growth accelerated dramatically and has now reached a worrisome level.

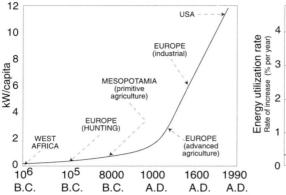

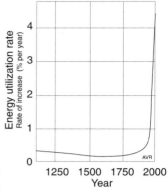

Figure 1.2   A very rough plot of the historical increase in the per capita energy utilization rate.

Figure 1.3   The annual rate of increase of per capita energy utilization was small up to the 19th century.

One driving force behind the increasing worldwide per capita energy utilization was the low cost of oil before 1973 when the price of oil was substantially lower than what it is currently.[†] Perez Alfonso, the Venezuelan Minister of Oil in 1946, was among those who recognized that this would lead to future difficulties. He was instrumental in creating OPEC in 1954, not as a cartel to squeeze out higher profits but to "reduce the predatory oil consumption to guarantee humanity enough time to develop an economy based on renewable energy sources." Alfonso also foresaw the ecological benefits stemming from a more rational use of oil.

OPEC drove the oil prices high enough to profoundly alter the world economy. The result was that the overall energy utilization rate slowed its increase. Owing to the time delay between the price increase and the subsequent response from the system, several years elapsed before a new equilibrium was established in the oil markets. The result was a major overshooting of the oil producing capacity of OPEC and the softening of prices that we witnessed up to the 1991 Iraqi crisis.

The recent effort of less developed countries (LDCs) to catch up with developed ones has been an important factor in the increase in energy demand. Figure 1.4 shows the uneven distribution of energy utilization rate throughout the world. 72% percent of the world population uses less than 2 kW/capita whereas 6% of the population uses more than 7 kW/ capita.

[†] In 1973, before the OPEC crisis, petroleum was sold at between $2 and $3 per barrel. The price increased abruptly traumatizing the economy. In 2000 dollars, the pre-1973 petroleum cost about $10/bbl (owing to a 3.8-fold currency devaluation), a price that prevailed again in 1999. However, in 2004, the cost had risen to over $50/bbl.

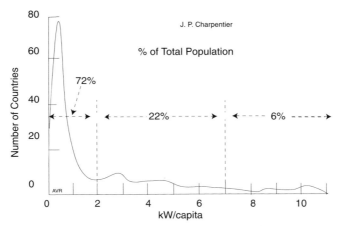

Figure 1.4   Most countries use little energy per
capita while a few developed ones use a lot.

There is a reasonable correlation between the total energy utilization rate of a country and its corresponding annual gross national product. About 2.2 W are used per dollar of yearly GNP. Thus, to generate each dollar, 69 MJ are needed. These figures, which are based on 1980 dollars, vary with time, in part owing to the devaluation of the currency, but also due to changing economic circumstances. It fact, it has been demonstrated that during an energy crisis, the number of megajoules per dollar decreases, while the opposite trend occurs during financial crises.

Further industrialization of developed countries may not necessarily translate into an increase of the per capita energy utilization rate—the trend toward higher efficiency in energy use may have a compensating effect. However, in the USA, the present decline in energy utilization[†] is due mainly to a change in the nature of industrial production. Energy intensive primary industries (such as steel production) are phasing out owing to foreign competition, while sophisticated secondary industries (such as electronics and genetic engineering) are growing.

Technological innovation has resulted in more efficient use of energy. Examples of this include better insulation in houses and better mileage in cars. Alternate energy sources have, in a small measure, alleviated the demand on fossil fuels. Such is the case of using ethanol from sugar cane for the propulsion of automobiles. It is possible that the development of fusion reactors will, one day, bring back the times of abundant energy.

Introduction of a more efficient device does not immediately result in energy economy because it takes a considerable time for a new device to

---

[†] The use of energy by the American industry was less in 1982 than in 1973.

be widely accepted. The reaction time of the economy tends to be long. Consider the privately owned fleet of cars. A sudden rise in gasoline price has little effect on travel, but it increases the demand for fuel efficiency. However, car owners don't rush to buy new vehicles while their old ones are still usable. Thus, the overall fuel consumption will only drop many years later, after a significant fraction of the fleet has been updated.

Large investments in obsolete technologies substantially delay the introduction of more desirable and efficient systems. A feeling for the time constants involved can be obtained from the study of the "market penetration function," discussed in Section 1.7.

1.6   The Population Explosion

In the previous section we discussed the *per capita* energy utilization rate. Clearly the total rate of energy utilization is proportional to the planetary population which has been growing at an accelerated rate.[†]

The most serious problem that confronts mankind is the rapid growth in population. The planet has a little more than 6 billion inhabitants, and the growth rate these last few decades has been around 1.4% per year. Almost all projections predict a population of about 7 billion by the year 2010. This will be the case even if, right now, everyone were to agree on a limit of two children per family. Under present-day actuarial conditions, the population would eventually stabilize at around 11 billion by the year 2050. Thus, population growth alone could account for 1.4% a year increase in energy demand, in the next few decades.

If, in 2050, all the estimated 11 billion inhabitants of Earth were to use energy at the present day USA level (11 kW/capita), the world energy utilization rate would reach 122 TW—a 16-fold increase over the present 7.6 TW. Such a rate is probably one order of magnitude higher than can be supplied unless fusion energy becomes practical and inexpensive.

A more modest scenario views the worldwide energy utilization rate stabilizing at the present level of Eastern Europe: 5 kW per capita. This would lead to an overall rate of 65 TW in 2050, which is still too high. Finally, if the world average kept its present 2 kW per capita, the rate would grow to 26 TW by the middle of next century. Clearly, it is difficult to provide adequate energy for 11 billion people. This is one more reason for attempting to limit the planetary population growth.

The constant population increase has its Malthusian side. About 10% of the world's land area is used to raise crops—that is, it is **arable**

[†] On 10/12/99, a 3.2 kg baby was born in Bosnia. Kofi Annan, General Secretary of the United Nations was on hand and displayed the new Bosnian citizen to the TV cameras because, somewhat arbitrarily, the baby was designated as the 6,000,000,000th inhabitant of this planet.

land, (See "Farming and Agricultural Technology: Agricultural Economics: Land, output, and yields." Britannica Online.) This means that roughly 15 million km^2 or 1.5×10^9 hectares are dedicated to agriculture. Up to the beginning of the 20th century, on average, each hectare was able to support 5 people (*Smil*), thus limiting the population to 7.4 billion people. More arable land can be found, but probably not enough to sustain 11 billion people. What limits agricultural productivity is nitrogen, one kilogram of which is (roughly) needed to produce one kilogram of protein. Although it is the major constituent of air, it is, in its elemental form, unavailable to plants and must either be "fixed" by appropriate micro-organisms or must be added as fertilizer.

Nitrogen fertilizers are produced almost exclusively from ammonia, and when used in adequate amounts can increase land productivity by nearly an order of magnitude. The present day and the future excess population of the planet can only exist if sufficient ammonia is produced. Although there is no dearth of raw materials for this fertilizer (it is made from air and water), its intensive use has a serious adverse environmental effect as discussed in the article by Smil.

1.7   The Market Penetration Function

A new technology, introduced in competition with an established one, may take over a progressively larger fraction of the market. Is it possible to forecast the rate at which such penetration occurs?

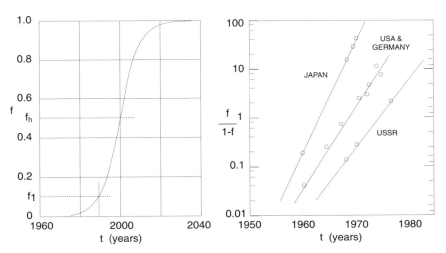

Figure 1.5   Left: Typical penetration function vs time.   Right: The penetration function of oxygen steel technology fits accurately the Fisher-Pry rule.

Let f be the fraction of the total market captured by the new technology. As time progresses, f grows from 0 to some value equal or less than 1. The latter corresponds to the new technology having totally replaced all competition. In due time, f may decrease again when a even newer technologies is introduced.

An empirical plot of the ascending phase of f vs time, t, has an "S" shape as exemplified by Figure 1.5 (left). A **market penetration time** is defined as $\Delta T \equiv (t_h - t_1)$, where t_h is the time at which $f = 0.5 \equiv f_h$, and t_1 is the time at which $f = 0.1 \equiv f_1$. ΔT may be negative if the technology in question is being replaced. It is then called the **abandonment time**. Fisher and Pry (1971) and Pry (1973) showed that when $\ln \frac{f}{1-f}$ is plotted versus time, a straight line results. Figure 1.5 (right) illustrates an example of how the Fisher-Pry equation provides an excellent fit to the empirical data. The data show how, in four different countries, the use of oxygen in steel converters is gradually substituted for the older open-hearth and Bessemer technologies. The straight lines in the plots correspond to a regression of the type:

$$\ln \frac{f}{1-f} = at + b. \tag{1}$$

Constants a and b characterize the market and the particular technology considered. One would expect that the fractional rate of technology penetration of the market, $\frac{1}{f}\frac{df}{dt}$, is proportional to the fraction, $(1-f)$, of the market that has not yet been penetrated:

$$\frac{1}{f}\frac{df}{dt} = a(1-f). \tag{2}$$

The empirical evidence of Figure 1.5 (right) and of Equation 1 supports the model of Equation 2, because the former is the integral of the latter.

The quantities, a and b depend on the nature of the technology and on the specific location where the technology is being introduced. It is possible to generalize the Fisher-Pry equation by making it independent of these parameters.

For $t = t_h, f = 0.5$ and

$$\ln \frac{f}{1-f} = at_h + b = 0 \qquad \therefore \qquad b = -at_h. \tag{3}$$

For $t = t_1, f = 0.1$ and

$$\ln \frac{f}{1-f} = at_1 + b = -2.2. \tag{4}$$

Subtracting one equation from the other,

$$2.2 = a(t_h - t_1) = a\Delta t \qquad \therefore \qquad a = \frac{2.2}{\Delta t}. \tag{5}$$

Thus, the market penetration formula can be written as:

$$\ln \frac{f}{1-f} = 2.2 \frac{(t - t_h)}{\Delta t}. \tag{6}$$

Equation 6 is a function of only the normalized independent variable, $(t - t_h)/\Delta t$. This permits presenting data with different a's and b's in a single graph. An example of such a plot is shown in Figure 1.6, prepared by Fisher and Pry. Data for 17 different cases of technology penetration are shown, with a surprisingly small scatter of points.

The Fisher-Pry model is insensitive to the overall market volume. Many factors that affect the market as a whole don't appear to influence its distribution among different technologies.

Figure 1.5 shows that the take over time for oxygen steel differed among countries: in Japan it was 5 years, in West Germany and in the USA 6, and in the Soviet Union 8 years. The rapid penetration of the technology was partially due to the fast depreciation of plants allowed by law.

Marchetti (1978) showed that the market penetration law is also applicable to energy. Figure 1.7 illustrates the fraction of the market supplied by a particular energy source as a function of time. The data are for the USA. The graph shows how energy from wood started abandoning the market in the 19th century owing to the introduction of coal as a source of fuel.

Coal, after penetrating the market for half a century, was forced out by oil and natural gas. Owing to the dispersed nature of the market, the time constants of both penetration and abandonment of energy products is much longer than that of most other technologies. Table 1.3 lists the different takeover times (abandonment times have a "minus" sign).

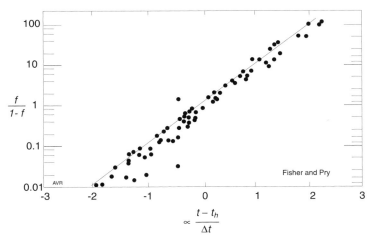

Figure 1.6   Fisher-Pry plot for 17 different substitutions.

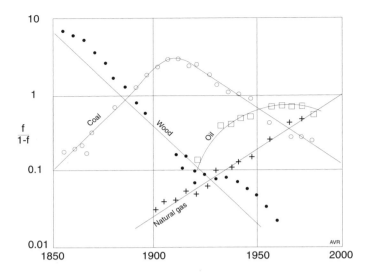

Figure 1.7   Long before the OPEC intervention, the
Fisher-Pry model would have predicted the current decline
in oil consumption.

<table>
<tr><td colspan="2">**Table 1.3**</td><td colspan="3">**Table 1.4**</td></tr>
</table>

Table 1.3		Table 1.4		
Take-over and Abandonment Times.		**Fisher-Pry Coefficients for Wood, Coal, and Gas.**		
			a	b
Wood	−38 years	Wood	−0.0585	110.20
Coal	38 years	Coal	−0.0439	85.18
Coal	−50 years	Gas	0.0426	−84.64
Oil	30 years			

Examine the period beginning in 1920. Wood, coal, and natural gas
seem to have behaved according to the Fisher-Pry model. During this
period, hydroelectric energy made a constant contribution of about 3.6%
of the total. The regression coefficients for wood, coal and gas are shown
in Table 1.4.

Since $\sum f = 1$, the fraction of the energy market supplied by oil can be
calculated by subtracting from 1 the fractional contributions of the remaining fuels. When this is done, one arrives at the curve for oil penetration
shown in Figure 1.7. It can be seen that it matches reasonably well the
actual data (open squares).

The regression coefficients were obtained from data for 1920 through
1950 only; the rest of the information for these items resulted from extending the straight lines in the graph. Yet, the derived oil penetration curve

shows a decline starting around 1970, which, in fact, did occur. The recent decline in relative oil consumption could have been predicted back in 1950, years before the creation of OPEC! One can therefore conclude that the reduction in relative oil usage would have occurred regardless of the actions of OPEC. All OPEC did was to affect the overall price of energy.

1.8   Planetary Energy Resources

In Section 1.5, we pointed out that the rate of per capita energy utilization rose rapidly in the last century. This, combined with the fast increase in population mentioned in Section 1.6, leads one to the inescapable conclusion that we are facing a serious challenge if we hope to maintain these trends in the foreseeable future. To investigate what can be done to resolve this difficulty we must first inquire what energy resources are available (Section 1.8) and next (Section 1.9) how we are using the resources at present.

Figure 1.8 shows the planetary energy resources. These can be renewable or nonrenewable.

Geothermal energy has been used for a very long time in Iceland and more recently in Italy, New Zealand, and the United States. Great expansion of its contribution to the total energy supply does not seem probable.

Gravitational energy—that is, energy from tides (see Chapter 16) has been used in France. Tides can only be harnessed in certain specific localities of which there is a limited number in the world.

Of the renewable resources, solar energy is by far the most abundant. A small part of it has been absorbed by plants and, over the eons, has been stored as coal, oil, and gas. Estimates of fossil reserves (as well as of nuclear fuel reserves) are extremely uncertain and are sure to be greatly underestimated because of incomplete prospecting. Table 1.5 gives us an idea of our fossil fuel reserves and Table 1.6 shows roughly estimated reserves of fissionable materials. These estimates do not include the old Soviet Union and China.

The values given in the table are very far from precise. They may, however, represent a *lower* limit. People who estimate these numbers tend to be conservative as testified by the fact that there is actually a secular *increase* in proved reserves. As an example, the proved reserves of dry natural gas, 2200 EJ in 1976, rose to 5500 EJ in 2002 not withstanding the substantial consumption of gas in the intervening years.

For oil and gas, the table lists the sum of proved reserves, reserve growth and undiscovered reserves.

Proved reserves are fuels that have been discovered but not yet produced. Proved reserves for oil and gas are reported periodically in the *Oil and Gas Journal*.

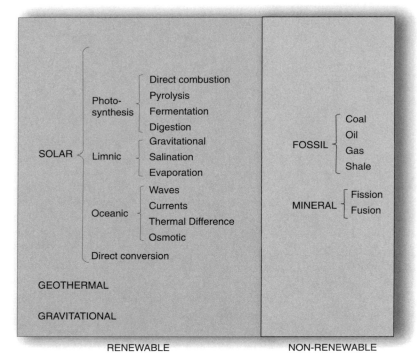

Figure 1.8   The energy resources of Earth.

<table>
<tr><td colspan="2">**Table 1.5**
Known Fossil Fuel Reserves</td></tr>
<tr><td>Methane clathrate</td><td>>100,000 EJ (1998)</td></tr>
<tr><td>Coal</td><td>39,000 EJ (2002)</td></tr>
<tr><td>Oil</td><td>18,900 EJ (2002)</td></tr>
<tr><td>Gas</td><td>15,700 EJ (2002)</td></tr>
<tr><td>Liquefied gas</td><td>2,300 EJ (2002)</td></tr>
<tr><td>Shale</td><td>16,000 EJ (?)</td></tr>
</table>

Table 1.6	
Known Reserves of	
Fissionable Materials[†]	
^{235}U	2,600 EJ
^{238}U	320,000 EJ
^{232}Th	11,000 EJ

[†]Does not include the former USSR and China.

Reserve growth represents the increase in the reserves of existing fields owing to further development of these fields and to the introduction of better technology for their extraction.

Undiscovered reserves represent the best possible guess of possible new discoveries.

Reserve growths and undiscovered reserves are estimated by the US Geological Survey (<http://greenwood.cr.usgs.gov/energy/WorldEnergy/DDS-60/>). For example, in 2002 the Oil and Gas Journal reported proved reserves of oil of 7280 EJ and the USGS estimated a growth of 4380 EJ and undiscovered oil reserves amounting to 5630 EJ adding up to the total of 18,900 EJ listed in the Table.

The indicated reserves also include 3000 EJ of proved dry natural gas that is currently too far from pipe lines to be economically transported to consumers.

In addition to the dry natural gas (mostly methane), a well will also produce other gases (propane, for example) that can be liquefied and shipped. The table lists a worldwide reserve of 2300 EJ in 2002.

For coal, the table shows only proved reserves. The total reserves for this fuel are, thus, substantially larger than listed.

One number in the table that is particularly uncertain is that referring to hydrated methane. William P. Dillon, a geologist of the USGS, testified in the U.S. House of Representatives in 1998, that "the amount of methane contained in the world's gas hydrate accumulations is enormous, but estimates of the amounts are speculative and range over three orders-of-magnitude from about 100,000 to 270,000,000 trillion cubic feet [100,000 to 270,000,000 EJ] of gas." We, being ultraconservative, listed the lower figure.

Methane Clathrate

Clathra is the Latin word for "bar" or "cage".

Atoms in a number of molecules group themselves in such a fashion that a cavity (or cage) is left in the center. The most famous of these arrangement is the "buckyball," a molecule consisting of 60 carbon atoms arranged as a hollow sphere capable of engulfing a number of substances. Buckyballs, discovered in the early 1980s, are not alone among "hollow" molecules. Under appropriate circumstances, water will freeze forming a cage consisting, sometimes, of 20 water molecules, but more commonly, of 46 water molecules. The configuration is unstable (it decays into a common ice crystal) unless certain gases become trapped in the central cage of the large molecule. Gases commonly trapped are methane, ethane, propane, iso-butane, n-butane, nitrogen, carbon dioxide, and hydrogen sulfide.

The ice crystal consisting of 46 water molecules is able to trap up to 8 "guest" gas molecules (a water-to-gas ratio of 5.75:1). In natural deposits, methane is by far the most abundant and the one of greatest interest to the energy field. Usually, up to 96% of the cages are fully occupied. These solid hydrates are called **clathrates**.

(continues)

(continued)

The density of the clathrate is about 900 kg/m^3. This means that the methane is highly compressed. See Problem 1.28. Notwithstanding its low density, water ice clathrate does not float up from the bottom of the ocean because it is trapped beneath the ocean sediment.

Clathrates form at high pressure and low temperature under sea and are stable at sufficient depth. The methane is the result of anaerobic digestion of organic matter that continuously rains down on the ocean floor. See Chapter 13.

There is no mature technology for the recovery of methane from clathrates. Proposed processes all involve destabilizing the clathrate and include:

1. Raising the temperature of the deposits.
2. Depressurization the deposits.
3. Injecting methanol or other clathrate inhibitor.

The latter process may be environmentally undesirable.

There are dangers associated with methane clathrate extraction. The most obvious ones are the triggering of seafloor landslides and the accidental release of large volumes of methane into the Earth's atmosphere where it has a powerful greenhouse effect.

Read more about clathrates in *Clathrates: little known components of the global carbon cycle* <http://ethomas.web.wesleyan.edu/ees123/clathrate.htm>

1.9 Energy Utilization

Most of the energy currently used in the world comes from non-renewable sources as shown in Figures 1.9 and 1.10, which display energy sources in 2001 for the whole world and for the United States, respectively. The great similarity between these two charts should not come as a surprise in view of the US using such a large fraction of the total world consumption.

What may be unexpected is that most of the renewable resources (geothermal, biomass, solar and wind) make such a small contribution to the overall energy picture. Figure 1.11 shows that as late as 1997 only 12% of the energy used to generate electricity in the USA came from renewable sources. Of these, 83% came from hydroelectrics. Thus, only 2% of the total came from the remaining renewables.

Disappointingly, so far, the contribution of solar and wind energy has been very small, much less than that of geothermal. Most of the renewable energy comes from hydro electric plants and some, from biomass.

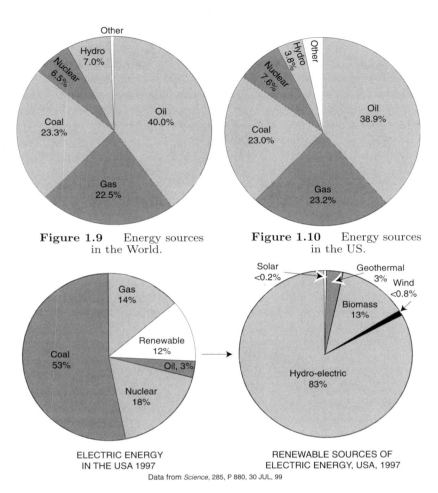

Figure 1.9   Energy sources in the World.

Figure 1.10   Energy sources in the US.

ELECTRIC ENERGY
IN THE USA 1997

RENEWABLE SOURCES OF
ELECTRIC ENERGY, USA, 1997

Data from *Science*, 285, P 880, 30 JUL, 99

Figure 1.11   Sources of electric energy in the United States.

Table 1.7
Energy Use, USA 2001

Source	Used (EJ)	Capacity (GW)	Utilization factor
Thermal	9.69	600	51.1%
Nuclear	2.71	98	87.6%
Hydro	0.97	99	31.1%
Geothermal	0.26		
Wind	0.0208	4.28[†]	15.4%
Other	0.015		

[†]This datum is from AWEA (American Wind Energy Association), all other are from EIA (Energy Information Administration.)

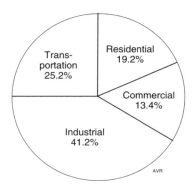

Figure 1.12   The different users of energy in the USA.

For all sources of energy, the cost of the plant is proportional to the installed capacity, while the revenue is proportional to the energy generated. The **plant utilization factor** is the ratio of the energy produced to that which would be produced if the plant operated uninterruptedly at full capacity (Table 1.7).

Observe the extremely high utilization factor of nuclear plants and the rather small factor of wind generators, the latter resulting from the great variability of wind velocity. Although specific data for solar plants are not available, they also suffer from a low utilization factor owing the day/night cycle and the vagaries of meteorological conditions.

It is of interest to know which are the main users of energy in the United States.

American residences account for nearly 20% of all energy used. Most of it is used for ambient heating, an area in which considerable economy can be realized, especially through better home design.

Waste heat from electric power plants can be used for ambient heating in homes and offices. "District heating" is common In Sweden. Thermal power plants in that country, operate with an average 29% efficiency but 24% of the total fuel energy (from the heat rejected by the stem plant) is piped, as hot water, to buildings in the neighborhood. Thus, only 47% of the available combustion energy is wasted. In contrast, in the United States, a total of 68% of the combustion energy is wasted in spite of the larger average steam plant efficiency (32%). District heating requires the location of power plants in densely populated areas. This is, of course, inadvisable in the case of nuclear plants and large fossil fueled installations. However, fuel cell plants (see Chapter 9), being noiseless and pollution free, can be placed in a downtown area.

It is probably in the transportation sector (25% of the total energy use) where modern technology can have the most crucial impact. Fuel cell cars promise to increase automobile efficiency while reducing pollution.

1.10   The Ecology Question

We have shown that there is an almost unavoidable trend toward increasing energy utilization. We have also pointed out that at present the energy used is at least 85% of fossil origin. Finally, we have shown that the fossil fuel reserves seem ample to satisfy our needs for a good fraction of the next millennium. So, what is the problem?

Most of the easily accessible sources of oil and gas have already been tapped. What is left is getting progressively more expensive to extract. Thus, one part of the problem is economical. Another is political—most of the fuel used by developed nations is imported (using the large American reserves is unpopular and politicians hesitate to approve such exploration). This creates an undesirable vulnerability. There are also technological difficulties associated with the identification of new reserves and the extraction of fuels from more remote locations. The major obstacle, however, is ecological. Fossil fuels are still the most inexpensive and most convenient of all energy resources, but their use pollutes the environment, and we are quickly approaching a situation in which we can no longer dismiss the problem or postpone the solution.

By far, the most undesirable gas emitted is carbon dioxide whose progressively increasing concentration in the atmosphere (from 270 ppm in the late 1800 to some 365 ppm at present) constitutes a worrisome problem. It is sad to hear influential people (among them, some scientists) dismiss this problem as inconsequential, especially in view of the growing signs of a possible runaway ecological catastrophe. For instance, in the last few decades, the thickness of the north polar ice has decreased by 40% and on the first year of the current millennium, a summertime hole appeared in the polar ice. Since increased concentrations of CO_2 can lead to global warming, some people have proposed increasing the emission of SO_2 to stabilize the temperature because of the cooling effect of this gas. Even ignoring the vegetation-killing acid rain that would result, this proposal is equivalent to balancing a listing boat by piling stones on the other side.

The lack of public concern with the CO_2 problem may be due to the focus on planetary temperature rise. Although the growth in CO_2 concentration is very easily demonstrated, the conclusion that the temperature will rise, although plausible, is not easy to prove. There are mechanisms by which an increase of greenhouse gases would actually result in a *cooling* of Earth. For instance, increasing greenhouse gases would result in enhanced evaporation of the tropical oceans. The resulting moisture, after migrating toward the poles, would fall as snow thereby augmenting the albedo of the planet and, thus, reducing the amount of heat absorbed from the sun.

Some scientist and engineers who are less concerned with political correctness, are investigating techniques to reduce (or at least, to stabilize) the concentration of atmospheric carbon dioxide. This can, in principle, be accomplished by reducing emissions or by disposing carbon dioxide in such a way as to avoid its release into the air. Emissions can be reduced by diminishing overall energy consumption (an utopian solution), by employing alternative energy sources, by increasing efficiency of energy use, and by switching to fuels that yield more energy per unit amount of carbon emitted. 1 kmole of methane, CH_4, when burned yielding liquid water and carbon dioxide, releases 889.6 MJ and emits 1 kilomole of carbon—it generates heat at a rate of 889.6 MJ per kilomole of carbon. n-heptane, C_7H_{16}, which can represent gasoline, releases 4820 MJ of heat per kilomole burned and emits 7 kilomoles of CO_2—a rate of 688.6 MJ per kilomole of carbon. Clearly, the larger the number of carbon atoms in the hydrocarbon molecule, the lower the ratio of the heat of combustion to the amount of carbon dioxide emitted because the ratio of hydrogen to carbon decreases. This is one reason for preferring methane to oil and oil to coal.

Alternative forms of energy are attractive but, at least for the present, are too expensive to seriously compete with fossil fuels.

In order to select a carbon dioxide disposal technique, it is important to inquire where nature stores the existing carbon.

Table 1.8 shows the estimated amount of carbon stored in different places.

Methods to dispose of CO_2 could include:

1.10.1 Biological

Photosynthesis removes carbon dioxide from the air. The biomass produced must be preserved if it is to permanently affect the CO_2 concentration. This means it cannot be burned or allowed to rot. There seems to be limited capacity for this method of CO_2 disposal. It should be noted that the biological uptake rate of carbon is, at present, only 0.002×10^{15} kg year.

1.10.2 Mineral

CO_2 is removed naturally from the air by forming carbonates (principally of magnesium and calcium). The gas is removed by reacting with abundant silicates. However, this process is too slow to cope with manmade emissions.

Table 1.8
Stored Carbon on Earth.

Oceans	45×10^{15} kg
Fossil fuels	10×10^{15} kg
Organic matter	2.4×10^{15} kg
Atmosphere	0.825×10^{15} kg

Carbon in the atmosphere

How much carbon is there in the atmosphere?

The surface area of earth is 510×10^{12} m^2, while the scale height of the atmosphere is around 8800 m (see the section on Boltzmann's law in Chapter 2). Consequently the volume of air (all of it compressed to 1 atmosphere pressure) is $510 \times 10^{12} \times 8800 = 4.5 \times 10^{18}$ m^3.

Present day atmospheric CO_2 concentration is 13.5×10^{-6} kmol/m^3. Thus, the atmosphere contains $13.5 \times 10^{-6} \times 4.5 \times 10^{18} = 61 \times 10^{12}$ kmol of CO_2 and, therefore, 61×10^{12} kmol of carbon. Since the atomic mass of carbon is 12 daltons, the mass of carbon in the atmosphere is 0.73×10^{15} kg. Compare with the 0.825×10^{15} kg of the table.

A simpler way to achieve about the same result is to consider that the atmospheric pressure at sea level is 1 kg/cm^2 or 10^4 kg/m^2. Consequently, the total mass of the atmosphere is $510 \times 10^{12} \times 10^4 = 510 \times 10^{16}$. Of this 360×10^{-6} is carbon dioxide and 12/44 of this is carbon. Carbon content of the atmosphere is $510 \times 10^{16} \times 365 \times 10^{-6} \times 12/44 = 0.51 \times 10^{15}$ kg, a result comparable with the previous one.

Ziock et al. propose the use magnesium silicates to sequester carbon dioxide at the point where fossil fuels are burned. Enormous deposits of magnesium oxide-rich silicates exist in the form of olivines and serpentines.

For serpentine, the net reaction involved is

$$Mg_3Si_2O_5(OH)_4 + 3CO_2 \rightarrow 3MgCO_3 + 2SiO_2 + 2H_2O$$

Notice that the end products are materials that already exist naturally in great abundance.

Substantial additional research is needed to improve the proposed disposal system and to make it economical.

1.10.3   Subterranean

CO_2 can be sequestered underground as the oil industry has been doing (for secondary oil recovery) for more than 50 years. The volume of the exhaust gases of a combustion engine is too large to be economically stored away. It is necessary to separate out the carbon dioxide, a task that is not easy to accomplish. One solution is proposed by Clean Energy Systems, Inc. of Sacramento, CA. The suggested equipment extracts oxygen from air (a well developed process) and mixes this gas with the fuel. Combustion produces steam and CO_2 at high temperature and pressure and drives several turbines at progressively lower temperatures. The water in the final exhaust is condensed and recycled leaving the carbon dioxide to be pumped, at 200 atmospheres, into an injection well. At present, no turbines exist capable of operating at the high temperature (over 3000 C) of the combustion products. See *Anderson, et al.*

1.10.4 Undersea

The Norwegian government imposes a stiff carbon dioxide emission tax that has made it economical to install disposal systems. They pump the gas deep into the ocean. It appears that liquid carbon dioxide can be injected into the seas at great depth and that it will stay there for a long time. More work is required to see if such scheme is indeed feasible and economical.

1.11 Nuclear Energy

Chemical fuels, such as oil or methane, release energy when the atoms in their molecules are rearranged into lower energy configurations. The energies involved are those of molecular binding and are of the order of some tens of MJ/kmol. When the *components* of an atom are arranged into lower energy configurations, then the energy released is orders of magnitude larger (GJ/kmole) because of the much larger intra-atomic binding energies.

The internal structure of atoms can be changed in different ways:

1. An atomic nucleus can be bombarded with a neutron, absorbing it. A different atom emerges.
2. An atom can spontaneously change by emitting either electrons (beta-rays) or helium nuclei (alpha-rays). Such radioactive decay releases energy which can be harvested as, for instance, it is done in **Radioisotope Thermal Generators** (RTG).
3. Atoms with large atomic number can be made to break up into smaller atoms with the release of energy. This is called **nuclear fission** and requires that the atomic number, Z, be larger than 26.
4. Atoms with low atomic numbers can be assembled into a heavier one, releasing energy. This is called **nuclear fusion** and requires that the final product have an atomic number smaller than 26.[†]

Nuclear energy has developed a bad reputation especially after the Chernobyl accident. Nevertheless it is still a source of substantial amounts of energy in many countries. In 2001, the US led the world in installed capacity—98 GW, followed by France (60 GW) and Japan (42 GW).[††]

The utilization factor of nuclear plants was excellent. In the US, the plants generated 87.6% of the energy they would have delivered if they had operated uninterruptedly at full power. In France, this number was 69.5% and in Japan, 75.4%.[††]

[†] All are transmutations, the age-old dream of medieval alchemists.

[††] The French and the Japanese data are for 1996.

Of the total electricity generated, nuclear plants in the US contributed a relatively modest 18%, while in France, heavily reliant on this form of energy, the contribution was 76.1%. In Japan, it was 33.4%. In 2000, Germany decided to phase out its 19 nuclear power plants. Each one was assigned a 32-year life after which they would be deactivated. Many plants have already operated more than half of their allotted life time.

The cost of nuclear electricity is high, about double of that from fossil fuel. In the US (1996) it was 7 cents/kWh, while that of a state of the art natural gas plant it was 3 cents/kWh (Sweet, William #1). Advanced reactor designs may bring these costs down considerably while insuring a greater safety in the operation of the plants (Sweet, William #2). This promised reduced cost combined with the ecological advantage of no greenhouse gas emission—a growing concern—may lead to renewed popularity of nuclear generators.

The major objection to fission type reactors is not so much the danger of the operation of the power plants, but rather the problem of disposing of large amounts of long-lived radioactive by-products. If the need for such disposal can be avoided, then there is good reason to reconsider fission generators as an important contributor to the energy supply system.

Specifications of new generation nuclear fission reactors might include (not necessarily in order of priority), the following items:

1. Safety of operation (including resistance to terrorist attacks)
2. Affordability
3. Reliability
4. Absence of weaponizable sub-products
5. Absence of long-lived waste products
6. Ability to transmute long-lived radioactive waste products from old reactors into short-lived radioactive products

The U.S. Department of Energy is funding research (2004) in several technologies that might realize most of the specifications above. One of these is the **heavy metal nuclear reactor** technology. Although the technology is complicated, it appears that this type of reactor may be able to not only produce wastes with relatively short half-lives (100 years contrasted with 100,000 years of the current waste), but in addition may be able to use current type waste as fuel thus greatly alleviating the waste disposal problem. Furthermore, because heavy-metal reactors operate at high temperatures (yet at low pressures), the thermolytic production of hydrogen (see Chapter 10) for use in fuel cell-driven automobiles looms as a good possibility. For further reading on this topic see Loewen.

The waste disposal problem is absent in fusion devices. Unfortunately, it has been impossible to demonstrate a working prototype of a fusion machine, even after several decades of concerted research.

Table 1.9
Masses of Some Particles Important to Nuclear Energy

Particle	Symbol	Mass (daltons[†])	Mass (kg)
electron ♣	e	0.00054579903	$9.1093897 \times 10^{-31}$
muon •	μ	0.1134381	1.883566×10^{-28}
proton ♣	p	1.007276467	1.672648×10^{-27}
neutron ♣	n	1.008664909	$1.6749286 \times 10^{-27}$
$^1_1$H		1.007825032	$1.673533967 \times 10^{-27}$
$^2_1$D		2.014101778	$3.344496942 \times 10^{-27}$
$^3_1$T		3.016049278	$5.008271031 \times 10^{-27}$
$^3_2$He		3.016029319	$5.008237888 \times 10^{-27}$
$^4_2$He		4.002603254	$6.646483555 \times 10^{-27}$
alpha ♣	α	4.001506175	$6.644661810 \times 10^{-27}$
$^5_3$Li		5.01254	$8.323524107 \times 10^{-27}$
$^6_3$Li		6.015122794	$9.988353127 \times 10^{-27}$
$^7_3$Li		7.01600455	$1.165035751 \times 10^{-26}$
$^{10}_5$B		10.012937	$1.662688428 \times 10^{-26}$
$^{11}_5$B		11.009305	1.82814×10^{-26}

[†]The dalton is not yet the official name for the atomic mass unit.

To do even a superficial analysis of the technical aspects of nuclear reactions, we need to know the masses of some of the atoms involved. See table above. Most of mass values are from Richard B. Firestone. Those marked with a ♣ are from Audi and Wapstra (1993), and the one marked with a • is from a different source. It can be seen that the precision of the numbers is very large. This is necessary because, in calculating the energy released in a nuclear reaction, one uses the small difference between large numbers which is, of course, extremely sensitive to uncertainties in the latter.

The listed values for the masses of the nucleons (the proton and the alpha, in the table) are nearly the values of the masses of the corresponding atoms minus the mass the electron(s). On the other hand, there is a large difference between the the mass of a nucleon and the sum of the masses of the component protons and neutrons. Indeed, for the case of the alpha, the sum of the two protons and the two neutrons (4.03188278 daltons) exceeds the mass of the alpha (4.001506175 daltons) by 0.030376606 daltons—about 28 MeV of mass. This is, of course, the large **nuclear binding energy** necessary to overcome the great electrostatic repulsion between the protons.

1.11.1   Fission

There are at least three fissionable elements of practical importance: ^{235}U, ^{239}Pu and ^{233}U. Of these, only ^{235}U is found in nature in usable quantities; ^{239}Pu and ^{233}U must be created by transmutation from "fertile" materials, respectively ^{238}U and ^{232}Th.

A nuclear fission reaction (with a corresponding release of energy) occurs when a fissionable material interacts with thermal, i.e., low energy, neutrons. The collision of high energy neutrons with ^{235}U, for example, is elastic, whereas low energy neutrons are captured:

$$^{235}_{92}\text{U} + ^{1}_{0}\text{n} \rightarrow ^{236}_{92}\text{U}. \tag{7}$$

The resulting ^{236}U decays with the emission of alpha-particles (lifetime 7.5 seconds). More importantly, the uranium also suffers spontaneous fission:

$$^{236}_{92}\text{U} \rightarrow 3^{1}_{0}\text{n} + \text{fission products} + ... \quad 3 \times 10^{-11} \text{ joules.} \tag{8}$$

Thus, under the proper circumstances, $^{235}_{92}$U absorbs a neutron and the resulting atom splits into smaller nuclei simultaneously releasing 3 neutrons and about 3×10^{-11} joules of energy:

$$^{235}_{92}\text{U} + ^{1}_{0}\text{n} \rightarrow 3\,^{1}_{0}\text{n} + \text{fission products} + 3 \times 10^{-11} \text{ J.} \tag{9}$$

Per kilogram of $^{235}_{92}$U, the energy released is

$$\frac{3 \times 10^{-11}\frac{\text{J}}{\text{atom}} \times 6 \times 10^{26}\frac{\text{atoms}}{\text{kmol}}}{235\frac{\text{kg}}{\text{kmol}}} = 77 \quad \text{TJ/kg.}$$

Compare this with the energy from chemical reactions which is frequently of the order of a few tens of MJ/kg.

When Otto Hahn, in 1939, demonstrated uranium fission, it became immediately obvious that a sustained "chain" reaction would be achievable. To such an end, all that was needed was to use one of the emitted neutrons to split a new uranium atom. In trying to build such a fission reactor, a number of problems had to be overcome.

1. The $^{235}_{92}U + ^{1}_{0}$n reaction requires slow (thermal) neutrons. The high energy neutrons emitted will not do. Thus, these neutrons must be made to transit through some material that has the property of slowing the particle down without absorbing it. Examples of such "moderating" substances are heavy water and graphite.

2. Fast neutrons may be absorbed by impurities in the fuel or in the moderator. The fuel is a mixture of $^{235}_{92}$U and $^{238}_{92}$U. The latter is an abundant "impurity" that absorbs fast neutrons but not slow ones. To reduce neutron losses, it may be necessary to "enrich" the fuel, i.e, increase the $^{235}_{92}$U/$^{238}_{92}$U ratio.[†] It is also necessary to place the fuel into a number of long rods embedded in a mass of moderator. This configuration allows most of the fast neutrons to escape the fuel region and reach the moderator where they are slowed and may eventually reenter one of the fuel rods. They now have insufficient energy to interact with the $^{238}_{92}$U but will do so with $^{235}_{92}$U perpetuating the reaction.

Clearly, it is essential that exactly one of the released neutrons is, on average, used to trigger a new fission. If more than one, the reaction will grow exponentially, if less, it will die out. Control systems are use to adjust this number to precisely one. Fortunately, the process is, to a degree, self adjusting—if the reaction rate rises, so will the temperature, and this reduces the probability of neutron capture.

Uranium isotopes cover the range from 227 to 240 in atomic mass, but natural uranium consists chiefly of:

Table 1.10
Uranium Isotopes

Isotope	Abundance (%)	Lifetime (years)
^{238}U	99.283	4.5×10^9
^{235}U	0.711	7.1×10^8
^{234}U	0.005	2.5×10^5

It is estimated that in the Western World there are reserves of uranium oxide (U_3O_8) amounting to some 6×10^9 kg, but only 34×10^6 kg are fissionable, corresponding to an available energy of 2600 EJ. Compare this with the 40,000 EJ of available coal energy.

The relatively modest resources in fissionable uranium led to "breeder reactors" in which fertile materials are transformed into fissionable ones.

Take ^{238}U, which suffers inelastic collisions with high energy neutrons (neutrons from fission):

$$^{238}_{92}U + ^1_0n \rightarrow ^{239}_{92}U \rightarrow ^{~~0}_{-1}e + ^{239}_{93}Np \rightarrow ^{~~0}_{-1}e + ^{239}_{94}Pu \qquad (10)$$

[†] No enrichment is needed if the moderator is heavy water (D_2O) as used in the CANDU reactor. This is the **CAN**adian **D**euterium **U**ranium, pressurized heavy water reactor that uses natural (unenriched) uranium and heavy water as both moderator and coolant.

or take ^{232}Th:

$$\ce{^{232}_{90}Th} + \ce{^{1}_{0}n} \rightarrow \ce{^{233}_{90}Th} \rightarrow \ce{^{0}_{-1}e} + \ce{^{233}_{91}Pa} \rightarrow \ce{^{0}_{-1}e} + \ce{^{233}_{92}U}. \qquad (11)$$

By creating plutonium in a breeder reactor, all uranium can be made to yield energy: 320,000 EJ become available. Even larger amounts of energy could be derived from thorium.

1.11.2 Fusion

The main objections to fission reactors are related to

1. lack of safety,
2. dangerously radioactive "ashes," and
3. scarcity of fuel.

Fusion reactors may overcome all of the above objections.
The reaction that is, by far, the easiest to ignite is[†]

$$\ce{^{3}_{1}T} + \ce{^{2}_{1}D} \rightarrow \ce{^{4}_{2}He} + \ce{^{1}_{0}n} \qquad (12)$$

To estimate the reaction energy released, one calculates the amount of mass lost. The mass of $\ce{^{3}_{1}T}$ is 5.00827×10^{-27} kg as given by the table at the beginning of this subsection. The mass of the deuterium is 3.3434×10^{-27} kg so that the mass of the left side of the equation is 8.3526×10^{-27} kg. On the right-hand side of the equation the sum of the masses of the alpha particle (the helium ion) and the neutron is 8.3214×10^{-27} kg, a deficit of 3.12×10^{-29} kg. When multiplied by c^2, this yields an energy of 2.80×10^{-12} joules per deuterium/tritium pair. The correct value is very slightly larger (it is nearer 2.81×10^{-12} J). The small discrepancy is mainly due to the fact that we used the mass of the atoms instead of that of the corresponding ions. The reaction yields 337 TJ per kg of tritium/deuterium alloy or 562 TJ per kg of tritium.

The energy released by the reaction is carried by both the alphas and the neutrons. The conversion of the neutron energy to usable forms has an efficiency of only some 40% because the particles are uncharged and heat management and mechanical heat engines are involved. On the other hand, the alphas can be directly converted to electricity at a much higher efficiency ($\approx$ 90%). See Rostoker, Monkhorst, and Binderbauer (1997); Moir and Barr (1973); Momota et al. (1998); Yoshikawa et al. (1991); Bloch and Jeffries (1950). In addition, the heavy neutron flux creates serious radioactivity and material destruction problems. Consequently, it is

---

[†] The larger the atomic number, Z, the greater the difficulty of reaction owing to the large electrostatic repulsion between nuclei.

important to know how the released energy is divided between the alphas and the neutrons. This can be done by assuming that the momenta are equally divided between the two types of particles:

$$m_\alpha v_\alpha = m_n v_n, \tag{13}$$

and combining this with with the energy equation,

$$\frac{1}{2}m_\alpha v_\alpha^2 + \frac{1}{2}m_n v_n^2 = W = W_\alpha + W_n. \tag{14}$$

Here, m_α is the mass of the alpha, m_n is the mass of the neutron, v_α is the velocity of the alpha, v_n is the velocity of the neutron, and W is the energy released by one pair of reacting atoms. Solving these simultaneous equations leads to

$$W_\alpha = \frac{W}{\frac{m_\alpha}{m_n} + 1} \tag{15}$$

and

$$W_n = \frac{W}{\frac{m_n}{m_\alpha} + 1}. \tag{16}$$

For the reaction under consideration, it is found that neutrons carry about 14 MeV, while the more massive alphas carry only some 3.5 MeV.

The T+D reaction is popular because of its high reactivity, which should facilitate ignition, and because the atomic number of the fuel is $Z = 1$, thus minimizing radiation losses. This is because radiation is a function of Z^2. However, it has drawbacks:

1. One neutron is emitted for each 2.8×10^{-12} J generated, whereas, in fission, the rate is one neutron per 10^{-11} J. Thus, the neutron bombardment is serious: it radioactivates substances and weakens structures by causing dislocations in the crystal lattice and by generating hydrogen bubbles inside materials.
2. As pointed out before, most of the energy is in the neutron stream reducing the recovery efficiency.
3. Although deuterium is not radioactive, tritium is radioactive with a lifetime of 12 years. It has the tendency to stick around by replacing normal hydrogen in water molecules.
4. There is no natural source of tritium; it must be obtained from lithium:

$$_3^6\text{Li}^6 + {}_0^1\text{n} \rightarrow {}_1^3\text{T} + {}_2^4\text{He} + ...   7.7 \times 10^{-13} \text{ joules} \tag{17}$$

Thus, each lithium atom yields $2.8 \times 10^{-12} + 7.7 \times 10^{-13} = 3.57 \times 10^{-12}$ J. One kg of lithium yields 350 TJ.

The world reserves of lithium are not known accurately. Conservative estimates are of 10^{10} kg. However, most of this is ^{7}Li. The desired isotope, ^{6}Li, has a relative abundance of 7.4%. Consequently, one can count on only 740×10^6 kg of this material or 260,000 EJ of energy.

In order of ease of ignition, the next two reactions are

$$\ce{_1^2D + _1^2D \rightarrow _2^3He + _0^1n} + ... \quad 5.23 \times 10^{-13} \text{ joules,} \tag{18}$$

$$\ce{_1^2D + _1^2D \rightarrow _1^3T + _1^1H} + ... \quad 6.45 \times 10^{-13} \text{ joules.} \tag{19}$$

The above reactions have equal probability of occurring.

The tritium produced will react with the deuterium according to Reaction 12. The average energy of the D+D reaction is

$$\frac{(5.23 + 6.45 + 28.0) \times 10^{-13}}{5} = 7.94 \times 10^{-13} \text{ J per D atom.} \tag{20}$$

The D+D reaction is still dirty (neutronwise) and still involves a radioactive gas (tritium). However, it does not use a fuel of limited abundance, such as lithium. It uses only deuterium, which is available in almost unlimited amounts. In common water, there is one D_2O molecule for every 6700 H_2O molecules. One can estimate roughly how much deuterium is available:

The oceans cover about 2/3 of the Earth's surface, which is 5.1×10^{14} m^2. Assuming an average depth of 3000 m, the ocean has a volume of 10^{18} m^3 and a mass of 10^{21} kg. Of this, 1/9 is the mass of hydrogen, and 2/6700 of the latter is the mass of deuterium, amounting to some 3.3×10^{16} kg or about 10^{31} J—an amount of energy that, for practical purposes, can be considered unlimited.

Next, in order of ignition difficulty is the ^{2}D + ^{3}He reaction that burns cleanly: no radioactive substances are involved and no neutrons are generated. Also clean is the ^{3}H + ^{3}He reaction.

The catch in these reactions is that there is no natural ^{3}He on earth; it must be made from the (dirty) fusion of Li and H. However, it is estimated that over a billion tons of the material exists on the moon. This may, one day, justify a mining operation on our satellite.

The ^{3}H on the moon comes from the solar wind that has, for billions of years, deposited it there. The ^{3}H on earth is trapped by the atmosphere and is eventually evaporated away.

An interesting reaction involves ^{11}B, the common isotope of boron:

$$\ce{_5^{11}B + _1^1H \rightarrow _6^{12}C^* \rightarrow _2^4He + _4^8Be.} \tag{21}$$

$\ce{_6^{12}C^*}$ is nuclearly excited carbon which spontaneously decays into an alpha and $\ce{_4^8Be}$ a very unstable atom with a lifetime of 2×10^{-16} seconds. Fortunately, it is an alpha-emitter:

$$\frac{8}{4}\text{Be} \rightarrow 2\frac{4}{2}\text{He}. \tag{22}$$

The overall reaction is:

$$^{11}_{5}\text{B} + ^{1}_{1}\text{H} \rightarrow 3^{4}_{2}\text{He}. \tag{23}$$

or, using a different notation

$$^{1}_{1}\text{H} + ^{11}_{5}\text{B} = 3\alpha. \tag{24}$$

It appears that this **triple alpha** reaction can be made to sustain itself in a **colliding beam fusion reactor** (see Rostoker, Binderbauer, and Monkhorst, 1997) but this has not yet been demonstrated. If it does work, we would have a clean fusion reactor using abundantly available fuel and capable of operating in units of moderate size, in contrast with the T+D reaction in a Tokamak with must be 10 GW or more if it can be made to work at all.

It should be noticed that ^{10}B will also yield a triple-alpha reaction when combining with a deuteron:

$$^{10}_{5}\text{B} + ^{2}_{1}\text{D} \rightarrow 3^{4}_{2}\text{He}. \tag{25}$$

Both isotopes of boron considered above are abundant, stable, and nonradioactive. Natural boron consists essentially of 20% ^{10}B and 80% ^{11}B.

The triple alpha reaction may also be an important player in the cold fusion process (if such process exists at all). See the next subsection.

Table 1.11 lists the percentage of the energy of a reaction that is carried away by neutrons.

Although fusion reactors have not yet been demonstrated,[†] there is a possibility that they will become the main source of energy some 50 years from now. If so, they may provide the bulk of the energy needed by humanity and the energy crunch will be over.

Table 1.11
Neutron Yields

Reaction	% of energy Carried by Neutrons
D + T	65...75
D + T	65...75
D + D	20...45
B + H	< 0.1

[†] Fusion research dates back to at least 1938 when Jacobs and Kantrowitz built a first magnetic confinement fusion reactor at NACA's (now NASA) Langley Memorial Aeronautical Laboratory.

1.11.3   Cold Fusion

At the beginning of the millennium, when this subsection was being rewritten, the cold fusion question remained unresolved. So far, no one has been able to reproduce the claims of Pons and Fleishmann, but, on the other hand, no one has been able to disprove the existence of cold fusion. As a matter of fact, cold fusion can and has been demonstrated. Let us review what we know for sure of this topic.

As indicated in Subsection 1.11.2, deuteron will react spontaneously with deuteron in one of these two reactions:

$$\ _1^2D + \ _1^2D \rightarrow \ _1^3T + \ _1^1H \tag{26}$$

$$\ _1^2D + \ _1^2D \rightarrow \ _2^3He + \ _0^1n \tag{27}$$

These reactions have about the same probability of occurrence and they produce a substantial amount of energy. The problem is that the probability of occurrence (under normal conditions) is extremely small, of the order of one fusion per galaxy per century according to a good-humored scientist,

It is easy to understand the reluctance of the ^2D atoms to get together: they carry positive charges and therefore repel one another. This can be overcome by imparting sufficient kinetic energy to the atoms, as, for instance, by heating them to extreme temperatures as in thermonuclear fusion.

There is a neat trick suggested by Alvarez (late professor of the University of California at Berkeley and Nobel Prize winner) that increases by 85 orders of magnitude the reaction cross-section (read probability). Replacing the orbital electron of the deuterium by a muon which is 207 times heavier, collapses the orbital by a large factor.[†] Muon mediated fusion can be observed in the laboratory as Jones (Brigham Young) demonstrated. The catch is that it takes more energy to create the muon than what one gets from the fusion.

Thus, cold fusion certainly does occur. More than that, cold fusion occurs (almost certainly) even when not mediated by muons.

Jones described an experiment that appears to prove just that. He used an electrolytic cell consisting of a platinum positive electrode and a palladium (sometimes, titanium) negative electrode. The electrolyte was D_2O (heavy water). Since water is a poor conductor of electricity, salts had to be added to the solution. Here is Jones's extraordinary recipe:

[†]A deuteron/electron molecule is about 74,000 fm in size, while the deuteron/ muon molecule is only 250 fm in size. The nuclei are, therefore, 300 times closer together and this raises enormously the probability of fusion.

"The electrolyte is a mixture of about 160 g of deuterium oxide (D_2O) plus various metal salts in about 0.2 g amounts each: $FeSO_4$, $NiCl_2$, $PdCl_2$, $CaCO_3$, $LiSO_4$, $NaSO_4$, $CaH_4(PO_4)_2$, $TiOSO_4$, and a very small amount of $AuCN$."

A chemist might be horrified by the cocktail above—it would be hard to tell what is going on.[†]

When a current was forced through the cell, a small flux of neutrons with a characteristic energy of 2.5 Mev was observed. Jones, a physicist, did a good job of neutron detection. Since 2.5 Mev is the energy of the neutrons in Reaction 27, this experiment tends to show that indeed fusion is going on.

Jones observed that some 8 hours after start of operation, the neutron "signal" turned off by itself. This effect was attributed to the poisoning of the palladium electrode by deposition of metals from the solution. In fact, etching the electrode revived the cell.

The reaction rate observed by Jones was small, perhaps 10^{-20} fusions per deuterium pair per second. This could be explained if the deuterium molecules were somehow squeezed from 74,000 fm to half this distance by their residence in the palladium lattice.[††] Jones dubs this **piezonuclear fusion**.

Pons and Fleishmann ran similar experiments but, being chemists not physicists, adopted a simpler electrolyte: a LiOH solution in D_2O (heavy water). They also failed to make careful neutron measurements. What they reported is that, after a prolonged pre-cooking, some cells suddenly developed a great deal of heat, billions of time more than in the Jones experiment. Unfortunately, these results were never reproduced by other experimenters and this casts severe doubts on their validity. Here is where I will don my devil's advocate mantle and, just for the fun of it, will defend the P&F results.

In a lecture delivered at the Utah University on March 31, 1989, Stanley Pons relates the most spectacular of his results. "A cube of palladium with a volume of 1 cm^3 was used as cathode of an electrolyzer with lithium hydroxide dissolved in D_2O as an electrolyte. A current of 250 mA/cm^2 was applied for several weeks/months [sic] with nothing remarkable happening. A Geiger counter detected no radiation. The current was cut to 125 mA/cm^2 late one day, and next morning the cube of palladium and the electrolysis cell were gone. A nearby Geiger counter was also ruined."[†††]

[†] Jones was trying to create a chemical environment somewhat like the one in the soil because he was trying to show that some of the internal heat of our planet is generated by deuterium fusion.

[††] A possible cause of the squeezing would be the increase of the electron mass to a few times its free mass.

[†††] As related by Patrick Nolan, 1989 (paraphrased).

There was a long delay (several days, at least) before heat evolved. Since the Jones cell poisons itself in 8 hours, this cell will never reach the primed state and no heat can be observed.

Why such a delay? Hydride hydrogen storage systems (see Chapter 11) are well known and are commercially available. One popular system uses a TiFe alloy to absorb H_2. Many other metals and alloys will do the same. Palladium, in particular, is a notorious H_2 absorber. It is not used commercially owing to its high price.

When TiFe powder (after duly activated) is exposed to hydrogen, it will form a (reversible) hydride, TiFeH. If the amount of hydrogen is small, there will be a mixture of TiFe and TiFeH in the powder. This mixture, called β-phase, has the empirical formula $TiFeH_x$, where x becomes 1 when all the material has been hydrided.

After full hydridization, addition of more hydrogen will cause the formation of a di-hydride, $TiFeH_2$ (γ-phase). Clearly, the hydrogen is more densely packed in the (di-hydride) γ-phase than in the β-phase. It is, therefore, plausible that the fusion will proceed faster once the γ-phase is reached. How long does it take to reach this γ-phase?

In the described experiment, Pons used a current density of 250 mA/cm^{-2}, a total current of 1.5 A, added over the six sides of the cubic cathode. This corresponds to a production of 9.4×10^{18} deuterons/second. Each cubic centimeter of palladium contains 68×10^{21} atoms. Thus, it takes 7,200 seconds or 2 hours) for the palladium, in this particular experiment, to start becoming di-hydrided. This assumes that all the deuterons produced are absorbed by the palladium and, thus, the time calculated is a rough lower limit.

Could the heat have resulted from a chemical reaction? The highest enthalpy of formation of any palladium salt seems to be 706 MJ/kmole, for palladium hydroxide. Atomic mass of palladium is 106 daltons and density is 12 g cm^{-3}. This means that one gets 80 kJ cm^{-3} chemically. Pons and Fleishmann have (they say) gotten 5 MJ cm^{-3}, two orders of magnitude more than chemistry allows.

Conclusion:

1. The heat produced cannot be due to classical fusion reaction (insufficient neutrons, tritium, and γ-rays).
2. The heat produced cannot be due to chemical reaction.
3. Then, simplistically, the heat was not produced.

There is at least one more possible reaction which occurs very rarely:

$$^2_1D + {}^2_1D \rightarrow {}^4_2He \tag{28}$$

As written above, this reaction cannot take place because two particles are converted into a single particle and it is impossible to conserve simultaneously energy and momentum under such conditions. For the reaction to proceed, it is necessary to shed energy and, in classical physics, this is done by emitting a 16 MeV γ-ray. Pons did not report γ-rays. There is still an outside possibility that the energy can be shed by some other mechanism such as a phonon, although physicists tell me that this is nonsense. Observe that Reaction 28 produces one order of magnitude more energy per fusion than do Reactions 26 and 27.

So far, we have attempted to explain the hypothetical cold fusion as the result of deuteron-deuteron reaction. It has been difficult to account for the absence of the expected large fluxes of neutrons or gamma rays. It is even more difficult to imagine such reaction proceeding when common water is used in place of heavy water. Nevertheless, some experimentalists make exactly such a claim.

There have been suggestions that cold fusion actually involves nuclear reactions other than those considered so far. Let us recapitulate what has been said about cold fusion.

1. The results, if any, are not easily reproduced.
2. No substantial neutron flux has been detected. This seems to eliminate the deuteron-deuteron reactions of Equations 26 and 27.
3. No substantial gamma ray flux has been detected. This eliminates the classical form of the deuteron-deuteron reaction of Equation 28.
4. Reactions are reported to be highly dependent on the exact nature of the palladium electrode.
5. Reactions have been reported with an H_2O instead a D_2O electrolyte.

The following cold fusion mechanism fitting the above observations has been recently proposed.

Boron is a common impurity in palladium. Natural boron exists in the form of two isotopes with the relative abundance of 20% for ^{10}B and 80% for ^{11}B. Thus, under some special circumstances, the two triple-alpha reactions of Equations 24 and 25 might occur. They emit neither neutrons nor gamma rays and can occur with either normal water or heavy water.

The boron impurity may be interstitial or it may collect in grain boundaries. The reaction may only occur if the boron is in one or the other of these distributions. It may also only occur when the amount of impurity falls within some narrow range. Thus, a palladium rod may become "exhausted" after some time of operation if the boron concentration falls below some given limiting concentration.

Perhaps the worst indictment of the P&F experiment is its irreproducibility. No one has claimed to have seen the large heat production reported from Utah. Pons himself states that his experiment will only work occasionally—he claims that there is *live* palladium and *dead* palladium.

This could be interesting. Hydrogen absorbed in metals is known to accumulate in imperfections in the crystal lattice. It is possible that such
defects promote the high concentrations of deuterium necessary to trigger
the reaction.

I still have an old issue of the CRC handbook that lists the thermoelectric power of silicon as both $+170$ mV/K and -230 mV/K. How can it
be both positive and negative? Notice that the determination of the sign
of the Seebeck effect is trivial; this cannot be the result of an experimental
error. In both cases "chemically pure" silicon was used. So, how come?
We have a good and classical example of irreproducibility. That was back
in the 1930s. Now any EE junior knows that one sample must have been
p-silicon, while the other, n-silicon. Both could be "chemically pure"—to
change the Seebeck sign, all it takes is an impurity concentration of 1 part
in 10 million. Is there an equally subtle property in the palladium that will
allow fusion in some cases?

In April 1992 Akito Takahashi of Osaka University revealed that his
cold fusion cell produced an average excess heat of 100 W over periods
of months. The electric power fed to the cell was only 2.5 W. The main
difference between the Takahashi cell and that of other experimenters is
the use of palladium sheets (instead of rods) and of varying current to
cause the cell to operate mostly under transient conditions. The excess
heat measured is far too large to be attributed to errors in calorimetry.
Disturbing to theoreticians is the absence of detectable neutrons. See D.
H. Freedman's (1992) report.

In spite of being saddled with the stigma of "pseudo-science," cold
fusion does not seem to go away. The September 2004 issue of *IEEE Spectrum*, published a report titled "Cold Fusion Back from the Dead," in which
recent work on cold fusion by reputable laboratories is mentioned. It quotes
the US Navy as revealing that the Space and Naval Systems Center (San
Diego) was working on this subject.[†] It also mentioned the Tenth International Conference on Cold Fusion that took place in Cambridge, MA, in
August 2003.

It appears that by 2004, "a number of groups around the world have
reproduced the original Pons-Fleishmann excess heat effect ... " Mike McK-
ubre of SRI International maintains that the effect requires that the palladium electrode be 100% packed with deuterium (one deuterium-to-one-
palladium atom). This coincides with our wild guess at the beginning of
this subsection.

At the moment, cold fusion research has gone partially underground,
at least as far as the media are concerned. Yet, the consensus is that it

---

[†] It is reported that Stanislaw Szpak, of the SNSC, has taken infrared pic-
tures of miniexplosions on the surface of the palladium, when cold fusion appears
to be taking place.

merits further study. This is also the opinion of independent scientists such as Paul Chu and Edward Teller who have been brought in as observers. It may be that cold fusion will one day prove practical. That is almost too good to be true and, for the classical fusion researchers, almost too bad to be true.

1.12 Financing

Some of the proposed alternative energy sources, such as the fusion reactor, require, for their implementation, a scientific breakthrough. Others need only technological development, as is the case of wind turbines or of ocean thermal energy converters. Still others have reached a fairly advanced stage of development, but their massive implementation awaits more favorable economic conditions, such as further increase in the price of oil. The production of synthetic fuel from coal falls in this category, as does the utilization of shale.

Finding new sources of energy is not difficult. What is difficult is finding new sources of *economically attractive* energy. It is, therefore, important to estimate the cost of the energy produced by different methods. One of the main ingredients of the cost formula is the cost of financing, examined below.

Frequently, the financing of the development is borne by the government, especially during the early high-risk stages of the work. It is an important political decision for the nation to finance or not to finance the development of a new energy source. For instance, the Solar Power Satellite scheme is one that has possibilities of being economical. However, its development costs, estimated as nearly 80 billion dollars, are too high to be funded by private corporations. Thus, the SPS system will be implemented only if the government feels justified in paying the bill.

Financing the implementation is simpler. Engineers can estimate roughly how the investment cost will affect the cost of the product by using a simple rule of thumb:

"The yearly cost of the investment can be taken as 20%[†] of the overall amount invested."

Thus, if a 1 million dollar power plant is to be built, one must include in the cost of the generated energy, a sum of $200,000 per year.

To allow a comparison of the costs of energy produced by different alternative sources, the Department of Energy has recommended a standard method of calculating the cost of the capital investment.

[†] This percentage is, of course, a function of the current interest rate. In the low interest rate regimen of the early years of this millennium, the percentage is lower than 20%.

We will here derive an expression for the cost of a direct reduction loan.

Assume that the payment of the loan is to be made in N equal install-ments. We will consider a \$1.00 loan. Let x be the interest rate of one payment period (say, one month) and let p be the value of the monthly payment. At the end of the first month, the amount owed is

$$1 + x - p \tag{29}$$

and, at the end of the second month, it is

$$(1 + x - p)(1 + x) - p = (1 + x)^2 - p(1 + 1 + x) \tag{30}$$

and, at the end of the third month, it is

$$[(1 + x)^2 - p(1 + 1 + x)](1 + x) - p$$
$$= (1 + x)^3 - p[1 + (1 + x) + (1 + x)^2]. \tag{31}$$

At the end of N months, the amount owed is zero because the loan has been repaid. Thus,

$$(1 + x)^N - p[1 + (1 + x) + (1 + x)^2 + ... + (1 + x)^{N-1}] = 0, \tag{32}$$

whence

$$p = \frac{1}{(1 + x)^1 + (1 + x)^{-2} + ... + (1 + x)^{-N}} = \left[\sum_{\gamma=1}^{N} z^\gamma\right]^{-1}, \tag{33}$$

where

$$z \equiv (1 + x)^{-1}. \tag{34}$$

But

$$\sum_{\gamma=1}^{N} z^\gamma = \frac{1 - z^{N+1}}{1 - z} - 1, \tag{35}$$

hence

$$p = \frac{1 - z}{z - z^{N+1}} = \frac{x}{1 - (1 + x)^{-N}}. \tag{36}$$

The formula above yields the magnitude of the monthly payment as a function of the interest rate (per month) and the number of payments.

As an example, consider a small entrepreneur who owns a Diesel-electric generating plant in which he has invested \$1000 per kW. The utilization factor is 50%—that is, 4380 kWh of electricity are produced yearly for each kW of installed capacity. Taxes and insurance amount to

$50 year^{-1} kW^{-1}. Fuel, maintenance, and personnel costs are $436 kW^{-1} year^{-1}. In order to build the plant, the entrepreneur borrowed money at 12% per year and is committed to monthly payments for 10 years. What is the cost of the generated electricity?

The monthly rate of interest is

$$(1 + x)^{12} = 1.12 \qquad \therefore \qquad x = 0.009489. \tag{37}$$

The number of payments is

$$N = 10 \text{ years} \times 12 \text{ months/year} = 120. \tag{38}$$

The monthly payment is

$$p = \frac{0.009489}{1 - (1 + 0.009489)^{-120}} = \$0.013995 \text{ month}^{-1}. \tag{39}$$

The yearly payment is

$$P = 12p = \$0.167937 \text{ year}^{-1}. \tag{40}$$

If there were no interest, the yearly payment would be $0.1. Thus, the yearly cost of interest is $0.067937.

All the above is on a loan of $1.00. Since the plant cost $1000 kW^{-1}, the cost of the investment is $167.94 kW^{-1} year^{-1}. But, on a per kW basis, there is an additional expense of $50 for taxes and insurance, raising the yearly total to $217.94. Thus, in this example, the yearly investment cost is 21.79% of the total amount.

A total of 4380 kWh per kW installed are generated (and sold) per year. The fixed cost per kWh is, therefore

$$\frac{217.94}{4380} = 0.0497 \text{ \$ kWh}^{-1}, \tag{41}$$

whereas, the fuel, maintenance, and personnel cost is

$$\frac{436}{4380} = 0.0995 \text{ \$ kWh}^{-1}. \tag{42}$$

Total cost is 0.1492 $ kWh^{-1}. This is commonly expressed as 149.2 mils/kWh, an awkward unit. It is better to use 149.2 $/MWh or, to stick to the conventional SI units of measure, $41.4 GJ^{-1}.

When the loan is paid after 10 years, does the entrepreneur own the plant? Maybe. The Diesel-generator may have only a 10-year life and a new one may have to be acquired.

CHAPTER 1   Generalities   39

References

Anderson, R, H. Brandt, H. Mueggenburg, J. Taylor, and F. Viteri, A power plant concept which minimizes the cost of carbon dioxide sequestration and eliminates the emission of atmospheric pollutants, *Clean Energy Systems, Inc., 1812 Silica Avenue, Sacramento, CA 95815*, **1998**.

Audi, G., and A. H. Wapstra, The 1993 atomic mass evaluation, *Nuclear Physics A 565* **1993**.

Bloch, F., and C. D. Jeffries, *Phys. Rev.* **77**, 305, **1950**.

Casson, Lionel, Godliness & Work, *Science 81*, 2, 36, **1981**.

Firestone, Richard B., <http://ie.lbl.gov/toi2003/MassSearch.asp>

Fisher, J. C., and R. H. Pry, A simple substitution model of technological change, *Report 70-C-215, General Electric, R. & D. Center*, June, **1970**.

Fleishmann, M., and S. Pons, Electrochemically induced nuclear fusion of deuterium, *J. Electroanal. Chem., 261*, 301–308, **1989**.

Freedman, D. H., A Japanese claim generates new heat, *News and Comments, Science, 256*, 24 April **1992**.

Jones, S. E., et al., Observation of cold nuclear fusion in condensed matter, *reprint from Brigham Young University*, March 23, **1989**.

Hafele, W., and W. Sassin, Resources and endowments. An outline on future energy systems, *IIASA, NATO Science Comm. Conf.*, Brussels, April **1978**.

Loewen, Eric P., Heavy-metal nuclear power, *American Scientist*, November-December **2004**.

Marchetti, C., Primary energy substitution models, *Int.Inst.Appl.Syst. An.(IIASA)* internal paper WP-75-88, June **1975**.

Moir, R. W., and W. L. Barr, *Nucl. Fusion* **13**, 35, **1973**.

Momota, H. et al. *Fusion Technol.* **21**, 2307, **1992**.

Nolan, Patrick, *e-mail circular*, 31 March **1989**.

Peterka, V., Macrodynamics of technological change: Market penetration by new technologies, *Int. Inst. Appl. Syst. An. (IIASA)*, RR-77-22, November **1997**.

Rafelski, J., et al., Theoretical limits on cold fusion in condensed matter, *AZPH-TH/89-19*, March 27, **1989**.

Rostoker, N., H. Monkhorst, and M. Binderbauer, *Office of Naval Research reports*, February, May, and August **1997** (available upon request).

Rostoker, Norman, Michl W. Binderbauer, and Hendrik J. Monkhorst, *Science* **278**, 1419, 21 Nov. **1997**.

Smil, Vaclav, Global population and the nitrogen cycle, *Scientific American*, p. 76, July **1997**.

Sweet, William #1, A nuclear reconnaissance, *IEEE Spectrum* 23, Nov. **1997**.

Sweet, William #2, Advanced reactor development rebounding, *IEEE Spectrum* 23, Nov. **1997**.

Yoshikawa, K., T. Noma, and Y. Yamamoto *Fusion Technol.* **19**, 870, **1991**.

Ziock, Hans-J., Darryl P. Butt, Klaus S. Lackner, and Christopher H. Wendt, *Reaction Engineering for Pollution Prevention, Elsevier Science*, **2000**.

Abundant statistical information on energy: http://www.eia.doe.gov/

For more detailed information on some topics in this chapter, read:
Sørensen, Bent, Renewable energy, *Academic Press* **2003**.

PROBLEMS

1.1   Assume that from 1985 on the only significant sources of fuel are:

    1. coal (direct combustion),
    2. oil,
    3. synthetic liquid fuel (from coal), and
    4. natural gas.

    Sources a, b, and c are assume to follow the market penetration rule:

$$ln\frac{f}{1-f} = at + b$$

where f is the fraction of the market supplied by the fuel in question and t is the year (expressed as 1988, for instance, not as simply 88). The coefficients are:

	a	b
for coal:	−0.0475,	92.14;
for oil:	−0.0436,	86.22.

    The above coefficients are derived from historical data up to 1975.

    The objective of this exercise is to predict what impact the (defunct) federal coal liquefaction program would have had on the fuel utilization pattern.

    According to the **first in, first out** rule, the "free" variable, i.e., the one that does not follow the market penetration rule, is the natural gas consumption fraction, f_{ng}. The questions are:
– in what year will f_{ng} peak?
– what is the maximum value of f_{ng}?

    Assume that f_{syn} (the fraction of the market supplied by synthetic fuel) is 0.01 in 1990 and 0.0625 in 2000. Please comment.

1.2   The annual growth rate of energy utilization in the world was 3.5% per year in the period between 1950 and 1973. How long would it take to consume all available resources if the consumption growth rate of 3.5% per year is maintained?

    Assume that the global energy resources at the moment are sufficient to sustain, at the current utilization rate

    a. 1000 years,
    b. 10,000 years.

1.3   A car moves on a flat horizontal road with a steady velocity of 80 km/h. It consumes gasoline at a rate of 0.1 liters per km. Friction of the tires on the road and bearing losses are proportional to the velocity and, at

80 km/h, introduce a drag of 222 N. Aerodynamic drag is proportional to the square of the velocity with a coefficient of proportionality of 0.99 when the force is measured in N and the velocity in m/s.

What is the efficiency of fuel utilization? Assuming that the efficiency is constant, what is the "kilometrage" (i.e., the number of kilometers per liter of fuel) if the car is driven at 50 km/h?

The density of gasoline is 800 kg per cubic meter and its heat of combustion is 49 MJ per kg.

1.4   Venus is too hot, in part because it is at only 0.7 AU from the sun. Consider moving it to about 0.95 AU. One AU is the distance between Earth and Sun and is equal to 150 million km.

To accomplish this feat, you have access to a rocket system that converts mass into energy with 100% efficiency. Assume that all the energy of the rocket goes into pushing Venus. What fraction of the mass of the planet would be used up in the project? Remember that you are changing both kinetic and potential energy of the planet.

1.5   Consider the following arrangement:

A bay with a narrow inlet is dammed up so as to separate it from the sea, forming a lake. Solar energy evaporates the water causing the level inside the bay to be h meters lower than that of the sea.

A pipeline admits sea water in just the right amount to compensate for the evaporation, thus keeping h constant (on the average). The inflow water drives a turbine coupled to an electric generator. Turbine plus generator have an efficiency of 95%.

Assume that there is heat loss neither by conduction nor by radiation. The albedo of the lake is 20% (20% of the incident radiation is reflected, the rest is absorbed). The heat of vaporization of water (at STP) is 40.6 MJ per kilomole. Average solar radiation is 250 W/square meter.

If the area of the lake is 100 km^2, what is the mean electric power generated? What is the efficiency? Express these results in terms of h.

Is there a limit to the efficiency? Explain.

1.6   The thermonuclear (fusion) reaction

$$^{11}_{5}B + ^{1}_{1}H \rightarrow 3^{4}_{2}He,$$

is attractive because it produces essentially no radiation and uses only common isotopes.

How much energy does 1 kg of boron produce? Use the data of Problem 1.11.

1.7   The efficiency of the photosynthesis process is said to be below 1% (assume 1%). Assume also that, in terms of energy, 10% of the biomass produced is usable as food. Considering a population of 6 billion people, what percentage of the **land** area of this planet must be planted to feed these people.

1.8   Each fission of ^{235}U yields, on average, 165 MeV and 2.5 neutrons. What is the mass of the fission products?

1.9   There are good reasons to believe that in early times, the Earth's atmosphere contained no free oxygen.

Assume that all the oxygen in the Earth's atmosphere is of photosynthetic origin and that all oxygen produced by photosynthesis is in the atmosphere. How much fossil carbon must there be in the ground (provided no methane has evaporated)? Compare with the amount contained in the estimated reserves of fossil fuels. Discuss the results.

1.10   What is the total mass of carbon in the atmosphere?

CO_2 concentration is currently 330 ppm but is growing rapidly!

If all the fossil fuel in the estimated reserves (see Section 1.8) is burned, what will be the concentration of CO_2

1.11   Here are some pertinent data:

Particle	Mass (daltons)	Particle	Mass (daltons)
electron	0.00054579903	alpha	4.001506175
muon	0.1134381	^5_3Li	5.01254
proton	1.007276467	^6_3Li	6.015122794
neutron	1.008664909	^7_3Li	7.01600455
^1_1H	1.007825032	$^{10}_5B$	10.012937
^2_1D	2.014101778	$^{11}_5B$	11.009305
^3_1T	3.016049278		
^3_2He	3.016029319		
^4_2He	4.002603254		

Constants	
c	2.998×10^8 m/s
h	6.625×10^{-34} joule-sec

To convert daltons to kg, divide by $6.02213670 \times 10^{26}$.

Deuterium is a very abundant fusion fuel. It exists in immense quantities in earth's oceans. It is also, relatively easy to ignite. It can undergo three different reactions with itself:

$$^2_1D + ^2_1D \rightarrow ^3_1T + ^1_1H \tag{1}$$

$$^2_1D + ^2_1D \rightarrow ^3_2He + _0n^1 \tag{2}$$

$$^2_1D + ^2_1D \rightarrow ^4_2He + h\nu \tag{3}$$

For each reaction, calculate the energy released and, assuming equipartition of momenta of the reaction products, the energy of each product. What is the energy of the photon released in Reaction 3?

1.12   Random access memories (RAMs) using the "Zing Effect" were first introduced in 1988 but only became popular in 1990 when they accounted for 6.3% of total RAM sales. In 1994 they represented $712 million of a total of $4.75 billion. Sales of all types of RAMs reached $6 billion in 1997.

A company considering the expansion of Z-RAM production needs to have an estimate of the overall (all manufacturers) sales volume of this type of memory in the year 2000. Assume that the growth rate of the overall dollar volume of RAM sales between 1900 and 2000 is constant (same *percentage* increase every year)

1.13   A 1500-kg Porsche 912 was driven on a level highway on a windless day. After it attained a speed of 128.7 km/h it was put in neutral and allowed to coast until it slowed down to almost standstill. The coasting speed was recorded every 10 seconds and resulted in the table below.

From the given data, derive an expression relating the decelerating force to the velocity.

Calculate how much horse power the motor has to deliver to the wheel to keep the car at a constant 80 mph.

Coasting time (s)	Speed (km/h)	Coasting time (s)	Speed (km/h)
0	128.7	100	30.6
10	110.8	110	25.9
20	96.2	120	20.4
30	84.0	130	16.2
40	73.0	140	12.2
50	64.2	150	9.2
60	56.4	160	5.1
70	48.0	170	2.0
80	41.8	180	0
90	35.8		

1.14   The California Air Resources Board (CARB) mandated, for 1995, an upper limit of 200 g/km for the emission of CO_2 from a minivan.

This could be achieved by bubbling the exhaust through a $Ca(OH)_2$ bath or through a similar CO_2 sequestering substance. However, this solution does not seem economical. Assume that all the produced CO_2 is released into the atmosphere.

What is the minimum mileage (miles/gallon) that a minivan had to have by 1995. Assume gasoline is pentane (C_5H_{12}) which has a density of 626 kg m^{-3}. A gallon is 3.75 liters and a mile is 1609 meters. The atomic mass of H is 1, of C is 12, and of O is 16.

1.15   A geological survey revealed that the rocks in a region of Northern California reach a temperature of 600 C at a certain depth. To exploit this geothermal source, a shaft was drilled to the necessary depth and a spherical cave with 10 m diameter was excavated. Water at 30 C is injected into the cave where it reaches the temperature of 200 C (still in liquid form, owing to the pressure) before being withdrawn to run a steam turbine.

Assume that the flow of water keeps the cave walls at a uniform 200 C. Assume, furthermore that, at 100 m from the cave wall, the rocks are at their 600 C temperature. Knowing that the heat conductivity, λ, of the rocks is 2 W m^{-1}K^{-1}, what is the flow rate of the water?

The heat capacity of water is 4.2 MJ m^{-3}K^{-1} and the heat power flux (W m^{-2}) is equal to the product of the heat conductivity times the temperature gradient.

1.16   The following data are generally known to most people:

a. The solar constant, C (the solar power density), at earth's orbit is 1360 W m^{-2};
b. the astronomical unit (AU, the average sun-earth distance) is about 150 million km;
c. the angular diameter of the moon is 0.5°.

Assume that the sun radiates as a black body. From these data, estimate the sun's temperature.

1.17   Using results from Problem 1.16, compare the sun's volumetric power density (the number of watts generated per m^3) with that of a typical Homo sapiens.

1.18   Pollutant emission is becoming progressively the limiting consideration in the use of automobiles. When assessing the amount of pollution, it is important take into account not only the emissions from the vehicle but also those resulting from the fuel production processes. Gasoline is a particularly worrisome example. Hydrocarbon emission at the refinery is some 4.5 times larger than that from the car itself. Fuel cell cars (see Chapter 9) when fueled by pure hydrogen are strictly a zero emission vehicle. However, one must inquire how much pollution results from the production of the hydrogen. This depends on what production method is used (see Chapter 10). The cheapest hydrogen comes from reforming fossil fuels and that generates a fair amount of pollution. A clean way of producing hydrogen is through the electrolysis of water; but, then, one must check how much pollution was created by the generation of the electricity. Again, this depends on how the electricity was obtained: if from a fossil fuel steam plant, the pollution is substantial, if from hydroelectric plants, the pollution is zero.

The technical means to build and operate a true zero emission vehicle are on hand. This could be done immediately but would, at the present stage of the technology, result in unacceptably high costs.

Let us forget the economics and sketch out roughly one possible ZEV combination. Consider a fuel-cell car using pure hydrogen (stored, for instance, in the form of a hydride—Chapter 11). The hydrogen is produced by the electrolysis of water and the energy required for this is obtained from solar cells (Chapter 14). Absolutely no pollution is produced. The system is to be dimensioned so that each individual household is independent. In other words, the solar cells are to be installed on the roof of each home.

Assume that the car is to be driven an average of 1000 miles per month and that its gasoline driven equivalent can drive 30 miles/gallon. The fuel cell version being much more efficient, will drive 3 times farther using the same energy as the gasoline car..

How many kilograms of hydrogen have to be produced per day?

How large an area must the solar cell collector have?

You must make reasonable assumptions about the solar cell efficiency, the efficiency of the electrolyzer and the amount of insolation (Chapter 12).

1.19   From a fictitious newspaper story:

A solar power plant in the Mojave Desert uses 1000 photovoltaic panels, each "40 meters square." During the summer, when days are invariably clear, the monthly sale of electricity amounts to $22,000. The average price charged is 3 cents per kWh. The plant is able to sell all the electricity produced.

There is an unfortunate ambiguity in the story: "40 meters square" can be interpreted as a square with 40 meters to its side or as an area of 40 m^2.

From the data in the story, you must decide which is the correct area.

1.20   Sport physiologists have a simple rule of thumb: Any healthy person uses about 1 kilocalorie per kilometer per kilogram of body weight when running.

It is interesting to note that this is true independently of how well trained the runner is. A trained athlete will cover 1 km in much less time than an occasional runner but will use about the same amount of energy. Of course, the trained athlete uses much more power.

The overall efficiency of the human body in transforming food intake into mechanical energy is a (surprisingly high) 25%!

A good athlete can run 1 (statute) mile in something like 4 minutes and run the Marathon (42.8 km) in a little over 2 hours.

1. Calculate the power developed in these races. Repeat for a poor performer who runs a mile in 8 minutes and the Marathon in 5 hours. Assume a body weight of 70 kg.
2. Evaporation of sweat is the dominant heat removal mechanism in a human body. Is this also true for a dog? For a horse?
3. Assuming that all the sweat evaporates, i.e., none of it drips off the body, how much water is lost by the runners in the four cases above? The latent heat of vaporization of water is 44.1 MJ/kmole.

1.21   One major ecological concern is the emission of hot-house gases, the main one being CO_2.

A number of measures can be taken to alleviate the situation. For instance, the use of biomass derived fuels does not increase the carbon dioxide content of the atmosphere.

Fossil fuels, on the other hand are a major culprit. Suppose you have the option of using natural gas or coal to fire a steam turbine to generate electricity. Natural gas is, essentially, methane, CH_4, while coal can be taken (for the purposes of this problem only) as eicosane, $C_{20}H_{42}$. The higher heat of combustion of methane is 55.6 MJ/kg and that of eicosane is 47.2 MJ/kg.

For equal amounts of generated heat, which of the two fuels is preferable from the CO_2 emission point of view? What is the ratio of the two emission rates?

1.22   A planet has a density of 2500 kg/m^3 and a radius of 4000 km. Its "air" consists of 30% ammonia, 50% carbon dioxide, and 20% nitrogen.

Note that the density, δ_{earth}, of Earth is 5519 kg/m^3.

What is the acceleration of gravity on the surface of the planet?

1.23   At 100 million km from a star, the light power density is 2 kW/m^2. How much is the total insolation on the planet of Problem 1.22 if it is 200 million km from the star. The total insolation on earth is 173,000 TW.

1.24 $^3_2$He can be used as fuel in "dream" fusion reactions—that is, in reactions that involve neither radioactive materials nor neutrons. Two possible reactions are

$$^2_1D + {}^3_2He \rightarrow {}^1_1H + {}^4_2He \tag{1}$$

and

$$^3_2He + {}^3_2He \rightarrow {}^1_1H + {}^1_1H + {}^4_2He \tag{2}$$

1 For each of the above reactions, calculate the energy (in kWh) released by 1 kg of $_2^3$He.

On earth, $_2^3$He represents 0.00013% of the naturally occurring helium. The US helium production amounts, at present, to 12,000 tons per year.

2. If all this helium were processed to separate the helium-three, what would be the yearly production of this fuel?

There are reasons to believe that there is a substantial amount of $_2$He^3 on the moon. Let us do a preliminary analysis of the economics of setting up a mining operation on our satellite.

One of the advantages of using "dream" reactions is that only charged particles (protons and alphas) are produced. The energy associated with charged particles can be more efficiently transformed into electricity than when the energy is carried by neutrons which must first produce heat that is then upgraded to mechanical and electric energy by inefficient heat engines. Thus, it is not necessarily optimistic to assign a 30% efficiency for the conversion of fusion energy into electricity.

3. How many kWh of electricity does 1 kg of $_2^3$He produce? Use the most economical of the two reactions mentioned.

Assume that the plant factor is 70% (the reactor delivers, on average, 70% of the energy it would deliver if running constantly at full power). Assume further that the cost of the fusion reactor is $2000/kW and that the cost of borrowing money is 10% per year. Finally, the cost of running the whole operation is $30 kW^{-1}year^{-1}.

4. How much would the electricity cost (per kWh) if the fuel were free?
5. How much can we afford to pay for 1 kg of $_2^3$He and still break even when electricity is sold at 5 cents per kWh?

1.25   Between 1955 and 1995, the ocean temperature (Atlantic, Pacific, and Indian) increased by 0.06 C.

Estimate how much energy was added to the water.

What percentage of the solar energy incident on earth during these 40 years was actually retained by the ocean?

1.26   There seems to be a possibility that climate changes will cause the polar ice caps to melt. The amount of ice in Antarctica is so large that if it were to melt, it would submerge all port cities such as New York and Los Angeles.

Estimate by how much the sea level would rise if only the north pole ice is melted leaving Greenland and Antarctica untouched.

1.27   Refueling a modern ICV with 50 liters of gasoline may take, say, 5 minutes. A certain amount of energy was transferred from the pump to the car in a given time. What is the power represented by this transfer? Assume that the overall efficiency of a gasoline car is 15% and that of an electric car is 60%. How much power is necessary to charge the batteries of the electric car in 5 minutes (as in the ICV case)? Assume that the final drive train energy is the same in both the ICV and the EV. Is it practical to *recharge* a car as fast as *refueling* one?

1.28   Some of the more attractive fuels happen to be gases. This is particularly true of hydrogen. Thus, storage of gases (Chapter 11) becomes an important topic in energy engineering. Lawrence Livermore Labs, for instance, has proposed glass micro-balloons, originally developed for housing minute amounts of tritium-deuterium alloy for laser fusion experiments. When heated, the glass becomes porous and hydrogen under pressure can fill the balloons. Cooled, the gas is trapped.

Clathrate is one of nature's ways of storing methane, even though no one is proposing it as a practical method for transporting the gas.

Methane clathrate frequently consists of cages of 46 H_2O trapping 8 CH_4 molecules.

1. What is the gravimetric metric concentration, GC, of methane in the clathrate? Gravimetric concentration is the ratio of the mass of the stored gas to the total mass of gas plus container.

Consider a hermetic container with 1 m^3 internal volume filled completely with the clathrate described, which has a density of 900 kg/m^3. Assume that by raising the temperature to 298 K, the material will melt and methane will evolve. Assume also (although this is not true) that methane is insoluble in water.

2. What is the pressure of the methane in the container?

1.29   A Radioisotope Thermal Generator (RTG) is to deliver 500 W of dc power to a load at 30 V. The generator efficiency (the ratio of the dc power out to the heat power in) is 12.6%. The thermoelectric generator takes heat in at 1200 K and rejects it at 450 K. The heat source is plutonium-241. This radioactive isotope has a half-life of 13.2 years and decays emitting α and β^- particles. These particles have an aggregate energy of 5.165 MeV.

Only 85% of the power generated by the plutonium finds its way to the thermoelectric generator. The rest is lost.

How many kilograms of plutonium are required? Note that radioactive substances decay at a rate proportional to the amount of undecayed substance and to a constant decay rate, λ:

$$\frac{dN}{dt} = -\lambda N.$$

1.30   In the USA we burn (very roughly) an average of 150 GW of coal, 40 GW of oil and 70 GW of natural gas.

Assume that

Coal is (say) $C_{20}H_{44}$ and that it yields 40 MJ per kg,
Oil is (say) $C_{10}H_{22}$ and yields 45 MJ per kg,
Natural gas is CH_4 and yields 55 MJ per kg.

How many kg of carbon are released daily by the combustion of coal alone? (Clearly, after you have handled coal, the other two fuels can be handled the same way. But, for the sake of time, don't do it.)

Part I

Heat Engines

Chapter 2
A Minimum of Thermodynamics and of Kinetic Theory of Gases

2.1 The Motion of Molecules

A gas is a collection of particles (molecules) that to a first approximation interact with one another solely through elastic collisions—in other words, through collisions that conserve both energy and momentum. If molecules were dimensionless, point-like objects, their thermal energy would be only that of linear motion in three dimensions—they would have only *three degrees of freedom*. In reality, molecules are more complicated. Even a simple monatomic one, such as helium, may be able to spin (because it has a finite dimension) and may, therefore, have more than three degrees of freedom. Multiatomic molecules can also vibrate and this confers to them additional degrees of freedom.

At a given moment, some molecules have large kinetic energy, while others have little. However, over a sufficiently long period of time, each has the same *average kinetic energy*, $<W_{mol}>$. This intuitive result is called the **principle of equipartition of energy**. What is not so immediately obvious is that the principle applies even to a collection of molecules with different masses: the more massive ones will have smaller average velocities than the lighter ones, but their average energy will be the same. According to this principle, the energy associated with any degree of freedom is the same. The instantaneous velocities have a **Maxwellian** distribution, as discussed in Section 2.28.

2.2 Temperature

It is useful to distinguish the two components of the average molecular energy: $<W_{mol,\ linear}>$ and $<W_{mol,\ spin\ \&\ vibr.}>$,

$$<W_{mol}>=<W_{mol,\ linear}> + <W_{mol,\ spin\ \&\ vibr.}> \qquad (1)$$

The pressure a gas exerts on an obstacle is the result of molecules colliding with the obstacle. Clearly, only the *linear* motion of the molecules can contribute to pressure; spin and vibration do not.

Temperature is a measure of $<W_{mol,\ linear}>$. It is defined by

$$T = \frac{2}{3k} <W_{mol,\ linear}>. \qquad (2)$$

The factor $\frac{1}{3}$ in the proportionality constant, $\frac{2}{3k}$, results from the three degrees of freedom. k is **Boltzmann's constant** and has, of course, the dimensions of energy per temperature (in the SI, $k = 1.38 \times 10^{-23}$ joules/kelvin).

In terms of temperature, the average energy of linear molecular motion in a three-dimensional gas is

$$<W_{mol,\ linear}>= 3\frac{k}{2}T \tag{3}$$

and, per degree of freedom,

$$<W_{mol,\ linear,\ per\ deg.\ of\ freed.}>= \frac{k}{2}T \tag{4}$$

Since each degree of freedom (associated with linear motion, spin, or with vibration) has the same energy, the average total molecular energy is

$$<W_{mol}>= \frac{\nu}{2}kT, \tag{5}$$

where ν is the number of degrees of freedom.

2.3   The Perfect-Gas Law

It is obvious that a simple relationship must exist between pressure and temperature. Consider motion in a single dimension normal to a surface. Upon impact, the molecule deposits a momentum of $2mv$ on the wall (the factor 2 accounts for the impinging velocity being v and the reflected velocity being another v). The flux of molecules (i.e., the number of molecules moving through a unit area in unit time) is $\frac{1}{2}nv$, where n is the **concentration** of molecules (i.e., the number of molecules per unit volume). Here the $\frac{1}{2}$ accounts for half the molecules moving in one direction, while the other half moves in the opposite direction because we are assuming that there is no net flux—that is, no bulk gas motion. The rate of change of momentum per unit area per unit time, i.e., the pressure exerted by the gas is, thus,

$$p = nmv^2. \tag{6}$$

Since the kinetic energy of the molecules moving in the direction being considered is $\frac{1}{2}mv^2$ and since this energy is $\frac{1}{2}kT$,

$$p = nkT. \tag{7}$$

The pressure is proportional to both the gas concentration and to its temperature. This is the **perfect-gas law** applicable to ideal gases, in which particles neither attract nor repel one another. Practical gases may not follow this law exactly owing to weak Van der Waals' forces between molecules and to the finite size of molecules (which causes the volume available to the gas to be smaller than the volume of the vessel containing it).

The higher the concentration of the gas, the greater the error when using the perfect gas law. But in many situations the error tends to be small—the volume of air at 300 K and 100 kPa is overestimated by only 0.07%.

It proves convenient to count particles in terms of *kilomoles* (just as it might be useful to count loaves of bread in terms of *dozens*). While a dozen is equal to 12, a kilomole is 6.022×10^{26}. This latter quantity is called **Avogadro's number, N_0**.[†]

Observe that $n = \mu N_0 / V$ (μ is the number of kilomoles and V is the volume of the container).

$$p = \mu \frac{N_0 kT}{V} = \mu \frac{RT}{V}. \tag{8}$$

This is another form of the perfect gas law.

$$R \equiv kN_0 = 1.38 \times 10^{-23} \times 6.022 \times 10^{26} = 8314 \quad \text{J K}^{-1}\text{kmole}^{-1}. \tag{9}$$

R is the **gas constant**.

2.4 Internal Energy

The total **internal energy**, U, of a gas is the sum of the energy of all molecules.

$$U \equiv \sum_i W_{mol_i} = \mu N_0 <W_{mol}> = \mu N_0 \frac{\nu}{2} kT = \mu \frac{\nu}{2} RT. \tag{10}$$

Thus, the internal energy, U, of a quantity, μ, of gas depends only on the temperature, T, and on the number, ν, of degrees of freedom of its constituent molecules.

2.5 Specific Heat at Constant Volume

When, in a fixed amount of gas, the temperature is changed, the internal energy also changes. When the volume is kept constant (as is the case of a mass of gas confined inside a rigid container), the rate of change of its internal energy with a change of temperature (per kilomole of gas) is called the **specific heat at constant volume, c_v**:

$$c_v = \frac{1}{\mu} \frac{dU}{dT} = \frac{\nu}{2} R. \tag{11}$$

Heat energy added to a gas is equally divided among the various degrees of freedom—hence the larger the number of degrees of freedom—the

---

[†] If Avogadro's number is taken as the number of molecules per mole (instead of kilomoles as one does when using the SI), then its value is 6.022×10^{23}.

more energy is necessary to increase the energy of linear motion—that is, to increase the temperature. For this reason, the specific heats of a gas are proportional to ν.

We can see from Equation 11, that when a quantity μ of gas changes its temperature, then its internal energy changes by

$$\Delta U = \mu \int_{T_0}^{T} c_v dT. \tag{12}$$

Notice that we have left c_v inside the integral to cover the possibility that it may be temperature dependent, although this is not obvious from our derivations so far.

2.6 The First Law of Thermodynamics

Introduce an amount, Q, of heat energy into an otherwise adiabatic gas-filled cylinder equipped with a frictionless piston. **Adiabatic** means that no heat is exchanged between the gas in the cylinder and the environment. If the piston is allowed to move, it can do external work, W, by lifting a weight. If held immobile, no work will be done. In general, $W \neq Q$. In fact, since energy cannot be created from nothing, $W \leq Q$. What happens to the excess energy, $Q - W$?

The principle of conservation of energy requires that the internal energy of the system increase by just the correct amount. A "bookkeeping" equation is written:

$$\Delta U = Q - W, \tag{13}$$

where ΔU is the increase in internal energy. In differential form, the change in internal energy can be related to the incremental heat added and to the incremental work done by the system.

$$dU = dQ - dW. \tag{14}$$

Equations 13 and 14 are the mathematical statement of the **first law of thermodynamics**. It is a statement of conservation of energy. In all cases of interest here, the internal energy is the energy associated with the random motion—the **thermal** energy. A more complicated system may increase its internal energy through such additional mechanisms as atomic or molecular excitations, ionization, and others.

When heat is added at constant volume, there is no external work $(dW = 0)$ and, consequently, $dU = dQ$. Since the specific heat at constant volume is dU/dT (per kilomole),

$$\mu c_v = \frac{dU}{dT} = \frac{dQ}{dT}. \tag{15}$$

Figure 2.1   Cylinder with frictionless piston.

2.7 The Pressure-Volume Work

We mentioned, in the previous subsection, that work can be extracted from a closed cylinder-with-piston system. How much work is generated?

The force on the piston is pA, where A is the area of the piston face. If the piston moves a distance, dx, it does an amount of work:

$$dW = pA dx \qquad (16)$$

The volume of the cylinder is changed by:

$$dV = A dx. \qquad (17)$$

Thus,

$$dW = p\, dV \qquad (18)$$

and

$$W = \int p\, dV. \qquad (19)$$

2.8 Specific Heat at Constant Pressure

c_v is the amount of heat that has to be delivered to one kilomole of gas to increase its temperature by 1 K provided that the *volume* is kept unaltered. In a system like the one depicted in Figure 2.1, this corresponds to immobilizing the piston. On the other hand, if the *pressure* is kept constant, then, in order to increase the temperature by the same 1 K, more energy is needed. The extra energy is required because in addition to increasing the internal energy, heat must also do work lifting the piston. This work (per unit temperature rise) is pdV/dT, or

$$p\frac{dV}{dT} = p\frac{d}{dT}\frac{RT}{p} = R \qquad (20)$$

because p is constant. It follows that

$$c_p = c_v + R = \frac{\nu}{2}R + R = \left(1 + \frac{\nu}{2}\right)R \qquad \text{J K}^{-1}\text{kmole}^{-1}. \qquad (21)$$

2.5

The ratio of the two specific heats is

$$\gamma \equiv \frac{c_p}{c_v} = \frac{R(1 + \nu/2)}{R\nu/2} = 1 + \frac{2}{\nu}. \tag{22}$$

2.9 Adiabatic Processes

In the closed system we have considered so far, we described the interplay between the internal energy, U, the work, W, and the heat, Q. The simplest possible system is one in which the cylinder is so well insulated that heat can neither enter nor leave. In such an **adiabatic** system, $\Delta Q = 0$. As the piston moves down, the work it does is entirely transformed into an increase in internal energy: $\Delta U = W$. The compression can be accomplished in a gradual manner so that at any given instant the pressure exerted by the piston is only infinitesimally larger than that of the gas—the compression is a succession of quasi-equilibrium states and the pressure is always uniform throughout the gas. Such is the case, for instance, when the piston is pressed down by the connecting rod of a mechanical heat engine, even though the action may appear to be very rapid. It is also possible to compress a gas abruptly as when an immobilized piston loaded with a heavy weight is suddenly released. In this case, the pressure of the gas immediately under the piston will rise rapidly but there is no time to transmit this change to the rest of the gas. A nonequilibrium situation is created. The former case—gradual compression—is by far the most common and most important. Nevertheless, we will consider first the abrupt compression, because gradual compression can be treated as an infinite succession of infinitely small abrupt steps.

2.9.1 Abrupt Compression

We will start with a qualitative description of what happens and then will examine an example.

Assume that a cylinder-and-piston system is in equilibrium. The piston, with a face area, A, is at a height, h_0 above the bottom of the cylinder enclosing a volume, $V_0 = h_0 A$. The force on the piston is F_0,[†] so that the pressure of the gas is $p_0 = F_0/A$. Next, the piston is clamped into place so that it cannot move and an additional mass is added to it. This increases the force to a value, F_1. At the very instant the piston is released, it will exert a pressure, $p_1 = F_1/A$, on the gas, but the latter is still at the substantially lower pressure, p_0. The piston will descend explosively to a height, h_1, and, after a while, will settle at a new height, h_1, when the gas pressure has risen to p_1. An amount of work, $W_{0\to1} = F_1(h_0 - h_1)$ has

---

[†] The force on the piston is the sum of the force exerted by the atmosphere plus the force owing to the weight of the piston.

been done on the gas and, owing to the adiabatic conditions, this work is entirely translated into an increase, $\Delta U = \mu c_v (T_1 - T_0)$, in internal energy. The compression caused a reduction in volume and an increase in pressure and temperature of the gas.

If next, the force on the piston is returned to its original value, $F_2 = F_0$, the piston will shoot up, and it is found that it will settle at a height, $h_2 > h_0$. The temperature will fall from the value, T_1, after the compression, to a new value, $T_2 > T_0$. The system does not return to its original state and the reason is obvious: The compression was caused by a force, F_1, but the expansion was against a smaller force, F_0. Thus, an amount of energy, $W_{0 \rightarrow 1} - W_{1 \rightarrow 2}$ was left over. This particular cycle extracted some energy from the environment. Hence, by definition, it is an irreversible process.

Example:

Consider the adiabatic cylinder-and-piston system shown in Figure 2.1. In our example, it contains $\mu = 40.09 \times 10^{-6}$ kilomoles of a gas whose $\gamma = 1.4$, independently of temperature. Its temperature is $T_0 = 300$ K. The cross-sectional area of the cylinder is $A = 0.001$ m^2.

The piston slides with no friction, exerting a force, $F_0 = 1000$ N. Consequently, the piston causes a pressure, $p_0 = \frac{F_0}{A} = \frac{1000}{0.001} = 10^6$ Pa.

The volume of the gas is

$$V_0 = \frac{\mu R T_0}{p} = \frac{40.09 \times 10^{-6} \times 8314 \times 300}{10^6} = 0.0001 \text{ m}^3. \qquad (23)$$

The piston hovers at $h_0 = V_0/A = 0.0001/0.001 = 0.1$ m above the bottom of the cylinder.

At equilibrium, the pressure of the gas is equal to the pressure the piston exerts. The specific heat at constant volume is

$$c_v = \frac{R}{\gamma - 1} = 20,785 \quad \text{J K}^{-1}\text{kmole}^{-1}. \qquad (24)$$

The internal energy of the gas is

$$U_0 = \mu c_v T_0 = 40.09 \times 10^{-6} \times 20{,}785 \times 300 = 250 \text{ J}. \qquad (25)$$

and the $p_0 V_0^\gamma$ product is

$$p_0 V_0^\gamma = 10^6 \times 0.0001^{1.4} = 2.51. \qquad (26)$$

(continued)

(continued)

The fixed characteristics of the system are,

Area, $A = 0.001$ m^2,
Gas amount, $\mu = 40.09 \times 10^{-6}$ kmoles,
Gamma, $\gamma = 1.4$,
Specific heat at constant volume, $c_v = 20{,}785$ JK^{-1}kmole^{-1},

and the initial data are

Force, $F_0 = 1000$ N,
Volume, $V_0 = 0.0001$ m^3,
Pressure, $p_0 = 10^6$ Pa,
Temperature, $T_0 = 300$ K,
Internal energy, $U_0 = 250$ J,
Height, $h_0 = 0.1$ m,
$p_0 V_0^\gamma = 2.51$.

Data on all the phases of this exercise are displayed in Table 2.1 at the end of the next subsection.

What happens if the force exerted by the piston is abruptly increased so that F_1 is now 10,000 N? The piston will go down stopping at a height, h_1. (Actually the piston will initially overshoot its mark and then oscillate up and down until the internal losses of the gas dampen out these oscillations.) When at equilibrium, the pressure of the gas is

$$p_1 = \frac{F_1}{A} = \frac{10{,}000}{0.001} = 10^7 \text{ Pa.} \tag{27}$$

In moving from h_0 to h_1, the piston did an amount of work, $W_{0\to1}$,

$$W_{1\to2} = F_1(h_0 - h_1). \tag{28}$$

Since the cylinder is adiabatic, the internal energy of the gas must increase by an amount,

$$\Delta U_0 = \mu c_v (T_1 - T_0) = F_1(h_0 - h_1), \tag{29}$$

where T_1 is the temperature of the gas after compression. It is

$$T_1 = \frac{p_1 V_1}{\mu R} = \frac{p_1 A h_1}{\mu R} = \frac{F_1}{\mu R} h_1. \tag{30}$$

Introducing Equation 30 into Equation 29,

$$F_1(h_0 - h_1) = \frac{c_v}{R} F_1 h_1 - \mu c_v T_0. \tag{31}$$

(continued)

(continued)
Solving for h_1,

$$h_1 = \frac{\gamma - 1}{\gamma}\left(h_0 + \mu c_v \frac{T_0}{F_1}\right) \tag{32}$$

Using the values of the example, $h_1 = 0.0357$ m.
The volume of the gas is now

$$V_1 = Ah_1 = 0.001 \times 0.0357 = 35.7 \times 10^{-6} \text{ m}^3. \tag{33}$$

The gas temperature is

$$T_1 = \frac{p_1 V_1}{\mu R} = \frac{10^7 \times 35.7 \times 10^{-6}}{40.09 \times 10^{-6} \times 8314} = 1071.4 \text{ K.} \tag{34}$$

and the pV^γ product is

$$p_1 V_1^\gamma = 10^7 \times (35.7 \times 10^{-6})^{1.4} = 5.94. \tag{35}$$

Collecting these data, we obtain the following values after the abrupt compression:

$F_1 = 10{,}000$ N,
Volume, $V_1 = 35.7 \times 10^{-6}$ m^3,
Pressure, $p_1 = 10^7$ Pa,
Temperature, $T_1 = 1071.4$ K,
Height, $h_1 = 0.0357$ m,
$p_1 V_1^\gamma = 5.94$.

The amount of energy the piston delivered to the gas is

$$W_{0\to 1} = F_1(h_0 - h_1) = 10{,}000(0.1 - 0.0357) = 643 \quad \text{J.} \tag{36}$$

In this example, the sudden application of 10,000 N (an increase of 9000 N) resulted in a strongly nonequilibrium situation. At the moment this additional force was applied, the piston exerted a pressure of 10^7 Pa while the opposing pressure of the gas was only 10^6 Pa. The piston descended explosively seeking a new equilibrium. We will attempt to reverse the situation, starting with the values above, also listed in the second column ("abrupt compression") of Table 2.1. We suddenly remove 9000 N (leaving the 1000 N we had originally). Calculations entirely parallel to the one we just did would lead to the new final values below:

(continued)

(continued)

$F_2 = 1000$ N.
Volume, $V_2 = 268 \times 10^{-6}$ m^3.
Pressure, $p_2 = 10^6$ Pa.
Temperature $T_2 = 795.6$ K.
Height $h_2 = 0.265$ m.
$p_2 V_2^\gamma = 9.98$.

The force on top of the piston is back to its original value of 1000 N, but the state of the gas is very far from that at the beginning of the experiment. This, as we pointed out, is to be expected. We compressed the gas with a 10,000 N force and then lifted the piston against a much smaller 1000 N force. Although the final height is larger than it was initially, some energy is left over. Indeed, the internal energy of the gas is now $U_2 = \mu c_v T_2 = 663$ J, an increase of 413 J over the initial value of $U_0 = 250$ J. This is, of course, the difference between the mechanical input energy, $W_{0 \to 1}$ and the mechanical output energy, $W_{1 \to 2}$.

2.9.2 Gradual Compression

Can a gas be compressed adiabatically in such a way that when expanded it returns exactly to the same state? In other words, can an adiabatic compression be reversible? The answer is yes, provided the force is applied gradually. The compression (or expansion) must proceed in a number of steps each of which maintains the gas in quasi-equilibrium. This can be demonstrated numerically in a simple way by using a spread sheet such as Excel.

Starting with the conditions we had at the beginning of the experiment, increment the force by a small amount, ΔF, so that $F_i = F_{i-1} + \Delta F$. Calculate h_i from

$$h_i = \frac{\gamma - 1}{\gamma} \left(h_{i-1} + \mu c_v \frac{T_{i-1}}{F_i} \right). \tag{37}$$

Calculate T_i from

$$T_i = \frac{F_1 h_i}{\mu R}. \tag{38}$$

Iterate until $F_i = F_{final}$. Here, F_{final}(10,000 N, in this example) is the final value of the force.

For sufficiently small ΔF, it is found that $h_{final} = 0.0193$ m and $T_{final} = 579.2$ K. It is also found that if we decompress in the same manner, we return to the original values of h, V, and T. The process is reversible.

In addition, it turns out that the final value of pV^γ is the same as the initial one. Indeed, pV^γ is the same in all steps of the calculation. This is not a coincidence. Later on in this chapter, we will demonstrate that in a reversible adiabatic process, pV^γ is constant. In Chapter 4 we will demonstrate that this so called **polytropic law** applies to all **isentropic processes**. The use of the polytropic law allows the calculation of reversible adiabatic processes in a simple way, not requiring the iteration technique mentioned above.

The results of the reversible compression are listed below:

Volume, $V_{reversible} = 19.3 \times 10^{-6}$ m^3.

Pressure, $p_{reversible} = 10^7$ Pa.

Temperature $T_{reversible} = 579.2$ K.

Height $h_{reversible} = 0.0193$ m.

$(pV^\gamma)_{reversible} = 2.51$.

$F_{reversible} = 10{,}000$ N.

Table 2.1
Variables in Different Phases of the Compression Experiment

Phase	Subscript	Vol. liters	Press. MPa	Temp. K	Height cm	pV^γ	Force N
Initial	"0"	100	1	300	10	2.51	1000
Abrupt compression	"a"	35.7	10	1071	3.57	5.94	10000
Abrupt expansion	"$a\ rever.$"	265	1	796	26.5	9.98	1000
Gradual compression	"$reversible$"	19.3	10	579	1.93	2.51	10000

2.9.3 p-V *Diagrams*

It is easier to understand the process by plotting the pressure versus volume behavior of the gas as illustrated in Figure 2.2.

Since reversible compression is the result of a large number of succeeding equilibrium steps, we can calculate the pressure and volume after each step. This is indicated by the smooth (exponential looking) curve in the figure. Notice that the shaded area under the curve represents the amount of work done during the compression.

The compression in the experiment starts from 1 MPa (when the gas volume was 100 liters) and ends when the pressure reaches 10 MPa (at a volume of 19.3 liters).

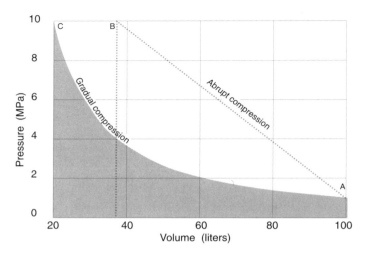

Figure 2.2   For a reversible adiabatic compression, the progression from the initial state, A, to the final state, C, is smooth and predictable. For an abrupt compression, it is impossible to specify the path of the gas.

In the case of the abrupt compression, the initial state, (A), is the same as in the previous case. The final state, (B), has the same pressure (10 MPa) as before but, since the gas is at a higher temperature, the volume is larger (35.7 liters versus 19.3 liters). However, the path from one state to the other is unknown, a fact indicated by the dotted line. During such a rapid compression, the pressure and the temperature cannot be specified because they are not uniform throughout the mass of the gas. As the piston presses down, the gas piles up in the vicinity of the piston, not having time to spread out uniformly.

The terms "gradual" and "abrupt" are relative. In most machines, even at high speed operation, the compression (or expansion) can be taken (with little error) as gradual.

2.9.4 Polytropic Law

Assume that the cylinder with piston used in our discussion of the pressure-volume work is insulated so that no heat can be exchanged between the gas inside and the environment outside. We have adiabatic conditions, and the heat exchanged is

$$dQ = 0. \tag{39}$$

Consider an infinitesimal step in the compression process. The work is

$$dW = p \, dV \tag{40}$$

From the first law of thermodynamics, we have

$$dQ = dU + dW = 0, \tag{41}$$

and from the perfect gas law,

$$p = \mu \frac{RT}{V}. \tag{42}$$

It follows that

$$dW = \mu RT \frac{dV}{V} \tag{43}$$

and

$$dU = \mu c_v dT, \tag{44}$$

$$\mu c_v \frac{dT}{T} + \mu R \frac{dV}{V} = 0, \tag{45}$$

$$c_v \ln T + R \ln V = constant. \tag{46}$$

But $R = c_p - c_v$, hence

$$\ln T + (\gamma - 1) \ln V = constant, \tag{47}$$

$$T \, V^{\gamma-1} = constant. \tag{48}$$

Since $pV = \mu RT$,

$$pV^\gamma = constant. \tag{49}$$

This is called the **polytropic** law and is the characteristic of a reversible adiabatic compression or expansion.

2.10 Isothermal Processes

In the previous sub-section, we discussed a particularly important thermodynamic transformation—the adiabatic process. We will now examine another equally important transformation—the isothermal process.

Under all circumstances, the work done when a gas expands from volume V_0 to V_1 (see the Section 2.7) is,

$$W_{0,1} = \int_{V_0}^{V_1} pdV. \tag{50}$$

Under isothermal conditions,

$$pV = p_o V_0 \qquad \therefore \qquad p = p_o \frac{V_0}{V} \tag{51}$$

$$W_{0,1} = p_0 V_0 \int_{V_0}^{V_1} \frac{dV}{V} = p_0 V_0 \ln \frac{V_1}{V_0} = p_0 V_0 \ln \frac{p_0}{p_1} = \mu RT \ln \frac{p_0}{p_1}. \tag{52}$$

We will re-derive this result using a more detailed procedure[†] in the hope that this will bring out more clearly the basic underlying mechanism of the process.

Assume that the cylinder is no longer thermally insulated; it is in thermal contact with a bath maintained at a constant 300 K. The frictionless piston is held in place by a weight, so that the initial pressure is p_0. We want to know how much energy can be extracted by allowing the expansion of the gas to a final pressure, p_f. This requires removing some weight from the piston. We assume that there is no outside atmospheric pressure.

Imagine the original weight had a mass of 10 kg and that, suddenly, 9 kg are removed. The piston will shoot up lifting 1 kg. Since there is no friction, the system oscillates, damped only by the internal dissipation of the gas which, cooled by expansion, warms up again as heat from the bath is conducted in. After a new equilibrium is established, a mass of 1 kg will have been lifted by, say, 1 meter. The work done is $W = mgh = 1 \times 9.8 \times 1 = 9.8$ J. Although the final temperature is the same as the initial one, *the processes is not isothermal*, because during the expansion the temperature first decreased and then rose to its original value. If isothermal, the temperature must not change throughout the whole process.

Let us rerun the experiment leaving this time 2 kg on the piston. The mass will rise and the oscillations will eventually settle down. Now an additional 1 kg is removed and the process repeats itself. The final state is the same as in the first experiment, but the work done is now $2 \times 9.8 \times 0.444 + 1 \times 9.8 \times 0.556 = 14.2$ J because the mass was raised 0.444 m in the first step (see Problem 2.3). The obvious reason for this larger amount of work is that, although only 1 kg reached the 1 m height, a total of 2 kg was lifted part of the way.

Maximum work is done by using an infinite number of steps, each one removing infinitesimally small amounts of mass. Under such circumstances, the expansion is isothermal because heat from the bath flows into the gas after each infinitesimal cooling, thus keeping the temperature at a constant 300 K. What is this maximum work?

Define a decompression ratio, r

$$r \equiv \frac{p_0}{p_f}. \tag{53}$$

Let the decompression proceed by (geometrically) uniform steps and let n be the number of steps. Then,

$$r^{1/n} = \frac{p_{i-1}}{p_i}. \tag{54}$$

The work done in step i is

$$W_i = (h_i - h_{i-1})F_i \tag{55}$$

[†] Suggested by Prof. D. Baganoff of Stanford University.

where F_i is the weight lifted in step i and is

$$F_i = p_i A, \tag{56}$$

A being the area of the piston.

The height reached by the weight after step i is

$$h_i = \frac{V_i}{A}, \tag{57}$$

hence,

$$W_i = \frac{V_i - V_{i-i}}{A} p_i A = V_i p_i - V_{i-1} p_i = V_0 p_0 - V_{i-1} p_i, \tag{58}$$

because after each step the temperature is returned to its original value, T_0, and consequently $pV = p_0 V_0$.

$$W_i = V_0 p_0 \left(1 - \frac{p_i}{p_{i-1}}\right) = V_0 p_0 \left(1 - r^{-1/n}\right). \tag{59}$$

The total work is

$$W = V_0 p_0 \sum_{i=1}^{n} \left(1 - r^{-1/n}\right) = V_0 p_0 n \left(1 - r^{-1/n}\right), \tag{60}$$

$$W = V_0 p_0 \ln r. \tag{61}$$

This is the formula derived earlier. An expanding mass of gas does maximum work when the expansion is isothermal.

2.11 Functions of State

The **state** of a given amount of perfect gas is completely defined by specifying any two of the following three variables: pressure, volume, or temperature. The third variable can be derived from the other two (if the number of kilomoles of the gas is known) by applying the perfect gas law.

When a gas changes from one state to another, it is possible to calculate the change in its internal energy. To do this, it is sufficient to know the amount of gas, its specific heat at constant volume and the initial and final temperatures. It is completely irrelevant what intermediate temperatures occurred during the change.

On the other hand, to calculate the amount of work done during a change of state, it is necessary to know exactly which way the change took place. Knowledge of the initial and final states is not sufficient. In the experiment described in the preceding subsection, the work to achieve the final state depended on which way the mass was raised. One cannot tell how much work is required to go from one state to another just by knowing what these states are. If one sees a person on top of a hill, one cannot know how much effort was made to climb it. The person may have taken an easy, paved path or may have traveled a longer, boulder strewn route. In other words, energy is **not** a function of the state of gas.

2.12 Enthalpy

So far we have considered only **closed systems** in which a fixed mass of gas is involved. Many devices, such as a turbine, are **open systems** involving a flow of gas. Instead of fixing our attention on a given mass of gas, we must consider a given volume through which a fluid flows suffering some thermodynamic transformation. In order to quantify the changes in energy in such an open system, one must account for the energy that the fluid brings into the system and removes from it.

To force the flow of the fluid into the open system depicted in the Figure 2.3, imagine a (fictitious) piston exerting a pressure, p, to press a volume, V, of gas into the device. The piston exerts a force, pA, and, to push the gas a distance, L, uses an energy, $pAL = pV$. The flow of energy into the device is $p_{in}V_{in}$. Exiting, the gas carries an energy, $p_{out}V_{out}$. The net energy deposited by the flow is $p_{in}V_{in} - p_{out}V_{out}$. If the gas also changed its internal energy, then the total work that the device generates is

$$W = \Delta(pV) + \Delta U = \Delta(pV + U) \equiv \Delta H. \tag{62}$$

This assumes that the device is adiabatic (heat is not exchanged with the environment through the walls).

The combination, $pV + U$, occurs frequently in thermodynamics and it becomes convenient to define a quantity, called **enthalpy**,

$$H \equiv U + pV. \tag{63}$$

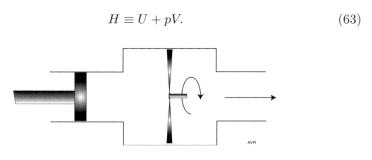

Figure 2.3   An open system in which
the flow of gas does some work.

H, U, and pV, being energies, are relative—that is, they must be referred to some arbitrary level. Their magnitudes are of little importance; what is of interest is their *change*:

$$\Delta H = \Delta U + \Delta(pV) = \Delta U + p\Delta V + V\Delta p. \tag{64}$$

At constant pressure, ΔH is simply $\Delta U + p\Delta V$ and is therefore equal to the heat, ΔQ, added to the system:

$$\Delta H = \Delta U + W = \Delta Q \quad \text{(at constant pressure)}. \tag{65}$$

For this reason, enthalpy is sometimes called the **heat content**. It is a quantity commonly used by chemists because reactions are frequently carried out in open vessels—that is, at constant pressure.

From Equation 12 (per kilomole)

$$\Delta U = \int_{T_0}^{T} c_v \, dT. \tag{66}$$

Using the perfect gas law (again, per kilomole), and Equation 64,

$$\Delta H = \Delta U + \Delta(pV) = \Delta U + \Delta(RT) = \int_{T_0}^{T} c_v \, dT + R \int_{T_0}^{T} dT$$

$$= \int_{T_0}^{T} (c_v + R) \, dT = \int_{T_0}^{T} c_p \, dT \tag{67}$$

Compare Equations 66 and 67.

2.13   Degrees of Freedom

The formulas we developed for the specific heats and for their ratio, γ, require knowledge of the number of degrees of freedom, ν, of the molecules. From our derivation, this number should be an integer and it should be independent of temperature. What do experimental data have to say?

Consider monatomic gases whose molecules have three translational degrees of freedom. If this were all, their γ would be exactly $1+2/3 = 1.667$ (Equation 22). Table 2.2 shows that the measured value of γ for helium, argon, and krypton is *approximately* (but not exactly) the expected one. However, since the molecules of these gases have some volume, they must be able to spin. At least one additional degree of freedom must be assigned to this spin motion, and ν should not be smaller than 4. This would lead to a γ of $1 + 2/4 = 1.5$, a value significantly below the observed one.

Take now the diatomic molecule, H_2. It should have 3 translational and 2 rotational degrees of freedom. In addition, it can vibrate, which should contribute another 2 degrees of freedom (one for the kinetic energy of vibration and one for the potential). Thus, the total number of degrees of freedom should be at least 7 and the value of γ should be $1+2/7 = 1.286$.

2.17

At 2300 K, hydrogen does have a γ of 1.3—corresponding to a $\nu = 6.67$, and as can be seen from Figure 2.4, it appears that at even higher temperature, the number of degrees of freedom might reach 7. At very low temperatures, ν tends toward 3 and H_2 behaves as a pointlike monatomic molecule.

Between these temperature extremes, ν varies smoothly, assuming fractional, non-integer, values. Clearly, any single molecule can only have an integer number of degrees of freedom, but a gas, consisting of a large collection of molecules, can have an *average* ν that is fractional. At any given temperature, some molecules exercise a small number of degrees of freedom, while others exercise a larger one. In other words, the principle of equipartition of energy (which requires equal energy for all degrees of freedom of all molecules) breaks down. Thus, for real gases, the specific heats and their ratio, γ, are temperature dependent, although not extremely so. See the graphs in "Free Energy Dependence on Temperature" in Appendix A to Chapter 9.

For some estimates it is sufficient to assume that these quantities are constant. One can assume $\nu = 5$ for H_2 and O_2, and $\nu = 7$ for the more complicated molecule H_2O, all at ambient temperature. More precise calculations require looking up these values in tables. See, for instance, a listing of observed values of c_p and of γ for H_2, O_2, and H_2O in Table 9.5.

Some general trends should be remembered: More complex molecules or higher temperatures lead to a larger number of degrees of freedom and consequently larger specific heats and smaller γ.

Table 2.2

Ratio of the Specific Heats of Some Monatomic Gases

Gas	Temp. (K)	γ
Helium	93	1.660
Argon	298	1.668
Krypton	298	1.68

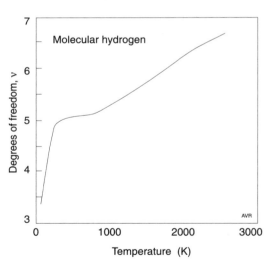

Figure 2.4   The number of degrees of freedom of molecular hydrogen as a function of temperature.

2.14 Entropy

When one considers different forms of energy, one can intuitively rank them in order of their "nobility." Electric energy must be quite "noble"—it can easily be transformed into any other kind of energy. The same is true of mechanical energy because it can (theoretically) be transformed into electricity and vice versa without losses. Heat, however, must be "degraded" energy—it is well known that it cannot be entirely transformed into either electric or mechanical energy (unless it is working against a heat sink at absolute zero). It turns out that chemical energy has a degree of "nobility" lower than that of electricity, but higher than that of heat.

Still, intuitively, one can feel that the higher the temperature, the higher the corresponding "nobility" of the heat—that is, the more efficiently it can be transformed into some other form of energy.

Let us try to put these loose concepts on a more quantitative basis.

Consider two large adiabatic reservoirs of heat: one (which we shall call the source) at a temperature, T_H, and one (the sink) at a lower temperature, T_C. The reservoirs are interconnected by a slender metal rod forming a thermally conducting path between them. We shall assume that, for the duration of the experiment, the heat transferred from source to sink is much smaller than the energy stored in the reservoirs. Under such circumstances, the temperatures will remain unaltered.

Assume also that the rod makes thermal contact with the reservoirs but not with the environment. The amount of heat that leaves the source must, then, be exactly the same as that which arrives at the sink. Nevertheless, the heat loses part of its "nobility" because its arrival temperature is lower than that at the departure. "Nobility" is lost in the conduction process.

Form the Q/T ratio at both heat source and at the heat sink. Q is the amount of heat transferred. Clearly, $Q/T_H < Q/T_C$. We could use this ratio as a measure of "ignobility" (lack of "nobility"), or, alternately, the ratio $-Q/T$ as a measure of "nobility." Loosely, **entropy** is what we called "ignobility":

$$S \equiv \frac{Q}{T}. \tag{68}$$

It is important to realize that in the above experiment, energy was conserved but "nobility" was lost; it did not disappear from the experimental system to emerge in some other part of the universe—it was lost to the universe as a whole. There is no law of conservation of "ignobility" or entropy. In any closed system, at best, the entropy will not change, but if it does, it always increases.[†] This is a statement of the **second law of thermodynamics**.

[†] However, a given system does not always tend toward maximum randomness. Systems may spontaneously create complicated structures such as life forms emerging from some primeval soup.

Since there is no heat associated with electric or mechanical energy, these forms have zero entropy.

We have seen that entropy is a measure of both the *quality* of heat and of the randomness of an arrangement. The higher the temperature associated with a quantity, Q, of heat, the lower the corresponding entropy, Q/T. On the other hand, the higher the temperature of a gas, the higher the randomness of its molecules—hence the higher its entropy. To resolve this apparent paradox, consider an amount of heat, Q_{in}, flowing, at temperature, T_{in}, into a volume of gas otherwise adiabatic. We have an input into the gas of both energy and entropy. Clearly, the increase in energy causes the gas temperature to rise, and, at the same time, the incoming entropy causes an increase in the entropy of the gas, because, although entropy is not conserved, it cannot, under the second law of thermodynamics, simply disappear. Hence raising the temperature of the gas also raised its entropy.

2.14.1 *Changes in Entropy*

Returning to the question of functions of state, it is important to know that entropy is such a function. To determine the change in entropy in any process, it is sufficient to determine the entropies of the final and the initial states and to form the difference.

Some processes can drive a system through a full cycle of changes (pressure, volume, and temperature) in such a way that, when the cycle is complete, the system is returned to the initial state. Such processes are **reversible**. To be reversible, the net heat and the net work exchanged with the environment must be zero. In any reversible process, the change in entropy of a substance owing to a change from State "1" to State "2" is

$$\Delta S = \int_{"1"}^{"2"} \frac{dQ}{T} \tag{69}$$

Here, S is the entropy and ΔS is the change in entropy.

In an **adiabatic** processes, $dQ = 0$. Hence

$$S = constant. \tag{70}$$

In an **isothermal** process, $\Delta S = Q/T$ because T is constant. Notice, however, that according to the first law of thermodynamics, $\Delta U = Q - W$, but, in an isothermal change, $\Delta U = 0$—hence $W = Q$ and, since in such a change, $W = p_0 V_0 \ln p_1/p_2$,

$$\Delta S = \frac{V_0 p_0}{T} \ln \frac{p_1}{p_2} = \mu R \ln \frac{p_1}{p_2}. \tag{71}$$

In an **isobaric** process,

$$\Delta S = \mu \int_{"1"}^{"2"} c_p \frac{dT}{T} \tag{72}$$

and, if c_p is constant,

$$\Delta S = \mu c_p \ln \frac{T_2}{T_1}. \tag{73}$$

In an **isometric** process,

$$\Delta S = \mu \int_{"1"}^{"2"} c_v \frac{dT}{T} \tag{74}$$

and if c_v is constant,

$$\Delta S = \mu c_v \ln \frac{T_2}{T_1}. \tag{75}$$

The change in entropy of μ kilomoles of a substance owing to an isobaric change of phase is

$$\Delta S = \mu \frac{Q_L}{T}, \tag{76}$$

where Q_L is the **latent heat of phase change** (per kilomole) and T is the temperature at which the change takes place.

Collecting all these results:

Table 2.3

Changes in Entropy

Process	ΔS	Equation
Adiabatic	zero	(70)
Isothermal	$\mu R \ln \frac{p_1}{p_2}$	(71)
Isobaric	$\mu \int_{"1"}^{"2"} c_p \frac{dT}{T}$	(72)
Isobaric (const. c_p)	$\mu c_p \ln \frac{T_2}{T_1}$	(73)
Isometric	$\mu \int_{"1"}^{"2"} c_v \frac{dT}{T}$	(74)
Isometric (const. c_v)	$\mu c_v \ln \frac{T_2}{T_1}$	(75)
Phase change	$\mu \frac{Q_L}{T}$	(76)

2.15 Reversibility

In the experiment of Section 2.14, it is impossible to reverse the direction of heat flow without a gross change in the relative temperatures, T_H and T_C. It is an **irreversible** process.

Let us examine another case (suggested by professor D. Baganoff of Stanford University) designed to illustrate reversibility:

Consider again an adiabatic cylinder with a frictionless piston. It contains 10 m^3 of an ideal monatomic gas ($\gamma = 1.67$) under a pressure of 100 kPa at 300 K. We will follow the behavior of the gas during compression and subsequent expansion by means of the p versus V and the T versus V diagrams in Figure 2.5. The initial state is indicated by "1" in both graphs.

When the piston is moved so as to reduce the volume to 2 m^3, the pressure will rise along the adiabatic line in such a way that the pV^γ product remains constant. When State "2" is reached, the pressure will have risen to 1470 kPa and the temperature to 882 K. The work done during the compression is $\int p\,dV$ between the two states and is proportional to the area under the curve between "1" and "2."

If the piston is allowed to return to its initial position, the gas will expand and cool, returning to the initial state, "1." The process is completely reversible. Now assume that inside the cylinder there is a solid object with a heat capacity equal to that of the gas. If we compress the gas rapidly and immediately expand it, there is no time for heat to be exchanged between the gas and the solid, which will remain at its original temperature of 300 K. The process is still reversible.

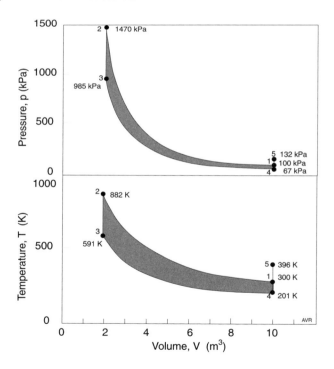

Figure 2.5   Pressure-volume and temperature-volume diagrams for the experiment described in the text.

However, if one compresses the gas and then waits until thermal equilibrium is established, then half of the heat generated will be transferred to the solid and the gas will cool to $(882 + 300)/2 = 591$ K, while its pressure falls to 985 kPa according to the perfect gas law. This is State "3."

When the gas expands again to 10 m^3, it cools down to 201 K and the pressure falls to 67 kPa (State "4"). Immediately after this expansion, the solid will be at 591 K. Later, some of its heat will have been transferred back to the gas whose temperature rises to 396 K and the pressure will reach 132 kPa ("5"). The gas was carried through a full cycle but did not return to its initial state: its temperature (and internal energy) is now higher than initially. The increase in internal energy must be equal to the work done on the gas (i.e., it must be proportional to the shaded area in the upper portion of Figure 2.5 which is equal to the area under curve "1" to "2" minus that under curve "3" to "4." The process is irreversible.

What happens if the compression and the expansion are carried out infinitely slowly? Does the process become reversible? You will find that it does when you do Problem 2.2. An electric analogy may clarify the situation. When a real battery (represented by a voltage source, V, with an internal resistance R) is used to charge an ideal energy accumulator, part of its energy will dissipate as heat through the internal I^2R losses, leaving only part to reach the accumulator. Clearly, the relative loss decreases as the current decreases—that is, as the charge time increases. If the energy is transferred from the battery to the accumulator infinitely slowly ($I \to 0$) there are no losses. The system is reversible in the sense that all the energy transferred to the accumulator can later be returned to the battery.

2.15.1 Causes of Irreversibility

Among the different phenomena that cause thermodynamic processes to become irreversible one can list:

2.15.1.1 Friction

Of all the causes of irreversibilities, friction is perhaps the most obvious. For example, in the cylinder-piston case, if some energy is lost by friction during compression, it is not returned during expansion; on the contrary, additional losses occur during this latter phase.

2.15.1.2 Heat Transfer Across Temperature Differences

Consider a metallic wall separating a source of heat—say a flame—from the input of a heat engine. All the heat, Q, absorbed from source is transmitted without loss through the wall, yet for this heat to flow there must be a temperature difference across the wall. The source side is at T_1, the engine side is at T_2, and T_1 must be larger than T_2. The entropy on the source side is Q/T_1 and on the engine side is Q/T_2, which is, of course larger than Q/T_1. So, in passing through the wall, the entropy was increased—the heat became less "noble" on the engine side.

In Chapter 3, we will show that the maximum efficiency of a heat engine is $\eta = \frac{T_H - T_C}{T_H}$. If the engine could have operated without the wall, its efficiency could have reached $\frac{T_1 - T_C}{T_1}$ and would be larger than when operated on the other side of the wall when it would be limited to $\frac{T_2 - T_C}{T_2}$.

2.15.1.3   Unrestrained Compression or Expansion of a Gas

In the subsection on adiabatic processes, we dealt with an example of abrupt expansion of a gas and found that it led to irreversibilities.

2.16   Negentropy

We have stressed that energy cannot be consumed. The conservation of energy is one of the laws of nature. When we use energy, we degrade it, so all energy we use is eventually degraded to heat and, one hopes, is radiated out into space in the form of long-wave infrared radiation.

Consider an example. An engine produces energy by extracting heat from the warm surface waters of the ocean (at some 300 K) and rejecting a smaller amount of heat to the cold waters near the bottom at, say, 275 K. All energy produced is used to compress adiabatically 10,000 kilomoles of a gas from 10^5 to 10^7 pascals (from about 1 to 100 atmospheres). If the initial temperature of the gas was 300 K, what is its final temperature?

Let p_0 and p_H be, respectively, the initial and the final pressures, and T_0 and T_H be the corresponding temperatures. Then,

$$\frac{T_H}{T_0} = \left(\frac{p_H}{p_0}\right)^{\frac{\gamma - 1}{\gamma}}, \tag{77}$$

where γ is taken, in this example, as 1.4 (equivalent to 5 degrees of freedom). The corresponding specific heat at constant volume is 20.8 kJ K^{-1}kmole^{-1}.

For a 100:1 adiabatic compression, the temperature ratio is 3.7, and the T_H will be $3.7 \times 300 = 1110$ K. After compression, the volume is

$$V_H = \mu \frac{R T_H}{p_H} = 10{,}000 \frac{8314 \times 1110}{10^7} = 9230 \text{ m}^3. \tag{78}$$

The work required for such a compression is

$$W = 20.8 \times 10^3 \times 10{,}000 \times (1110 - 300) = 168 \text{ GJ}. \tag{79}$$

Let the gas cool to 300 K—the temperature of the ocean surface. All the thermal energy (168 GJ) is returned to the waters. If the canister with gas is towed to the beach, we will have not removed any energy from the ocean, yet we will have 10,000 kilomoles of gas at a pressure of

$$p = \mu \frac{R T}{V} = 10{,}000 \frac{8314 \times 300}{9230} = 2.7 \text{ MPa}. \tag{80}$$

The internal energy of the compressed gas is no larger than that of an equal amount of the gas at the same temperature but at a much lower pressure; the internal energy is independent of pressure. Let the gas expand through a 100% efficient turbine, allowing its pressure to fall to 10^5 Pa. The work done by the turbine must come from the internal energy of the gas; this means that the gas must cool down. The temperature ratio (Equation 77) will be 2.56. Thus the gas exhausted from the turbine will be at 117 K.

The energy delivered by the turbine to its load is

$$W = 20.8 \times 10,000 \times (300 - 117) = 38 \text{ GJ.} \tag{81}$$

We have "generated" 38 GJ that certainly did not come from the ocean. Where did it come from? It came from the gas itself that had its internal energy reduced by cooling to 117 K. Actually, it came from the ambient air that was itself cooled by the exhaust from the turbine. The work delivered by the turbine will eventually be degraded to "waste" heat, reheating the environment by exactly the same amount it was cooled. Thus, in this whole process, energy was conserved, but we were able to perform useful work (the generator output) by removing something from the ocean. What was it?

We did randomize the ocean by mixing cold bottom water with warm surface water—we increased the entropy of the ocean. The canister carried away some **negentropy**. Here we used the semantically more acceptable concept of **negative entropy**, a quantity that can be consumed. Marchetti (1976), who coined the word "negentropy," proposed the above system as a method for "supplying energy without consuming energy." This is, of course, what is done in all cases in which energy is utilized. Although the system is technically feasible, it is commercially unattractive because the canister has but a small negentropy-carrying capacity compared with the large requirements of material for its construction. The main point to be learned from this example is that the consumable is negentropy, not energy.

2.17   How to Plot Statistics

The World Almanac lists the U.S. population distribution by age (for 1977) in the following manner:

Table 2.4
U.S. Population Distribution by Age

Age Interval (years)	Number of People (millions)
Under 5	15.2
5–13	32.2
14–17	16.8
18–20	12.8
21–44	72.0
45–64	43.8
65 and over	23.5

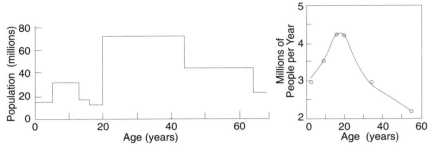

Figure 2.6   Inappropriate way of plotting population versus age.   The area under the plot does not have the dimension of number of people.

Figure 2.7   Number of people plotted in 1-year age interval.

Table 2.5
U.S. Population Distribution by Age

Age Interval (years)	Mean Age (years)	Age Interval (years)	People (millions)	(people/year)
0–5	2.5	5	15.3	3.07
5–3	9.5	9	32.2	3.58
14–17	16	4	16.8	4.20
18–20	19,5	3	12.8	4.26
21–44	33.0	24	72.0	3.00
45–64	55.0	20	43.8	2.19
over 64	?	?	23.5	—

In search of a meaningful way of graphically presenting the data above, one can build the histogram shown in Figure 2.6. However, owing to the disparate age intervals chosen in tabulating the data, the histogram is not very enlightening. It would be better to use uniform age intervals—the smaller, the better. In the limit, the best would be to plot $\partial N/\partial A$ versus A, where N is the number of people and A is the age. To do this, we will have to construct another table, derived from the data in Table 2.4.

The data in the last column of Table 2.5 are plotted in Figure 2.7. A continuous line joined the data points. The area under the resulting curve is proportional to the population of the country. In illustrating the energy distribution of, for instance, gas molecules, it is informative to present plots of $\partial N/\partial W$ versus W or versus v (either energy or velocity) but not a plot of N versus W, as one is sometimes tempted to do.

2.18   Maxwellian Distribution

If molecules have a uniform velocity distribution, how many molecules have velocities less than a given value $|v|$? To answer this question, let us remember that at any time, each molecule is at some position $(x,\ y,\ z)$

and has some velocity $(v_x,\ v_y,\ v_z)$. As far as the energy in the gas is concerned, the exact position of the molecules is irrelevant but their velocity is not. Although the individual velocities are changing all the time, in a gas at constant temperature, any instant in time is statistically equivalent to any other. In other words, any instantaneous picture of the velocities is adequate to describe the statistical behavior of the gas.

Let us plot the velocities of the molecules in a system of orthogonal coordinates, v_x, v_y, and v_z—that is, in **velocity space**. Alternatively, we could plot the momenta, mv_x, mv_y, and mv_z, in the **momentum space**. Since we are assuming that molecules have uniform velocity (or momentum) distribution, the velocity (or the momentum) space is uniformly populated. Thus, the number of molecules that have less than a certain velocity, $|v|$, is proportional to the volume of a sphere of radius v (or p) in the space considered. This means that the number of molecules with velocity less than $|v|$ (or momenta less than $|p|$) must be proportional to v^3 (or p^3). Hence, the number of molecules with velocity between v and $v + dv$ (momenta between p and $p + dp$) must be proportional to $\partial v^3/\partial v$—that is, to v^2 (or p^2). In real systems, uniform velocity distribution is unusual. In common gases, a distribution that fits experimental observation is one in which the probability, f, of finding a molecule with a given energy, W, is

$$f = \exp\left(-\frac{W}{kT}\right).\tag{82}$$

Under such conditions, the number of molecules with velocities between v and $v + dv$ is

$$\frac{\partial N}{\partial v} = \Lambda v^2 \exp\left(-\frac{mv^2}{2kT}\right)\tag{83}$$

or

$$\frac{\partial N}{\partial W} = \frac{2^{1/2}}{m^{3/2}}\Lambda W^{1/2} \exp\left(-\frac{W}{kT}\right)\tag{84}$$

where Λ is a constant and $W = mv^2/2$.

This is the so-called **Maxwellian distribution**.

Clearly,

$$N = \int_0^\infty \frac{\partial N}{\partial v}dv = \Lambda \int_0^\infty v^2 \exp\left(-\frac{mv^2}{2kT}\right)dv\tag{85}$$

where N (the total number of molecules) does not change with temperature.

It turns out that

$$\int_0^\infty v^2 \exp\left(-\frac{mv^2}{2kT}\right)dv = \frac{\pi^{1/2}}{4}\left(\frac{2kT}{m}\right)^{3/2}.\tag{86}$$

Thus,

$$N = \Lambda \frac{\pi^{1/2}}{4} \left(\frac{2kT}{m} \right)^{3/2} \qquad \therefore \qquad \Lambda = 4N\pi^{-1/2} \exp \left(\frac{m}{2kT} \right)^{3/2}, \quad (87)$$

$$\frac{\partial N}{\partial v} = 4N\pi^{-1/2} \left(\frac{m}{2kT} \right)^{3/2} v^2 \exp \left(-\frac{mv^2}{2kT} \right) \qquad (88)$$

and

$$\frac{\partial N}{\partial W} = 2N\pi^{-1/2} \frac{W^{1/2}}{(kT)^{3/2}} \exp \left(-\frac{W}{kT} \right). \qquad (89)$$

The shape of the $\partial N/\partial v$ versus v plot depends, of course, on the temperature, as shown in Figure 2.8, where T_0 is an arbitrary reference temperature. However, the area under the curve, being a measure of the total number of molecules in the gas, is independent of temperature.

The peak value of $\partial N/\partial v$ is

$$\frac{\partial N}{\partial v} = \frac{2N}{e} \left(\frac{2m}{\pi kT} \right)^{1/2}, \qquad (90)$$

and occurs when $v = \sqrt{2kT/m}$ or, equivalently, when $W = kT$.

As T approaches 0, $\partial N/\partial v_{[max]}$ approaches ∞ and occurs for $v = 0$. The distribution becomes a delta function at $T = 0$.

This means that according to this classical theory, at absolute zero, all the molecules have zero velocity and zero energy.

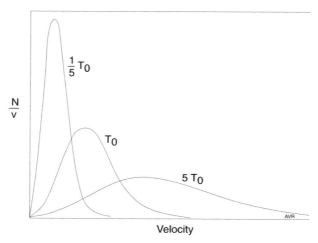

Figure 2.8   The Maxwellian velocity distribution at three different temperatures.

2.19 Fermi-Dirac Distribution

Electrons in metals do not behave in a Maxwellian way. They are governed by the **Pauli exclusion principle**, which states that in a given system, no two electrons can have the same quantum numbers, and, consequently, they cannot all have zero energy at absolute zero. Rather, at absolute zero electrons must be distributed uniformly in energy up to a given energy level. No electron has energy above this level, called the **Fermi level**. Thus, the probability, f, of finding electrons at a given energy level is:

$$f = 1, \quad \text{for } W < W_F \quad \text{and} \quad f = 0 \quad \text{for } W > W_F. \quad (91)$$

As before, the number of allowed states with momenta less than $|p|$ (notice the change in terminology) is proportional to the volume of a sphere with radius $|p|$ in the momentum space:

$$N \text{ (with momenta less than } |p|) = \frac{2}{h^3} \left(\frac{4}{3} \pi p^3 \right).$$

We have used $2/h^3$ as a proportionality constant. Although we will not present a justification for this, it should be noticed that the factor 2 is the result of 2 possible spins of the electron and that the dimensions come out right: N is the number of electrons per cubic meter.

Since $p = mv$, it follows that $p^2 = 2mW$ and

$$N \text{ (with momenta less than } |p|) = \frac{2}{h^3} \times \frac{4}{3} \pi (2mW)^{3/2}. \quad (92)$$

Thus

$$\frac{\partial N}{\partial W} = \frac{8\sqrt{2}\pi}{h^3} m^{3/2} W^{1/2}. \quad (93)$$

When the temperature is larger than zero, the probability that a given state be occupied is given by

$$f = \frac{1}{1 + \exp\left(\frac{W - \mu}{kT}\right)}. \quad (94)$$

The quantity μ is called the **chemical potential**. When $T = 0$, $\mu = W_F$ and the function above has the property that $f = 1$ for $W < W_F$ and $f = 0$ for $W > W_F$, as required. When $T \neq 0$, μ is slightly smaller than W_F.

The **density of states**, using this probability function becomes

$$\frac{\partial N}{\partial W} = \frac{8\sqrt{2}\pi}{h^3} m^{3/2} \frac{W^{1/2}}{1 + \exp\left(\frac{W - \mu}{kT}\right)} \quad (95)$$

and

$$N = \frac{8\sqrt{2}\pi}{h^3} m^{3/2} \int_0^\infty \frac{W^{1/2}}{1 + \exp\left(\frac{W-\mu}{kT}\right)} dW. \tag{96}$$

Since N is independent of T, the integral must itself be independent of T, which means that the chemical potential, μ, must depend on T in just the correct manner.

For $kT \ll \mu$,

$$\int_0^\infty \frac{W^{1/2}}{1 + \exp\left(\frac{W-\mu}{kT}\right)} dW \approx \frac{2}{3}\mu^{3/2} = \frac{2}{3}W_F^{3/2}, \tag{97}$$

that is, the integral does not depend on T and $\mu = W_F$.

At higher temperatures, to insure the invariance of the integral, μ must change with T approximately according to

$$\mu = W_F - \frac{(\pi kT)^2}{12W_F}. \tag{98}$$

The difference between μ and W_F is small. For Fermi levels of 5 eV and T as high as 2000 K, it amounts to only some 0.1%. Therefore, for most applications, the chemical potential can be taken as equal to the Fermi level. One exception to this occurs in the study of the thermoelectric effect. Figure 2.9 shows a plot of the density of states, $\partial N/\partial W$, versus W for three temperatures. Notice that the chemical potential, which coincided with the Fermi level at $T = 0$, has shifted to a slightly lower energy at the higher temperature.

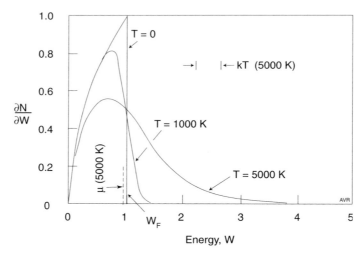

Figure 2.9   Energy distribution of electrons in a metal at three different temperatures.

2.20 Boltzmann's Law

A very useful result from statistical mechanics is **Boltzmann's law**, which describes the concentration of particles in a gas as a function of their potential energy and their temperature. This law is used in a number of chapters in this book.

Consider a force, F (derived from a potential), acting on each atom and aligned along the x-direction. nF is, of course, the total force acting on a cubic meter of gas and, if we restrict ourselves to a rectangular prism of base area, A, and height, dx, the force is $nFdxA$, and the pressure on the base is $nFdx$. In equilibrium, this pressure must balance the gas pressure, $kTdn$.

$$F \, dx = kT \frac{dn}{n}. \tag{99}$$

The potential energy, $W_{pot} = -\int F \, dx$, is

$$W_{pot} = -kT \ln n, \tag{100}$$

from which, we obtain Boltzmann's equation

$$n \propto \exp\left(-\frac{W_{pot}}{kT}\right). \tag{101}$$

Example
Atmospheric pressure

Each molecule in air has a potential energy, $mg\Delta h$, relative to a plane Δh meters closer to the ground.

According to Boltzmann's law, the concentration of molecules must vary as

$$n = n_0 \exp\left(-\frac{mg\Delta h}{kT}\right) \tag{102}$$

Note that kT/mg has the dimensions of length. It is called the **scale height**, H.

$$n = n_0 \exp\left(-\frac{\Delta h}{H}\right) \tag{103}$$

and, if T is independent of h,

$$p = p_0 \exp\left(-\frac{\Delta h}{H}\right). \tag{104}$$

(continued)

2.31

(continued)

In an isothermal atmosphere, the air pressure falls exponentially with height.

Taking the mean mass of the molecules in air as 29 daltons, and knowing that to convert daltons to kg, it suffices to divide by Avogadro's number, we find that the mass of a representative air molecule is about 48×10^{-27} kg. Consequently, the scale height of earth's atmosphere (isothermal at 300 K) is

$$H = \frac{1.38 \times 10^{-23} \times 300}{48 \times 10^{-27} \times 9.8} = 8{,}800 \text{ m.} \tag{105}$$

Appendix

Symbology

We will try to adopt the following convention for representing thermo-dynamic quantities such as

G, free energy,
H, enthalpy,
Q, heat,
S, entropy, and
U, internal energy:

1. Capital letters will indicate the quantity associated with an arbitrary amount of matter or energy.
2. Lowercase letters indicate the quantity per unit. A subscript may be used to indicate the species being considered. For example, the free energy per kilomole of H_2 will be represented by $\overline{g_{H_2}}$.

$g =$ free energy per kilogram.
$\overline{g} =$ free energy per kilomole.
$g^* =$ free energy per kilogram, at 1 atmosphere pressure.
$\overline{g}^* =$ free energy per kilomole, at 1 atmosphere pressure.
$\overline{g}_f =$ free energy of formation per kilomole.
$\overline{g}_f^\circ =$ free energy of formation per kilomole, at 298 K, 1 atmo-sphere, i.e., at RTP (Standard Free Energy of Formation).

For more information on some topics in this chapter, read:
Çengel, Y. A., and M. A. Boles, *Thermodynamics, an engineering approach*, McGraw-Hill, **1994**.

PROBLEMS

2.1   10 kg/s of steam ($\gamma = 1.29$) at 2 MPa is delivered to an adiabatic turbine (100% efficient). The exhaust steam is at 0.2 MPa and 400 K.

1. What is the intake temperature?
2. What power does the turbine deliver?

2.2   Show that the cylinder and piston experiment of Section 2.15 (with the solid object inside) is reversible provided the compression is carried out infinitely slowly. Do this numerically. Write a computer program in which compression and expansion take place in suitably small steps and are, in each step, followed by an equalization of temperature between the gas and the solid object within the cylinder.

2.3   Refer to the experiment described in Section 2.10. Show that the work done in lifting the 1 kg mass in two steps (first 2 kg, then 1 kg) is 14.2 J. Show that the 2 kg mass rises 0.444 m. Assume that the steps occur slowly enough so that the gas cooled by the expansion returns to the original temperature after each step.

2.4   Consider 10 m^3 (V_0) of gas ($\gamma = 1.6$) at 10^5 Pa (p_0) and 300 K (T_0).

1. How many kilomoles, μ, of gas are involved?
2. The gas is compressed isothermally to a pressure, p_f, of 1 MPa.

   2.1 What is the new volume, V_f?
   2.2 How much energy was used in the compression?

3. Now, instead of compressing the gas isothermally, start again (from V_0, p_0, and T_0) and compress the gas adiabatically to a pressure, p_2. The gas will heat up to T_2. Next, let it cool down isometrically (i.e., without changing the volume) to $T_3 = 300$ K and a pressure, p_3, of 1 MPa. In other words, let the state return to that after the isothermal compression.

   3.1 What is the pressure, p_2?
   3.2 What is the temperature, T_2, after the adiabatic compression?
   3.3 What is the work done during the adiabatic compression?
   3.4 Subtract the heat rejected during the isometric cooling from the work done during the adiabatic compression to obtain the net energy change owing to the process described in Item 3.3.

2.5   When a gas expands, it does an amount of work

$$W = \int_{V_0}^{V_1} p\,dV.$$

2.34

If the expansion is adiabatic, the polytropic law is observed and the integral becomes (see Chapter 6)

$$W = \frac{p_0 V_0^\gamma}{\gamma - 1} \left(V_1^{1-\gamma} - V_0^{1-\gamma} \right).$$

Show, by using the definitions of c_v and of γ, that this work is equal to the energy needed to raise the temperature of the gas from T_0 to T_1 under constant volume conditions.

2.6   The domains in a nonmagnetized ferromagnetic material are randomly oriented, however, when magnetized, these domains are reasonably well aligned. This means, of course, that the magnetized state has a lower total entropy than the nonmagnetized state.

There are materials (gadolinium, Gd, for example) in which this effect is large. At 290 K, polycrystalline gadolinium (atomic mass 157.25, density 7900 kg/m^3) has a total entropy of 67.6 kJ K^{-1} kmole^{-1} when unmagnetized and 65.6 kJ K^{-1} kmole^{-1} when in a 7.5 tesla field.

Assume that 10 kg of Gd are inside an adiabatic container, in a vacuum, at a temperature of 290 K. For simplicity, assume that the heat capacity of the container is negligible. The heat capacity of Gd, at 290 K, is 38.4 kJ K^{-1} kmole^{-1}.

Estimate the temperature of the gadolinium after a 7.5 T field is applied.

2.7   The French engineer, Guy Negre, invented an "eco-taxi," a low pollution vehicle to be built in Mexico beginning in the year 2001. Its energy storage system consists of compressed air tanks that, on demand, operate an engine (it could be a turbine, but in the case of this car, it is a piston device).

There are several problems to be considered. Let us limit ourselves to the turn around efficiency of the energy storage system. For comparison, consider that a lead-acid battery has a turn around efficiency of somewhat over 70% and fly wheels, more than 90%.

A very modern compressed gas canister can operate at 500 atmospheres.

1. Calculate the energy necessary to compress 1 kilomole of air ($\gamma = 1.4$) *isothermally* from 1 to 500 atmospheres. The temperature is 300 K.
2. One could achieve the same result by compressing the air adiabatically and then allowing it to cool back to 300 K. Calculate the energy necessary to accomplish this.
3. The compressed air (at 300 K) is used to drive a turbine (in the French scheme, a piston engine). Assume that the turbine is ideal—**isentropic**—and it delivers an amount of mechanical energy equal to the change of enthalpy the gas undergoes when expanding. How much

energy does 1 kilomole of air deliver when expanding under such conditions?

To solve this problem, follow the steps suggested below.

3.1. Write an equation for the change of enthalpy across the turbine as a function of the input temperature (300 K) and the unknown output temperature.

3.2. Using the polytropic law, find the output temperature as a function of the pressure ratio across the turbine. Assume that the output pressure is 1 atmosphere.

If you do this correctly, you will find that the temperature at the exhaust of the turbine is below the liquefaction point of the gases that make up air. This would interfere with the turbine operation, but in the present problem, disregard this fact.

3.3. Once you have the exhaust temperature, calculate the mechanical energy generated by the turbine.

4. What is the turn around efficiency of the compressed air energy storage system under the (optimistic) assumptions of this problem—that is, what is the ratio of the recovered energy to the one required to compress the air?

2.8   The cylinder in the picture has, initially, a 1 liter volume and is filled with a given gas at 300 K and 10^5 Pa. It is perfectly heat insulated and is in a laboratory at sea level. The frictionless piston has no mass and piston and cylinder, as well as the 1-ohm electric resistor, installed inside the device have negligible heat capacity.

At the beginning of the experiment, the piston is held in place, so it cannot move.

A 10-amp dc current is applied for 1 second. causing the pressure to rise to 1.5×10^5 pascals. Next, the piston is released and rises.

What is the work done by the piston?

Chapter 3
Mechanical Heat Engines

3.1 Heats of Combustion

The driving agent of a heat engine is a temperature differential. A heat engine must have a source and a sink of heat. The heat source may be direct solar radiation, geothermal steam, geothermal water, ocean water heated by the sun, nuclear energy (fission, fusion, or radioactivity), or the combustion of a fuel. In developed countries, over 90% of the energy is derived from combustion of fuels, almost all of which are of fossil origin.

When carbon burns completely in an oxygen atmosphere, the product is carbon dioxide which, under all normal circumstances in this planet, is a gas. However, most fuels contain hydrogen and, thus, when burned, they also produce water. The resulting water may leave the engine in liquid or in vapor form. In the former case it releases its latent heat of vaporization to the engine. For this reason, hydrogen-bearing fuels can be thought of as having two different heats of combustion: one, called the **higher heat of combustion**, corresponds to the production of liquid water, whereas the other, corresponding to the formation of water vapor, is called the **lower heat of combustion**.

The higher heat of combustion of hydrogen is 143 MJ/kg, while the lower is 125 MJ/kg. The heat of combustion of carbon is 32.8 MJ/kg. Thus, one could expect that the heat of combustion of a hydrocarbon, $C_n H_m$, is roughly

$$\Delta H = \frac{12n \times 32.8 + 143m}{12n + m} \tag{1}$$

for the case in which the product is liquid water.

Table 3.1
Comparison Between Estimated and Measured Higher Heats of Combustion

	ΔH (MJ/kg) estimated	ΔH (MJ/kg) measured	Error %
CH_4	60.4	55.6	8.3
$C_2 H_6$	54.8	52.0	5.3
$C_3 H_8$	52.8	50.4	4.6
$C_{10} H_{22}$	49.9	47.4	5.0
$C_{20} H_{42}$	49.2	47.2	4.0

Table 3.2

**Higher Heats of Combustion of Hydrocarbons
of the Form C_nH_{2n+a}**

$n \downarrow$	$a \rightarrow 2$	0	-2	-4	-6	-8	-10
0	143						
1	55.6						
2	52.0						
3	50.5		MJ/kg				
4	50.2						
5	49.1						
6	48.5						
7	48.2						
8	47.9						
10	47.4	47.3	46.2	45.3	44.4	42.9	41.8
16	47.4						
20	47.2						

Table 3.1 compares some higher heats of combustion calculated from our simplified formula with those actually measured. Naturally, our formula overestimates these heats because it does not take into account the binding energies between C and H and between C and C. However, our reasoning predicts the regular behavior of the heats as a function of molecular mass.

The higher the order of the hydrocarbon, the larger the relative amount of carbon compared with hydrogen: thus, the smaller the heat of combustion. Methane, the first of the aliphatic hydrocarbons, has the largest heat of combustion: 55.6 MJ/kg. As the order increases, the heat of combustion decreases tending toward 47 MJ/kg. This is illustrated in Figure 3.1. A similar regularity is observed as one moves from one series of hydrocarbons to the next. Table 3.2 demonstrates this effect.

An alcohol results when an OH radical replaces a hydrogen in a hydrocarbon (Figure 3.2). Gaseous hydrocarbons form liquid alcohols. Thus, for vehicles (where a liquid is more useful than a gas), it is advantageous to partially oxidize hydrocarbons transforming them into alcohols. This partial oxidation causes the alcohol to have a smaller heat of combustion than the parent hydrocarbon:

CH_4 (55.6 MJ/kg) yields CH_3OH (22.7 MJ/kg),
C_2H_6 (52.0 MJ/kg) yields C_2H_5OH (29.7 MJ/kg).

However, 1 kmole (16 kg) of methane yields 1 kmole (32 kg) of methanol. Hence, 1 kg of methane can be transformed into 2 kg of methanol with an efficiency of

$$\eta = \frac{2 \times 22.7}{55.6} = 0.82 \tag{2}$$

Table 3.3

**Efficiency of Conversion of
Hydrocarbons into Alcohols**

Methane into methanol	0.82
Ethane into ethanol	0.88
Propane into propanol	0.91

Table 3.3 shows the *theoretical* efficiencies of converting hydrocarbons into their corresponding alcohols.

In addition to hydrocarbons and alcohols, there are a number of other fuels to be considered.

Fuels can also be derived from biomass (Chapter 13). They all contain oxygen in their molecule and, for this reason, have lower heats of combustion than hydrocarbons. Cellulose, for instance, has a heat of combustion of 17.4 MJ/kg.

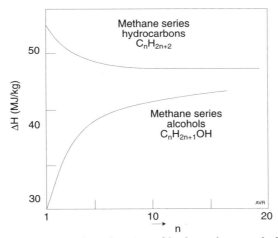

Figure 3.1   Heats of combustion of hydrocarbons and alcohols.

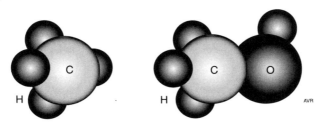

Figure 3.2   An alcohol results from the substitution of a hydrogen atom by a hydroxyl radical. The figure shows the structures of methane and its corresponding alcohol, methanol.

3.3

3.2 Carnot Efficiency

Mechanical and electrical energy are "noble" forms of energy: no entropy is associated with them. Consequently, it is theoretically possible to transform one into another without losses—in other words, without having to reject heat. Big machines can make this transformation with over 99% efficiency.

Heat engines transform a degraded form of energy—heat—into a noble form: either electrical or mechanical. This cannot be done without rejecting part of the input heat (unless the engine works against a sink at absolute zero). Thus, even theoretically, the achievable efficiency is smaller than 1.

Figure 3.3 indicates how a heat engine operates. The input is an amount of heat, Q_{in}, the useful output is a quantity, W (of mechanical or electrical energy), and Q_{out} is an amount of heat that must be rejected. The efficiency of such an engine is the ratio of the output energy to the heat input:

$$\eta = \frac{W}{Q_{in}} = \frac{Q_{in} - Q_{out}}{Q_{in}} \tag{3}$$

The entropy in the system must increase or, at best, suffer no change. Hence, the largest possible efficiency—**the Carnot efficiency**—corresponds to the situation in which the entropy at the output, Q_{out}/T_C, is equal to that at the input, Q_{in}/T_H. Of course, there is no entropy associated with the "noble" work output, W. Consequently,

$$\frac{Q_{in}}{T_H} = \frac{Q_{out}}{T_C} \tag{4}$$

$$Q_{out} = \frac{T_C}{T_H} Q_{in} \tag{5}$$

$$\eta_{CARNOT} = \frac{T_H - T_C}{T_H} \tag{6}$$

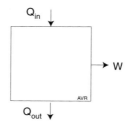

Figure 3.3   The operation of a heat engine.

Thus, the Carnot efficiency depends only on the temperatures between which the engine operates.

In steady state conditions,

$$\frac{Q_{out}}{W} = \frac{Q_{in} - W}{W} = \frac{Q_{in}}{W} - 1 = \frac{1}{\eta} - 1 \tag{7}$$

As η decreases, the ratio above increases rapidly. A modern fossil-fueled steam turbine with $\eta \approx 0.4$ rejects 1.5 joules of heat for each joule of useful energy produced. In a thermal nuclear reactor such as all reactors in the United States, the efficiency of the turbine is lower (some 28%) because of limitations in T_H: the rejected heat is then 2.6 times larger than the generated energy.[†] The cooling tower of nuclear plants must be substantially larger than that of a fossil-fueled plant.

Automotive engines have small efficiencies ($\eta \approx 0.2$) and the large amount of heat they reject can pose a serious problem when special equipment is needed for this purpose. Fortunately, Otto and diesel engines do not need such equipment. The radiator found in most cars rejects only the heat from component inefficiency, not the thermodynamically rejected heat.

3.3 Engine Types

All mechanical heat engines involve four processes:

1. compression (or pumping),
2. heat addition,
3. expansion,
4. heat rejection.

Figure 3.4 illustrates the four processes as they occur in a **close-cycle Rankine** engine. The expansion can be accomplished through a turbine or through a cylinder with a piston, as in the locomotives one still sees in period movies.

Since steam engines at the beginning of the 20th century turbine had less than 10% efficiency, the amount of heat that the condenser (in the close-cycle machines) had to reject was large: some 9 joules for each joule of mechanical energy produced (Equation 7).

To avoid unwieldy condensers, early locomotives released the spent vapor exhausting from the cylinder directly into the air (thus leaving to the environment the task of heat removal). See Figure 3.5.

[†] This is not true of fast reactors such as the heavy-metal ones that operate at high temperatures.

3.5

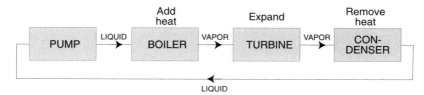

Figure 3.4   The four processes in the Rankine cycle. In the close cycle configuration, the working fluid is condensed for reuse.

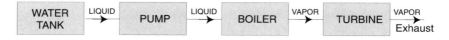

Figure 3.5   In the open-cycle Rankine, exhaust vapor is released directly into the environment. A large water tank is required.

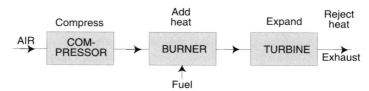

Figure 3.6   The Brayton cycle used in jet planes is an internal combustion, open-cycle engine.

Water could not be recovered in such open-cycle machines, the locomotive had to be equipped with a large water tank whose size limited the range of the locomotive.

Heat must be added from an external source, as the Rankine cycle is an **external heat source** (most frequently, an **external combustion**) cycle.

Another example of an open-cycle engine (in this case an **internal combustion cycle**) is the Brayton turbine used in jet and turbojet planes, illustrated in Figure 3.6.

Examples of close-cycle engines are:

> the Rankine or "vapor"-cycle engine,
> the Stirling engine.

Examples of open-cycle engines are:

> the Otto (spark-ignition) engine,
> the diesel (compression-ignition) engine,
> the open-cycle Brayton engine.

The Rankine engine is distinguished by having the working fluid change phase during the cycle. Different working fluids can be employed. For operation at very low temperatures (say, boiler temperature of some 25 C, as is the case in ocean thermal energy converters), the most appropriate fluid for the Rankine-cycle is ammonia. For low operating temperatures such as those in some direct solar engines, one of several available freons may work best. Most fossil-fueled Rankine engines use water vapor, although mercury may be effective at higher temperatures.

Steam engines are ill suited for automobiles—they either require a large condenser or a large water tank. In part, because no heat rejection device is required, the open-cycle Brayton turbine is universally used in jet planes. Various engines that use only gases as working fluids differ in the nature of the processes used, as illustrated in Table 3.4. For good efficiency, mechanical heat engines must operate with high compression ratios.

In gas turbines, two different types of compressors can be used: radial or axial. Radial compressors can be built with compression ratios of about 3:1 per stage. Thus, for a 9:1 compression, two stages are sufficient. The disadvantage is that it is difficult to obtain high efficiencies with this type of compressor, especially when stages are ganged. In radial (centrifugal) compressors, the air is accelerated towards the rim and must be redirected to the intake at the hub of the next stage. However, channeling air around involves losses. Furthermore, part of the energy imparted to the air is rotational and therefore has to be converted to translational energy by means of (lossy) **diffusers**.

Axial compressors yield only a small compression ratio per stage, say, 1.2:1. Thus, a large number of stages must be employed. The Rolls-Royce "Tyne" aircraft turbine achieves an overall 13.5:1 compression using 15 stages, with a mean compression ratio of 1.189:1 per stage. The large number of stages is practical because of the simplicity with which the output of one stage can be fed to the input of the next. The trouble with axial compressors is that their efficiency falls rapidly with decreasing size owing mostly to blade tip leakage, something that is difficult to avoid in small machines. For this reason, Brayton turbines of relatively low power are still of the radial type.

Table 3.4
Nature of the Four Processes in Different Engines

Cycle	Compression	Heat addition	Expansion	Heat Rejection
Carnot	adiabatic	isothermal	adiabatic	isothermal
Otto	adiabatic	isometric	adiabatic	isometric
Diesel	adiabatic	isobaric	adiabatic	isometric
Brayton	adiabatic	isobaric	adiabatic	isobaric
Stirling	isothermal	isometric	isothermal	isometric
Ericsson	isothermal	isobaric	isothermal	isobaric

Gas turbines may be at the verge of opening a new significant market as bottoming cycle for solid oxide fuel cells whose exhaust consists of high temperature gases well suited to drive this class of turbines.

In both Rankine and Brayton engines, each of the processes is carried out in a different part of the equipment. Compression and expansion are accomplished by separate devices and combustion occurs in its special chamber. This allows optimization of each part.

In Otto and diesel engines, three processes are carried out in the same engine part—the cylinder. The fourth process—heat rejection—is carried out in the environment. The multiple functions of the cylinder require design compromises with resulting lower component efficiency.

While Rankine and Brayton engines are of the continuous combustion types, Otto and diesel operate intermittently and are able to tolerate higher working temperatures, since the combustion phase lasts only briefly. Also, heat is generated directly in the working fluid, thereby doing away with heat transfer problems encountered in all external combustion engines.

The Stirling engine, which will be examined in greater detail toward the end of this chapter, holds considerable promise, but, although quite old in concept, it has never achieved the popularity that it appears to deserve. It is, at least theoretically, the most efficient of the mechanical heat engines. It is less polluting than the Otto and the diesel because it burns fuel externally. Additionally, since the Stirling engine involves no explosions, it runs more smoothly and with less noise than other engines. Finally, in spite of being a close-cycle device, it requires only moderate heat rejection equipment.

The Stirling cycle also finds application in some refrigeration equipment. Since it operates with helium, hydrogen, or air, it employs no freons or other ecologically damaging fluids.

D.G. Wilson (1978) summarized the characteristics of the most common mechanical heat engines, as shown in Table 3.5.

Table 3.5

Some Characteristics of the More Common Combustion Engines

	Temper. ratio	Compon. effic.	Relative thermodyn. EFFIC.	Heat rejection (Thermodyn.)
Otto	High	Fair	Poor	None
Diesel	High	Fair	Poor	None
Rankine	Low	Good	Fair	Large
Stirling	Moderate	Good	Very Good	Medium
Brayton (o.c.)	Mod./high†	Very Good	Good	None
Brayton (c.c.)	Moderate	Very Good	Good	Medium

† Development of more modern materials is expected to permit use of higher temperature ratios.

3.8

3.4 Efficiency of an Otto Engine

In an ideal Otto cycle, a fuel/air mixture is compressed adiabatically from a volume, V_1, to a volume, V_2, during the compression phase. The gas traces out the line between the points 1 and 2 in the p-V diagram of Figure 3.7. Thus,

$$\frac{p_2}{p_1} = \left(\frac{V_1}{V_2}\right)^\gamma \tag{8}$$

From the perfect gas law

$$\frac{p_2 V_2}{T_2} = \frac{p_1 V_1}{T_1} \tag{9}$$

$$T_2 = T_1 \frac{p_2 V_2}{p_1 V_1} = T_1 \left(\frac{V_1}{V_2}\right)^{\gamma-1} \tag{10}$$

$$T_2 = T_1 r^{\gamma-1} \tag{11}$$

where r is the **compression ratio**, V_1/V_2. The compression is adiabatic because, owing to the rapidity with which it occurs, no heat is exchanged between the interior of the cylinder and the outside. Consequently, the work done is equal to the increase in internal energy of the gas:

$$W_{1,2} = \mu c_v (T_2 - T_1) = \mu c_v T_2 (1 - r^{1-\gamma}). \tag{12}$$

At the end of the compression phase, Point 2, a spark ignites the mixture, which, ideally, is assumed to burn instantaneously. Both temperature and pressure rise instantaneously, the latter reaching the value indicated in Point 3 of the diagram. There is no change in volume during this heat addition or **combustion** phase. The amount of heat added is

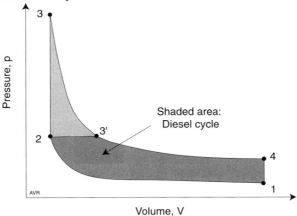

Figure 3.7   The p-V diagram of an Otto cycle.

3.9

$$Q_{2,3} = \mu c_v (T_3 - T_2). \tag{13}$$

The behavior during expansion is similar to that during compression:

$$W_{3,4} = \mu c_v T_3 (1 - r^{1-\gamma}). \tag{14}$$

The useful output of the engine is $W_{3,4} - W_{1,2}$, while the energy input is $W_{2,3}$. Hence the efficiency is

$$\eta = \frac{W_{3,4} - W_{1,2}}{Q_{2,3}} = \frac{T_3(1 - r^{1-\gamma}) - T_2(1 - r^{1-\gamma})}{T_3 - T_2} = 1 - r^{1-\gamma}. \tag{15}$$

Notice that from Equation 11, $r^{1-\gamma} = T_1/T_2$. Thus, Equation 15 can be written as

$$\eta = 1 - \frac{T_1}{T_2} = \frac{T_2 - T_1}{T_2} = \eta_{CARNOT}. \tag{16}$$

Thus, the ideal Otto cycle achieves the Carnot efficiency of an engine working between the maximum, precombustion, temperature and the intake temperature. The ideal Otto cycle cannot achieve the Carnot efficiency determined by the highest and lowest temperature during the cycle.

Theoretical efficiency is plotted versus the compression ratio in Figure 3.8. The diesel efficiency is lower than the Otto because, in the former, the combustion is supposed to occur slowly, at constant pressure (see darker shaded area in Figure 3.7), not at constant volume as in spark ignition engines. Thus, the total area enclosed by the p-V trace is smaller. In fact, the ideal efficiency of the diesel cycle is

$$\eta = 1 - r^{1-\gamma} \left[\frac{r_c^{\gamma} - 1}{\gamma (r_c - 1)} \right], \tag{17}$$

where $r_c \equiv V_{3'}/V_2$ (Figure 3.7) is called the **cutoff ratio** and is the expansion ratio during the combustion period.

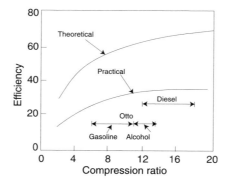

Figure 3.8   The efficiency of an Otto and a diesel engine
as a function of their compression ratios.

In practice, diesel engines operate with higher efficiency than those based on the Otto cycle, because the latter must operate at lower compression ratios to avoid **knocking** (Section 3.6).

Efficiency can be improved by

1. raising γ, and
2. raising the compression ratio.

As γ is larger for air than for fuel vapors, the leaner the mixture, the higher the efficiency. This is in part counterbalanced by the tendency of lean mixtures to burn slowly causing a departure from the ideal Otto cycle. In addition, if mixtures become too lean, ignition becomes erratic—the engine runs "rough" and tends to backfire.

The stoichiometric air/fuel ratio for gasoline is 14.7:1. However, maximum power is achieved with a very rich mixture (12:1 to 13:1), while maximum efficiency requires lean mixtures (16:1 to 18:1). The **stratified combustion** engine achieves an interesting compromise by injecting fuel and air tangentially into the cylinder so that, owing to the resulting centrifugal force, the mixture is richer near the cylinder wall and becomes progressively leaner toward the axis. Combustion is initiated in the rich region and propagates inward. While the mean mixture is sufficiently lean to ensure high efficiency, ignition is still reliable. Incidentally, diesel engines can operate with leaner mixtures than spark-ignition engines.

As we are going to see, the nature of the fuel used in Otto cycle engines limits their maximum usable compression ratio. Too high a ratio causes the engine to "ping," "knock," or "detonate," particularly during acceleration. Compression ratios that allow a car to accelerate without detonation are too small for efficiency at cruising speeds. This can be ameliorated by changing the ignition timing. The spark is retarded during acceleration and advanced for cruising. An ingenious variable compression ratio engine was developed by Daimler-Benz. The cylinder has two pistons, one connected to the crankshaft as usual and an additional one that sits freely above the first. A variable volume of oil can be inserted between these two pistons regulating their spacing thereby adjusting the compression ratio.

Consider a gasoline engine with a 9:1 compression ratio using a fuel-air mixture with a γ of 1.3. Its ideal efficiency is about 50%.

The efficiency is reduced by departures from the ideal cycle, such as

1. failure to burn fast enough. The combustion does not occur at constant volume. On the other hand, high speed diesels tend to burn not at constant pressure, but at a somewhat rising pressure.
2. heat loss through the cylinder wall and through the piston and connecting rod. Thus, the heat generated by the burning mixture does not all go into expanding the gas. Design efforts have focused on producing a more nearly adiabatic cylinder.

Typically, these departures reduce the efficiency to some 80% of ideal. The engine in the example would then have a 40% efficiency.

It can be seen that for high efficiency one needs:

1. high compression ratio,
2. fast combustion,
3. lean mixture, and
4. low heat conduction from cylinder to the exterior.

There are numerous losses from friction between solid moving parts (**rubbing friction**) and from the flow of gases (**pumping friction**). Rubbing friction can be reduced by clever design, use of appropriate materials, and good lubricants. Pumping friction can be managed by adequate design of input and exhaust systems. Losses can be reduced by increasing the number of valves (hence the popularity of cars with 4 valves per cylinder). Also, part of the power control of an engine can be achieved by adjusting the duration of the intake valve opening, thereby avoiding the resistance to the flow of air caused by the throttle. Cars may soon be equipped with electronic sound cancellation systems, a technique that dispenses with the muffler and therefore permits a better flow of the exhaust gases.

An engine must use ancillary devices whose efficiency influences the overall performance. These include the alternator, the water pump, and the radiator cooling fan. In modern cars, the latter typically operates only when needed instead of continuously as in older vehicles. About half of the engine output is consumed by these devices. In the example we are considering only some 20% of the combustion energy is available to the accessories (e.g., air conditioner and power steering) and to the transmission. The latter is about 90% efficient, so the residual power available to the propulsion of the car could be as little as 18% of the fuel energy.

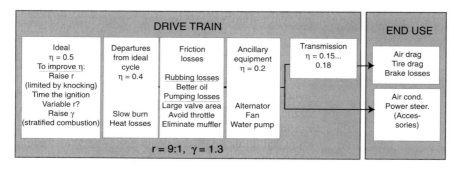

Figure 3.9   Drive train and end use in a spark-ignition vehicle. With a compression ratio of 9:1 and with an air/fuel mixture having a gamma of 1.3, the ideal engine has 50% efficiency. Assorted losses and ancillary equipment reduce the final efficiency to less than 18%.

Engine, transmission, and ancillary parts constitute the **drive train**. The load on the power train (e.g., tire drag, aerodynamic losses, brake losses, and accessories) constitutes the **end use load.** We saw that, roughly, the end use load is 18% or less of the fuel energy.

3.5 Gasoline

Without a doubt the most popular automotive fuel currently is gasoline. Gasoline is not a chemically unique substance—it's composition has been continuously improved since its introduction and is also adjusted seasonally. It is a mixture of more than 500 components dominated by hydrocarbons with 3 to 12 carbon atoms. Most are branched (see discussion of the difference between *octane* and *isooctane* in the next section). For this book, the two main characteristics of gasoline are the following.

3.5.1 Heat of Combustion

Since the composition of gasoline is variable, its heat of combustion is not a fixed quantity. One may as well use the values for heptane or octane (≈ 45 MJ/kg) as a representative higher heat of combustion.

3.5.2 Antiknock Characteristics

As far as energy content is concerned, gasoline has a decisive advantage over alcohol. However, there is no point in using a high energy fuel if this leads to a low engine efficiency. As discussed previously, the efficiency is determined in part by the compression ratio which, if too high, causes knocking (see next section). Alcohols tolerate substantially higher compressions than most gasolines and therefore lead to greater engine efficiencies. This somewhat compensates for the lower specific energy of these fuels. Gasolines with better antiknock characteristics (higher octane rating) do not necessarily have higher energy content, in fact, they tend to have lower energy. It makes sense to use the gasoline with the lowest possible octane rating (i.e., the cheapest) compatible with the engine being fueled. On the other hand, cheaper gasolines, independent of their octane rating, may cause gum formation and other deposits in the engine and may result in more exhaust pollution.

3.6 Knocking

The efficiency of an engine increases when the compression ratio increases. In spark ignition engines, the compression must be limited to that tolerated by the fuel used. Compressions that are too high cause **detonation** or **knocking**, a condition that, in addition to being destructive to the pistons, leads to a reduction in the power of the motor.

The difference between **explosion** and **detonation** is the rapidity of combustion. Gunpowder, for instance, will explode when confined but will only burn with a hiss when ignited in free air. On the other hand, substances such as nitrogen triiodide will decompose so fast that, even unconfined, will make a loud noise.

An explosion within a cylinder exerts a steady force on the piston analogous to the force a cyclist puts on the pedal of his bike. A detonation is as if the cyclist attempted to drive his vehicle with a succession of hammer blows.

The ability of a fuel to work with a high compression ratio without detonating is measured by its **octane rating**. A fuel is said to have an octane rating, O_f if it behaves (as far as detonation is concerned) like an isooctane/n-heptane mixture containing $O_f\%$ octane. The fuel need contain neither octane nor heptane.

A fuel may have an octane rating larger than 100%—a result of extrapolation.

Experimentally, the octane rating of fuels with $O_f > 100$ is determined by comparing with isooctane to which a fraction, L (by volume), of tetraethyl lead, $(C_2H_5)_4Pb$, has been added. The octane rating is given by

$$O_f = 100 + \frac{107,000L}{(1 + 2786L) + \sqrt{1 + 5572L - 504,000L^2}}. \tag{18}$$

An addition of 0.7% (i.e., of a fraction of 0.007) of tetraethyl lead to isooctane leads to a 120 octane ratio, a value common in aviation gasoline.

Notice that the compound used is *isooctane*. The reason is that n-octane—normal or unbranched octane—has extremely poor antiknocking behavior whereas isooctane resists knocking well.

The critical compression ratio of hydrocarbons[†] decreases rapidly with the number of carbons in the molecule. Thus, methane may have a CCR of 13 while that of n-heptane is down to 2.2 and that of n-octane is even lower.

However, if the structure of the molecule is changed (i.e., if a different isomer is used), the knock resistance may increase substantially: Isooctane has a CCR of 6, benzene, the basic aromatic hydrocarbon, has a CCR of 15, For this reason, unleaded gasoline may contain considerable amounts of aromatics to insure a high octane rating.

Compare the structure of two isomers of octane (C_8H_{18})

[†] The critical compression ratio, CCR, is the compression ratio that just causes knocking to begin under given experimental condition, such as 600 rpm and 450 K coolant temperature.

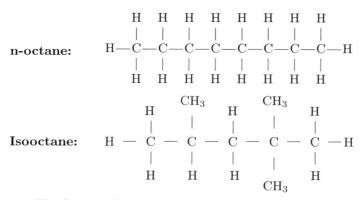

The formula above shows that isooctane is technically pentane in which three hydrogens have been replaced each by a methyl (CH_3) radical. Two of the substitutions occur in position 2 and one in position 4 of the molecule. Hence isooctane is *2,2,4-trimethylpentane*.

The effective octane rating of a fuel depends on the conditions under which it operates. For this reason, more than one octane rating can be associated with any given fuel. The rating displayed on the gas station pump is usually an average of two differently measured values. A more complete discussion of this topic can be found in a book by J. B. Heywood.

Additives increase the octane rating of gasoline. Iodine can be used, but is expensive. Up to a few years ago, tetraethyl lead was the standard additive in **leaded** gasolines. Environmental concerns have eliminated this type of fuel. High octane rating is now achieved by increasing the percentage of cyclic (benzene series) hydrocarbons. Thus, one avoids poisoning by increasing the risk of cancer and, incidentally, paying more for fuel.

The presence of ethanol in gasoline increases its resistance to detonation as indicated in Figure 3.10. It can be seen that the addition of 30% ethanol to low grade gasoline raises its octane rating from 72 to 84. The octane rating, O_m, of **gasohol** (gas/alcohol mixture) can be calculated from the octane rating, O_g, of the original gasoline and from the **blending octane value**, B, of the alcohol:

$$O_m = Bx + O_g(1 - x), \tag{19}$$

where x is the ratio of the additive volume to that of the gasoline. Depending on the initial quality of the fuel, the blending octane value of ethanol can be as high as 160. Methanol has a B of 130, although, when used alone, its rating is only 106. Gasohol can achieve high octane ratings without the use of lead and with only moderate addition of cyclic hydrocarbons. Thus, gasohol brings substantial public health advantages.[†]

[†] This may not be quite true when the additive is methanol because of the formaldehyde in the exhaust. With ethanol, the exhaust contains, instead, some acetaldehyde, a relatively innocuous substance.

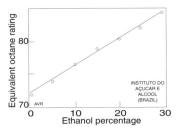

Figure 3.10   Addition of ethanol to gasoline results
in a mixture with higher octane rating.

Since 1516, Brazil has been the world's leading sugar cane producer. The widely fluctuating international price of sugar prompted Brazil to develop gasohol as a means of disposing of excess production. In years when the price was low, the alcohol percentage in Brazilian gasoline was high (typically, 24%). When sugar prices were high, much less ethanol found its way into automotive fuel (say, 5%). Starting in the 1970s Brazil decided to sell pure (hydrated) alcohol as fuel for its fleet of specially designed cars thus achieving a certain independence from the importation of oil.

Alcohol is more than an additive—it is, itself, a fuel. However, its energy content is lower than that of gasoline (Table 3.6). Per unit volume, ethanol contains only 71% of the energy of heptane, the main constituent of gasoline. Nevertheless, Brazilian alcohol-driven cars (using gasoline-free ethanol) have a per liter kilometrage that approaches that of gasoline engines. This is due to the higher efficiency of the high compression ethanol motors. However, at present ethanol is, per liter, more expensive than gasoline and the Brazilian program requires a substantial government subsidy to make the use of pure alcohol acceptable to the public.

Because water can be mixed with alcohol—inviting the "stretching" of the fuel sold at refueling stations—Brazilian pumps are equipped with densitometers permitting the consumer to check on the quality of the product.

Table 3.6
Properties of Two Important Alcohols
Compared with Heptane and Octane.
Higher heats of combustion for fuels at 25 C

	Mol. mass	kg/ liter	MJ/ kg	MJ/ liter	MJ/ kg (rel. to octane)	MJ/ liter (rel. to octane)
Methanol	32	0.791	22.7	18.0	0.475	0.534
Ethanol	46	0.789	29.7	23.4	0.621	0.697
n-Heptane	100	0.684	48.1	32.9	1.00 6	0.979
iso-Octane	114	0.703	47.8	33.6	1.000	1.000

3.7 Hybrid Engines for Automobiles

Automobile emission standards are established individually by each state, but the leader is the California Air Resources Board (CARB) which has proposed the most stringent emission specifications in the country. These included a requirement that, by a given date, 2% of the vehicles sold in California be "zero emission vehicles (ZEV)." This requirement was later postponed. Automobile manufacturers have spent considerable effort in the exegesis of the expression, ZEV.

Clearly, a purely electric vehicle (EV) emits no noxious gases. Nevertheless, it consumes electricity that is generated in part by burning fossil fuels, which produces pollutants. An EV does pollute, albeit very little compared with a conventional internal combustion vehicle (ICV). Some argue that if an automobile equipped with an internal combustion engine emits the equivalent amount of pollution (or less) than the total emission from an EV, then such an ICV should also be considered a "zero emission" car. The general interest in this type of vehicle is attested by the great popularity of the Toyota Prius.

A hybrid vehicle is an electric car equipped with an additional fuel-driven power source. There are several reasons why hybrids lead to a substantial lowering of emission:

1. Whereas a normal automotive engine has to operate over a wide range of powers, from idling to full acceleration, the battery-charging engine of a hybrid is optimized for operation at constant power and can be fine-tuned for maximum efficiency and minimum pollution.
2. There is no waste during the frequent idling periods that occur in normal city driving.
3. Regenerative braking that returns power to the battery during deceleration can be implemented in a relatively simple manner.

There are two general categories of hybrid vehicles: series and parallel.

In series hybrid vehicles, the power applied to the wheels comes entirely from the electric motor(s). The fuel driven component simply recharges the battery.

In parallel hybrids, wheel power is derived from both electric and IC motors. Clutches are used to couple these different power plants to the wheels according to the requirements of the moment.

Series hybrids are relatively simple but require large electric motors capable of delivering full acceleration power. They must, in addition, have auxiliary systems to maintain battery charge. Thus series hybrids have large drive motors, a charging motor, and a generator. The sum of the powers of these three components substantially exceeds the power necessary to drive the vehicle. This can be expensive.

In parallel hybrids the electric motors can be much smaller, and the additional surge power comes from the IC power plant. However, the extra power plant in a hybrid does not have to be a heat engine. Fuel cells may prove ideal for such an application.

3.8 The Stirling Engine

Had the early automobile developers opted for a Stirling engine rather than an Otto, it is possible that present day combustion vehicles would be more efficient and less polluting. But a quirk of history tipped the scales away from the Stirling.

Stirlings have the following advantages:

1. They are more efficient than Otto and diesel engines.
2. They can operate with a wide variety of fuels.
3. Being an external combustion engine they tend to generate less pollutants. They still produce large amounts of carbon dioxide, but, owing to their greater efficiency, they produce less than current automotive engines of equivalent power. They can operate well with fuels having a low carbon-to-hydrogen ratio, thus producing more energy per unit amount of carbon emitted.
4. They are low-noise devices because no explosions are involved.

In addition to its application to engines, the Stirling cycle can be adapted for refrigeration without needing CFCs.

With the rapidly approaching era of fuel cells, the future of any new mechanical heat engine is, at best, questionable. Nevertheless, a more detailed examination of the Stirling cycle is an excellent tool for gaining a certain insight into the analysis of combustion engines in general.

The Stirling cycle consists of an isothermal compression, an isometric heat addition, an isothermal expansion, and an isometric heat rejection (c.f., Table 3.4). Its great efficiency results from the possibility of **heat regeneration** described in more detail later in this chapter. A number of Stirling engine configurations have been tried. Table 3.7 lists the most common.

Table 3.7

Several Stirling Engine Configurations

$$
\left\{
\begin{array}{l}
\text{Kinematic} \left\{
\begin{array}{l}
\text{Alpha (two cylinders, two pistons)} \\
\text{Beta (one cylinder with piston \& displacer)} \\
\text{Gamma (one cylinder with piston,} \\
\quad \text{another with displacer)}
\end{array}
\right. \\
\text{Free piston} \\
\text{Ringbom}
\end{array}
\right.
$$

In all configurations, two pistons are employed. In some cases one is the **power piston** and the other is the **displacer**. The distinction will become clear when we examine examples of the engine.[†] In **kinematic** engines, both pistons are driven by the crankshaft, in general by means of connecting rods. In the **free piston** configuration, the pistons are not mechanically connected to any part of the engine.

The **Ringbom** configuration uses one kinematic and one free piston.

Since the alpha configuration is the easiest to understand, we will examine it in more detail.

Consider two cylinders interconnected by a pipe as shown in Figure 3.11. One cylinder (labeled "HOT") is continuously heated by an external source, which can be a flame, a radioisotope capsule, concentrated solar energy, and so on. The temperature of the gas in this cylinder is T_H. The other cylinder (labeled "COLD") is continuously cooled by circulating cold water or blowing cool air or, in small engines, perhaps simply by convection. The temperature of the gas in this cylinder is T_C. At any rate, there is, as in any other heat engine, a **heat source** and a **heat sink**.

The space above the pistons is filled with a working gas (in practical engines, this may be hydrogen or helium). In order to follow the cycle, we will use a specific example. A gas with a $\gamma = 1.40$ is used. The volume of each cylinder can, by moving the piston, be changed from 10^{-3} m^3 to 0 m^3—that is, from 1 liter to 0 liters.

Initially (State 0), the "cold" piston is all the way down. The volume in this cylinder is $V_{C_0} = 10^{-3}$ m^3, the temperature is $T_{C_0} = T_C = 300$ K, and the pressure (in both cylinders) is $p_{C_0} = p_{H_0} = 10^5$ Pa or 1 atmosphere.

From the perfect gas law, $pV = \mu RT$, we can calculate that the amount of gas in the "cold" cylinder is 40.1×10^{-6} kilomoles.

The amount of gas in the connecting pipe and in the "hot" cylinder (V_{H_0}) is assumed to be negligible.

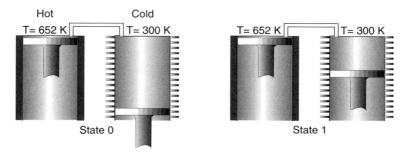

Figure 3.11   The first two phases of an alpha-Stirling cycle.

[†] Power pistons compress or do work on expansion. Displacers do no work. All they do is transfer fluid from one region of the machine to another.

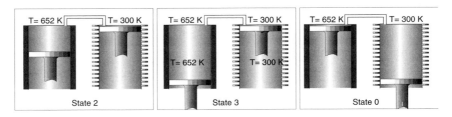

Figure 3.12   The final phases of an alpha-Stirling cycle.

Phase 0→1   (Isothermal compression)

The "cold" piston is moved partially up so that the gas volume is now $V_{C_1} = 10^{-4}$ m^3 (a compression ratio, $r = 10$). Since the cylinder is in contact with the heat sink, the heat generated by the compression is removed and the temperature remains unchanged. In other words, the compression is **isothermal**. The energy required is

$$W_{0\to1} \equiv W_{compress} = \mu R T_C \ln \frac{V_{C_0}}{V_{C_1}} = 230 \text{ J.} \tag{20}$$

The temperature did not change while the pressure increased 10-fold. State 1 of the gas is

$$V_{C_1} = 10^{-4} \text{ m}^3,$$

$$T_C = 300 \text{ K},$$

$$p_{C_1} = 10^6 \text{ Pa.}$$

Phase 1→2   (Gas transfer, followed by isometric heat addition)

The "cold" piston goes all the way up, and the "hot" piston goes partially down so that $V_{H_2} = 10^{-4}$, $V_{C_2} = 0$, and the total volume of the gas does not change. Theoretically, there is no energy cost to this gas transfer, but, the gas is now in contact with the hot source and will start heating up. Assume, arbitrarily, that the temperature rises to 652 K. For this to happen, the heat source must deliver to the gas an amount of heat, $Q_{1\to2}$.

A gas whose $\gamma = 1.4$ has c_v of 20.8 kJ K^{-1} kmole^{-1}. Hence, the heat necessary to raise the temperature from 300 to 652 K, while keeping the volume unchanged (**isometric** heat addition) is

$$Q_{1\to2} \equiv Q_{add} = \mu c_v \Delta T = \mu c_v (T_H - T_C)$$
$$= 40 \times 10^{-6} \times 20.8 \times 10^3 (652 - 300) = 293 \text{ J.} \tag{21}$$

Since the gas temperature went up without a change in volume, the pressure must have increased. State 2 of the gas is

$$V_{H_2} = 10^{-4} \text{ m}^3,$$

$$T_H = 652 \text{ K},$$

$$p_{H_2} = \frac{652}{300} \times 10^6 = 2.17 \times 10^6 \text{ Pa}.$$

Phase 2→3   (Isothermal expansion)

The high pressure pushes the "hot" piston down until the volume in the cylinder reaches 10^{-3} m^3. The corresponding 10:1 expansion would cool the gas down, but heat from the external source keeps the temperature constant—we have an **isothermal** expansion that delivers 500 J to the crankshaft:

$$W_{2\rightarrow3} \equiv W_{expan} = \mu R T_H \ln \frac{V_{H_3}}{V_{H_2}} =$$

$$40.1 \times 10^{-6} \times 8314 \times 652 \ln 10 = 500 \text{ J}. \qquad (22)$$

We have chosen, arbitrarily, a T_H of 652 K in the preceding phase, so that the energy delivered to the crankshaft comes out a round number. The heat input required is, of course, $Q_{2\rightarrow3} \equiv Q_{expan} = 500$ J.

State 3 of the gas is

$$V_{H_3} = 10^{-3} \text{ m}^3,$$

$$T_H = 652 \text{ K},$$

$$p_{H_3} = 2.17 \times 10^5 \text{ Pa}.$$

Phase 3→0   (Isometric heat rejection)

Finally, the pistons return to their initial position. The gas volume does not change but, owing to its transfer to the "cold" cylinder, it cools **isometrically** to 300 K and, thus, returns to State 0. This completes the cycle. The heat removed during this phase is

$$Q_{3\rightarrow0} = \mu c_v \Delta T = 40 \times 10^{-6} \times 20.8 \times 10^3 (652 - 300) = 293 \text{ J}. \qquad (23)$$

exactly the same as $Q_{1\rightarrow2}$.

In one cycle, the crankshaft receives 500 J from the "hot" piston ($W_{2\rightarrow3}$) and returns 230 J used in the compression phase ($W_{0\rightarrow1}$). A net mechanical energy of $500 - 230 = 270$ J constitutes the output of the machine. This happens at a cost of two heat inputs, $Q_{1\rightarrow2}$ and $Q_{2\rightarrow3}$, amounting to 793 J.

The efficiency of the device is

$$\eta = \frac{W_{expan} - W_{compress}}{Q_{add} + Q_{expan}} = \frac{500 - 230}{293 + 500} = \frac{270}{793} = 0.34.$$

The efficiency of a Carnot cycle working between 652 and 300 K is

$$\eta_{CARNOT} = \frac{652 - 300}{652} = 0.54. \qquad (24)$$

Thus, the Stirling engine, as described, realizes only a modest fraction of the Carnot efficiency. However, a relatively simple modification changes this picture.

We observe that $Q_{3 \to 0} = Q_{1 \to 2}$. The heat that has to be removed from the gas in the $3 \to 0$ phase can be stored in a **regenerator** inserted in the pipe connecting the two cylinders and it can be used to supply $Q_{1 \to 2}$. This means that the only heat required from the heat source is $Q_{2 \to 3}$ (500 J). Using a perfect, ideal, regenerator the efficiency of the Stirling cycle is

$$\eta = \frac{270}{500} = 0.54, \tag{25}$$

which is exactly the Carnot efficiency.

In practice regenerators can be realized by using, for example, steel wool whose large surface to volume ratio guarantees a speedy heat exchange.

In a beta-configured Stirling engine, a single cylinder is used. See the schematics in Figure 3.14. The lower piston is called the **power** piston and fits tightly in the cylinder so that gas can be compressed. The upper piston is the **displacer** and fits loosely so that it is quite leaky. The function of this displacer is to move the gas from the "cold" space (Space 2) just above the power piston to the "hot" space above the displacer (Space 1).

The phases of this cycle are the same as those of the alpha-cycle so we can use the same quantitative example. In Figure 3.14, we divide the $1 \to 2$ phase into two subphases: $1 \to 2^*$ in which the displacer comes down and moves the cold gas of Space 2 into the hot region of space 1. Again, ideally, no energy is consumed in this subphase because the gas leaks freely through the gap between displacer and cylinder. Isometric heat addition is performed in Subphase $2^* \to 2$.

The problem with this configuration is that there is no provision for the important regeneration function. This can be accomplished by using an external path connecting Spaces 1 and 2, as indicated in Figure 3.13.

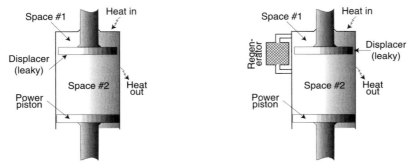

Figure 3.13   A regenerator can be fitted to a beta-Stirling engine in the manner indicated above.

3.22

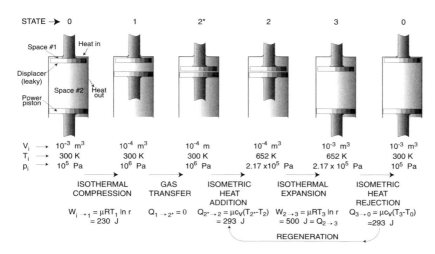

Figure 3.14   The various phases of a beta-Stirling cycle.

3.9   The Implementation of the Stirling Engine

One possible implementation of the Stirling engine is that shown in Figure 3.15. Air from a *combustion blower* is fed to a *burner* where it is mixed with fuel and ignited. The resulting hot gases are passed through an *input heat exchanger* that transfers heat to the working fluid in "Space #1" of the engine.

The gases that exit the input side of this heat exchanger still contain a considerable amount of heat that can be recovered by preheating the air coming from the combustion blower. This lowers the amount of fuel required to maintain the working fluid of the engine at the required temperature.

The cooling of the working fluid in "Space #2" is accomplished through a separate *cooling heat exchanger*.

As in all types of engines, a number of factors cause the specific fuel consumption of a practical Stirling engine to be much higher than the theoretical predictions and its pollutant emissions to be higher than desirable. Further development would be necessary if this category of engines were to become popular in the automotive world. This will probably not happen because it appears that combustion engines are approaching the end of their popularity. In all probability, fuel cells will gradually take over most areas where IC engines are now dominant. Owing to the enormous size of the automotive industry, the transition from IC to FC technology must, necessarily, proceed gradually. An important intermediate step is the use of hybrid cars.

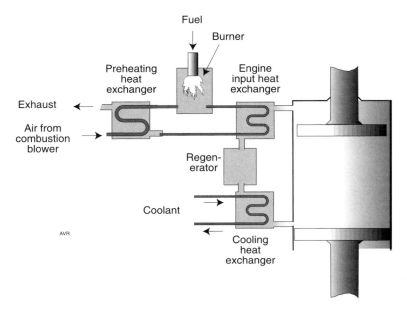

Figure 3.15   One possible implementation of a Stirling engine.

Real efficiency departs from from the theoretical because:

1. The efficiency of mechanical heat engines increases with increasing compression ratio. The Stirling engine is no exception (see Problem 3.1). The internal volume of the heat exchangers and the regenerator causes the compression ratio to be smaller than if heat could be applied directly to "Space #1."

2. The combustion gases leaving the input heat exchanger are still quite hot and carry away considerable energy. To recover part of the latter a preheater is used as explained previously.

3. The regenerator cannot operate ideally and return all the heat deposited in it during part of the cycle.

4. There are frictional losses when the working fluid is transferred from one space to the other through the heat regenerator. This is the main reason for using hydrogen as a working fluid, as it leads to minimum frictional losses.

5. Ideally, the motion of the pistons should be intermittent, but this is difficult to implement in machines operating at reasonably high rpm. Thus, one has to make a compromise in the piston programming.

6. The limited time for the heat exchanges in the working fluid leads to temperatures that never reach the desired steady state levels.

The power output of a Stirling engine can be controlled by either adding or removing working fluid. To reduce power, a compressor removes some working fluid from the engine and stores it at high pressure in a holding tank. To increase power, gas from this tank can be quickly delivered back to the engine. The input heat exchanger temperature is continuously monitored and this information is used to control the fuel flow.

3.10 Cryogenic Engines

As pointed out, a heat engine depends for its operation on a heat source and a sink which, by definition, must be at a lower temperature. Almost invariably, the sinks are at nearly the ambient temperature while the source is driven by combustion, nuclear reaction, solar energy, and so forth, to a relatively elevated temperature. However, it is possible to use a source that is at ambient temperature, while the temperature of the sink is lowered by a supply of a cryogenic substance.

Ordonez and his collaborators at the University of North Texas have proposed such cryogenic engines. Their experimental engines operate on an open Rankine cycle and use liquid nitrogen for the sink. Figure 3.16 shows a schematic representation of the setup.

The liquid nitrogen in the cryogenic reservoir is under the (adjustable) pressure, p_1. The liquid is fed to a heat exchanger to which enough heat is added (from the outside air) to produce gaseous N_2 at the pressure, p_1, and the temperature, T_1.

The exhaust valve is closed and the intake valve is opened so that μ kilomoles of nitrogen enter the cylinder. This causes the piston to be pushed down until the volume reaches a value, v_2. The process is isobaric $(p_2 = p_1)$ and isothermic $(T_2 = T_1)$ because sufficient heat is added to the heat exchanger.

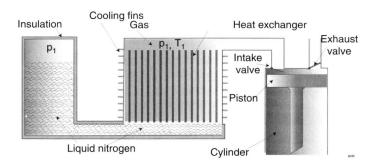

Figure 3.16   A cryogenic heat engine using an open Rankine cycle.

We have:

$$v_2 = \mu \frac{RT_2}{p_2} = \mu \frac{RT_1}{p_1}. \tag{26}$$

The work produced by the piston is

$$W_{12} = p_1 v_2 = \mu RT_1. \tag{27}$$

Next, the intake valve is closed. The high pressure nitrogen continues to expand until the pressure inside the cylinder is p_3 and the volume reaches v_3. This process may be carried out isothermally ($T_3 = T_1$) or adiabatically ($T_3 < T_1$), depending on the type of hardware used. Consider the more favorable isothermal case:

$$W_{23} = \mu RT_1 \ln \frac{p_3}{p_1}. \tag{28}$$

The total work done by the piston during the 1-to-3 phase is

$$W_{13} = \mu RT_1 + \mu RT_1 \ln \frac{p_3}{p_1}. \tag{29}$$

However, this is not the available work. Part of the total work is done by the piston on the atmosphere:

$$W_{atm} = p_{atm} v_3, \tag{30}$$

where

$$v_3 = \mu \frac{RT_1}{p_3}, \tag{31}$$

hence, assuming $T_1 = T_{atm}$,

$$W_{atm} = \mu RT_1. \tag{32}$$

The net useful work done is

$$W_{net} = W_{13} - W_{atm} = \mu RT_1 \ln \frac{p_3}{p_1}. \tag{33}$$

This means that the net work is exactly that owing to the isothermal expansion of the gas.

Example

Choose $p_1 = 1$ MPa and assume that $T_1 = T_{atm} = 298$ K and $p_3 = p_{atm} = 0.1$ MPa. The expansion ratio, $r = p_1/p_3$, is 10:1. The net work per kilomole of "fuel" is 5.7 MJ/kmole of N_2 or 204 kJ/kg of N_2.

It may be difficult to achieve isothermal expansion—it would require an additional heat exchanger to warm up the nitrogen as it forces the piston down. However, something similar is required in a Stirling engine, so it is possible.

Let us examine the case of adiabatic expansion that is easier to achieve. If no heat is added during the expansion, the temperature will fall:

$$T_3 = T_1 \left(\frac{p_1}{p_3}\right)^{\frac{1}{\gamma}-1}. \tag{34}$$

Here, $\gamma = 1.4$ for nitrogen. The work during the expansion is

$$W_{23} = \mu c_v (T_1 - T_3), \tag{35}$$

where $c_v \approx \frac{R}{\gamma-1} = 20.8$ kJ K^{-1} kmole^{-1}.

For the example with $p_1 = 1$ MPa, we get $T_3 = 154$ K and $W_{23} = 3$ MJ/kmole.

When the volume, v_3, is reached, the gas will have cooled to T_3 so that

$$v_3 = \mu \frac{RT_3}{p_3}, \tag{36}$$

consequently the work done on the atmosphere is

$$W_{atm} = p_{atm} v_3 = \mu R T_3. \tag{37}$$

The net work is now,

$$W_{net} = \mu R T_1 - \mu R T_3 + W_{23} = \mu R(T_1 - T_3) + \mu c_v(T_1 - T_3)$$
$$= \mu(T_1 - T_3) R \frac{\gamma}{\gamma - 1}. \tag{38}$$

Still for the example being considered, the specific net energy is 4.2 MJ/kmole or 150 kJ/kg. Compare with the 5.7 MJ/kmole or 204 kJ/kg for the isothermic case and with the 47,000 kJ/kg for the typical gasoline.

Clearly, the specific energy of the cryogen will increase with increasing operating pressure. The gain, however, is logarithmic. Thus, by raising the pressure to 10 MPa (a factor of 10) the specific energy rises to 11.4 MJ/kmole, a gain of 2. Observe that 10 MPa correspond to approximately 100 atmosphere and would lead to a rather heavy (and expensive) engine.

Gasoline has to be used in an internal combustion engine with some 20% efficiency, while the pneumatic motor used in the cryogenic engine can have very high component efficiency. This would reduce the practical specific energy advantage of gasoline to a factor of 40 over the nitrogen.

In practice, the Ordonez engine has yielded around 19 kJ/kg thus far. A demonstration car using the engine does 0.3 miles to the gallon, which is not practical. The efficiency will probably be improved.

Conclusions

Vehicles equipped with the cryogenic engine achieve only very modest mileage. Demonstrated mileage is 0.3 miles/gallon. Present day cost of liquid nitrogen is $0.50/kg or $1.52/gallon, essentially the same as that of gasoline. This means that fuel cost would be about 100 times larger than that of a gas driven car.

Assume a 10-fold improvement in performance for the cryogenic car— fuel cost would still be one order of magnitude larger than for the gasoline one. Add to this the need to carry huge amounts of cryogenic material, which would constitute a considerable hazard in case of accident.

The advantage of the system is its low pollution (but not necessarily zero pollution because there is need of energy to produce the liquid nitrogen). The question is whether these advantages compensate for the serious disadvantages that we discussed.

References

Heywood, John B., *Internal Combustion Engine Fundamentals*, McGraw-Hill, Inc., **1988**.

Ordonez, C. A., and M. C. Plummer, Cold thermal storage and cryogenic heat engine for energy storage applications, *Energy Sources*, 19:389–396, **1997**.

Ordonez, C. A., Cryogenic heat engine, *Am. J. Phys.* **64** (4), April **1996**.

Plummer, M. C., C. P. Koehler, D. R. Flanders, R. F. Reidy, and C. A. Ordonez, Cryogenic heat engine experiment, *Advances in Cryogenic Engineering* **43** 1245-1252 **1998**.

PROBLEMS

3.1

1. Demonstrate that the theoretical efficiency of a Stirling engine without regenerator is

$$\eta = \eta_{CARNOT} \left(1 + \frac{\nu \eta_{CARNOT}}{2 \ln r} \right)^{-1}$$

where η_{CARNOT} is the Carnot efficiency associated with the engine temperature differential, ν is the number of degrees of freedom of the working gas and r is the compression ratio.

2. What gas would you suggest as a working fluid? Why?

3. In the example in the text, a compression ratio of 10 was used. What would the efficiency of that engine be if the ratio were raised to 20? What are the disadvantages of using this higher compression ratio? Is it worth the effort?

3.2   Plot a pressure versus volume diagram and a temperature versus entropy diagram for the Stirling engine in the example given in the text. What do the areas under the pV and the TS lines represent?

3.3   Consider two cylinders, A and B, equipped with pistons so that their internal volume can be changed independently. The maximum volume of either cylinder is 10 m^3 and the minimum is zero. The cylinders are interconnected so that the gas is at the same pressure anywhere in the system. Initially, the volume of A is 10 m^3 and of B is 0. In other words, piston A is all the way up and B is all the way down. The system contains a gas with $\gamma = 1.4$.

1. If this is a perfect gas, what is the number of degrees of freedom of its molecules and what is its specific heat at constant volume?

2. The pressure is 0.1 MPa and the temperature is 400 K. How many kilomoles of gas are in the system?

3. Now, push piston A down reducing the volume to 1 m^3, but do not change the volume of cylinder B. What are the temperature and pressure of the gas assuming adiabatic conditions? What energy was expended in the compression?

4. Next, press A all the way down and, simultaneously, let B go up so that the volume in cylinder A is zero and that in cylinder B is 1 m^3. What are the pressure and temperature of the gas in B?

5. The next step is to add heat to the gas in B so that it expands to 10 m^3 at constant temperature. How much heat was added? How much work did Piston B deliver? What is the final pressure of the gas?

6. Now press B all the way down while pulling A up. This transfers gas from one cylinder to the other and (theoretically) requires no energy. Cylinder A rejects heat to the environment and the gas cools down to

400 K. The pistons are not allowed to move. The cycle is now complete. How much heat was rejected?

7. What is the efficiency of this machine—that is, what is the ratio of the net work produced to the heat taken in?

8. What is the corresponding Carnot efficiency?

9. Sketch a pressure versus volume and a temperature versus volume diagram for the process described.

10. Derive a formula for the efficiency as a function of the compression ratio, r. Plot a curve of efficiency vs r in the range $1 < r \leq 100$.

11. If this ratio were to reach the (unrealistic) value of 10,000, what would the efficiency be? Does this exceed the Carnot efficiency? Explain.

3.4   A car is equipped with a spark ignition engine (Otto cycle). It uses gasoline (assume gasoline is pure pentane) as fuel, and, for this reason, its compression ratio is limited to 9. The highway mileage is 40 miles/gallon.

Since pure ethanol is available, the car owner had the engine modified to a compression ratio of 12 and is using this alcohol as fuel. Assuming that in either case the actual efficiency of the car itself is half of the theoretical efficiency, what is the mileage of the alcohol car?

The lower heats of combustion and the densities are

Pentane: 28.16 MJ/liter, 0.626 kg/liter;
Ethanol: 21.15 MJ/liter, 0.789 kg/liter.

Do this problem twice, once using $\gamma = 1.67$ and once using $\gamma = 1.4$.

3.5   Consider a cylinder with a frictionless piston. At the beginning of the experiment it contains one liter of a gas ($\gamma = 1.4$, $c_v = 20$ kJ K^{-1} kmole^{-1}) at 400 K and 10^5 Pa.

1. How many kmoles of gas are in the cylinder?

2. What is the pV product of the gas?

Move the piston inwards reducing the volume of the gas to 0.1 liters. This compression is adiabatic.

3. What is the pressure of the gas after the above compression?

4. What is the temperature of the gas after the above compression?

5. How much work was done to compress the gas?

Now add 500 J of heat to the gas without changing the temperature.

6. After this heat addition phase, what is the volume of the gas?

7. After this heat addition phase, what is the pressure?

8. Since the gas expanded (the piston moved) during the heat addition, how much work was done?

Let the gas expand adiabatically until the volume returns to 1 liter.

9. After the expansion, what is the pressure of the gas?

3.30

10. After the adiabatic expansion what is the temperature of the gas?
11. How much work was done by the expansion?

Now remove heat isometrically from the gas until the pressure reaches $10^5 Pa$. This will bring the system to its original State 1.

12. What is the net work done by the piston on an outside load?
13. What is the total heat input to the system (the rejected heat cannot be counted)?
14. What is the efficiency of this machine?
15. What is the Carnot efficiency of this machine?
16. Sketch a pressure versus volume diagram of the cycle described in this problem statement.

3.6   Assume that gasoline is pure pentane (actually, it is a complicated mixture of hydrocarbons best represented by heptane, not pentane). Consider a 1:4 ethanol-gasoline mixture (by volume). The gasoline has an 86 octane rating. The blending octane rating of ethanol is 160. Use $\gamma = 1.4$.

1. What is the energy per liter of the mixture compared with that of the pure gasoline?
2. What is the octane rating of the mixture?

Assume that the maximum tolerable compression ratio is $r = 0.093 \times O_r$ where O_r is the octane rating.

3. What is the highest compression ratio of the gasoline motor? Of the mixture motor?
4. What is the relative efficiency of the two motors?
5. What is the relative kilometrage (or mileage) of two identical cars equipped one with the mixture motor and the other with the gasoline motor?

3.7   An open-cycle piston type engine operates by admitting 23×10^{-6} kmoles of air at 300 K and 10^5 Pa. It has a 5.74 compression ratio.

Compression and expansion are adiabatic. Heat is added isobarically and rejected isometrically.

Heat addition is of 500 J.

Air has $c_v = 20{,}790$ J K^{-1} kmole^{-1} and a $\gamma = 1.4$.

What is the theoretical efficiency of this engine? Compare with its Carnot efficiency.

Proceed as follows:

Calculate the initial volume of cylinder (at end of admission).

Compress adiabatically and calculate the new V, p, T, and the work required.

Add heat and calculate new state variables.

Expand and calculate the work done.

3.8   A certain Stirling engine realizes one half of its theoretical efficiency and operates between 1000 K and 400 K. What is its efficiency with

1. a perfect heat regenerator, argon working fluid, and 10:1 compression?
2. the same as above but with a 20:1 compression ratio?
3. the same as in (1.) but without the heat regenerator?
4. the same as in (2.) but without the heat regenerator?

3.9   Rich mixtures reduce the efficiency of Otto engines, but mixtures that are too lean do not ignite reliably. The solution is the "stratified combustion engine."

Consider an engine with a 9:1 compression ratio. A rich mixture may have a gamma of 1.2 while, in a lean one it may be 1.6. Everything else being the same, what is the ratio of the efficiency with the lean mixture to that with the rich mixture?

3.10   Consider an Otto (spark-ignition) engine with the following specifications:

Maximum cylinder volume, V_0: 1 liter (10^{-3} m^3).
Compression ratio, r: 9:1.
Pressure at the end of admission, p_0: 5×10^4 Pa.
Mixture temperature at the end of admission, T_0: 400 K.
Average ratio, γ, of specific heats of the mixture: 1.4.
Specific heat of the mixture (constant volume), c_v,: 20 kJ K^{-1} kmole^{-1}

The calculations are to be based on the ideal cycle—no component losses. At maximum compression, the mixture is ignited and delivers 461 J to the gas.

If the engine operates at 5000 rpm, what is the power it delivers to the load?

3.11   If pentane burns stoichiometrically in air (say, 20% O$_2$ by volume), what is the air-to-fuel mass ratio?

Atomic masses:

H: 1 dalton.
C: 12 daltons.
N: 14 daltons.
O: 16 daltons.

Neglect argon.

3.12   The higher heat of combustion of n-heptane (at 1 atmosphere, 20 C) is 48.11 MJ/kg. What is its lower heat of combustion?

3.32

3.13   One mole of a certain gas ($\gamma = 1.6$, $c_v = 13{,}860$ J K^{-1}kmole^{-1}) occupies 1 liter at 300 K. For each step, below, calculate the different variables of state, p, V and T.

Step $1 \rightarrow 2$
Compress this gas adiabatically to a volume of 0.1 liters.
How much of the energy, $W_{1,2}$, is used in this compression?

Step $2 \rightarrow 3$
Add isothermally 10 kJ of heat to the gas.
What is the external work, if any?

Step $3 \rightarrow 4$
Expand this gas adiabatically by a ratio of 10:1 (the same as the compression ratio of Step $1 \rightarrow 2$).

Step $4 \rightarrow 1$
Reject heat isothermally so as to return to State 1. What is the energy involved?

What is the overall efficiency of the above cycle?
What is the Carnot efficiency of the cycle?
What power would an engine based on the above cycle deliver if it operated at 5000 rpm (5000 cycles per minute)?

3.14   The Stirling engine discussed in the example in the text uses an isothermal compression, followed by an isometric heat addition, then by an isothermal expansion, and, finally, by an isometric heat rejection.

Isothermal compression may be difficult to achieve in an engine operating at high rpm. Imagine that the engine actually operates with an adiabatic compression.

Assume that the other steps of the cycle are the same as before. Thus, the isometric heat addition still consists of absorption of 293 joules of heat. This means that the "hot" cylinder is no longer at 652 K, it is at whatever temperature will result from adding isometrically 293 J to the gas after the initial adiabatic compression.

Calculate the theoretical efficiency of the engine (no heat regeneration) and compare it with the Carnot efficiency.

Calculate the power produced by this single cylinder engine assuming that the real efficiency is exactly half of the ideal one and that the engine operates at 1800 rpm. Each full rotation of the output shaft corresponds to one full cycle of the engine. Use $\gamma = 1.4$.

3.15   Assume an engine (Engine #1) that works between 1000 K and 500 K has an efficiency equal to the Carnot efficiency.

The heat source available delivers 100 kW at 1500 K to the above engine. A block of material is interposed between the source and the engine to lower the temperature from 1500 K to the 1000 K required. This block of material is 100% efficient—the 100 kW that enter on one side are all delivered to the engine at the other side.

What is the Carnot efficiency of the system above? What is the power output of the system?

Now, replace the block by a second engine (Engine #2) having a 10% efficiency. The heat source still delivers 100 kW. What is the overall efficiency of the two engines working together?

3.16   A boiler for a steam engine operates with the inside of its wall (the one in contact with the steam) at a temperature of 500 K, while the outside (in contact with the flame) is at 1000 K.

1 kW of heat flows through the wall for each cm^2 of wall surface. The metal of the wall has a temperature dependent heat conductivity, λ, given by $\lambda = 355 - 0.111T$ in MKS units. T is in kelvins.

1. Determine the thickness of the wall,
2. Determine the temperature midway between the inner and outer surface of the wall.

3.17   A 4-cycle Otto (spark ignition) engine with a total displacement (maximum cylinder volume) of 2 liters is fueled by methane (the higher heat of combustion is 55.6 MJ/kg, however, in an IC engine, what counts is the lower heat of combustion). The compression ratio is 10:1. A fuel injection system insures that, under all operating conditions, the fuel-air mixture is stoichiometrically correct. The gamma of this mixture is 1.4.

Owing to the usual losses, the power delivered to the load is only 30% of the power output of the ideal cycle. Because of the substantial intake pumping losses, the pressure of the mixture at the beginning of the compression stroke is only 5×10^4 Pa. The temperature is 350 K.

If the engine operates at 5000 rpm, what is the power delivered to the external load?

3.18   Consider a spark-ignition engine with a a 9:1 compression ratio. The gas inside the cylinder has a $\gamma = 1.5$.

At the beginning of the compression stroke, the conditions are:

$V_1 = 1$ liter,
$p_1 = 1$ atmosphere,
$T_1 = 300$ kelvins.

At the end of the compression, 10 mg of gasoline are injected and ignited. Combustion is complete and essentially instantaneous.

Take gasoline as having a heat of combustion of 45 MJ/kg.

1. Calculate the ideal efficiency of this engine.
2. Calculate the Carnot efficiency of an engine working between the same temperatures as the spark-ignition engine above.

3. Prove that as the amount of fuel injected per cycle decreases, the efficiency of the Otto cycle approaches the Carnot efficiency.

3.19   In a diesel engine, the ignition is the result of the high temperature the air reaches after compression (it is a **compression ignition** engine). At a precisely controlled moment, fuel is sprayed into the hot compressed air inside the cylinder and ignition takes place. Fuel is sprayed in relatively slowly so that the combustion takes place, roughly, at constant pressure. The compression ratio, r, used in most diesel engines is between 16:1 and 22:1. For diesel fuel to ignite reliably, the air must be at 800 K or more.

Consider air as having a ratio of specific heat at constant pressure to specific heat at constant volume of 1.4 ($\gamma = 1.4$). The intake air in a cold diesel engine may be at, say, 300 K. What is the minimum compression ratio required to start the engine?

3.20   We have a machine that causes air ($\gamma = 1.4$) to undergo a series of processes. At the end of each process, calculate the state of the gas (pressure, volume, and temperature) and the energy involved in the process.

The initial state (State #1) is

$p_1 = 10^5$ Pa,
$V_1 = 10^{-3}$ m^3,
$T_1 = 300$ K.

1. 1st Proc. (Step 1→2): Compress adiabatically, reducing the volume to 10^{-4} m^3.
2. 2nd Proc. (Step 2→3): Add 200 J of heat isobarically.
3. 3rd Proc. (Step 3→4): Expand adiabatically until $V_4 = 10^{-3}$ m^3.
4. List all the heat and mechanical inputs to the machine and <u>all</u> the mechanical outputs. From this, calculate the efficiency of the machine. (Hint: Don't forget to add all the processes that deliver energy to the output.)

3.21   A--Adiabatic   B--Isothermal

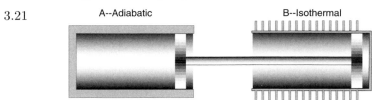

A crazy inventor patented the following (totally useless) device: Two geometrically identical cylinders (one adiabatic and the other isothermal) have rigidly interconnected pistons as shown in the figure.

The system is completely frictionless, and at the start of the experiment (State #0), the pistons are held in place so that the gases in the cylinders are in the states described below:

Cylinder A   Cylinder B
(Adiabatic)   (Isothermal)
$V_{A_0} = 1$ m^3, $V_{B_0} = 0.1$ m^3,
$p_{A_0} = 10^5$ Pa, $p_{B_0} = 10^6$ Pa,
$T_{A_0} = 300$ K. $T_{B_0} = 300$ K.

1. Now, the pistons are free to move. At equilibrium, what is the temperature of the gas in Cylinder A? The γ of the gas is 1.5.

An external device causes the pistons to oscillate back and forth 2500 times per minute. Each oscillation causes V_B to go from 0.1 m^3 to 1 m^3 and back to 0.1 m^3.

2. How much power is necessary to sustain these oscillations?

Consider the same oscillating system as above with the difference that in each compression and each expansion 1% of the energy is lost. This does not alter the temperature of the isothermal cylinder because it is assumed that it has perfect thermal contact with the environment at 300 K. It would heat up the gas in the adiabatic cylinder that has no means of shedding heat. However, to simplify the problem assume that a miraculous system allows this loss-associated heat to be removed but not the heat of compression (the heat that is developed by the adiabatic processes).

3. How much power is needed to operate the system?

3.22   In a diesel cycle one can distinguish the following different phases:

Phase 1 $\rightarrow$ 2 An adiabatic compression of pure air from Volume V_1 to Volume V_2.

Phase 2 $\rightarrow$ 3 Fuel combustion at constant pressure with an expansion from Volume V_2 to Volume V_3.

Phase 3 $\rightarrow$ 4 Adiabatic expansion from Volume V_3 to Volume V_4.

Phase 4 $\rightarrow$ 1 Isometric heat rejection causing the state of the gas to return to the initial conditions.

This cycle closely resembles the Otto cycle with a difference that in the Otto cycle the combustion is isometric while in the diesel it is isobaric.

Consider a cycle in which $V_1 = 10^{-3}$ m^3, $V_2 = 50 \times 10^{-6}$ m^3, $V_3 = 100 \times 10^{-6}$ m^3, $p_1 = 10^5$ Pa, $T_1 = 300$ K, and (for all phases) $\gamma = 1.4$.

1. Calculate the theoretical efficiency of the cycle.
2. Calculate the efficiency by using the efficiency expression for the diesel cycle given in Chapter 4 of the text.
3. Calculate the efficiency by evaluating all the mechanical energy (compression and expansion) and all the heat inputs. Be specially careful with what happens during the combustion phase (2 $\rightarrow$ 3) when heat from the fuel is being used and, simultaneously, some mechanical energy is being produced.

You should, of course, get the same result from 2 and 3.

Chapter 4
Ocean Thermal Energy Converters

4.1 Introduction

The most plentiful renewable energy source in our planet by far is solar radiation: 170,000 TW fall on Earth. Harvesting this energy is difficult because of its dilute and erratic nature. Large collecting areas and large storage capacities are needed, two requirements satisfied by the tropical oceans. Oceans cover 71% of Earth's surface. In the tropics, they absorb sunlight and the top layers heat up to some 25 C. Warm surface waters from the equatorial belt flow poleward melting both the arctic and the antarctic ice. The resulting cold waters return to the equator at great depth completing a huge planetary thermosyphon.

The power involved is enormous. For example, the Gulf Stream, has a flow rate of 2.2×10^{12} m^3 day^{-1} of water, some 20 K warmer than the abyssal layers. A heat engine that uses this much water and that employs as a heat sink the cool ocean bottom would be handling a heat flow of $\Delta T c \dot{V}$, where ΔT is the temperature difference, c is the heat capacity of water (about 4 MJ m^{-3}K^{-1}), and $\dot{V}$ is the flow rate.[†] This amounts to 1.8×10^{20} J day^{-1} or 2100 TW. The whole world uses energy at the rate of only ≈ 8 TW. These order of magnitude calculations are excessively optimistic in the sense that only a minuscule fraction of this available energy can be practically harnessed. Nevertheless, ocean thermal energy holds some promise as an auxiliary source of energy for use by humankind.

Figure 4.1 shows a typical temperature profile of a tropical ocean. For the first 50 m or so near the surface, turbulence maintains the temperature uniform at some 25 C. It then falls rapidly reaching 4 or 5 C in deep places. Actual profiles vary from place to place and also with the seasons.

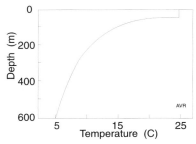

Figure 4.1   Typical ocean temperature profile in the tropics.

[†] The expression *flow rate* is redundant. The word *flow* is defined as *the volume of fluid flowing through a tube of any given section in a unit of time* (OED).

It is easier to find warm surface water than sufficiently cool abyssal waters, which are not readily available in continental shelf regions. This limits the possible sitings of ocean thermal energy converters.

4.2 OTEC Configurations

Two basic configurations have been proposed for OTECs:

1. those using hydraulic turbines, and
2. those using vapor turbines.

The first uses the temperature difference between the surface and bottom waters to create a hydraulic head that drives a conventional water turbine. The advantages of this proposal include the absence of heat exchangers.

Consider a hemispherical canister as depicted in the left hand side of Figure 4.2. A long pipe admits cold water, while a short one admits warm water. The canister is evacuated so that, in the ideal case, only low-pressure water vapor occupies the volume above the liquid surface. In practice, gases dissolved in the ocean would also share this volume and must be removed. This configuration was proposed by Beck (1978).

At a temperature of 15 C, the pressure inside the canister is about 15 kPa (0.017 atmospheres). At this pressure, warm water at 25 C will boil and the resulting vapor will condense on the parts of the dome refrigerated by the cold water. The condensate runs off into the ocean, establishing a continuous flow of warm water into the canister. The incoming warm water drives a turbine from which useful power can be extracted. The equivalent hydraulic head is small and turbines of large dimensions would be required.

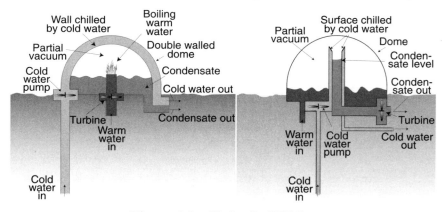

Figure 4.2   Hydraulic OTECs.

To increase the hydraulic head, Zener and Fetkovich (1975) proposed the arrangement of Figure 4.2 (right). The warm surface water admitted to the partially evacuated dome starts boiling. The resulting vapor condenses on a funnel-like surface that seals one of the two concentric cylinders in the center of the dome. This cylinder receives cold water pumped from the ocean depths, which chills the steam-condensing surface. The collected condensed water subsequently flows into the central pipe creating a head that drives the turbine. The efficiency of the device is substantially enhanced by the foaming that aids in raising the liquid.

OTECs developed in the 1980s were of the vapor turbine type. They can use open cycles (Figures 4.3A and B), close cycles (Figure 4.3C) or hybrid cycles (Figure 4.3D). The open cycle avoids heat exchangers (or, if fresh water is desired, it requires only a single heat exchanger). However, the low pressure of the steam generated demands very large diameter turbines. This difficulty is overcome by using a close (or a hybrid) cycle with ammonia as a working fluid. Most work has been done on the close-cycle configuration, which is regarded as more economical. However, the costs of the two versions may turn out to be comparable.

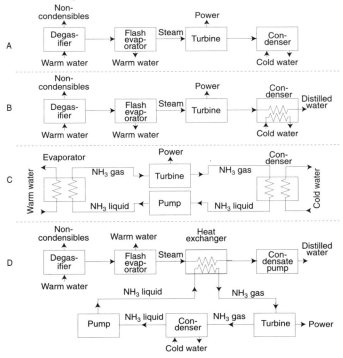

Figure 4.3   OTEC configurations include the open-cycle type without distilled water production (A), the open-cycle type with distilled water recovery (B), the close-cycle (C), and the hybrid-cycle (D).

4.3 Turbines

A turbine generates mechanical energy from a difference in pressure. Usually, the state of the gas at the inlet and the pressure of the gas at the exhaust are specified.

Let p_{in} and T_{in} be the pressure and the temperature at the inlet of the turbine and p_{out}, T_{out} the corresponding quantities at the exhaust.

The output of the turbine is the mechanical work, W. The heat, Q, is exchanged with the environment by some means other than the circulating gases. Most practical turbines are sufficiently well insulated to be assumed **adiabatic**—that is, a condition in which $Q = 0$.

The inlet gas carries an enthalpy, H_{in}, into the turbine, while the exhaust removes H_{out} from the device. Conservation of energy requires that[†]

$$W = H_{in} - H_{out}. \tag{1}$$

Expressing the quantities on a per kilomole basis (quantities per kilomole are represented by lower case letters), we can write

$$W = \mu_{in}h_{in} - \mu_{out}h_{out} = \mu(h_{in} - h_{out}), \tag{2}$$

because, under steady state conditions, $\mu_{in} = \mu_{out} \equiv \mu$.

In a perfect gas,

$$h_{in} - h_{out} = \int_{T_{out}}^{T_{in}} c_p dT. \tag{3}$$

Assuming a constant specific heat,

and
$$h_{in} - h_{out} = c_p(T_{in} - T_{out}), \tag{4}$$

$$W = \mu c_p \left(T_{in} - T_{out}\right). \tag{5}$$

$$W = \mu c_p T_{in} \frac{T_{in} - T_{out}}{T_{in}} = \mu c_p T_{in} \eta_{CARNOT} \tag{6}$$

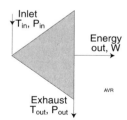

Figure 4.4   A turbine.

[†] Provided there is no appreciable change in kinetic, potential, magnetic, and other forms of energy.

The above equation looks similar to that which describes the behavior of a heat engine. However, the quantity, $\mu c_p T_{in}$, although having the dimensions of energy, is not the heat input to the device; rather it is the enthalpy input. For a given input state and a given exhaust pressure, the mechanical energy output increases with decreasing exhaust temperature. The lowest possible value of T_{out} is limited by the second law of thermodynamics that requires that the entropy of the exhaust gases be equal or larger than that of the inlet gases. The lowest exhaust temperature (highest output) is achieved by a turbine operating **isentropically**, one in which the entropy is not changed. Any deviation from this condition is due to irreversibilities (losses) in the device. These losses will generate heat and thus increase T_{out}.

Isentropic Processes

If there is no change in entropy in the gas that flows through the turbine, then we have an isentropic process.

From the first law of thermodynamics:

$$dQ = dU + pdV \tag{7}$$

and from the second law:

$$dQ = TdS \tag{8}$$

$$TdS = dU + pdV \tag{9}$$

From the definition of enthalpy, $H = U + pV$,

$$dU = dH - Vdp - pdV \tag{10}$$

Hence,

$$TdS = dH - Vdp, \tag{11}$$

$$dS = \frac{dH}{T} - \frac{V}{T}dp \tag{12}$$

But $dH = c_p dT$ and $V/T = R/p$, hence

$$dS = c_p \frac{dT}{T} - R\frac{dp}{p} \tag{13}$$

If the process is isentropic, then $dS = 0$, thus,

$$c_p \frac{dT}{T} = R\frac{dp}{p} = (c_p - c_v)\frac{dp}{p} \tag{14}$$

$$\frac{dT}{T} = \left(1 - \frac{1}{\gamma}\right)\frac{dp}{p}. \tag{15}$$

(continues)

4.5

(continued)

$$\ln T + \left(\frac{1-\gamma}{\gamma}\right) \ln p = constant \tag{16}$$

$$T p^{\frac{1-\gamma}{\gamma}} = constant \tag{17}$$

And, finally, from the perfect gas law,

$$p V^\gamma = constant. \tag{18}$$

Thus, the polytropic law derived for the case of adiabatic compression (see Chapter 2) applies to any isentropic process.

What is the exhaust temperature, $T_{out_{min}}$, in an isentropic turbine? Using the polytropic law,

$$p_{in} V_{in}^\gamma = p_{out} V_{out}^\gamma, \tag{19}$$

Applying the perfect gas law, we can eliminate the volumes:

$$p_{in}^{1-\gamma} \mu R T_{in}^\gamma = p_{out}^{1-\gamma} \mu R T_{out_{min}}^\gamma, \tag{20}$$

$$T_{out_{min}} = T_{in} \left(\frac{p_{out}}{p_{in}}\right)^{\frac{\gamma-1}{\gamma}}. \tag{21}$$

The energy delivered by the isentropic turbine is

$$W = \mu c_p T_{in} \left[1 - \left(\frac{p_{out}}{p_{in}}\right)^{\frac{\gamma-1}{\gamma}}\right] \tag{22}$$

A turbine may be considered adiabatic in the sense that it does not exchange heat with the environment except through the flowing gas. However, it may exhibit internal losses that cause the exhaust temperature to be larger than that calculated from Equation 21.

The **isentropic efficiency** of a turbine is the ratio between the actual work produced by the turbine to the work it would produce if the input and output had the same entropy.

4.4   OTEC Efficiency

The Carnot efficiency of an OTEC is low due to the small temperature difference that drives it. OTECs must abstract a large quantity of heat from the warm surface waters and reject most of it to the cold bottom waters. They handle great volumes of water. How do such volumes compare with those handled by a hydroelectric plant of the same capacity?

Let the temperature difference between the warm and the cold water be $\Delta T \equiv T_H - T_C$. Prof. A. L. London of Stanford University has shown that the minimum water consumption occurs when the temperature difference across the turbine (in a close-cycle system) is $\Delta T/2$, leaving the remaining $\Delta T/2$ as the temperature drop across the two heat exchangers. If half of this is the drop across the warm water exchanger, and if $\dot{V}$ is the flow rate of warm water, then the power abstracted is $\frac{1}{4}\Delta T\, c\, \dot{V}$. The Carnot efficiency is $\Delta T/2T_H$. Assuming that the cold water flow is equal to that of the warm water, i.e., that $\dot{V}_{TOT} = 2\dot{V}$, the power generated (with ideal turbines) is

$$P_{OTEC} = \frac{c}{16T_H}\Delta T^2 \dot{V}_{TOT} = 850\,\Delta T^2 \dot{V}_{TOT}. \tag{23}$$

We took $T_H = 296$ K. Notice that the power is proportional to ΔT^2. The power of a hydroelectric plant is

$$P_{HYDRO} = \delta g \Delta h \dot{V} \tag{24}$$

where δ is the density of water (1000 kg/m^3), g is the acceleration of gravity, and h is the height difference between the input and output water levels.

We ask now how large must Δh be for an hydroelectric plant to produce the same power as an OTEC that handles the same water flow.

$$\Delta h = \frac{\Delta T^2 c}{16 T_H \delta g} = 0.085\,\Delta T^2. \tag{25}$$

An OTEC operating with 20 K temperature difference delivers the same power as a hydroelectric plant with identical flow rate and a (moderate) 34 m head. Thus, the volumes of water required by an OTEC are not exorbitant.

The above calculations are quite optimistic; they failed, for instance, to account for the considerable amount of energy required by the various pumps, especially the cold water pump. Nevertheless, the main problem with OTECs is not the large volume of water but rather one of heat transfer. Compared with fossil-fueled plants of the same capacity, the heat exchangers of an OTEC are enormous. Up to half the cost of a close-cycle OTEC is in the heat exchangers.

4.5 Example of OTEC Design

OTECs are not designed for minimum water consumption, but rather, for minimum cost. This alters the temperature distribution in the system. Figure 4.5 shows the temperatures in a Lockheed project. It is of the close-cycle type, using ammonia as the working fluid. The overall temperature difference is 18.46 K. The warm water flow rate is 341.6 m^3/s.[†] It enters

---

[†] To gain an idea of how much water is pumped, consider a 25 × 12 m competition swimming pool. The warm water pump of the Lockheed OTEC under discussion would be able to fill such a pool in less than 2 seconds!

the heat exchanger at 26.53 C and exits at 25.17 C, having been cooled by 1.36 K. This warm water delivers to the OTEC a total thermal power of

$$P_{in} = 341.6 \times 4.04 \times 10^6 \times 1.36 = 1876.9^\dagger \qquad (26)$$

The ammonia at the turbine inlet is at 23.39 C while at the outlet it is at 11.11 C, which leads to a Carnot efficiency of

$$\eta_{CARNOT} = \frac{23.39 - 11.11}{273.3 + 23.39} = 0.041 \qquad (27)$$

Electricity is produced at 90.2% of the Carnot efficiency. However a great deal of the generated power is used by the different pumps as illustrated in Table 4.1. In fact, in this example, 28% of the total power generated is used internally just to run the system. In typical steam engines, pumps are mechanically coupled to the turbine, not electrically as in this OTEC.

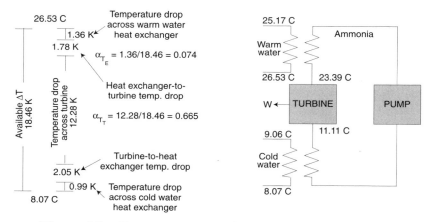

Figure 4.5   Temperatures in an OTEC designed by Lockheed.

Table 4.1
Internal Power Use in the Lockheed OTEC

Condensate pump (recirculates ammonia)	2.54 MW
Reflux pump (recirculates ammonia that failed to evaporate)	0.04 MW
Warm water pump	4.83 MW
Cold water pump	12.14 MW
TOTAL	19.55 MW

† MWt stands for thermal power, whereas MWe stands for electric power.

The main power consumer is the cold water pump not only because it handles the largest amount of water but also because it has to overcome the friction on the very long cold water pipe.

The electric power generated is

$$P_{gen} = 1876.9 \times 0.041 \times 0.902 = 69.4 \text{ MWe} \tag{28}$$

However, the electric power available at the output bus is only

$$P_{bus} = 69.4 - 19.6 = 49.8 \text{ MWe} \tag{29}$$

Thus, the overall efficiency of this OTEC is

$$\eta = \frac{49.8}{1876.9} = 0.0265. \tag{30}$$

This 2.65% efficiency is what can be expected of any well-designed OTEC.

The pressure of the ammonia at the inlet side of the turbine is 0.96 MPa (9.7 atmos) and, at the outlet, 0.64 MPa (6.5 atmos). An amount of heat equal to $1876.9 - 69.4 = 1807.5$ MW must be rejected to the cold water by the heat exchanger. This is done by taking in 451.7 m^{3}s^{-1} of water at 8.07 C and heating it up by 0.99 K to 9.06 C.

It is necessary to make the distance between the cold water outlet and the warm water inlet sufficiently large to avoid mixing. This gives the ocean currents opportunity to sweep the cold water away. In absence of currents, OTECs may have to move around "grazing" fresh warm water. Propulsive power for this grazing can easily be obtained from the reaction to the outlet water flow.

4.6 Heat Exchangers

The overall efficiency of an OTEC is small. The Lockheed OTEC of the example converts 1877 MWt into 49.8 MW of salable electricity—an efficiency of 2.6%. However, the "fuel" is completely free so the overall efficiency is of no crucial importance. What counts is the investment cost which greatly depends on the cost of the heat exchangers.

The power transferred through a heat exchanger is

$$P_{therm} = \gamma A \Delta T_{EXB} \tag{31}$$

where γ is the heat transfer coefficient of the exchanger, A is its area, and ΔT_{EXB} is the mean exchanger-to-boiler temperature difference.

Lockheed hoped to achieve a $\gamma = 2800$ W m^{-2}K^{-1}. With a ΔT_{EXB} of approximately 2.5 K and a P_{therm} of 1877 MW, the required area is about 270,000 m^2—that is, over 500×500 m.

Even minor fouling will considerably lower the value of γ and thus it is important to keep the heat exchanger surfaces clean and free from algae. It is possible to electrolyze a small fraction of the incoming water to liberate algae-killing chlorine.

Technically, the ideal material for OTEC heat exchangers seems to be titanium owing to its stability in seawater. Aluminum, being less expensive was also considered.

4.7   Siting

We saw that the economics of OTECs depend critically on ΔT. Consequently, a site must be found where a (comparatively) large ΔT is available. There is, in general, no difficulty in finding warm surface waters in tropical seas; the problem is to find cold water because this requires depths uncommon in the vicinity of land. For this reason, land based OTECs may be less common than those on floating platforms.

A large ΔT (some 17 K or more) is not the only siting requirement. Depth of a cold water layer at 6 C or less should be moderate, certainly less than 1000 m; otherwise the cost of the cold water pipe would become excessive. Anchoring problems suggest placing OTECs in regions where the total depth is less than some 1500 m. Surface currents should be moderate (less than 2 m/s).

Total ocean depth is, of course, of no importance if dynamically positioned OTECs are contemplated. This can be accomplished by taking advantage of the thrust generated by the seawater exhaust of the plant. As explained previously, if there are no currents, such thrusting may be necessary to keep the cooler exhaust water from mixing with the warm intake.

Another siting consideration is distance from the shore if one contemplates bringing in the generated electricity by means of electric cables. OTECs may be used as a self-contained industrial complex operating as floating factories producing energy intensive materials. Ammonia, for instance, requires for its synthesis, only water, air, and electricity and is almost the ideal product for OTEC manufacture. It is easier to ship the ammonia than to transmit electric power to shore. Once on shore, ammonia can be used as fertilizer or it can be converted back into energy by means of fuel cells (see Chapter 9).

One OTEC arrangement involves the use of only the cold ocean bottom water. Such water is first pumped through heat exchangers and then into shallow ponds, where it is heated by the sun. It can, in this manner, reach temperatures well above those of the ocean, leading to larger Carnot efficiencies.

OTECs using solar heated ponds can be combined with mariculture. Deep ocean waters tend to be laden with nutrients and, when heated by the sun, will permit the flourishing of many species of microscopic algae. The algarich water flows into a second pond where filter feeding mollusks are raised. Oysters, clams, and scallops are produced. The larger of these animals are either kept for reproduction or sold in the market. The smaller ones are destroyed and thrown into a third pond where crustaceans (shrimps, lobsters, etc.) feed on them. The effluent of this pond should not be returned directly to the ocean because the animal waste in it is a source of pollution. A fourth and final pond is used to grow seaweed that clean up the water and serve as a source of agar or carrageen or, alternatively, as a feedstock for methane-producing digesters (see Chapter 13). The warm water from this pond is used to drive the OTEC.

References

Beck, E. J., Ocean thermal gradient hydraulic power plant, *Science*, **189**, 293, **1975**.

Zener, C., and J. Fetkovich, Foam solar sea power plant, *Science*, **189**, 294, **1975**.

PROBLEMS

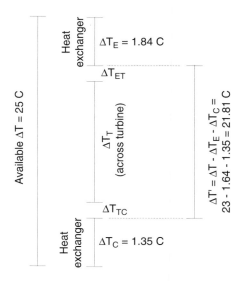

4.1   An OTEC is to deliver 100 MW to the bus bar. Its warm water comes from a solar heated pond that is kept at 33 C. The water exhausted from the heat exchanger is returned to this same pond at a temperature of 31 C. (There is a slight heat loss in the pipes.) To reestablish the operating temperature, the pond must absorb heat from the sun. Assume an average (day and night) insolation of 250 W/m^2 and an 80% absorption of solar energy by the water.

Cold water is pumped from the nearby abyss at a temperature of 8 C. Refer to the figure for further information on the temperatures involved.

The warm water loses 1.84 K in going through the heat exchanger, while the cold water has its temperature raised by 1.35 K in its exchanger.

Assume that 80% of the remaining temperature difference appears across the turbine and that the rest is equally distributed as temperature differentials between the colder side of the warm heat exchanger (whose secondary side acts as an evaporator) and the turbine inlet and between the warmer side of the cold heat exchanger (whose secondary side acts as a condenser) and the turbine outlet (ΔT_{ET} and ΔT_{TC}, in the figure).

Internal power for pumping and other ends is 40 MW. The efficiency of the turbine-generator combination is 90%. Estimate the rates of flow of warm and cold water.

What is the required surface of the heating pond, assuming no evaporation?

If the residence time of the water in the pond is 3 days, what depth must it have have?

4.12

4.2   The Gulf Stream flows at a rate of 2.2×10^{12} m^3/day. Its waters have a temperature of 25 C. Make a rough estimate of the area of the ocean that collects enough solar energy to permit this flow.

4.3   Assume that ammonia vaporizes in the evaporator of an OTEC at constant temperature (is this strictly true?). If the warm water enters the heat exchanger with a temperature ΔT_1 higher than that of the boiling ammonia and leaves with a ΔT_2, what is the mean ΔT? *To check your results:* If $\Delta T_1 = 4$ K and $\Delta T_2 = 2$ K, then $<\Delta T> = 2.88$ K.

4.4   A 1.2 GWe nuclear power plant is installed near a river whose waters are used for cooling. The efficiency of the system is 20%. This is the ratio of electric output to heat input.

Technical reasons require that the coolant water exit the heat exchangers at a temperature of 80 C. It is proposed to use the warm coolant water to drive an OTEC-like plant. Assume:

the river water is at 20 C,
the OTEC efficiency is one half of the Carnot efficiency,
half of the available ΔT is dropped across the turbine.

1. What is the flow rate of this water?
2. What is the maximum electric power that can be generated by such a plant?

4.5   An OTEC pumps 200 cubic meters of warm water per second through a heat exchanger in which the temperature drops by 1%. All the heat extracted is delivered to the ammonia boiler. The ammonia temperature at the turbine inlet is equal to the mean temperature of the water in the warm water heat exchanger minus 1 K. The condenser temperature is kept at 10 C by the cooling effect of 250 cubic meters per second of cold water. The efficiency of the turbine/generator system is 90%. 12 MW of the produced electricity is used for pumping.

What must the intake temperature of the warm water be, so that a total of 20 MW of electricity is available for sale?

What must the intake temperature be so that the OTEC produces only exactly the amount of power needed for pumping?

4.6   Consider an OTEC whose turbine/generator has 100% mechanical efficiency. In other words, the system operates at the Carnot efficiency. Input temperature to the turbine is T_H and the output is at T_C.

The input heat comes from a heat exchanger accepting water at $T_{H_{in}}$ and discharging it at a lower temperature, $T_{H_{out}}$. The flow rate of warm water through this heat exchanger is $\dot{V}_H$.

4.13

The heat sink for the turbine is another heat exchanger taking in water at $T_{C_{in}}$ and discharging it at a higher temperature, $T_{C_{out}}$. The flow rate of cold water through this heat exchanger is $\dot{V}_C$. Refer to the figure above.

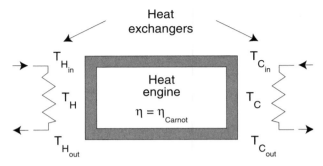

All the heat extracted from the warm water by the heat exchanger is transferred to the input of the turbine. All the heat rejected by the turbine is absorbed by the cold water heat exchanger and removed.

The following information is supplied:

$$T_{H_{in}} = 25 \text{ C.}$$
$$T_{C_{in}} = 8 \text{ C.}$$
T_H is the mean of $T_{H_{in}}$ and $T_{H_{out}}$.
T_C is the mean of $T_{C_{in}}$ and $T_{C_{out}}$.
$$\dot{V}_H = \dot{V}_C \equiv \dot{V} = 420 \text{ m}^3/\text{s.}$$
$T_{H_{out}}$ and $T_{C_{out}}$ are not given.

Clearly, if $T_{H_{out}}$ is made equal to $T_{H_{in}}$, no heat is extracted from the warm water, and the power output is zero. On the other hand, if $T_{H_{out}}$ is made equal to $T_{C_{in}}$, maximum power is extracted from the warm water but the output power is again zero because (as you are going to find out) the Carnot efficiency goes to zero.

Between these extremes, there must be a value of $T_{H_{out}}$ that maximizes power output. Determine this value, and determine the value of $T_{H_{out}}$ that maximizes the Carnot efficiency.

4.14

Chapter 5
Thermoelectricity

So far we have discussed heat engines in general and have examined in some detail the heat engine used in Ocean Thermal Energy Converters. All these engines convert heat into mechanical energy.

In this chapter and in the next two, we consider engines that transform heat directly into electricity: the thermoelectric, the thermionic, and the radio-noise converters. In the chapter on photovoltaic cells, we will discuss the thermophotovoltaic converter that transforms heat into radiant energy and then into electricity. Other engines such as magnetohydrodynamic engine that converts heat into kinetic energy of a plasma and then into electricity will not be covered here.

In most of the chapters of this book we have adopted the strategy of first introducing a simplified model that explains the phenomena that underlie the operation of the device being studied, and, from that, we deduce the way the device behaves. In this chapter, we will change the approach, describing first the behavior of thermocouples and only later will we examine the underlying phenomenology. The reason for this is that although the behavior of thermoelectric devices is easy to describe, there is no simple way of explaining how these effects come about. In fact, were we to use classical mechanics with Maxwellian electron energy distribution, we would prove that there is no Peltier effect and that the Thompson effect in metals should be two orders of magnitude larger than what it really is. The Peltier and Thompson effects are two of the effects associated with thermoelectricity.

5.1 Experimental Observations

Consider a heat-conducting bar whose ends are at different temperatures, T_H and T_C. Clearly a certain amount of heat power, P_{H_F}, will enter through one face and an amount, P_{C_F}, will leave through the opposite face. If the bar has its remaining sides perfectly insulated so no heat can exit through them, then

$$P_{H_F} = P_{C_F} = \Lambda(T_H - T_C) \tag{1}$$

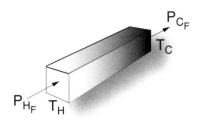

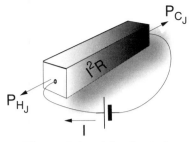

Figure 5.1   A bar with a temperature gradient.

Figure 5.2   A bar heated by a current.

Λ is the **heat conductance** of the bar, which, in the SI is measured in W/K. The subscript F is used because heat conduction is, sometimes, referred to as the **Fourier** effect in honor of Jean Baptiste Joseph Fourier's noted contributions to the study of heat diffusion.

If, on the other hand, the bar is at uniform temperature but is heated (by means of a current) to a higher temperature than that of the bodies in contact with the two end faces, then heat must flow out. The heat power generated by the current is I^2R, where R is the resistance of the bar. Half of this generated heat will flow out one end, and half the other.

$$P_{H_J} = P_{C_J} = \tfrac{1}{2}RI^2. \tag{2}$$

We used the subscript, J, to indicate that the heat is the result of the **Joule** effect (James Prescott Joule). Note that the arrow directions for heat flow are different in Figure 5.1 and Figure 5.2. Had we chosen to keep the same direction as in the former figure then P_{H_J} would be negative.

If the bar is simultaneously submitted to a temperature differential and a current, then the Fourier and Joule effects are superposed:

$$P_H = \Lambda(T_H - T_C) - \tfrac{1}{2}RI^2, \tag{3}$$

$$P_C = \Lambda(T_H - T_C) + \tfrac{1}{2}RI^2. \tag{4}$$

Here we used the directions of heat power flow indicated in the inset. However, in a more complicated structure, the heat power flow can depart surprisingly from expectation.

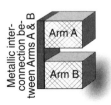

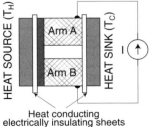

Figure 5.3   A simple thermocouple (left) and a test setup (right).

Consider a **thermocouple** consisting of two *dissimilar* materials (conductors or semiconductors) joined together at one end. The materials (arms A and B in Figure 5.3) may touch one another directly or may be joined by a metallic strip as indicated. As long as this metallic strip is at uniform temperature, it has no influence on the performance of the thermocouple (provided the strip has negligible electric resistance and essentially infinite heat conductivity). The free ends of arms A and B are connected to a current source.

Again, if the connecting wires are at uniform temperature, they exert no influence. Two blocks maintained at uniform temperature are thermally connected, respectively, to the junction and to the free ends. These blocks are electrically insulated from the thermocouple.

The block in contact with the junction is the **heat source** and is at the temperature T_H. The other block is the **heat sink** and is at T_C. The rate of heat flow, P_H, from the source to the sink is measured as explained in the box at the end of this subsection.

Assume the thermocouple is carefully insulated so that it can only exchange heat with the source and with the sink. If we measure P_H as a function of $T_H - T_C$ with no current through the thermocouple, we find that P_H is proportional to the temperature difference (as in Equation 1):

$$P_H = \Lambda(T_H - T_C). \tag{5}$$

As an example, take $\Lambda = 4.18$ W/K, then

$$P_H = 4.18(T_H - T_C). \tag{6}$$

If we force a current, I, through a thermocouple with an internal resistance, R, we expect, as explained, P_H to be given by (cf., Equation 3)

$$P_H = \Lambda(T_H - T_C) - \tfrac{1}{2}RI^2, \tag{7}$$

or, if the resistance of the thermocouple is 2.6×10^{-4} ohms,

$$P_H = 2090 - 1.3 \times 10^{-4}I^2, \tag{8}$$

where we used $T_H = 1500$ K and $T_C = 1000$ K, as an example.

The expected plot of P_H versus I appears as a dotted line in Figure 5.4. It turns out that a change in P_H, as I varies, is in fact observed but it is not independent of the sign of I. The empirically determined relationship between P_H and I is plotted, for a particular thermocouple, as a solid line. A second order regression fits the data well:

$$P_H = 2090 + 1.8I - 1.3 \times 10^{-4}I^2 \tag{9}$$

5.3

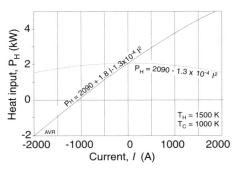

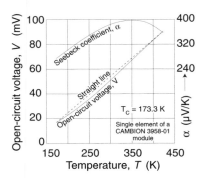

Figure 5.4   Heat input vs current characteristics of a thermocouple.

Figure 5.5   Open-circuit voltage and Seebeck coefficient.

In the above, we recognize the heat conduction term because it is independent of I. We also recognize the Joule heating term, $1.3 \times 10^{-4}\, I^2$.

In addition to these two terms, there is one linear in I.

This means that if the current is in one direction, heat is transported from the source to the sink and, if inverted, so is the heat transport. Evidently heat energy is carried by the electric current. This reversible transport is called the **Peltier** effect (Jean Charles Athanase Peltier, 1785/1845).

From empirical evidence, the Peltier heat transported is proportional to the current. We can, therefore, write

$$P_{Peltier} = \pi\, I \tag{10}$$

where π is the **Peltier coefficient.**

If we connect an infinite impedance voltmeter to the thermocouple instead of a current generator, we observe a voltage, V, that is dependent on the temperature difference, $\Delta T \equiv T_H - T_C$. The dependence is nonlinear as illustrated in Figure 5.5, which shows the relationship of the open-circuit voltage of a thermocouple to the temperature, T_H, T_C being held, in this case, at a constant 173.3 K.[†]

The **Seebeck coefficient**, α, is the slope of the V versus T_H plot and depends somewhat on the temperature:

$$\alpha \equiv \frac{dV}{dT}. \tag{11}$$

We are going to show later that there is a relationship between the Peltier and the Seebeck coefficients:

$$\pi = \alpha T. \tag{12}$$

The Peltier coefficient is a strong function of the temperature.

[†] The dependence of V on ΔT can be represented quite accurately by a power series, $V = a_0 \Delta T + a_1 \Delta T^2 + a_2 \Delta T^3 + \dots$ The values of the different coefficients, a_i, are slightly dependent on the reference temperature, T_C.

The open-circuit voltage developed by the thermocouple is

$$V = \int_{T_C}^{T_H} \alpha \, dT, \tag{13}$$

or, if a **mean Seebeck coefficient**, $<\alpha>$, is used,

$$V = <\alpha> (T_H - T_C). \tag{14}$$

In the balance of this chapter, although we will work with $<\alpha>$, we will represent it simply by α. The use of the mean value of α allows us to correctly described the thermocouple performance in terms of only the four effects mentioned so far: Fourier, Joule, Peltier, and Seebeck completely ignoring the heat convected by the carriers—that is, the **Thomson effect**. In the end of this chapter, we will justify this omission.

Thermocouples find use as

1. Thermometers.
2. Generators capable of transforming heat directly into electricity.
3. Heat pumps and refrigerators.

Temperatures as well as electric quantities can be easily measured with good accuracy (a moderately inexpensive voltmeter can have accuracies of 0.1% of full scale). On the other hand, it is difficult to determine the precise flow of heat.

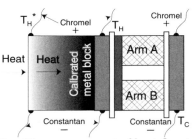

Figure 5.6   The measurement of heat flow in a thermocouple.

Conceptually, the heat flow can be measured by insulating the heat source except at its contact with the thermocouple. Heat can be supplied to this source by an electric resistor. The electric power necessary to keep T_H constant is a measure of the heat flow from source to couple.

(continues)

(continued)

More commonly, the flow of heat is measured by inserting a metallic block (whose conductivity as a function of temperature has been carefully measured) between the heat source and the thermocouple. If no heat leaks out through the side walls of the block, then the heat flow through the thermocouple can be determined from the temperature drop, $T_H^* - T_H$, along the metallic block.

T_H^* and T_H, as well as T_C, are each measured by attaching thermocouple wires to appropriate regions of the device.

5.2   Thermoelectric Thermometers

Since the open-circuit (Seebeck) voltage is a monotonic function of the temperature, thermocouples are an obvious choice as thermometers. When carefully calibrated, they will function with surprising accuracy over the temperature range from some 20 K to over 1700 K. However, to cover this full range, different units are required.

Because the Seebeck voltage is not a linear function of temperature. this voltage must be translated to temperature by means of look-up tables.

A very large number of combinations of materials have been considered for thermometry, but only a few have been standardized. Nine of these combinations are designated by identifying letters as shown in Table 5.1.

The first material in each pair in the table above is the positive leg, the second the negative leg. Composition is by weight.

Table 5.1
Standardized Thermocouple Pairs

Type	Material	Recommended Range (K)
B	Pt+30%Rh vs Pt+6%Rh	1640–1970
C	W+5%Re vs W+26%Re	1920–2590
E	90%Ni+10%Cr vs 55%Cu+45%Ni	370–1170
J	99.5%Fe vs 55%Cu+45%Ni	370–1030
K	90%Ni+10%Cr vs 95%Ni+2%Al+2%Mn+1%Si	370–1530
N	84%Ni+14%Cr+1.5%Si vs 95%Ni+4.5%Si+0.1 Mg	920–1530
R	Pt+13%Rh vs Pt	1140–1720
S	Pt+30%Rh vs Pt	1250–1720
T	Cu vs 55%Cu+45%Ni	70–620

Many of the alloys are better known by their commercial name:

55%Cu+45%Ni: Constantan, Cupron, Advance, ThermoKanthal JN.
90%Ni+10%Cr: Chromel, Tophel, ThermoKanthal KP, T-1.
99.5%Fe: ThermoKanthal JP.
95%Ni+2%Al+2%Mn+1%Si: Alumel, Nial, ThermoKanthal KN, T-2.
84%Ni+14%Cr+1.5%Si: Nicrosil.
95%Ni+4.5%Si+0.1 Mg: Nisil.

To identify a single-leg thermoelement, a suffix ("P" or "N") is attached to the type letter to indicate the polarity of the material. Thus,
for instance, EN—usually constantan—is the negative leg of Type E thermocouples. Materials must be selected taking into account a number of
characteristics, including:

1. Stability. The properties of the material should not change significantly with their use. The materials must be be chemically stable in
   the environment in which they are to operate. This is particularly true
   for high temperature devices operating in an oxidizing atmosphere, and
   is the reason for using noble metals instead of base ones.

   The material must be physically stable. It should not experience
   phase changes (especially in cryogenic applications). It should not be
   altered by mechanical handling (must, for example, not lose ductility).

   The material should be insensitive to magnetic fields.
2. Homogeneity. It is important that the material be homogeneous not
   only along a given sample but also from sample to sample so that a
   single calibration be valid for an entire batch.
3. Good thermoelectric power. The lower temperature limit for practical use of thermocouple thermometers (about 20 K) is the result of
   insufficient values of α as absolute zero is approached.
4. Low thermal conductivity. This is important at cryogenic temperatures.

The choice of the type of thermocouple depends mainly on the temperature range desired, as indicated in Table 5.1.

To achieve the highest precisions, it is essential to perform very careful
calibration. The extremely meticulous procedures required are described
in detail (with ample references) by Burns and Scroger. As an example, a
NIST (National Institute of Standards and Technology) publication reports
the calibration of a thermocouple (submitted by a customer) as having
uncertainties not exceeding 3 μV in the range of 0 C to 1450 C. Since the
Seebeck voltage for this particular sample was 14,940 μV at the highest
temperature mentioned, the uncertainty at that temperature was less than
0.02%.

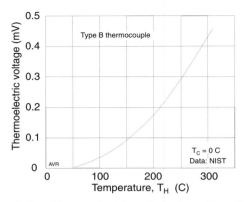

Figure 5.7   Thermoelectric voltage for a Type B thermocouple.

Very complete data of thermoelectric voltages versus temperature, tabulated at intervals of 1 celsius, for various types of thermocouples are found in NIST ITS-90 Thermocouple Database at <http://srdata.nist.gov/its90/main/>. An example of such data appears in Figure 5.7. Observe that the voltages developed by a thermocouple thermometer are small compared to those developed by a thermoelectric generator. See Section 5.3.

5.3   The Thermoelectric Generator

The ability of a thermocouple to generate a voltage when there is a temperature difference across it suggests its use as a heat engine capable of producing electricity directly. As a heat engine, its efficiency is limited by the Carnot efficiency, and therefore must be of the form

$$\eta = \frac{T_H - T_C}{T_H}\eta^* \tag{15}$$

where η^* depends on the geometry of the device, on the properties of the materials used, and on the matching of the generator to the load.

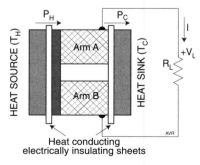

Figure 5.8   The thermocouple as a generator.

Consider a thermocouple represented by the simplified sketch in Figure 5.8. The electric resistance, R, and the heat conductance, Λ, are:

$$R = \frac{\ell_A}{A_A \sigma_A} + \frac{\ell_B}{A_B \sigma_B} \tag{16}$$

$$\Lambda = \frac{A_A \lambda_A}{\ell_A} + \frac{A_B \lambda_B}{\ell_B} \tag{17}$$

where A is the area of the cross section of each arm, ℓ is the length of each arm, σ is the electric conductivity, and λ is the thermal conductivity. Note that the two arms of the device are in parallel in so far as heat conduction is concerned but are electrically in series.

In the presence of a current, I, the heat power supplied by the heat source is, as we have seen from our experiment,

$$P_H = \Lambda(T_H - T_C) + \pi I - \frac{1}{2}RI^2. \tag{18}$$

π can be eliminated by using Equation 12. The Peltier term at the source is proportional to the temperature, T_H, at that point. Thus,

$$P_H = \Lambda(T_H - T_C) + \alpha T_H I - \frac{1}{2}RI^2. \tag{19}$$

The current through the load is

$$I = \frac{\alpha(T_H - T_C)}{R + R_L}, \tag{20}$$

and, consequently, the power delivered to the load is

$$P_L = \frac{\alpha^2(T_H - T_C)^2}{(R + R_L)^2} R_L. \tag{21}$$

The efficiency of the device is

$$\eta = \frac{P_L}{P_H} = \frac{T_H - T_C}{T_H} \times$$
$$\left[\frac{(R + R_L)^2}{R_L}\frac{\Lambda}{\alpha^2}\frac{1}{T_H} + \frac{1}{2}\frac{R}{R_L} + 1 + \frac{1}{2}\frac{R}{R_L}\frac{T_C}{T_H}\right]^{-1}. \tag{22}$$

It is convenient to write $R_L = m\,R$. The efficiency formula then becomes

$$\eta = \eta_{CARNOT} \times \left[1 + \frac{1}{2m}\left(1 + \frac{T_C}{T_H}\right) + \frac{(m+1)^2}{m}\frac{1}{T_H Z}\right]^{-1} \equiv \eta_{CARNOT} \times \eta^* \tag{23}$$

where a **figure of merit**, Z, of the thermocouple is defined as

$$Z \equiv \frac{\alpha^2}{\Lambda R}. \tag{24}$$

5.9

The dimension of Z is [temperature^{-1}]; therefore, in the SI, the unit of measure is K^{-1}. In the expression for η^*, all parameters other than Z are externally adjustable. The characteristics of the thermocouple are all contained in Z. *The larger Z, the greater the efficiency.*

In order to obtain a large Z, one must choose materials for the thermocouple arms that have large Seebeck coefficients. One must also make the ΛR product as small as possible. This can be achieved, again, by proper choice of materials and of the geometry of the device. If the arms are short and have a large cross section, then the resistance, R, tends to be small, but the heat conductance, Λ, tends to be correspondingly large. Likewise, if the arms are long and have a small cross section, the heat conductance, tends to be small but the resistance, R, tends to be correspondingly large. As it happens, there is a geometry that minimizes the ΛR product. This minimum occurs when the length, ℓ, and the cross-sectional area, A, of arms A and B satisfy the relationship (see derivation in Appendix 1)

$$\frac{\ell_A A_B}{\ell_B A_A} = \sqrt{\frac{\lambda_A \sigma_A}{\lambda_B \sigma_B}}. \tag{25}$$

Under the above conditions, the value of the ΛR product is

$$\Lambda R = \left[\left(\frac{\lambda_A}{\sigma_A} \right)^{1/2} + \left(\frac{\lambda_B}{\sigma_B} \right)^{1/2} \right]^2. \tag{26}$$

If one wants to maximize the efficiency of a thermoelectric generator, one has to choose the appropriate value for the load resistance, R_L. Remembering that we wrote $R_L = m\,R$, this means that we have to choose the appropriate value of m:

$$\frac{d}{dm} \left[\frac{(m+1)^2}{m} \frac{1}{ZT_H} + \frac{1}{2m} \left(1 + \frac{T_C}{T_H} \right) + 1 \right] = 0 \tag{27}$$

which leads to

$$m = \sqrt{1 + <T> Z} \tag{28}$$

where

$$<T> = \frac{T_H + T_C}{2}. \tag{29}$$

When this value of m is introduced into the expression for η^*, one obtains

$$\eta^* = \frac{(1 + <T> Z)^{1/2} - 1}{(1 + <T> Z)^{1/2} + T_C/T_H} = \frac{m-1}{m + T_C/T_H}. \tag{30}$$

Summing up, there are three different considerations in optimizing the efficiency of a thermocouple:

1. Choice of appropriate materials in order to maximize Z.
2. Choice of the best geometry in order to minimize ΛR.
3. Choice of the proper value of the load resistance relative to the internal resistance of the device, i.e., selection of the best value for m. Equation 23 shows that for $Z \to \infty$,

$$\eta^* = \frac{1}{1 + \frac{1}{2m}\left(1 + \frac{T_C}{T_H}\right)} \tag{31}$$

Maximum η^* is obtained by using $m = \infty$. Having an infinite m means that there is infinitely more resistance in the load than in the thermocouple. In other words, the couple must have zero resistance, which can only be achieved by the use of superconductors. Semiconductors, unfortunately, have inherently zero Seebeck coefficients. Consequently, $\eta^* = 1$ cannot be achieved even theoretically. Indeed, with present day technology, it is difficult to achieve Zs in excess of 0.004 K^{-1}. This explains why thermocouples have substantially less efficiency than thermomechanical engines.

Figure 5.9 plots η^* versus T_H (for $T_C = 300$ K), using two different values of Z, in each case for optimum m. From this graph one can see that thermocouples created with existing technology can achieve (theoretically) some 30% of the Carnot efficiency. Compare this with a modern steam generator:

In Paradise, Ky, TVA operates a 1150 MW power plant with a net overall efficiency of 39.4%. Input steam temperature is 812 K and heat is rejected at 311 K. The Carnot efficiency is, therefore, 61.7%. The system realizes 64% of the Carnot efficiency.

5.4 Figure of Merit of a Material

The figure of merit, Z, that we have used so far refers to a pair of materials working against each other. It would be convenient if a figure of merit could be assigned to a material by itself. This would help in ranking its performance in a thermocouple. To be able to develop such a figure of merit, we need to extend the definition of the Seebeck coefficient.

Measurements and theoretical predictions show that the Seebeck effect of any junction of superconductors is zero. This permits the definition of an absolute Seebeck coefficient of a normal conductor: it is the coefficient of the material working against a superconductor.

The absolute coefficient can, of course, be measured only at temperatures low enough to allow superconductivity. The coefficient of lead was measured between 7.2 and 18 K. Below 7.2 K, lead becomes itself a superconductor and the effect disappears; above 18 K there were no superconductors available when these measurements were made.

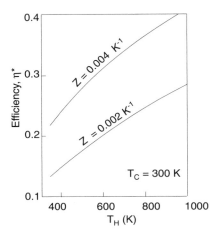

Figure 5.9 Efficiency of thermocouples vs. temperature.

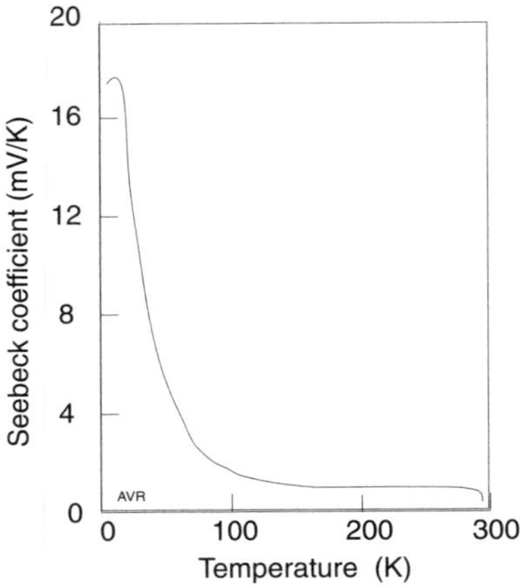

Figure 5.10 The Seebeck coefficient of germanium is large at low temperatures.

Seebeck coefficients for lead above 18 K were calculated from accurate measurements of the Thompson effect (see Section 5.15) using the Equation

$$\frac{d\alpha}{dT} = \frac{\tau}{T} \qquad \text{(from 138)}$$

The absolute α's for other materials were determined by measuring their thermoelectric voltage against lead.

Once the Seebeck coefficient is known, the figure of merit can be written as

$$Z = \frac{\alpha^2}{\Lambda R} = \alpha^2 \frac{\sigma A}{\ell} \frac{\ell}{\lambda A} = \alpha^2 \frac{\sigma}{\lambda}. \qquad (32)$$

In the above formula, all parameters refer to a single material.

It is clear that to maximize the figure of merit, one has to choose a material with the highest α and the smallest possible λ/σ. Unfortunately, this ratio is approximately the same for all metals. We will examine this question in some detail in the next section.

5.5 The Wiedemann-Franz-Lorenz Law

In the 19th century, physicists were having difficulties measuring the thermal conductivity of materials. Gustave Heinrich Wiedemann (1826–1899) observed in 1853 that, at least for metals, the ratio, λ/σ, appeared to be constant. If so, the thermal conductivity could be inferred from the easily measure electric conductivity. Eventually, a "law" was formulated in collaboration with Rudolf Franz (1827–1902) and Ludwig Valentine Lorenz

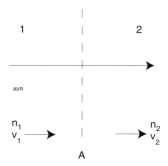

Figure 5.11   Thermal conductivity in a unidimensional gas.

(1829–1891) expressing the relationship between the thermal conductivity, the electric conductivity and the absolute temperature. This is Wiedemann-Franz-Lorenz law, which can be justified based on a simple classical model of electric conduction.

Consider the heat conduction in a unidimensional gas along which there is a temperature gradient (Figure 5.11). A surface, A, at the origin of coordinates is normal to the molecular motion. We assume that there is no net mass flow. Then,

$$n_1 v_1 + n_2 v_2 = 0. \tag{33}$$

We now use the symbol, ℓ, as the mean free path (not as the length of the arms, as before). The molecules that cross A coming from the left originate, on the average, from a point of coordinate $-\ell/2$.

Their kinetic energy will be $U_{(-\ell/2)}$. Those that come from the right have an energy $U_{(\ell/2)}$. Half the molecules move to the left and half to the right. The net energy flux, i.e., the power density, is

$$P = \frac{nv}{2}\left[U_{(-\ell/2)} - U_{(\ell/2)}\right]$$

$$= -\frac{nv}{2}\frac{\partial U}{\partial x}\ell \quad \text{W/m}^2. \tag{34}$$

The energy of each molecule is $\frac{1}{2}kT$, therefore

$$P = -\frac{nv}{4}\ell k\frac{\partial T}{\partial x} \equiv -\lambda\frac{\partial T}{\partial x}. \tag{35}$$

Here, the quantity, λ, is the thermal conductivity as before.

If we assume that the electrons are the only conveyers of heat in a metal (a reasonable assumption) and that they act as a gas with each electron carrying $\frac{3}{2}kT$ units of energy, then the heat conductivity of a metal should be

$$\lambda = \frac{3nv\ell k}{4}. \tag{36}$$

The factor 3 was included to account for the three degrees of freedom of electrons in a 3-dimensional gas. Actually, this overestimates the thermal conductivity because we did not correctly consider the statistical number of electrons of a 3-dimensional gas that cross a given surface per unit time. Although the numerical results are incorrect, the influence of the different physical parameters on the conductivity is correctly represented.

Let us now examine the electric conductivity, σ,

$$\sigma = qn\mu \tag{37}$$

where μ, the mobility, is the velocity a **carrier** (a mobile charge) attains under the influence of a unit electric field; it is the ratio of the **drift velocity**, v_d, to the electric field, E:

$$\mu = \frac{v_d}{E}. \tag{38}$$

Under the usual assumption that collisions are isotropic, after each collision, the velocity of the electron is statistically zero (because it has equal probability of going in any direction). This being the case, the average drift velocity of an electron is $\frac{1}{2}at$, where a is the acceleration, qE/m, and t is the mean free time, ℓ/v. Remember that v is the *thermal* velocity of the electron and is generally much larger than its *drift* velocity, v_d.

$$\mu = \frac{q\ell}{2mv} \tag{39}$$

and

$$\sigma = \frac{q^2 n\ell}{2mv}. \tag{40}$$

The λ/σ ratio becomes

$$\frac{\lambda}{\sigma} = \frac{3mv^2 k}{2q^2} = \frac{3k^2 T}{2q^2} \tag{41}$$

because

$$mv^2 = kT. \tag{42}$$

The correct ratio is

$$\frac{\lambda}{\sigma} = \frac{\pi^2}{3}\frac{k^2 T}{q^2} \equiv LT = 2.44 \times 10^{-8} T. \tag{43}$$

This expression is called the **Wiedemann-Franz-Lorenz** law and the constant, $L = 2.44 \times 10^{-8}$ $(V/K)^2$, is the **Lorenz number**.

If one attempted to reduce the heat conductance by using materials of low thermal conductivity, one would automatically increase the electric resistance of the thermocouple because of the proportionality between σ and λ. Thus, no improvement can be expected from manipulating these

5.14

properties and therefore, in most metallic conductors, the figure of merit depends only on the Seebeck coefficient and the temperature:

$$Z = \frac{\alpha^2}{L}\frac{1}{T} = 4.1 \times 10^7 \frac{\alpha^2}{T}. \quad K^{-1} \qquad (44)$$

Observe that the larger L, the smaller Z.

The Wiedemann-Franz-Lorenz law is not actually a law; it is rather an approximation that applies approximately to many metals but not to materials in general. Likewise, the Lorenz number is not constant as predicted by the theory but varies from metal to metal as shown in Table 5.2.

The precision of the numbers in Table 5.2 is questionable as they were calculated from separate measurements of the thermal conductivity and the electric resistivity published in the *Handbook of Chemistry and Physics* (CRC). Since different samples were likely used in the two measurements and because the conductivities are very sensitive to the degree of impurity, one cannot have great confidence in the ratios displayed.

Additionally, the measurement of thermal conductivity is probably not accurate beyond the second significant figure. More importantly, the assumption that electrons are the only carriers of heat in a solid is certainly only an approximation for metals and is not valid for most other materials, as we will discuss in more detail later in this text.

Table 5.2
The Lorenz Number for Some Metals

Metal	$\dfrac{L}{(V/K)^2 \times 10^8}$	Metal	$\dfrac{L}{(V/K)^2 \times 10^8}$
Ag	2.29	Na	2.18
Al	2.10	Ni	2.03
Au	2.53	Os	3.00
Be	1.60	Pb	2.51
Cd	2.44	Pd	2.62
Co	2.11	Pt	2.57
Cr	4.56	Sn	2.75
Cu	2.13	Ta	2.37
Fe	2.68	Ti	3.45
Gd	5.07	W	3.24
Hg	2.82	Zn	2.32
Ir	2.65	Zr	3.10
K	2.33		
Mg	1.71		
Mo	2.65		

5.6 Thermal Conductivity in Solids

In solids, heat propagates via two different mechanisms:

1. It is conducted by the same carriers that transport electric charges. The corresponding thermal conductivity, λ_C, is called **carrier conductivity**. It is related to the electric conductivity by the Wiedemann-Franz-Lorenz law discussed above.
2. It is also conducted by thermal vibrations of the crystalline lattice (i.e., by phonons) propagating along the material. This conductivity is called the **lattice conductivity**, λ_L.

The first mechanism is dominant in metals as they have an abundance of carriers and have relatively soft lattices. However, the opposite phenomenon characterizes semiconductors. For example, diamond has negligible carrier concentration and consequently has negligible electric conductivity. Nevertheless, diamond exhibits a thermal conductivity 11 times greater than that of aluminum and 30 times that of iron. In fact, it is the best heat conductor of all naturally occurring substances.[†]

The total thermal conductivity is the sum of the carrier and the lattice conductivities,

$$\lambda = \lambda_C + \lambda_L. \tag{45}$$

Therefore, the figure of merit becomes

$$Z = \frac{\alpha^2 \sigma}{\lambda_C + \lambda_L} = \frac{\alpha^2}{\lambda_C/\sigma + \lambda_L/\sigma} = \frac{\alpha^2}{LT + \lambda_L/\sigma} \tag{46}$$

As a consequence, the value of Z in Equation 44 is an upper limit.

It can be seen from Table 5.3 that mechanically hard semiconductors (e.g., silicon and germanium) have lattice conductivities that are orders of magnitude larger than their respective carrier conductivities. For example, the lattice conductivity for silicon is 400 times larger than its carrier conductivity. On the other hand, the ratio of these conductivities in soft semiconductors tends to be smaller (e.g., approximately 4, for BiTe).

[†] Synthetic crystals, such as silicon nitride (SiN) and aluminum nitride (AlN), when carefully prepared, may have thermal conductivities exceeding that of diamond.

Diamonds prepared by the chemical vapor deposition method (CVD), which are available from Fraunhoff IAF in wafers of up to 15 cm diameter and more than 2 mm thickness, have a conductivity of 5300 W/(m K) at 118 K and of 2200 W/(m K) at 273 K. Compare with copper that, over this same range of temperatures, has about 380 W/(m K).

Table 5.3
Thermal Conductivities
of Some Semiconductors

	W/(m K) Room temperature Doping corresponding to $\alpha = 200\ \mu V/K$		
Material	λ_L	λ_C	λ_L/λ_C
Silicon	113	10.3	377
Germanium	63	0.6	105
InAs	30	1.5	20
InSb	16	1	16
BiTe	1.6	0.4	4

Compared with metals, semiconductors have an unfavorable λ/σ ratio, but have such a large advantage in α that they are the material invariably used in thermoelectric generators, refrigerators, and heat pumps. Metals are exclusively used in thermometry. It is possible that superlattices having negligible lattice conductivity can be created. Superlattices consist of alternating layers of two different semiconductors. Layer spacing can be made smaller than the mean free path of conduction electrons yet larger than the mean free path of phonons. Consequently, electric and heat carrier conductivities are (relatively) undisturbed, but heat lattice conductivity is reduced. See Venkatasubramanian et al. If such low heat conduction lattices can be made with αs of 350 $\mu V/K$, then thermocouples would reach the golden target of having a ZT product of 5, which, according to experts, would open an immense field of practical applications for these devices.

5.7   Seebeck Coefficient of Semiconductors

No metal has a Seebeck coefficient larger than 100 $\mu V/K$. The great majority has coefficients much smaller than 10 $\mu V/K$ as can be seen from Table 5.4. Some semiconductors have coefficients of some 300 $\mu V/K$ at usable temperatures. Since the figure of merit depends on the square of the Seebeck coefficient, one can see that some semiconductors have an order of magnitude advantage over metals.

The data in Table 5.4 were taken from the CRC *Handbook of Thermoelectronics*, which, in turn, obtained the values from different sources. All elements in the table are metals. The Seebeck coefficient for semiconductors depends critically on the doping level. The polarity of the Seebeck effect depends on whether the semiconductor is of the p or n type. Intrinsic semiconductors have zero Seebeck coefficients. For small doping concentrations, the Seebeck effect grows rapidly with the doping level reaching a peak and the decreasing again as depicted in Figure 5.12.

Table 5.4
Seebeck Coefficient for Most Metals
μV/K, Temperature: 300 K

Ag	1.51	Eu	24.5	Nb	−0.44	Sr	1.1
Al	−1.66	Fe	15	Nd	−2.3	Ta	−1.9
Au	1.94	Gd	−1.6	Ni	−19.5	Tb	−1
Ba	12.1	Hf	5.5	Np	−3.1	Th	−3.2
Be	1.7	Ho	−1.6	Os	−4.4	Ti	9.1
Ca	10.3	In	1.68	Pb	−1.05	Tl	0.3
Cd	2.55	Ir	0.86	Pd	−10.7	Tm	1.9
Ce	6.2	K	−13.7	Rb	−10	U	7.1
Co	−30.8	La	1.7	Re	−5.9	V	0.23
Cr	21.8	Lu	−4.3	Rh	0.6	W	0.9
Cs	−0.9	Mg	−1.46	Ru	−1.4	Y	−0.7
Cu	1.83	Mn	−9.8	Sc	−19	Yb	30
Dy	−1.8	Mo	5.6	Sm	1.2	Zn	2.4
Er	−0.1	Na	−6.3	Sn	−1	Zr	8.9

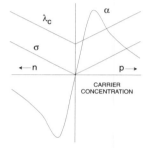

Figure 5.12   Seebeck coefficient and electric and heat conductivities of a semiconductor as a function of carrier concentration.

5.8   Performance of Thermoelectric Materials

One should not select thermoelectric materials based solely on their figure of merit. A given material with good Z may be of little value because it has too low a melting point. Table 5.5 shows the figure of merit, Z, the maximum operating temperature, T_H, the ZT_H product and the "efficiency," η, for several materials. The "efficiency" was calculated from

$$\eta = \frac{(1 + Z <T>)^{1/2} - 1}{(1 + Z <T>)^{1/2} + T_C/T_H} \frac{T_H - T_C}{T_H} \tag{47}$$

where T_C was 300 K. Although $BiSb_4Te_{7.5}$ has the highest figure of merit of the materials in the table, its low maximum operating temperature makes it relatively inefficient. On the other hand, lead telluride, with a much smaller figure of merit, can operate at 900 K and can, therefore, yield 12.6% efficiency when its cold side is at 300 K.

Table 5.5
Performance of Some Thermoelectric Materials

Material	Z $(K^{-1} \times 10^3)$	T_H (K)	ZT_H	η
Bi_2Te_3	2.0	450	0.9	5.4%
$BiSb_4Te_{7.5}$	3.3	450	1.5	7.6%
Bi_2Te_2Se	2.3	600	1.38	11.1%
PbTe	1.2	900	1.08	12.6%
CeS(+Ba)	0.8	1300	1.04	14.3%

The product, ZT, is the **dimensionless figure of merit**. Each semiconductor has a range of temperature over which its ZT is best. For low temperatures around 100 C, Bi_2Te_3 works best. At much higher temperatures around 500 C, PbTe, popular in the 1960s, was the choice. For this range the so called TAGS (tellurium/antimony/germanium/silver) is now preferred. JPL recently (2003) developed Zn_4Sb_3, an excellent solution for the intermediary range around 350 C. The data in Figure 5.13 are for p-type semiconductors. PbTe, for instance, when doped with sodium, forms a p-type semiconductor and when doped with lead iodide, forms an n-type semiconductor.

Much larger values of ZT at room temperature have been demonstrated in the laboratory. Indeed, Venkatasubramanian et al. have reported a p-type Bi_2Te_3/Sb_2Te_3 superlattice device with a ZT of 2.4 at room temperature.

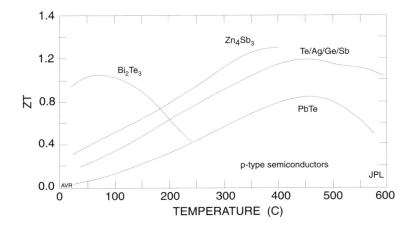

Figure 5.13   For each semiconductor, ZT is best over a given temperature range.

5.9   Some Applications of Thermoelectric Generators

The unparalleled reliability and simplicity of the thermoelectric generators makes them the preferred device in applications in which unattended operation is more important than efficiency. Its applications include:

1. Power supplies for spacecraft that operate too far from the Sun to take advantage of photovoltaics.
2. Topping cycles for stationary power plants (potentially).
3. Generators for oil producing installations, including ocean platforms.
4. Electric power providers for air circulating fans in residential heating systems that otherwise would not operate during periods of electric power failures.
5. Power supplies for automotive use that take advantage of the heat that the engines shed.
6. Generators that produce the energy necessary to open the main valve in gas heating systems. The heat of the pilot flame activates the generator. The main gas valve will not open unless the pilot is ignited.

The heat necessary to drive thermoelectric generators may come from any number of sources. For example, it can come from the burning of fuel, from radioactive decay or from reject heat (such as exhaust gases in an automobile).

Radionuclide decay is a heat source for generating electricity in space and in remote locations. Table 5.6 from a University of Stony Brook Web page (Mechanical Engineering) lists some of the radionuclides used.

Long duration space missions invariably use plutonium-238 owing to its long half-life although this is an extremely high cost fuel amounting to many million dollars per RTG. For ground use, strontium-90 is preferred. Strontium-90 is, indeed, the radionuclide that powers the controversial 500 or so RTGs installed by the Soviet Union along the coast of the Kola Peninsula (bordering on Finland and Norway).

Table 5.6
Radionuclides Used in RTGs

Element	Half-life (years)	Specific power (kW/kg) (thermal)	Specific cost $/watt
Cesium-144	0.781	25	15
Curium-242	0.445	120	495
Plutonium-238	86.8	0.55	3000
Polonium-210	0.378	141	570
Strontium-90	28.0	0.93	250

Table 5.7
Specifications for the Radioisotope Thermal Generator (RTG)
Used in the 1981 "Galileo" Mission to Jupiter

	BOL	EOM	
Heat furnished by isotope source	2460	2332	W
Heat into converter	2251	2129	W
Heat into thermocouples	2068	1951	W
Hot junction temperature	1133	1090	K
Cold junction temperature	433	410	K
Thermoelectric efficiency	11.1	10.8	%
Generator efficiency	9.4	8.6	%
Power output	230	201	W
Weight	41.7	41.7	kg
Output voltage	30	30	V
Specific power	5.52	4.82	W/kg

The Radioisotope Thermal Generator (RTG) that powered the "Galileo" missions to the outer planets represented the state-of-the-art for thermoelectric power sources in 1978. It used the then novel selenium based semiconductors. The specifications for this RTG listed in Table 5.7 correspond to both the beginning of life (BOL) and the end of mission (EOM) conditions. The EOM conditions are not the same as end of life, which is many years longer. In this particular case, BOL is 1000 hours after fueling and EOM is 59,000 hours (almost 7 years) after fueling.

A ΔT of about 700 K was kept throughout the mission. The thermoelectric efficiency degraded only slightly in the 7 years of operation.

The advantage of an RTG is that it is extremely light if one considers that it includes both the electrical generator and the fuel for many years of operation. Even the lightest possible gasoline engine would be orders of magnitude heavier. A large airplane gasoline engine may deliver 1500 W/kg, but one must add the mass of the fuel and of the oxygen needed for longtime operation. The specific consumption of a gasoline engine is about 0.2 kg hp^{-1}h^{-1}.[†] For each kilogram of gasoline, the engine uses 3.1 kg of O_2. The specific consumption of *fuel plus oxygen* is 0.8 kg hp^{-1}h^{-1}. Since 200 W correspond to 0.27 hp, the hourly consumption of a gasoline driven generator that delivers 200 W is 0.24 kg of consumables. During the 59,000 hours of the mission, 14,000 kg of consumables would be used up. Thus, these consumables alone would mass over 3000 times more than the whole RTG. The latter, having no moving parts, requires no maintenance, while, on the other hand, it is inconceivable that a gasoline engine could possible operate unattended for 7 long years.

[†] One of the most economical aircraft engines ever built was a "turbo compound" engine that powered the Lockheed Constellation. Its specific consumption was 0.175 kg hp^{-1}h^{-1}

5.10   Design of a Thermoelectric Generator

Example:

A thermoelectric generator is to furnish 100 kW at 115 V. Input temperature is 1500 K, while the output is at 1000 K. This output temperature is high enough to drive a steam plant—the thermoelectric generator is to serve as a **topping cycle**. See Chapter 3.

The characteristics of materials of the thermocouple are:

Seebeck coeff., averaged over the temp. range of interest:	0.0005 V/K
Electric resistivity of arm A:	0.002 Ωcm
Electric resistivity of arm B:	0.003 Ωcm
Thermal conductivity of arm A:	0.032 W cm^{-1}K^{-1}
Thermal conductivity of arm B:	0.021 W cm^{-1}K^{-1}
Maximum allowable current density:	100 A cm^{-2}

To simplify the construction, arms A and B must have equal length (but not necessarily equal cross section). Calculate:

1. The maximum thermal efficiency,
2. the number of thermocouples in series,
3. the dimensions of the arms,
4. the open-circuit voltage, and
5. the heat input and the rejected heat at
    5.1.   full load, and
    5.2.   no load.

Solution:

$V_{oc} \equiv$ open-circuit voltage per thermocouple.

$$V_{oc} = \alpha(T_H - T_C) = 0.0005 \times (1500 - 1000) = 0.25 \text{ V}. \qquad (48)$$

$I \equiv$ current through each thermocouple (same as the total current through the battery because all elements are in series).

$$I = \frac{100{,}000 \text{ W}}{115 \text{ V}} = 870 \text{ A}. \qquad (49)$$

If there are n thermocouples, each with a resistance, R, then

$$n\,V_{oc} - nRI = 115 \text{ V}. \qquad (50)$$

(continued)

(continued)

To find n, we must know R. For maximum efficiency, the load resistance, R_L must be equal to mR_{batt}, or $R_L = mnR$, where $m = \sqrt{1+ <T> Z}$ (see Equations 28 and 29). Here, R is the resistance of each thermocouple and R_{batt} is the resistance of the whole battery—that is, it is nR.

$$R_L = \frac{115 \text{ V}}{870 \text{ A}} = 0.132 \ \Omega, \tag{51}$$

$$Z = \frac{\alpha^2}{\Lambda R}, \tag{52}$$

$$\Lambda R = \left[\sqrt{\lambda_A \rho_A} + \sqrt{\lambda_B \rho_B}\right]^2 = \left[\sqrt{0.032 \times 0.002} + \sqrt{0.021 \times 0.003}\right]^2$$
$$= 254 \times 10^{-6} \text{ V}^2/\text{K}, \tag{53}$$

$$Z = \frac{0.0005^2}{254 \times 10^{-6}} = 980 \times 10^{-6} \text{ K}^{-1}, \tag{54}$$

$$<T>= \frac{1500 + 1000}{2} = 1250 \text{ K}, \tag{55}$$

$$m = \sqrt{1 + 980 \times 10^{-6} \times 1250} = 1.49, \tag{56}$$

$$nR = \frac{R_L}{m} = \frac{0.132}{1.49} = 0.0886 \ \Omega, \tag{57}$$

$$n = \frac{115 + nRI}{V_{oc}} = \frac{115 + 0.0886 \times 870}{0.25} = 768.3. \tag{58}$$

We will need a total of 768 thermocouples.

$$V_{OC} \equiv \text{open-circuit voltage of the battery}$$
$$= nV_{oc} = 0.25 \times 768 = 192 \text{ V}, \tag{59}$$

$$P_{H_{no \ load}} = \Lambda_{batt}(T_H - T_C). \tag{60}$$

$\Lambda_{batt} = n\Lambda$ because all the thermocouples are thermally in parallel.

$$\Lambda = \frac{\Lambda R}{R} = \frac{254 \times 10^{-6}}{0.0886/768} = 2.20 \text{ W/K}, \tag{61}$$

$$\Lambda_{batt} = 768 \times 2.20 = 1690 \text{ W/K}, \tag{62}$$

$$P_{H_{no \ load}} = 1690(1500 - 1000) = 846 \text{ kW}, \tag{63}$$

(continued)

5.23

(continued)

$$P_{C_{no\ load}} = 846 \text{ kW}, \tag{64}$$

$$\begin{aligned} P_{H_{full\ load}} &= 846 + n\alpha T_H I - \tfrac{1}{2} I^2 nR \\ &= 846 + 768 \times 0.0005 \times 1500 \times 870 \times 10^{-3} \\ &- \tfrac{1}{2} \times 870^2 \times 0.0886 \times 10^{-3} = 1310 \text{ kW}, \end{aligned} \tag{65}$$

$$\eta = \frac{100}{P_H} = \frac{100}{1310} = 0.076 \tag{66}$$

$$P_{C_{full\ load}} = P_H - 100 = 1310 - 100 = 1210 \text{ kW}. \tag{67}$$

Since the length of the two arms is the same, Equation 25 simplifies to

$$\frac{A_B}{A_A} = \sqrt{\frac{\lambda_A \rho_B}{\lambda_B \rho_A}} = \sqrt{\frac{0.032 \times 0.003}{0.021 \times 0.002}} = 1.51, \tag{68}$$

For $J_{max} = 100$ A cm^{-2}, the smaller of the two cross sections, A_A, must be equal to $870/100 = 8.7$ cm^2. The larger cross section must be $A_B = 1.51 \times 8.7 = 13.1$ cm^2. The resistance of each individual thermocouple is

$$R = \frac{nR}{n} = \frac{0.0886}{768} = 0.000115 \ \Omega \tag{69}$$

$$0.000115 = \rho_A \frac{\ell}{A_A} + \rho_B \frac{\ell}{A_B} = \left(\frac{0.002}{8.7} + \frac{0.003}{13.2} \right) \ell \tag{70}$$

which leads to $\ell = 0.36$ cm.

The two arms have a rather squat shape.

If the heat rejected at 1000 K is used to drive a steam turbine having 30% efficiency, the electric power generated by the latter will be $0.3 \times 1210 = 363$ kW. Adding to this the 100 kW from the thermocouple, we will have a total of 463 kW and an overall efficiency of

$$\eta = \frac{463}{1310} = 0.35. \tag{71}$$

It can be seen that thermocouples can be used as acceptable topping engines.

5.11 Thermoelectric Refrigerators and Heat Pumps

The Peltier effect is reversible: the direction of the heat transport depends on the direction of the current. Heat can be transported from the cold to the hot side of the thermocouple, which, consequently, can act as a heat pump or a refrigerator. We will investigate how much heat can be transported. For the sake of simplicity we will make the (not completely realistic) assumption that α, R, and Λ are all temperature independent.

5.11.1 Design Using an Existing Thermocouple

If a given thermocouple battery is available, then, presumably, the values of α, R, and Λ are known. Assume, as an example, that $\alpha = 0.055$ V/K, $R = 4.2\ \Omega$, and $\Lambda = 0.25$ W/K. Assume also that heat is to be pumped from $T_C = 278$ K to $T_H = 338$ K, a ΔT of 60 K.

Let P_C be the heat power transported from the cold source to the cold end of the thermocouple.

$$P_C = -\Lambda \Delta T + \alpha T_C I - \tfrac{1}{2}RI^2. \tag{72}$$

For the current example,

$$P_C = -15.0 + 15.29I - 2.1I^2 \tag{73}$$

The electrical energy required to do this pumping is

$$P_E = \alpha \Delta T I + RI^2 \tag{74}$$

The ratio between the pumped heat and the required electric power is called the **coefficient of performance**, ϕ_C, of the heat pump,

$$\phi_C = \frac{-\Lambda \Delta T + \alpha T_C I - \tfrac{1}{2}RI^2}{\alpha \Delta T I + RI^2}. \tag{75}$$

In a loss-less thermocouple ($R = 0$ and $\Lambda = 0$), $\phi_C = T_C/\Delta T$, which is the **Carnot efficiency**, $\phi_{C_{Carnot}}$, of the heat pump. See Problem 5.40. For our example $\phi_{C_{Carnot}} = 4.63$. The actual thermocouple does not come anywhere close to this value.

Figure 5.14, shows how the power, P_C, pumped from the cold source varies with the current. If $I < 0$ (not shown in the figure), the heat is being pumped *into* the cold source. At $I = 0$ (also not shown), there is no Peltier effect and heat still flows *into* the cold side by conduction. As I increases, some Peltier pumping starts to counteract this heat conduction and, eventually (in a properly designed device), heat will actually begin flowing *from* the cold side to the hot side. This amount of heat will initially increase as I increases, but eventually Joule losses will begin to generate so much heat that the Peltier pumping is overwhelmed; further increases in I will result in a reduction of the heat extracted from the cold side.

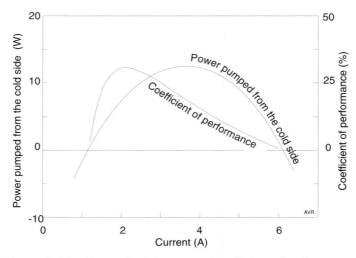

Figure 5.14 Pumped cold power and coefficient of performance as a function of current for the thermocouple of the example. Observe that the current that maximizes the pumped power is not the same as the one that maximizes the coefficient of performance.

It is easy to calculate what current causes maximum heat pumping:

$$\frac{dP_C}{dI} = \alpha T_C - RI = 0 \tag{76}$$

from which

$$I_{max\ cooling} = \frac{\alpha T_C}{R}. \tag{77}$$

The cold power pumped when this current is used is

$$P_{C_{max}} = -\Lambda \Delta T + \frac{\alpha^2 T_C^2}{2R}. \tag{78}$$

In our example, the current that maximizes the pumping is 3.64 A and the maximum power pumped is 12.83 W.

The lowest temperature that can be reached is that at which heat just ceases to be pumped—that is, $P_C = 0$:

$$\frac{\alpha^2 T_C^2}{2R} = \Lambda(T_H - T_C) \tag{79}$$

from which,

$$T_{C_{min}} = \frac{-1 + \sqrt{1 + 2ZT_H}}{Z}. \tag{80}$$

$$Z = \frac{\alpha^2}{\Lambda R} \tag{81}$$

which leads, for our example, to $Z = 0.00288\ \mathrm{K}^{-1}$, and $T_{C_{min}} = 249\ \mathrm{K}$. The current to achieve this is 3.26 A, but the pumped power is zero. However, any temperature above 249 K can be achieved.

5.26

With slightly more complicated math, one can find the current that maximizes the coefficient of performance:

$$\frac{d\phi_C}{dI} = (-\Lambda\Delta T + \alpha T_C I - \tfrac{1}{2}RI^2)(-1)(\alpha\Delta TI + RI^2)^{-2}(\alpha\Delta T + 2RI)$$
$$+ (\alpha\Delta TI + RI^2)^{-1}(\alpha T_C - RI) = 0, \tag{82}$$

The current that maximizes ϕ_C is

$$I = \frac{\Lambda\Delta T}{\alpha <T>}(m+1). \tag{83}$$

This can also be written

$$I = \frac{a\Delta T}{R(m-1)}, \tag{84}$$

see Problem 5.39.

Introducing the value of I into Equation 75 (and after considerable algebra), one finds that the maximum value for the coefficient of performance for the thermoelectric refrigerator is

$$\phi_{C_{opt}} = \frac{T_C}{\Delta T}\left(\frac{m - T_H/T_C}{m+1}\right) \tag{85}$$

where $m = \sqrt{1 + Z <T>}$, as before and $T_C/\Delta T$ is, as stated, the Carnot efficiency of the refrigerator.

Applying this to our example,

$$m = \sqrt{1 + Z <T>} = \sqrt{1 + 0.00288 \times \left(\frac{338 + 278}{2}\right)} = 1.374 \tag{86}$$

$$\phi_{C_{opt}} = \frac{278}{338 - 278}\left(\frac{1.374 - 338/278}{1.374 + 1}\right) = 0.308. \tag{87}$$

To obtain this coefficient of performance, one must use a current of

$$I = \frac{a\Delta T}{R(m-1)} = \frac{0.055 \times (338 - 278)}{4.2 \times (1.374 - 1)} = 2.10 \ A. \tag{88}$$

Table 5.8 compares two batteries of identical thermocouples, both pumping 100 W of heat from 258 K to 323 K. One battery is adjusted to pump this heat with a minimum number of cells—that is, it operates with current that maximizes P_C. The other battery operates with the current that maximizes the coefficient of performance. The substantially larger efficiency of the second battery comes at a cost of the larger number of cells required.

Table 5.8

Thermocouples Operated at Maximum P_C
and at Optimum ϕ_C
$P_C = 100$ W

Point of operation	Number of cells	P_E (W)	P_H (W)	ϕ_C
Max P_C	100	540	640	18.5%
Opt. ϕ_C	161	336	436	29.7%

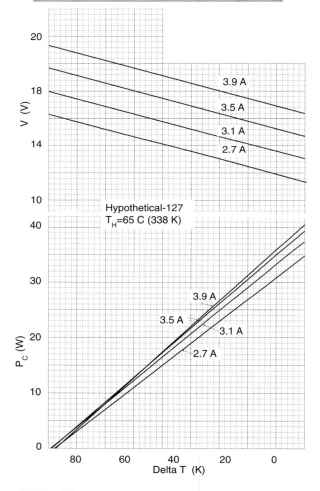

Figure 5.15   The characteristics of a given thermocouple are frequently displayed as shown.

Commonly, the characteristics of thermocouples used as heat pumps are displayed in graphs like the one in Figure 5.15, which corresponds to the unit in our example and is, roughly, similar to the Tellurex CZ1-1.0-127-1.27 unit—a battery consisting of 127 cells in series. This explains the large value of α—each cell has an α of $0.055/127 = 0.000433$ V/K. For an example of how to use such graphs, see Problem 5.33.

5.11.2 Design Based on Given Semiconductors

If semiconducting materials have been selected, but the exact dimensions of the thermocouples have not yet been determined, then although the values of α, ρ, and λ are known, those of R and Λ are not. Presumably, the dimensions of the arms of the thermocouple will be optimized,

$$\Lambda R = \left[\sqrt{\lambda_A \rho_A} + \sqrt{\lambda_B \rho_B} \right]^2 \equiv \beta, \tag{89}$$

thus establishing a known relationship between Λ and R.

Equation 72 becomes

$$P_C = -\beta \Delta T \frac{1}{R} + \alpha T_C I - \tfrac{1}{2} R I^2 \tag{90}$$

If a given cooling power is desired, what is the value of R that maximizes the coefficient of performance?

$$\phi_C = \frac{P_C}{P_E} = \frac{P_C}{\alpha \Delta T I + R I^2} \tag{91}$$

Solving Equation 90 for I,

$$I = \frac{\alpha T_C - \sqrt{\alpha^2 T_C^2 - 2\beta \Delta T - 2 P_C R}}{R} \tag{92}$$

We have selected the negative sign preceding the square root because we are searching for the *least* current.

$$\phi_C = \frac{P_C R}{\alpha \Delta T \left(\alpha T_C - \sqrt{\alpha^2 T_C^2 - 2\beta \Delta T - 2P_C R} \right) + \left(\alpha T_C - \sqrt{\alpha^2 T_C^2 - 2\beta \Delta T - 2P_C R} \right)^2} \tag{93}$$

Taking the derivative, $d\phi_C/dR$, setting it to zero, and solving for R,

$$R = \frac{-2\Delta T^2 \beta (2\beta + \alpha^2 T_A) + B(\alpha^2 \beta^{1/2} \Delta T T_C T_A - 2\beta^{3/2} \Delta T^2)}{\alpha^2 T_A^2 P_C} \tag{94}$$

where

$$T_A \equiv \Delta T + 2 T_C, \tag{95}$$

and

$$B \equiv \sqrt{4\beta + 2\alpha^2 T_A}. \tag{96}$$

As this result is sufficiently complicated to derive, it may be easier to solve the problem by trial and error using a spreadsheet.

Example:
We want a refrigerator capable of removing 10 W from a cold box at −5 C rejecting the heat to the environment at 30 C.

Owing to the temperature drops across the heat exchangers, the cold junction must be at –15 C and the hot one at 40 C.

The thermocouple materials have the following characteristics:

$$\alpha = 0.0006 \text{ V/K},$$
$$\lambda_A = 0.015 \text{ W cm}^{-1} \text{ K}^{-1},$$
$$\rho_A = 0.002 \ \Omega\text{cm},$$
$$\lambda_B = 0.010 \text{ W cm}^{-1} \text{ K}^{-1},$$
$$\rho_B = 0.003 \ \Omega\text{cm}.$$

The temperatures are
$$T_H = 313 \text{ K (40 C)},$$
$$T_C = 258 \text{ K } (-15 \text{ C}).$$

For optimum geometry,

$$\Lambda R \equiv \beta = \left[\sqrt{0.015 \times 0.002} + \sqrt{0.010 \times 0.003} \right]^2 = 120 \times 10^{-6} \text{ V}^2/\text{K}. \tag{97}$$

Applying Equations 94, 95, and 96,

$$T_A = 55 + 2 \times 258 = 571 \text{ kelvins} \qquad \text{(from 95)}$$

$$B = \sqrt{4 \times 120 \times 10^{-6} + 2 \times 0.0006^2 * 571} = 0.02985 \text{ V K}^{-1/2} \tag{96}$$

$$R = \left\{ -2 \times 55^2 \times 120 \times 10^{-6} \left(2 \times 120 \times 10^{-6} + 0.0006^2 \times 571 \right) \right.$$

$$+ 0.02985 \left[0.0006^2 \times (120 \times 10^{-6})^{1/2} \times 55 \times 258 \times 571 \right.$$

$$\left. \left. - 2 \times (120 \times 10^{-6})^{3/2} \times 55^2 \right] \right\} \bigg/ \left(0.0006^2 \times 571^2 P_C \right) = \frac{0.00335}{P_C} \tag{94}$$

(continued)

(continued)

A single thermocouple will, for this application, draw too much current and require an inconveniently low voltage. A better strategy would be to use 100 thermocouples, connected electrically in series and thermally in parallel. Hence, we want to pump 0.1 W per thermocouple. ($P_C = 0.1$ W),

$$R = 0.0335 \ \Omega. \tag{98}$$

The corresponding heat conductance is, from Equation 97

$$\Lambda = \frac{\beta}{R} = \frac{120 \times 10^{-6}}{0.0335} = 0.00358 \ \text{W/K} \tag{99}$$

The required current can be found from Equation 92:

$$I = \frac{0.0006 \times 258 - \sqrt{0.0006^2 \times 258^2 - 2 \times 120 \times 10^{-6} \times 55 - 2 \times 0.1 \times 0.0335}}{0.0335}$$
$$= 2.72 \ \text{A}. \tag{100}$$

The electric input power is

$$P_E = \alpha \Delta T I + R I^2 = 0.0006 \times 55 \times 2.72 + 0.00335 \times 2.72^2 = 0.337 \ \text{W} \tag{101}$$

And the coefficient of performance is

$$\phi_C = \frac{0.1}{0.337} = 0.296. \tag{102}$$

We can obtain this same value using Equation 93.

We now have the required values of R and of Λ. We must determine the geometry of the two arms. It facilitates the assembly of the thermocouple if both arms have the same length, ℓ—that is, if $\ell_A = \ell_B \equiv \ell$.

$$R = \rho_A \frac{\ell}{A_A} + \rho_B \frac{\ell}{A_B} \tag{103}$$

Using the values in our example,

$$\ell = \frac{0.0335}{\dfrac{0.002}{A_A} + \dfrac{0.003}{A_B}} \tag{104}$$

$$\Lambda = \lambda_A \frac{A_A}{\ell} + \lambda_B \frac{A_B}{\ell} \tag{105}$$

(continued)

(continued)

and

$$\ell = \frac{0.015A_A + 0.01A_B}{0.003580} \tag{106}$$

Equating equations 104 to 106, we obtain

$$A_A = \frac{3}{2}A_B \tag{107}$$

Next, we need to determine the maximum allowable current density, J_{max}. We can assume that $J_{max} = 300$ A/cm^2 and that the maximum allowable current though the thermocouple is 4 A (It is supposed to operate at 2.7 A.) This sets an approximate area for $A_A = 4/300 = 0.013$ cm^2. The value of A_B is 0.02 cm^2, and the length of each arm, from Equation 104, is 0.11 cm.

The required voltage to pump 10 W is

$$V = \frac{100P_E}{I} = \frac{100 \times 0.337}{2.72} = 12.4 \text{ V}. \tag{108}$$

5.12   Temperature Dependence

In Section 5.11, we made the assumption that α, Λ, and R are independent of temperature. In reality, this is not the case. If we take this dependence into account, we will greatly complicate the solution of the various design problems. We will not do this in this book and shall be satisfied with the approximate results of the preceding section. Nevertheless, it is useful to acknowledge that the parameters mentioned do vary when the temperature changes. This is illustrated in Figures 5.16 and 5.17.

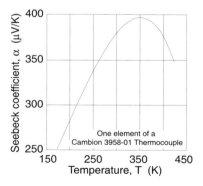
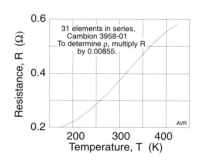

Figure 5.16   Representative behavior of the Seebeck coefficient and the resistance of a 1970 thermocouple as a function of temperature.

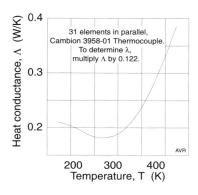

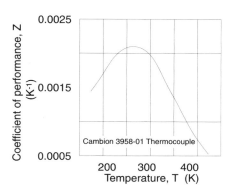

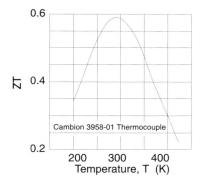

Figure 5.17   Representative behavior of the heat conductance, the coefficient of performance, and the ZT product of a thermocouple as a function of temperature. This corresponds to the state of the art in 1970. Since then, much progress has been made. The ZT values of modern materials comfortably exceeds unity, for example.

5.13   Battery Architecture

We saw that there are a number of characteristics of a thermocouple that have to be optimized for best performance: choice of material, proper geometry of the arms, and matching the load to the generator. There is an additional factor that needs to be considered in optimizing batteries of cooling devices. T_H and T_C are typically the same for all cells of a battery. However, the coefficient of performance can be improved by making different arrangements for the heat flow. Some of these alternative arrangements are described in an article by Bell (see references).

5.14   The Physics of Thermoelectricity†

In investigating the behavior of thermocouples we considered four different mechanisms:

---

† A fuller treatment of this subject is found in the book by Goldsmid.

1. **Heat conduction**, a topic with which most readers are thoroughly familiar. It was shown that heat is conducted by both the motion of carriers and by the vibration of the crystalline lattice. Heat conduction introduces losses in the performance of thermocouple.

2. **Joule losses**. As the current flows through the device, it encounters resistance and, consequently, generates heat. This constitutes the second loss mechanism in thermocouples. Joule losses result from the scattering of carriers by lattice imperfections (thermal vibrations, impurities, dislocations, etc.). Again, it is assumed that the reader is familiar with it.

3. **Seebeck effect**, the development of a voltage in a conductor as a result of a temperature differential. It is caused by increased carrier concentration in the cold regions of a conductor. The mechanism for this is examined in Subsection 5.14.1.

4. **Peltier effect**, the absorption or the release of heat at a junction of dissimilar conductors owing to the change in heat capacity of carriers when they leave one medium and enter a different one. In Section 5.1, we stated that there is a relationship between the Seebeck and the Peltier effects. We will derive this relationship in the present section. The mechanism for the Peltier effect itself will be discussed in Subsection 5.14.2.

So far, we have completely disregarded a fifth important effect—the Thomson effect. There is a very good reason for this omission as shall be explained in this section. We will also show how the Seebeck, Peltier, and Thomson effects are interrelated.

5. **Thomson effect**, the convection of heat by the flux of drifting carriers, will be discussed in Subsection 5.14.3.

5.14.1 The Seebeck Effect

Consider a length of pipe filled with a gas that is at uniform temperature. Clearly, both pressure and concentration are also uniform. However, if one end of the pipe is heated to a higher temperature than the other, the higher pressure on the hotter side will cause some flow of gas toward the colder end. When steady state is reestablished, the flow ceases and the pressure is again uniform. According to the perfect gas law, $p = nkT$, constant pressure means the nT product is also constant. Consequently, the concentration of the gas at the colder end will be higher than at the hotter end of the pipe.

Conduction electrons also behave as a gas. If there is a temperature gradient, their concentration in a conducting bar will be higher in the cold side, which, as a consequence, becomes negatively charged with respect to the hotter. The resulting electric potential is the **Seebeck voltage**, a quantity that depends both on the temperature difference and the nature of the of the conductor. If the carriers in the bar are holes, then the colder side will become positive with respect to the hotter. Thus, the polarity of the Seebeck voltage depends on the material of the conductor.[†]

An external connection is needed to tap into the voltage developed in the conducting bar. However, the external wire is submitted to the same temperature differential as the bar itself and will develop its own Seebeck voltage. If the external wire is made of the same material as the bar, the two voltages exactly cancel one another. If, however, the connection is made with a different material, then a net voltage may become available.

A thermocouple must always consist of two *dissimilar* materials, most often of opposing polarity so that the individual Seebeck voltages add up. Since the external connections wires are attached to the open ends of the thermocouple which, presumably, are at the same temperature, these wires contribute no additional thermoelectric voltage.

Thermocouples are low impedance devices (low voltage and large current) and, for many applications, must be connected in series forming a thermoelectric **battery**. See Figures 5.18 and 5.19.

In a thermocouple battery individual cells are electrically in series and thermally in parallel. Thus, if Λ is the heat conductance of one single cell, then $n\Lambda$ is the heat conductance of a battery of n cells. By the same token, the resistance of the battery is nR when R is the resistance of one cell. The voltage generated per unit temperature difference (equivalent to α in a single cell) is $n\alpha$. Consequently, the figure of merit of a battery of n identical cells is

$$Z = \frac{(n\alpha)^2}{n\Lambda \times nR} = \frac{\alpha^2}{\Lambda R} \tag{109}$$

the same as the figure of merit of each cell.

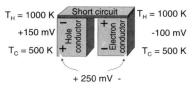

Figure 5.18   To obtain a useful output, two arms must be paired, forming a thermocouple.

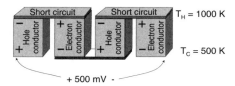

Figure 5.19   Thermocouples can be connected in series forming a battery.

[†] By convention, $\alpha < 0$ when the cold end is negative.

The above explanation of the Seebeck effect is oversimplified. It serves as a first order model to facilitate the understanding of the gross behavior of thermoelectric devices.

The migration of carriers to the cold side of a conductor creates an electric field that forces the electrons to drift back toward the hot side. A dynamic equilibrium is established when just as many carriers move under the influence of the pressure gradient as do (in the opposite direction) under the influence of the electric field.

At equilibrium, there is no *net* charge transfer from one end of the conductor to the other—the flux, nv, is the same in both directions. However, carriers moving down the temperature gradient, being more energetic, carry more heat than those moving in the opposite direction. Thus, even in the absence of a net particle flow, there is a net heat flow in the material. This explains the metallic heat conductivity.

Our simple model predicts that all electronic conductors have negative thermoelectric power—that is, a negative Seebeck coefficient. By the same token, p-type semiconductors (in which holes are the carriers) must have a positive Seebeck coefficient. The model is inadequate because, although most metal have negative αs, some, such as copper, do not.

To improve the theory, we have to consider the scattering of electrons as they move through the conductor. If the scattering cross section is temperature independent, the conclusions above hold because both fluxes (up and down the temperature gradient) are equally perturbed. However, if some mechanism causes hot electrons to be more severely scattered than cooler ones, then the flux of hot electrons is diminished and the negative thermoelectric power is reduced or even reversed. On the other hand, if the hot electrons are less scattered than the cold ones, then the negative thermoelectric power is enhanced.

Some materials exhibit a large Seebeck effect at low temperatures, as illustrated in Figure 5.10. This is due to phonon-electron interaction. When there is a temperature gradient in a material with sufficiently rigid crystalline lattice, heat is conducted as lattice waves as discussed in Section 5.4. Such waves can be interpreted as a flux of quasi-particles called **phonons**. Here, again, we use the duality of waves and particles. Phonons can interact with other phonons and also with electrons. At higher temperatures, phonon-phonon interaction is dominant, but at lower temperatures, phonon-electron interaction may become important. When this happens, the phonon flux (from the hot to the cold side) simply sweeps electrons along with it causing a large charge accumulation at the cold end enhancing the negative thermoelectric power. If the material is a p-type semiconductor, then it is holes that are swept along enhancing the positive thermoelectric power.

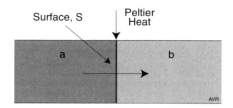

Figure 5.20   Heat convected by an electric current.

5.14.2 The Peltier Effect

Two different conductors, a and b, of identical cross section, A, are connected end-to-end as suggested in Figure 5.20. The surface, S, is the interface between them. Both conductors are at the same temperature, T, and a current, I, flows through them. Let the heat capacity of a typical conduction electron be c_a in conductor a and c_b in conductor b. The thermal energy associated with each (typical) electron is, respectively, $c_a T$ and $c_b T$. For instance, if the electron gas obeys the Maxwellian distribution, then $c_a T = c_b T = \frac{3}{2} kT$.

The current in either side of the interface is

$$I = qnvA. \tag{110}$$

In conductor a, the current convects thermal energy toward the interface at a rate

$$P_a = nvAc_a T \tag{111}$$

and in conductor b, it convects thermal energy away from the interface at a rate

$$P_b = nvAc_b T. \tag{112}$$

If $P_a > P_b$, then at the interface, energy must be rejected from the conductors into the environment at the rate

$$P = P_a - P_b = nvAT(c_a - c_v) = \frac{T}{q}(c_a - c_b)I \equiv \pi I. \tag{113}$$

For a Maxwellian electron gas, this model predicts a Peltier coefficient, $\pi = 0$, because, for such a gas, $c_a = c_b$. However, since there is a non-zero Peltier effect, one must conclude that in thermocouple materials, the electron gas does not behave in a Maxwellian manner. This agrees with the accepted non-Maxwellian model for electrons in metals. However, in lightly doped semiconductors, the conduction electrons are Maxwellian. Additional sophistication of our model is required to explain the Peltier effect in such materials.

5.14.3 The Thomson Effect

Consider again the unidimensional gas discussed in our derivation of thermal conductivity in Section 5.4. We want to derive a formula for the convective transport of heat. We will disregard heat conduction—its effect can simply be superposed on the results obtained here.

We assume that there is a net flux, nv, of molecules and that the temperature is not uniform along the gas column. Take three neighboring points: 1, 2, and 3. Each molecule that moves from 1 to 2 carries an energy cT_1. Here, c is the mean heat capacity of the molecule (i.e., c is $1/N$ of the heat capacity of N molecules). For each molecule that arrives at 2 coming from 1, another leaves 2 toward 3 carrying cT_2 units of energy. Thus, the increase in energy at 2 owing to the flow of gas must be $c(T_1 - T_2)nv$ joules per second per unit area.

If 1 and 2 are separated by an infinitesimal distance, then $T_1 - T_2 = -dT$ and the energy is transported at a rate

$$dP^* = -cnvdT \ \text{W/m}^2 \tag{114}$$

where P^* is the power *density*. If instead of a gas column, we have a free-electron conductor, then heat is convected by electrons and since $J = qnv$,

$$dP^* = -\frac{J}{q}c \ dT \ \text{W/m}^2 \tag{115}$$

or

$$dP = -\frac{I}{q}c \ dT \ \text{W}. \tag{116}$$

The above expression can be rewritten as

$$dP = \tau I dT \tag{117}$$

where τ is the **Thomson coefficient** and has the dimensions of V/K. Clearly,

$$\tau = -\frac{c}{q}. \tag{118}$$

For conductors in which the carrier distribution is Maxwellian, $c = \frac{3}{2}k$, and the Thomson coefficient is

$$\tau = -\frac{3}{2}\frac{k}{q} = -129 \ \mu V/K. \tag{119}$$

Many semiconductors do have a Thomson coefficient of about -100 μV/K. However, a more accurate prediction of the coefficient requires the inclusion of holes in the analysis.

5.38

Electrons in metals do not obey Maxwellian statistics. As discussed in Chapter 2, they follow the Fermi-Dirac statistics: only a few electrons at the high energy end of the distribution can absorb heat. Those that do, absorb energy of the order of kT units but they represent only a fraction of about kT/W_F of the total population. Hence, roughly, the mean heat capacity of the electrons is

$$c = \frac{\partial}{\partial T} \left(\frac{kT}{W_F} kT \right) = \frac{2k^2 T}{W_F}. \tag{120}$$

The ratio of the Fermi-Dirac heat capacity to the Maxwellian is about kT/W_F.

At room temperature, kT is some 25 meV, while a representative value for W_F is 2.5 eV. Therefore, the quantum heat capacity is approximately 100 times smaller than the classical one. For this reason, the Thomson coefficient of metals is small compared with that of semiconductors.

5.14.4 Kelvin's Relations[†]

When we derived the formulas for the performance of thermocouples, we stated that the Peltier coefficient, π, was equal to αT. We will now prove this assertion. Additionally, in developing the thermocouple formulas, we failed to introduce the Thomson effect discussed in the preceding subsection. Here, we will justify this omission. Heat flows from the hot source to the couple at a rate

$$P_H = \Lambda(T_H - T_C) + \pi_H I + I \int_{T_H}^{T_C} \tau_A dT - \tfrac{1}{2} I^2 R. \tag{121}$$

Notice that in this case we have included the Thomson heat convection. The Thomson coefficients of arms A and B are, respectively, τ_A and τ_B.

We will consider a hypothetical thermocouple that has neither electric resistance nor heat conductance. It is a reversible device with no losses. Since both Λ and R are zero,

$$P_H = \pi_H I + I \int_{T_H}^{T_C} \tau_A dT. \tag{122}$$

In the same manner, heat flows from the sink to the couple at a rate

$$P_C = -\pi_C I + I \int_{T_C}^{T_H} \tau_B dT. \tag{123}$$

[†] William Thomson was knighted Lord Kelvin in 1866, mainly in recognition of his work in trans-Atlantic telegraphy.

As there are no losses, the power, $V_L I$, delivered to the load is equal to the total heat power input, $P_H + P_C$. Moreover, because of the absence of resistance in the thermocouple, the load voltage, V_L, is equal to the open-circuit voltage, V_{oc}.

$$V_{oc}I = \pi_H I - \pi_C I + I \int_{T_C}^{T_H} (\tau_B - \tau_A)dT, \qquad (124)$$

P_H P_E P_C

$$V_{oc} = \pi_H - \pi_C + \int_{T_C}^{T_H} (\tau_B - \tau_A)dT. \qquad (125)$$

We want to find the Seebeck coefficient, α, at a given temperature. Let us hold T_C constant and see how V_{oc} varies with T_H:

$$\frac{\partial V_{oc}}{\partial T_H} = \frac{\partial \pi_H}{\partial T_H} + \frac{\partial}{\partial T_H} \int_{T_C}^{T_H} (\tau_B - \tau_A)dT$$

$$= \frac{\partial \pi_H}{\partial T_H} + \tau_B(T_H) - \tau_A(T_H). \qquad (126)$$

Since the above equation is valid for any T_H, we can replace the latter by the general temperature, T:

$$\alpha \equiv \frac{\partial V_{oc}}{\partial T} = \frac{\partial \pi}{\partial T} + \tau_B - \tau_A. \qquad (127)$$

The entropy entering the thermocouple is

$$S_{in} = \frac{P_H}{T_H}, \qquad (128)$$

and that leaving the thermocouple is

$$S_{out} = -\frac{P_C}{T_C}, \qquad (129)$$

hence, the entropy change in the device is

$$\Delta S = \frac{P_H}{T_H} + \frac{P_C}{T_C} \leq 0. \qquad (130)$$

The inequality is the result of the second law of thermodynamics. However, the thermocouple we are considering is a lossless one (isentropic), thus

$$\Delta S = \frac{P_H}{T_H} + \frac{P_C}{T_C} = 0, \qquad (131)$$

and $\frac{d}{dT}\Delta S = 0$, because Equation 131 holds for any value of T.

$$\frac{d\Delta S}{dT_H} = \frac{d}{dT_H}\left[\frac{\pi_H I}{T_H} - \frac{\pi_C I}{T_C} + I\int_{T_C}^{T_H}\frac{\tau_B}{T}dT - I\int_{T_C}^{T_H}\frac{\tau_A}{T}dT\right]$$

$$= I\left[\frac{1}{T_H}\frac{\partial\pi_H}{\partial T_H} - \frac{\pi_H}{T_H^2} + \frac{\tau_B(T_H) - \tau_A(T_H)}{T_H}\right] = 0. \tag{132}$$

We replaced the differential of the integral with respect to its upper limit by the value of the argument of the integral at this upper limit. See box below.

A Bit of Math

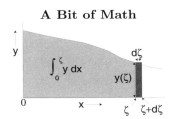

For those who have forgotten a little bit of their math, here is a simple derivation of what happens when one differentiates an integral with respect to one or both limits. Consider,

$$Int = \int_0^\zeta y\ dx, \tag{133}$$

where y is any well-behaved function of x. Refer to the figure. The integral corresponds to the light gray area and the upper limit is $x = \zeta$. We ask what happens when the limit is changed by a infinitesimal amount, $d\zeta$. Clearly, the area—the integral—increases by $y(\zeta)\ d\zeta$ which is represented by the small dark rectangle in the figure. $y(\zeta)$ is the value the function, y, assumes when $x = \zeta$ (the value of y at the upper limit). The change in the integral is

$$dInt = y(\zeta)\ d\zeta, \tag{134}$$

and the rate of change is

$$\frac{dInt}{d\zeta} = y(\zeta). \tag{135}$$

The value of the differential of an integral with respect to its upper limit is simple the value of the function at this upper limit.

Eliminating I from Equation 132, simplifying, and recognizing again that the result holds for any value of T_H,

$$\frac{\partial \pi}{\partial T} + \tau_B - \tau_A = \alpha = \frac{\pi}{T}. \tag{136}$$

The relationship between α and π used in the beginning of this chapter is thus proven correct.

The dependence of α on T is of interest:

$$\frac{\partial \alpha}{\partial T} = \frac{1}{T}\left(\frac{\partial \pi}{\partial T} - \frac{\pi}{T}\right) = \frac{1}{T}\left(\frac{\partial \pi}{\partial T} - \frac{\partial \pi}{\partial T} + \tau_A - \tau_B\right), \tag{137}$$

$$\frac{\partial \alpha}{\partial T} = \frac{\tau_A - \tau_B}{T}. \tag{138}$$

The above expression shows that when the two arms of a thermocouple have the same Thomson coefficients, the Seebeck coefficient is independent of temperature. This does not occur often.

The contribution of the Thomson effect to the thermocouple voltage is

$$V_\tau = \int_{T_C}^{T_H} (\tau_B - \tau_A) dT. \tag{139}$$

Using the relationship of Equation 138 (and ignoring the sign),

$$V_\tau = \int_{T_C}^{T_H} T\frac{d\alpha}{dT} dT = \int_{T_C}^{T_H} T d\alpha. \tag{140}$$

Integrating by parts,

$$V_\tau = (\alpha T)\Big|_{T_C}^{T_H} - \int_{T_C}^{T_H} \alpha dT. \tag{141}$$

The mean value of α in the temperature interval, T_C to T_H, is

$$<\alpha> = \frac{\int_{T_C}^{T_H} \alpha dT}{T_H - T_C}. \tag{142}$$

Hence,

$$V_\tau = \alpha_H T_H - \alpha_C T_C - <\alpha> (T_H - T_C), \tag{143}$$

$$V_\tau = \alpha(T_H - T_C) - <\alpha> (T_H - T_C). \tag{144}$$

If we use an average value for α—that is, if $\alpha_H = \alpha_C = <\alpha>$, then $V_\tau = 0$. Thus, it is possible to disregard the Thomson heat transport simply by using the **mean** Seebeck coefficient.

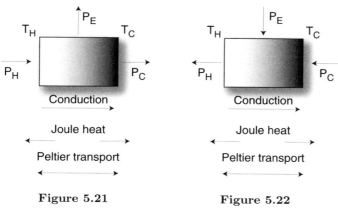

Figure 5.21   **Figure 5.22**
Directions of power flow in a thermocouple.

5.15   Directions and Signs

In the various examples discussed in this text, there has been no consistent definition of the direction in which the different powers flow into and out of a thermocouple. The directions, and the corresponding signs in the equations for P_H, P_C, and P_E, have been chosen in each case so as to best suit the problem being discussed.

While the choice of the directions for the flow of different powers is entirely arbitrary it must be consistent throughout a problem. When the thermocouple is used as a generator, the most intuitive direction is the conventional one shown in Figure 5.21 in which heat from a source flows into the couple and part of it is rejected to a sink, while electric power flows out of the device. This will cause all the powers to be larger than zero. For refrigerators and heat pumps, it is simpler to use the directions indicated in Figure 5.22 in which some heat is pumped away from a cold source by the thermocouple and is then rejected to a hotter sink. Naturally, to make heat flow from a colder to a hotter region, a certain amount of electric power is required to flow into the device.

Let us examine the equations for P_H and P_C. They consist of three terms: a conduction (Fourier) term, a Joule heating term, and a term that represents the Peltier transport of heat.

Clearly, the conduction term is positive when the flow is from T_H to T_C. In Figure 5.21, this flow is in the same direction as P_H and P_C and is therefore a positive term in the equations. In Figure 5.22, the conduction term opposes both P_H and P_C and is a negative term in the corresponding equations.

Half of the Joule heat flows out of the hot end and half out of the cold end. It is a negative term for P_H and a positive term for P_C in Figure 5.21 and it is positive for P_H and negative for P_C in Figure 5.22.

The Peltier heat can flow in either direction depending on the sign of the current applied to the thermocouple. Thus, in one of the equations (say in the equation for P_H) the sign of the Peltier term is arbitrary in both Figures 5.21 and 5.22. Once the sign is chosen in the P_H equation, then the same sign must be used in the P_C equation provided P_H and P_C are in the same direction. The reason for this becomes obvious if we look at the equation for $P_E = P_H - P_C$ in both figures.

In Figure 5.21,

$$P_H = \Lambda(T_H - T_C) + \alpha T_H I - \tfrac{1}{2}RI^2, \tag{145}$$

$$P_C = \Lambda(T_H - T_C) + \alpha T_C I + \tfrac{1}{2}RI^2, \tag{146}$$

$$P_E = \alpha(T_H - T_C)I - RI^2. \tag{147}$$

In Figure 5.22,

$$P_H = -\Lambda(T_H - T_C) - \alpha T_H I + \tfrac{1}{2}RI^2, \tag{148}$$

$$P_C = -\Lambda(T_H - T_C) - \alpha T_C I - \tfrac{1}{2}RI^2, \tag{149}$$

$$P_E = -\alpha(T_H - T_C)I + RI^2. \tag{150}$$

Notice that P_E of Figure 5.21 is the negative of that in Figure 5.22 (the arrows are inverted). Notice also that, to make the currents in the two figures consistent with one another, those in the latter figure are negative (they are in the direction of P_E, as are the currents in the former figure).

The thermocouple can be modeled as a voltage generator with an open-circuit voltage $\alpha(T_H - T_C)$ resulting from the Seebeck effect and an internal resistance, R. It can be seen that either equation for P_E represents such a model.

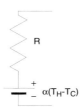

Figure 5.23   Electric model of a thermocouple.

Appendix

The figure of merit of a thermocouple is $Z = \alpha^2/\Lambda R$ and is maximized when the ΛR product is minimum. What is the corresponding thermocouple geometry?

$$\Lambda = \frac{A_A \lambda_A}{\ell_A} + \frac{A_B \lambda_B}{\ell_B},$$

$$R = \frac{\ell_A}{A_A \sigma_A} + \frac{\ell_B}{A_B \sigma_B},$$

$$\Lambda R = \frac{A_A \lambda_A}{\ell_A}\frac{\ell_A}{A_A \sigma_A} + \frac{A_B \lambda_B}{\ell_B}\frac{\ell_A}{A_A \sigma_A} + \frac{A_A \lambda_A}{\ell_A}\frac{\ell_B}{A_B \sigma_B} + \frac{A_B \lambda_B}{\ell_B}\frac{\ell_B}{A_B \sigma_B},$$

hence

$$\Lambda R = \frac{\lambda_A}{\sigma_A} + \frac{\lambda_B}{\sigma_B} + \frac{\ell_A A_B}{\ell_B A_A}\frac{\lambda_B}{\sigma_A} + \frac{\ell_B A_A}{\ell_A A_B}\frac{\lambda_A}{\sigma_B}$$

For a given choice of materials, $\frac{\lambda_A}{\sigma_A} + \frac{\lambda_B}{\sigma_B}$ is fixed and a minimum ΛR occurs when $\frac{\ell_A A_B}{\ell_B A_A}\frac{\lambda_B}{\sigma_A} + \frac{\ell_B A_A}{\ell_A A_B}\frac{\lambda_A}{\sigma_B}$ is minimum. Let

$$x \equiv \frac{\ell_A A_B}{\ell_B A_A}, \qquad a_1 \equiv \frac{\lambda_A}{\sigma_B}, \qquad a_2 \equiv \frac{\lambda_B}{\sigma_A}$$

then, the minimum occurs when $\frac{d}{dx}\left(a_2 x + \frac{a_1}{x}\right) = 0$ or $a_2 = \frac{a_1}{x^2}$ which leads to

$$\frac{\ell_A A_B}{\ell_B A_A} = \sqrt{\frac{\lambda_A \sigma_A}{\lambda_B \sigma_B}} \tag{25}$$

Putting Equation 25 into the expression for ΛR,

$$\Lambda R_{min} = \left(\sqrt{\frac{\lambda_A}{\sigma_A}} + \sqrt{\frac{\lambda_B}{\sigma_B}}\right)^2. \tag{26}$$

References

Bell, Lon E., Use of thermal insolation to improve thermoelectric system operating efficiency. International Thermoelectric Society conference (ITC2002), Long Beach, CA, Aug. 26–29, **2002**.

Burns, G. W., and M. G. Scroger, The calibration of thermocouples and thermocouple materials, National Institute of Standards and Technology (NIST), Special Publication 250-35, Apr. **1969**.

Goldsmid, H. J., *Electronic refrigeration*, Pion Limited, London, **1986**.

Venkatasubramanian, R., Edward Siivola, Thomas Colpitts, and Brooks O'Quinn, "Thin-film thermoelectric devices with high room-temperature figures of merit," *Nature. 413* 11, Oct. **2001**, pp. 597–602.

PROBLEMS

5.1   The Russians are quite advanced in thermoelectrics and have recently developed a secret material for such use. Data are, naturally, hard do obtain, but the CIA discovered that a thermoelectric cooler is being built capable of depressing the temperature by 100 K when the hot junction is at 300 K. What is the figure of merit of the thermocouple?

5.2   The Seebeck coefficient of a junction is

$$\alpha = 100 + T - 10^{-3}T^2 \ \mu\text{V/K}.$$

1. At what temperature is the Peltier coefficient maximum?
2. What is the coefficient at this temperature?
3. At what temperature are the Thomson coefficients of the two arms equal?

5.3   A thermoelectric cell has figure of merit of 0.002 K^{-1} and an internal resistance of 100 $\mu\Omega$. The average Seebeck coefficient is 200 μV/K. Heat flow meters are used to measure the flow of heat, P_H, from the hot source to the cell and the flow, P_C, from the cell to the cold sink. There is no other heat exchange between the cell and the environment.
The source is at 600 K and the sink at 300 K.

1. With no electric current flowing in the cell, what is the value of P_H and P_C?
2. A current, I, is now made to circulate. This modifies P_H. What currents cause P_H to go to zero?
3. What are the corresponding values of P_C?
4. Over what range of currents does the cell act as a power generator?

5.4   A block of metal, maintained at 850 K, is mounted on a pedestal above a platform at 350 K.
The experiment being conducted requires that there be absolutely no heat transfer from the metal block through the pedestal. To achieve this, the pedestal was made into a thermocouple and the appropriate current was driven through it. The thermocouple has a heat conductance of 1 W/K and a resistance of 1 milliohm. Its figure of merit is 0.001 K^{-1}. All these data are average values over the temperature range of interest.
What current(s) must be driven through the thermocouple? What is the voltage across its terminals? What is electric power required?

5.5   We want to pump 100 W of heat from 210 K to 300 K. Two stages of thermocouples must be used. The *second* stage pumps 100 W from 210 K to T_{H_2}. The first stage pumps the necessary energy from $T_{C_1} = T_{H_2}$ to 300 K.

5.47

All thermocouples are made of the same materials whose combined characteristics are:

$$\alpha = 0.001 \text{ V K}^{-1},$$
$$\Lambda R = 0.0005 \text{ K V}^2.$$

The geometry of each thermocouple is optimized. The current through the thermoelectric pair is adjusted for maximum cooling. The current through the first stage is not necessarily the same as that through the second.

Within each stage, all elements are connected electrically in series.

The total electric power input to the system depends critically on the choice of T_{H_2}. What value of T_{H_2} minimizes the electric power consumption, P_E? What is the value of P_E? What is the voltage applied to each stage?

5.6   A battery of thermocouples delivers 5 kW at 24.0 V to a load. The thermocouples were designed to operate at maximum efficiency under the above conditions.

The hot side of the battery is maintained at 1100 K and the cold side, at 400 K.

The figure of merit of each individual thermocouple cell is 0.0015 K^{-1}.

What is the efficiency of the system?

5.7   What is the heat conductance of a metal bar at uniform temperature ($T = 400$ K) if the bar has a resistance of 4 Ω?

5.8   To drive a current of 100 A through a thermocouple at uniform temperature (300 K), a power of 50 W is required.

The same thermocouple, when open-circuited and under a temperature differential from of 800 K to 300 K, has 0.5 V between the open terminals, on the cold side.

Assume that the resistance is independent of T.

How many watts are necessary to drive a current of 100 A through the thermocouple when the differential above is maintained? Is there a unique answer to this question?

5.9   A thermocouple works between 500 K and 300 K. Its resistance is 0.0005 Ω and its heat conductance is 0.2 W/K. The mean Seebeck coefficient (between 500 and 300 K) is 0.001 V/K.

What is the open circuit voltage generated by the thermocouple?

When there is no current, heat flows, of course, from the hot side to the cold side. Is it possible to make the heat that flows from the hot source to the thermocouple equal to zero? If so, what is the heat flow from the cold sink to the thermocouple? What is the electric power involved?

5.48

What is the voltage across the couple? Does the electric power flow into the thermocouple or out of it (i.e., does the couple act as a generator or a load)?

5.10   A thermocouple operates between 900 K and 300 K. When short-circuited, it delivers 212 A, and when open-circuited, 0.237 V. The dimensions of the arm have been optimized for the mean temperature of 600 K. The material in the arms are semiconductors that have negligible lattice heat conductivity.

1. Calculate the heat taken from the hot source when open-circuited?
2. What is the power delivered to a 500 $\mu\Omega$ load?
3. What is the efficiency of the device in Question 2?

5.11   A small thermoelectric generator (single pair) is equipped with two thermal sensors, one measuring T_H (in the hotter side) and one measuring T_C (in the colder side). An electric heater warms up the hotter side and all the heat thus generated is delivered to the thermocouple. A feedback system assures that T_H is kept at exactly 1000 K. The amount of electric power delivered to the heater can be measured.

On the colder side, a corresponding feedback system assures that T_C is kept at a constant 500 K.

The current, I, forced through the thermocouple can be adjusted to a desired value.

When $I = 0$, it takes 10 W of electric power to operate the heater at the hotter side of the thermocouple. Under such conditions, the thermocouple develops 0.50 V at its open-circuited terminals.

Now, a given current I is forced through the thermocouple so that the electric heater can be disconnected, while T_H still remains at 1000 K.

Next, the current I, above, is reversed and it is observed that it takes 18.3 W to maintain $T_H = 1000$ K.

1. What is the value of I?
2. What voltages are necessary to drive the currents I and $-I$.

5.12   A thermocouple is connected to an adjustable current source. Disregard the resistance of the wires connecting the current source to the thermocouple. These wires and the cold side of the thermocouple are at 300 K. Under all circumstances the current delivered has an absolute value of 100 A. Assume that the resistance of the thermocouple is temperature independent.

1. When $T_H = T_C = 300$ K, the absolute value of the voltage across the current source is 0.20 V. What is the voltage if the direction of the current is inverted?

2. When $T_H = 600$ K and $T_C = 300$ K, the absolute value of the voltage is 0.59 V. What are the voltages if the direction of the current is inverted?

5.13   Assume that the materials in the arms of a thermocouple obey strictly the Wiedemann-Franz-Lorenz law. The Seebeck coefficient, α, of the thermocouple is 150 μV/K independently of temperature and the electric conductivities, σ (the same for the two arms), are also temperature independent. Such unrealistic behavior has been specified to make the problem more tractable—both α and σ usually do vary with temperature.

The device is to operate between T_H and T_C. For the calculation of any temperature dependent quantity, use the arithmetic mean, T_m, of these two temperatures.

The geometry of the thermocouple has been optimized and the load connected to it always has the value that maximizes the efficiency of the system.

Show that, under the above circumstances, the electric power delivered to the load is independent of the choice of T_H provided that $\Delta T \equiv T_H - T_C$ is always the same.

At 800 K, what is the value of the ΛR product for this thermocouple?

5.14   A thermoelectric device, consisting of 100 thermocouples electrically in series and thermally in parallel, is being tested as a heat pump. One side is placed in contact with a cold surface so that it cools down to -3 C; the other side is maintained at 27 C.

The open-circuit voltage is measured by means of a high-impedance voltmeter and is found to be 900 mV.

Next, the electric output is shorted out and it is observed that a current of 9 A flows through the short.

The device is now removed from the cold surface and its cold end is insulated thermally so that absolutely no heat can flow in. A current of 50 A generated by an external source is forced through the device in such a direction that heat is pumped from the cold end to the warm end. A thermometer monitors the final temperature of the cold side. After steady state is reached, the temperature is 260 K. The hot side is still at 27 C.

Is this the lowest temperature that can be achieved? If not, what is the lowest temperature and what is the necessary current?

5.15   A thermoelectric device is being tested in a laboratory. It consists of a single thermocouple and has its hot side in intimate thermal contact with an electric heater whose total heat output is transferred to the thermocouple. In other words, the power delivered to the heater is equal to the heat power, P_H, that flows into the thermocouple. Under all circumstances, the hot side is maintained at 1000 K and the cold side, at 300 K.

5.50

The two temperatures, T_H and T_C, are monitored by thermometers.

The first step in the test reveals that when the device is open-circuited, it draws 14 watts of heat from the hot source and delivers a voltage of 0.28 V. When short-circuited, it delivers a current of 35 A.

1. What is the heat power input when the device is short circuited?
2. What is the heat power input when 0.4 V are applied in opposition to the open-circuit voltage?

5.16   A Radio-isotope Thermal Generator is used as an electric power source aboard a spacecraft. It delivers 500 W at 30 V to a load optimally matched to the thermoelectric generator. Under such circumstances, this generator operates at a 12.6% efficiency.

The hot side of the generator (T_H) is at 1300 K and the cold side (T_C) is at 400 K.

Assume that all the characteristics of the thermocouple $(\alpha, R, \text{and } \Lambda)$ are temperature independent.

1. What would be the efficiency of the generator if the load resistance were altered so that the power delivered fell to 250 W?

Assume now that the heat power source is a radionuclide which delivers a constant heat power independently of the demands of the load. Thus, the temperature becomes a function of the load power. If the radioactive material were inside an adiabatic container, the temperature would rise until the container was destroyed—a steady heat leak must be provided to limit the temperature.

The radionuclides in this problem release heat at a rate of 4984 W. The container, by itself, radiates enough energy to keep its outer skin at constant $T_L = 1000$ K under all circumstances, including when the thermoelectric generator is not installed, and, consequently, only the leakage path removes heat from the source. The cold side of the thermoelectric generator is always at $T_C = 400$ K.

When the thermoelectric generator is attached to the heat source and generates 500 W to a matched electric load, the temperature, T_S, falls to 1300 K (as in Question 1, this is T_H of the thermoelectric generator). In other words, the heat source delivers heat to two parallel paths: the leakage path and the thermoelectric generator.

2. Calculate the temperature of the heat source when no electric energy is being drawn from the thermoelectric generator.
3. When 250 W dc are drawn from the generator, what is the source temperature? There is more than one answer. Use the current with the smallest absolute value. Set up your equations and use a trial and error method.

5.17   Demonstrate that the voltage required to drive a thermoelectric heat pump is independent of the amount of power pumped and of the cold temperature, provided the current has been adjusted for maximum pumping.

5.18   Tungsten has an electric resistivity that (between 1000 and 3600 K) is given with acceptable precision by

$$\rho = -1.23 \times 10^{-7} + 3.49 \times 10^{-10}T,$$

where ρ is in Ω m.

Give me your best estimate for the thermal conductivity of tungsten at 1100 K and 1600 K.

5.19   A perfectly heat-insulated box is equipped with an electric heater, which allows the introduction of heat at an accurately measured rate. The only way heat can be removed is through a Peltier heat pump whose hot side is maintained at a constant 300 K.

The current, I, through the heat pump is controlled by a computer that senses the temperature inside the box and is set to keep it, if possible, at 280 K.

The experimenter chooses the amount of heat dissipated by the electric heater and tabulates the current the computer delivers to the heat pump. Here are the results:

Heat input (W)	Current (A)
0.50	7.382
1.00	13.964
1.25	21.225
1.00	32.702

For each case in the table, what is coefficient of performance (COP) of the heat pump?

5.20   At 300 K, the electric resistivity of a sample is 0.002 Ωcm and its heat conductivity is 0.03 W K^{-1} cm^{-1}. From these data, determine if the material is a metal or a semiconductor. Explain.

5.21   A thermoelectric battery consists of 1000 identical thermocouples electrically in series and thermally in parallel. It works between 1000 K and 500 K.

Each thermocouple has the following characteristics:

Heat conductance: 3 W/K,
Electric resistance: 200 $\mu\Omega$,
Seebeck coefficient: 0.0007 V/K,

How much power does this battery deliver to a 0.3 Ω load?

5.22   A thermoelectric generator consists of a number of series connected thermocouples. It operates between 1000 and 400 K.

When a 0.1 Ω load is used the current is 266.7 A, and the heat rejected to the cold sink is 48 kW. When open circuited, the voltage is 48 V.

What is the figure of merit (Z) of the device?

5.23   A prismatic block of pure sodium measures 1 by 1 by 10 cm. The two smaller faces (1 by 1 cm) are kept at a uniform temperature, one at 370 K and the other at 300 K.

Sodium has resistivity that can be represented (with fair accuracy) by

$$\sigma = 3.9 \times 10^7 - 6.6 \times 10^4 T \ \ S/m,$$

where T is in kelvins.

What is the heat power conducted by the sodium block?

Solve this problem twice:

First make some (drastic) simplifying assumption (use an average heat conductivity) to obtain an estimate of the heat power conducted.

Next, solve it by more realistically taking into account the fact that the heat conductivity varies along the length of the block.

5.24   A thermoelectric device, consisting of n series connected thermocouples, is tested as a heat pump. A certain current, I, is chosen and then the lowest attainable temperature, T_{Cmin}, is determined. This is done with a completely heat insulated cold end.

Next another value of I is used and another, different, T_{Cmin} is found.

In this manner, a tabulation of T_{Cmin} versus I is made. Throughout the whole experiment, T_H is maintained a constant 320 K. The combination that resulted in the lowest T_{Cmin} is

$I = 32.08$ A for a $T_{Cmin} = 244$ K.

The voltage necessary to drive this current through the thermoelectric device is 16.0 V.

What are the resistance, the heat conductance, and the α of the thermocouple battery?

5.25   The hot side of a thermocouple has, as a heat sink, a volume of water kept a 373 K under a pressure of 1 atmosphere. The water can only lose heat by evaporation (no conduction, except, of course, through the thermocouple itself). The heat of vaporization of water is 40 MJ/kmole. There is enough water so during the experiment the reservoir never runs dry. There is enough heat output from the hot side of the thermocouple to maintain the water at, at least, 373 K.

Assume that, under all temperatures considered in this problem, the device has

$$\Lambda R = 300 \times 10^{-6} \text{ V}^2/\text{K}.$$
$$\alpha = 0.002 \text{ V/K}.$$

Assume also that Λ does not change over the temperature range of interest.

Observe that this material has an extraordinarily high α. It does not exist at the moment. It will be invented in the year 2043. Just to emphasize a point, I specified this unrealistic Seebeck coefficient.

The cold side of the thermocouple is also in contact with a volume of liquid (not water), initially at 350 K. Again, the liquid (a total of 10 cubic centimeters) is in an adiabatic container and the heat loss by evaporation is negligible.

There is no temperature drop between the water and the hot side of the thermocouple and also between the cold side of the thermocouple and the cold liquid.

The current, $I = 11.67$ A, forced through the device has been chosen so that, under the initial conditions, it is the one that provides maximum heat pumping.

1. What is the heat power pumped initially from the cold liquid?
2. If the experiment runs long enough ($I = 11.67$ A), how cold will the liquid in the cold side get and how hot will the water on the hot side get?
3. What is the rate of hot water evaporation (in kg/s) when $T_C = 350$? What is the rate when the lowest temperature of the cold liquid is reached?
4. Is the electric power used in the preceding item larger, equal, or smaller than the heat power required to evaporate the water? Do this for the cool liquid temperature of 350 K and also for the lowest achievable temperature.
5. What is the voltage necessary to drive the 11.67 A current? Calculate this for each of the two coolant temperatures.

5.26   Usually the Seebeck coefficient, α, is a nonlinear function of the temperature, T. Under what circumstances is the coefficient independent of temperature?

5.54

5.27   Refer to the graph (next page) describing the performance of a
Peltier cooler, **Hypothetical-127**, which consists of 127 thermocouples
electrically in series and thermally in parallel. These data were generated
under the simplifying assumptions that the Seebeck coefficient, α, the resistance, R, and the thermal conductance, Λ, are all temperature independent.
In real life, the parameters actually vary with temperature.

1. Estimate the open-circuit voltage when $T_H = 65$ C and $T_C = -15$ C.
2. What is the figure of merit, Z, of the device when operating under the
   above conditions?
3. The figure of merit above is for the whole unit (127 thermocouples).
   What is the figure of merit for each individual thermocouple. Do not
   guess, show the result rigorously.
4. When the Peltier cooler, above, is pumping 22 W of heat from 298 K
   to 338 K (a ΔT of 40 K), how much heat is rejected at the hot side?
5. For the Peltier cooler, above, pumping heat from 298 K to 338 K, what
   would be the optimum current?
6. Assuming that α, Λ, and R are temperature independent, how much
   power does such a unit (used as generator) deliver to a load optimized
   for maximum efficiency? $T_H = 500$ K, $T_C = 300$ K.

5.28   A semiconductor can be doped so that it can have either p or n
conductivity. Its properties (all independent of temperature) are:

	p	n	
α	400	-400	μV/K
ρ	0.005	0.005	Ωcm
λ	0.02	0.02	W cm^{-1}K^{-1}
J_{max}	10	10	A cm^{-2}

Ammonia boils at -33.2 C under atmospheric pressure. The flask in
which the ammonia is kept (at 1 atmosphere), although insulated, allows
5 W of heat to leak in. To keep the ammonia from boiling away, we must
pump 5 W of heat from the flask. We propose to do this with thermocouples
using the materials described above. One single stage will be insufficient to
depress the temperature to the desired point—use two stages. The T_H of
the colder stage (Stage 2) is 261.2 K. The T_H of Stage 1 is 40 C. The current
through the two stages must be the same (they are connected in series) and
must be the maximum allowable current through the material. Eachach
stage must use the current that optimizes heat pumping. *The constraint
on the current of the two stages may appear to lead to an incompatibility.
Actually it does not.* Stage 2 consists of 10 thermocouples.

(Continues on Page 195.)

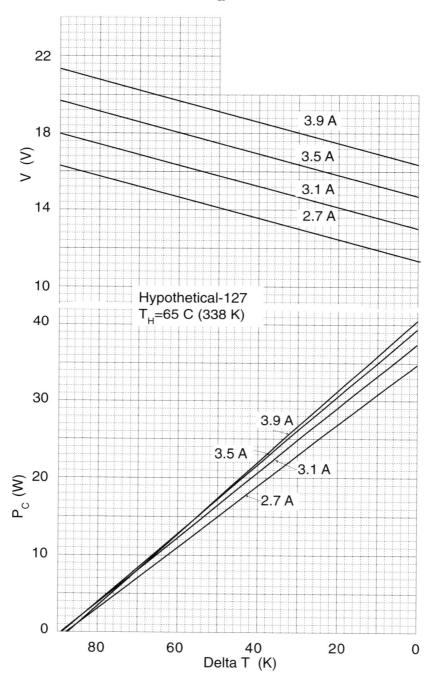

1. What is the electric power consumed? What voltage must be applied to the system assuming that all couples are electrically in series?
2. If power fails, the thermocouple will, unfortunately, provide a good path for heat leakage. Assuming the outside surface of the cooling system is still at 40 C, how many watts of heat leak in (as long as the ammonia is still liquid)? The heat of vaporization of ammonia is 1.38 MJ/kg. How long will it take to evaporate 1 kg of ammonia?

5.29   Three different materials, A, B, and C, were cut into rectangular prisms L cm long and A cm^2 in cross section and were sent to a lab to be tested.

The test consists in accurately measuring the heat power flow, P_H, between the square faces (the ones with 1 cm^2 area), when a temperature difference of exactly 1 K is maintained. The average temperature of each sample is T.

In addition, a current, I, is forced through the sample (it flows from one square face to the other). The voltage, V, developed across the sample is measured. The results are tabulated below.

Sample	V (V)	I (A)	P_H (W)	T (K)	A (cm^2)	L (cm)
A	0.00971	100	0.0257	500	10	1
B	0.00420	100	0.0290	50	2	5
C	22.20	0.001	4.17E-5	1000	8	3

The results for one or more of the samples above *may* be impossible. If so, identify which and explain why.

5.30   A thermoelectric unit consists of n cells. When there is no temperature difference across the leads, the resistance is 4 Ω. Λ is 0.3 W/K. When the unit is submitted to a 80 K temperature differential, it generates a 1.71 V across a 3-ohm load.

To simplify the problem, make the unrealistic assumption that α, R, and Λ are temperature independent.

1. What current pumps 10 watts of heat from 300 K to 350 K?

Define the coefficient of performance (COP) of a thermoelectric heating system as the ratio of the delivered heat power to the electric power required.

2. What is the coefficient of performance of the device of the previous problem when extracting heat from an environment at 300 K and delivering heat at 350 K. Express this as a function of I. Determine the current that optimizes the COP. Do not waste time and risk errors by taking derivatives. Use trial and error numerical solutions. How much heat is pumped under such circumstances?

5.31   A thermoelectric generator works between the temperatures 1000 K and 400 K. The open-circuit voltage of the generator is 70.0 V. The geometry of the individual cells has been optimized.

Measurements show that maximum efficiency is obtained when the generator delivers 5.22 A at 38.68 V.

1. What is the efficiency of the system?
2. What is the efficiency of the system when only 100 W are delivered to the load?
3. Why is the efficiency of the generator so much smaller when delivering 100 W compared to its efficiency at 200 W?

5.32   A single thermocouple is equipped with two heat exchangers through which water is forced to flow. The input water is at 300 K for both heat exchangers. The flow is adjusted so that the water exits the hot heat exchanger at 310 K and the cold heat exchanger at 290 K. For simplicity, these are also the temperatures of the two ends of the thermocouple. The current through the thermocouple is such that exactly 10 W of heat power are removed from the water in the cold heat exchanger. This current is the smallest current capable of pumping the 10 W from the cold side. The characteristics of the thermocouple are:

$\alpha = 0.0006$ V/K.
$\Lambda R = 120 \times 10^{-6}$ V^2/K.

1. What is the flow rate of the water in the cold heat exchanger?
2. What is the current through the thermocouple?
3. What is the flow rate of the water through the hot heat exchanger?
4. What is the coefficient of performance of this refrigerator? The coefficient of performance is defined as $\phi_C = P_C/P_E$.
5. Using this same thermocouple and the same T_H and T_C, it is possible to operate at a higher coefficient of performance. Clearly, the amount of heat pumped from the cold side would no longer be 10 W. What is the best coefficient of performance achievable?

5.58

5.33   Prove that the current that maximizes the coefficient of performance, ϕ_C,

$$I = \frac{\Lambda \Delta T}{\alpha <T>}(m+1).$$

can also be written

$$I = \frac{\alpha \Delta T}{R(m-1)}.$$

5.34   Demonstrate that the Carnot efficiency of a heat pump is $T_C/\Delta T$, where T_C is the temperature of the cold side and ΔT is the temperature difference between the hot and the cold side.

5.35   In the text, there is a description of the RTG used in the Galileo mission to Jupiter in 1981. From the data given, calculate the cost in dollars of the fuel used in this RTG.

5.36

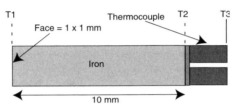

The figure illustrates an experimental set-up used to determine the characteristics of a thermocouple. It consists of a rectangular prism made of pure iron (conductivity = 80 Wm^{-1}K^{-1}) to which the thermocouple to be tested is attached. The iron slug is well insulated laterally so that heat can only enter and leave through the small faces. The temperatures of the two faces as well as that of the open end of the thermocouple can be accurately measured. All properties of the different elements are temperature independent,

Test #	T_1	T_2	T_3	Applied current	Observed voltage
	(K)	(K)	(K)	(A)	(V)
1		300	300	5	0.1
2	440.6	400	300	0	0.05

The table above lists two different tests carried out.

1. From the above, calculate the figure of merit of the thermocouple.
2. What was the temperature, T_1, in Test 1?

5.37   You have 2 liters of water in an adiabatic styrofoam container. You want to cool the water by means of a battery of thermocouples consisting of ten units, each one of which has the following (temperature independent) characteristics:

$\alpha = 0.055$ V/K.
$R = 4.2\ \Omega$.
$\Lambda = 0.25$ W/K.

The temperature of the hot side of the thermocouples is maintained at a constant 300 K.

1. How long does it takes to cool the water from 300 K to 285 K when you drive a 4 A constant current through each thermocouple unit.
2. Can you change the operating point of the thermocouple so as to shorten the cooling time? What is the shortest achievable cooling time?
3. How much electric energy is consumed in cooling the water when the current is that of the preceding question? Compare this with the heat energy extracted from the water. What is the coefficient of performance of the system?
4. Assume now that you adjusted the current to optimize the coefficient of performance. What is the current? What is the coefficient of performance?
5. What would be the coefficient of performance of an ideal (Carnot) heat pump cooling this amount of water?

5.60

Chapter 6
Thermionics

6.1 Introduction

Discovered in 1885 by Edison, thermionic emission achieved enormous industrial importance in the 20th century, but only in 1956 was the first thermionic heat-to-electricity converter demonstrated.

The operation of a vacuum thermionic generator is easy to understand, but the device is impractical owing to the space charge created by the electrons in the inter-electrode space. Introducing a plasma containing positive ions into this space overcame the space charge problem but created additional complications. It became necessary to use plasma, not only to cancel space charge but also as a source of additional electrons. Only this latter solution—the **ignited** plasma device—holds promise of becoming a useful thermionic generator.

In this text, we describe the fundamentals of the emission process, the transport of electrons through a vacuum, the operation of the vacuum diode converter, the creation of a cesium plasma and the behavior of the plasma when present in concentrations low enough to allow disregarding collision ionization, and, finally, a few words on the converter in which the collision ionization is of fundamental importance. The behavior of the plasma in the latter is too intricate to be discussed in detail in this book. Those who need a more complete explanation of the operation of plasma diodes, especially of the only practically important class operating in the ignited mode, should refer to an extensive article by Ned Razor. See References.

Clarifying some of the terminology used: Most electronic engineers think of "anode" as the positive electrode and "cathode" as the negative one. This is misleading. In Greek, anodos and kathodos mean, respectively, "the way up" and "the way down." In electrical engineering, these words mean simply the way in and the way out of a device (using the conventional direction of current flow). Thus, in a diode, the anode is the electrode through which electricity enters the device and the cathode is the exit electrode. It can be seen that here the anode is, indeed, the electrode connected to the positive terminal of the power supply. In a battery, however, electricity (conventionally) exits from the positive electrode, which is consequently called the cathode. To simplify the terminology, we will, when possible, use self-describing terms such as emitter (of electrons) for the cathode of a thermionic device and collector, for the anode.

Thermionic devices, once the very soul of electronics, still constitute a sizable market. Thermionics are the basis of **radio tubes** used in most high power radio and TV transmitters, of **cathode-ray tubes** employed in oscilloscopes, and in TV and computer monitors. In this latter function, they are being progressively displaced by other high resolution devices such as the much more compact LCD and plasma panels. Thermionics also play an essential role in most microwave tubes such as **klystrons**, **magnetrons**, and **traveling-wave tubes**.

The majority of the classical thermionic implements were **vacuum state** devices in which electrons flow through vacuum. Nevertheless, occasionally plasmas were used, as in the once popular mercury vapor rectifiers. Vacuum state devices are undergoing a revival in microelectronics. However they do not use thermionic emission, they take advantage instead of field emission phenomena.[†] This chapter, on the other hand, concerns itself with thermionic heat-to-electricity converters and excludes field emission.

In the simplest terms, a thermionic converter is a heat engine in which electrons are boiled off a hot surface (**emitter**) and are collected by a colder one (**collector**). There is an energy cost in evaporating electrons from a solid, just as there is a somewhat similar cost in evaporating a liquid. The *minimum* energy necessary to remove an electron from a metal or a semiconductor is called the **work function**. Dividing the work function by the electron charge defines a potential (voltage) that has the same numerical value as the work function itself when the latter is expressed in eV. In this text, we will use the symbol, ϕ, to represent the potential, and, consequently, the work function is $q\phi$. Although somewhat confusing, we will use the term "work function" to designate both the energy and the potential, as do most workers in thermionics. The reader will be able to distinguish the two meanings from the context.

The work function depends on the nature of the emitting substance; if the work function of the hot emitter surface is larger than that of the cold collector, then the difference may become available as electric energy in the load. Thus, if the emitter has a work function of 3 eV and the collector has one of 2 eV, a net potential of 1 volt could be available to the load.

Electrons boiled off from the emitter accumulate on the collector unless there is a provision to drain them off. This can be done by establishing an external path from collector to emitter. This path can include a load resistance, R_L, as depicted in Figure 6.1 through which a current, I, circulates. The external voltage drop creates an interelectrode electric field that opposes the electron flow. Notice that the collector or anode is the negative terminal of the thermionic generator. In the figure, the diode is represented by its conventional symbol in electronics.

[†] See the survey of vacuum microelectronics by Iannazzo.

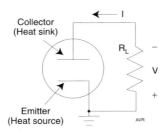

Figure 6.1   A thermionic generator.

The energy dissipated in the load comes from the emitter heat input. The required high temperature leads to substantial heat radiation losses. It is difficult to drive large currents across the vacuum gap in the emitter-to-collector space. To circumvent this difficulty, the vacuum can be replaced by a positive ion gas. Unfortunately, this introduces additional losses owing to collisions between electrons and ions.

Although thermionic converters are heat engines and are limited by the Carnot efficiency, they offer important advantages:

1. they yield electric energy directly, not mechanical energy as do mechanical heat engines;
2. they operate at high temperature, with corresponding high Carnot efficiencies;
3. they reject heat at high temperature, simplifying the design of heat sinks in space applications and making them useful as topping engines;
4. they operate at high power densities, leading to compact devices that use high-priced materials sparingly;
5. they operate at low pressures and have no moving parts.

In order to better understand the operation of a thermionic engine, it is useful to consider separately two of the basic processes involved: The thermal emission of electrons from a solid surface and the transport of such electrons across the interelectrode space.

6.2   Thermionic Emission

Qualitatively, it is easy to understand thermionic emission. In metals or semiconductors, at any temperature above absolute zero, free electrons move around in a random manner, their velocity distribution being a function of temperature. When the temperature is sufficiently high, some electrons have enough energy to overcome the forces that normally cause them to stay within the solid. However, it is not sufficient that the electrons have more than the escape energy; they must also have this excess kinetic energy associated with a velocity component normal to the emitting surface.

The thermionically emitted electron current is exponentially dependent on ϕ and is

$$J_0 = q\frac{4\pi}{h^3}mk^2T^2 \exp\left(-\frac{q\phi}{kT}\right) = A_{th}T^2 \exp\left(-\frac{q\phi}{kT}\right), \qquad (1)$$

where

$$A_{th} \equiv \frac{4\pi mqk^2}{h^3}. \qquad (2)$$

A_{th} is the **theoretical emission constant**. If the mass of the electron is taken as 9.1×10^{-31} kg, A_{th} has the value of 1.20×10^6 amperes m^{-2} K^{-2}. The actual mass of the electron may vary somewhat from material to material. Equation 2 is known as **Richardson's Equation** in honor of Owen Williams Richardson, who, for his work in thermionic emission, received the 1928 Nobel Prize in Physics.

The work function is a characteristic of the emitting surface. Since it is a temperature-dependent quantity, one can make the reasonable assumption that it might be a well-behaved function of T, and, as such, the corresponding potential can be expressed as

$$\phi = \phi_0 + a_1T + a_2T^2 + \dots \text{ eV}. \qquad (3)$$

The higher order terms are, presumably, small. Truncating after the linear term, Richardson's Equation becomes

$$J_0 = A_{th}T^2 \exp\left[-\frac{q}{kT}(\phi_0 + a_1T)\right] = A_{th}\exp\left(-\frac{q}{k}a_1\right)T^2\exp\left(-\frac{q\phi_0}{kT}\right)$$

$$= AT^2 \exp\left(-\frac{q\phi_0}{kT}\right). \qquad (4)$$

where

$$A \equiv A_{th}\exp\left(-\frac{q}{k}a_1\right). \qquad (5)$$

One has the choice of writing Richardson's equation (Equation 1) using a theoretical emission constant, A_{th}, which is the same for all emitters, together with a work function, ϕ, which not only depends on the nature of the emitting surface but is, in addition, dependent on the temperature of the emitter. Or one can opt for a Richardson's equation (Equation 4) in which the work function does not vary with temperature (but is still dependent on the nature of the emitter), together with an emission constant that, although temperature independent, also depends on the nature of the emitter. The first option describes better the physical nature of the emission phenomenon, but the second is easier to use. Experimental thermionic emission data are usually tabulated as values of A and ϕ_0. It is customary to drop the subscript "0," with the understanding that when using the second option, one must use the value of the work function at absolute zero.

6.4

Table 6.1
Properties of Some Thermionic Emitters

Material	Work function ϕ_0 (eV)	Emiss. Const, A (A m^{-2}K^{-2})	Melting Point (K)	Temperature coefficient eV/K
Pt	5.32	320,000	2045	0.000114
Ni	4.61	300,000	1726	0.000120
Cr	4.60	480,000	2130	0.000079
W	4.52	600,000	3683	0.000060
Mo	4.20	550,000	2890	0.000067
Ta	4.19	550,000	3269	0.000067
Th/W	2.63	30,000	- - -	
BaO + SrO	1.03	100	- - -	0.000318
Cs	1.81	- - -	302	0.000810

Table 6.1 shows properties of some materials typically used in thermionic devices. Work functions vary from 1 eV to a little over 5 eV, while the apparent emission constants, A, cover a much wider range, from 100 to 600,000 A m^{-2}K^{-2}, a result of the variation in the temperature coefficients of ϕ in the different materials.

It must be emphasized that ϕ is sensitive to the exact way in which the material was prepared and to the state of its surface. Some of the tabulated values correspond to single crystals of the material considered, having as clean a surface as possible.

Devices used in electronic engineering benefit from emitters made of materials with low work functions because this allows their operation at relatively low temperatures, saving heating energy and, sometimes, prolonging the useful life of the devices. Thus, oxide coated emitters (BaO + SrO) are popular in small vacuum tubes. However, the emitter life may still be limited by the weakness of the interfaces between the ceramic crystals.

Each emitter has a preferred operating temperature. Tungsten filaments work best at around 2500 K, thoriated tungsten at 1900 K, and oxides at 1150 K. Thoriated tungsten emitters are still used in high power vacuum tubes because of their lower work function compared with pure tungsten and their longer life. When pure tungsten is heated to its usual operating temperature, there is a tendency for single crystals to grow, and it is at the resulting interfaces that breaks occur. The presence of thorium retards the growth of these crystals.

Thermionic energy converters require that the work function of the emitter be larger than that of the collector. Since most converters use a cesium atmosphere to neutralize the space charge (see Subsection 6.7.1), and because cesium tends to condense on the cooler collector surface, the work function of the latter is usually near that of cesium: 1.81 eV. Thus, emitters must have work functions of more than 1.81 eV.

Owing to the exponential dependence on ϕ, the emitter current is more sensitive to ϕ than to A. For example, the current density of a BaO + SrO emitter at its normal operating temperature of 1150 K is

$$J_0 = 100 \times 1150^2 \times \exp -\frac{1.03 \ q}{1150 \ k} = 4090 \text{ A m}^{-2}. \tag{6}$$

At the same temperature, a tungsten emitter produces a much smaller current, notwithstanding having an emission constant, A, 6000 times larger:

$$J_0 = 600{,}000 \times 1150^2 \times \exp -\frac{4.52 \ q}{1150 \ k} = 1.3 \times 10^{-8} \text{ A m}^{-2}. \tag{7}$$

For this reason, tungsten emitters operate at high temperatures.

High temperatures are not the only cause of electron emission. Several other mechanisms can accomplish this, including the impact of photons (**photoelectric emission**), the impact of subatomic particles especially electrons themselves (**secondary emission**), and intense electric fields already alluded to in the previous section (**field emission**). Intense fields near the emitter of a thermionic device will alter the magnitude of the emitted current. In most thermionic generators, such fields are sufficiently small to allow their effect to be ignored.

6.3 Electron Transport

The simplest thermionic generator would consist of an emitting surface, (the **emitter**), heated to a sufficiently high temperature, T_H, and placed in the vicinity of a collecting surface (the **collector**), operating at a lower temperature, T_C. The space between these surfaces may be a vacuum. The heat source may be of any desired nature: a flame, a nuclear reactor, the heat from nuclear decay, concentrated sunlight, and so on.

The geometry of the device plays a role in its performance but here we are only going to consider two parallel plate electrodes because this is the easiest configuration to analyze.

Since electrons leave the emitter and travel to the collector through the interelectrode space, the conventional direction of an external current is, under all circumstances, out of the emitter into the collector. In other words, the collector is the anode and the emitter, the cathode of the device.

When operated as a generator, the device is connected to a load, as indicated in Figure 6.2. The voltage drop across the load causes the collector to become negatively polarized with respect to the emitter. In electronic applications, devices are usually operated with an externally applied bias that causes the collector to be positive with respect to the emitter. An **accelerating interelectrode potential** is established. See Figure 6.2.

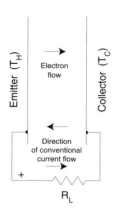

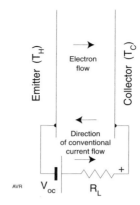

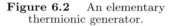

Figure 6.2   An elementary thermionic generator.

Figure 6.3   A forward-biased thermionic diode.

One might conclude that with an accelerating bias, all the emitted electrons should reach the collector, that is, that the interelectrode current density, J, should be equal to the emitted current, J_0, However, frequently J is smaller than J_0—the device is **space charge limited** or **unsaturated**, while when $J = J_0$, it is said to be **emission limited** or **saturated**.

Most thermionic *amplifiers* operate under space charge limitation; the current through the device depends on the applied voltage and is given by the **Child-Langmuir law** discussed later on. For maximum efficiency, thermionic *generators* must operate at the maximum possible current density and must, therefore, be emission limited.

In a vacuum, such as may exist in the interelectrode space of some devices, there is no mechanism for scattering the emitted electrons. As a consequence, the electron motion is dictated in a simple manner by the local electric field, or, in other words, by the interelectrode potential.

In the absence of charges in the interelectrode space of a parallel plate device—that is, if there are no emitted electrons between the plates—the electric field is constant and the potential varies linearly with the distance from the reference electrode (the emitter, in Figure 6.3). See curve "a" in Figure 6.4. A single electron injected into this space will suffer a constant acceleration. However, if the number of electrons is large, their collective charge will alter the potential profile causing it to sag—curve "b."

If the number of electrons in the interelectrode space is sufficiently large, the potential profile may sag so much that the electric field near the emitter becomes negative (curve "c"), thus applying a retarding force to the emitted particles. Only electrons that are expelled with sufficient initial velocity are able to overcome this barrier and find their way to the collector. This limits the current to a value smaller than that corresponding to the maximum emitter capacity.

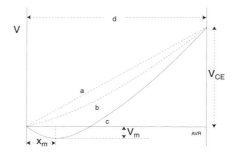

Figure 6.4   Potential across a planar thermionic diode.

6.3.1 The Child-Langmuir Law

Almost all thermionic generators operate under emission saturated conditions—that is, with no net space charges in the interelectrode space. Under such conditions the current does not depend on the voltage. However, to better understand how the presence of space charges limits the current, we will derive the equation that establishes the relationship between the applied voltage and the resulting current when the device is unsaturated.

Consider a source of electrons (emitter) consisting of a flat surface located on the $x = 0$ plane, and capable of emitting a current of density, J_0. See Figure 6.5 Another flat surface (collector) is placed parallel to the emitter at a distance, d, from it. A perfect vacuum exists in the space between the electrodes. Assume initially that the electrons are emitted with no kinetic energy—they just ooze out of the emitter.

A potential is applied to accelerate the electrons from emitter to collector. A current of density, J (where $J \leq J_0$), is established between the electrodes:

$$J = qnv, \tag{8}$$

where q is the charge of the electron, n is the electron concentration—the number of electrons per unit volume—and v is the velocity of the electrons. J is the flux of charges.

J must be constant anywhere in the interelectrode space, but both n and v are functions of x. We want to establish a relationship between J and the potential, V, at any plane, x. We will take $V_{(x=0)} = 0$. Then, owing to the assumed zero emission velocity of the electrons,

$$\tfrac{1}{2}mv^2 = qV, \tag{9}$$

from which

$$qn = J\sqrt{\frac{m}{2q}}\,V^{-1/2}. \tag{10}$$

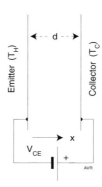

Figure 6.5   A vacuum diode.

However, the presence of electrons in transit between the electrodes establishes a space charge of density qn so that, according to Poisson's equation,

$$\frac{d^2V}{dx^2} = -\frac{nq}{\epsilon_0}, \tag{11}$$

where ϵ_0 is the permittivity of vacuum. Therefore,

$$\frac{d^2V}{dx^2} = -\frac{J}{\epsilon_0}\sqrt{\frac{m}{2q}}\,V^{-1/2} = -KV^{-1/2}, \tag{12}$$

where

$$K \equiv \frac{J}{\epsilon_0}\sqrt{\frac{m}{2q}}. \tag{13}$$

The solution of Equation 12 is

$$V = a + bx^\alpha. \tag{14}$$

The requirement that $V_{(x=0)} = 0$ forces $a = 0$. Thus,

$$\alpha(\alpha - 1)bx^{\alpha-2} = -Kb^{-1/2}x^{-\alpha/2}, \tag{15}$$

consequently

$$\alpha - 2 = -\frac{\alpha}{2} \qquad \therefore \qquad \alpha = \frac{4}{3}, \tag{16}$$

and

$$\frac{4}{9}b = -Kb^{-1/2} \qquad \therefore \qquad b^3 = \left(-\frac{9}{4}K\right)^2 = \left(\frac{9}{4}K\right)^2, \tag{17}$$

$$b = \left(\frac{9}{4}K\right)^{2/3}. \tag{18}$$

Thus

$$V = \left(\frac{9}{4}K\right)^{2/3} x^{4/3}. \tag{19}$$

Replacing K by its value from Equation 13 and solving for J,

$$J = \frac{4}{9}\epsilon_0 \sqrt{\frac{2q}{m}} \frac{V^{3/2}}{x^2} = \frac{2.33 \times 10^{-6}}{x^2} V^{3/2} \qquad \text{A m}^{-2}. \qquad (20)$$

When $V = V_{CE}$ (the collector-to-emitter or **anode** voltage), then $x = d$, where d is the interelectrode spacing. Hence,

$$J = \frac{2.33 \times 10^{-6}}{d^2} V_{CE}^{3/2} \qquad \text{A m}^{-2}. \qquad (21)$$

More commonly, the expression is written in terms of current, not current density, and is known as the **Child-Langmuir law**:

$$I = B \, V_{CE}^{3/2}. \qquad (22)$$

The **perveance**, B, is given by

$$B = \frac{2.33 \times 10^{-6}}{d^2} A, \qquad (23)$$

where A is the electrode area.

The Child-Langmuir law holds for diodes of any shape; however, the perveance depends on the particular geometry of the device.

In reality, electrons do not ooze out of the emitter; they are launched with a wide spectrum of velocities. As explained before, while in transit to the collector, they constitute a space charge that causes the potential in the interelectrode region to be lower than when electrons are absent. In fact, the potential of the region near the emitter may become more negative than that of the emitter itself creating a retarding electric field or "barrier."

The Child-Langmuir law is valid for such real diodes, but the perveance is now a function of the distance $d - x_m$ and the current is proportional to $(V + V_m)^{3/2}$. x_m is the distance from emitter to the plane where the potential reaches a minimum (see Figure 6.4). This plane represents a **virtual emitter**. V_m is the potential at $x = x_m$.

Even in the presence of space charge, a thermionic diode can become saturated if enough forward bias is applied. Consider a diode whose emitter is a tungsten plate heated to 2500 K. At this temperature, tungsten emits 3000 amperes for each square meter of surface: $J_0 = 3000$ A m^{-2}. If the collector-to-emitter voltage, V_{CE}, is small enough, then $J < J_0$ and the Child-Langmuir law is valid. At higher collector-to-emitter voltages, the law would predict currents larger than the emission capability of the cathode.

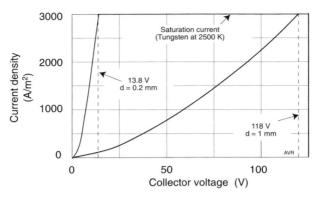

Figure 6.6   *V-J* characteristics of two vacuum thermionic diodes.

Since this is impossible, at large V_{CE}, J remains equal to J_0 independently of the value of V_{CE}. The voltage that just causes such saturation depends on the perveance of the device—that is, on the interelectrode spacing, d. The larger the spacing the higher the saturation voltage.

The *V-J* characteristics of two diodes with different perveances are shown in Figure 6.6. The diode with 1 mm interelectrode spacing will operate in the space charge limited region—that is, will obey the Child-Langmuir law as long as the collector voltage is 118 V or less. Above this voltage, it will saturate. The diode with a 0.2 mm spacing saturates at only 13.8 V.

Thermionic converters operate under reverse bias and their current is severely restricted by net space charges in the interelectrode region.

To reduce the total amount of charge, the electrodes must be placed very close together or, alternatively, a space charge neutralization scheme must be employed. Devices with spacing as small as 10 μm have been made, but they are difficult to build with large electrode areas.

It is possible to obtain a virtual zero interelectrode spacing (at least theoretically) by employing the configuration of Figure 6.7. Emitter and collector are co-planar and face a third electrode: an **accelerator**.

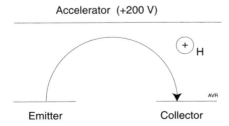

Figure 6.7   A vacuum diode with effectively zero interelectrode spacing.

A voltage is applied between emitter and accelerator—say, some 200 V. A magnetic field (normal to the plane of the figure) causes the path of the electrons to bend away from the accelerator toward the collector. The effective perveance of this device is infinite, that is, the current is always saturated. Since—in the ideal case—the electrons don't reach the accelerator, no power is used in the latter. In practice, however, a substantial number of electrons do reach this electrode using up a prohibitive amount of energy. Sufficiently reducing interelectrode spacing or using some scheme to circumvent the effect of space charges appears to be impractical.

In solids, the drift velocity of free electrons (or other carriers) is limited by frequent collisions with obstacles. The dissipated energy reveals itself as a resistance to the flow of current. Although there are no obstacles to scatter the electrons in a vacuum, there is nevertheless a severe impediment to the free flow of a current: the very electrons create, by their presence, a space charge that tends to oppose their motion. If instead of a vacuum the electrons move in a rarefied positive ion gas, there will be a certain scattering (with attending losses), but the space charge is neutralized and the electrostatic impediment to the flow of current is removed. The vast majority of thermionic generators use space charge neutralization. In the next section we will derive the behavior of a thermionic diode under the assumption that there is no space charge in the interelectrode space.

6.4 Lossless Diodes with Space Charge Neutralization

6.4.1 Interelectrode Potentials

To drive large currents through a thermionic diode, interelectrode space charges must be neutralized. Thus, we will restrict our attention to the space charge-free case. Under such circumstances, in a planar diode, the potential varies linearly with distance from the emitter.

If the electrodes are externally shorted out and if thermoelectric effects are neglected, then there is no potential difference between the collector and the emitter: the energy reference level (the **Fermi level**) is the same at the two electrodes. Figure 6.8 shows the voltages in a shorted diode.

ϕ_E stands for the work function of the emitter and ϕ_C for that of the collector. Notice that the larger the ordinate, the more negative the potential (voltage). To free itself from the emitter, an electron must have (at least) an energy, $q\phi_E$; in other words, the potential just outside the emitter (referred to the Fermi level) is $-\phi_E$. The figure represents a case in which $\phi_E > \phi_C$.

Since we are accustomed to plots in which the voltage increases upward, we have redrawn Figure 6.8 upside down in Figure 6.9. It can be seen that the potential just outside the collector is more positive than that just outside the emitter, and, for that reason (in absence of space charge), the diode current, J, is saturated, that is, $J = J_0$, provided there is negligible emission from the collector:

$$J_0 = AT^2 \exp - \frac{q\phi_E}{kT}. \qquad (24)$$

When a load is placed externally between collector and emitter as shown in Figure 6.1, the current from the generator will cause a voltage, V, to appear across this load in such a way that the terminal nearer the emitter becomes positive. The potential distribution then resembles that shown in Figure 6.10.

As long as $|V + \phi_C| < |\phi_E|$, the interelectrode potential will accelerate the electrons and the current density will remain equal to J_0.

However, if V becomes large enough to invalidate the above inequality, then a retarding potential, V_R, is developed between the electrodes, with a magnitude

$$|V_R| = |V + \phi_C - \phi_E| = |V - \Delta V|, \qquad (25)$$

where

$$\Delta V \equiv \phi_E - \phi_C. \qquad (26)$$

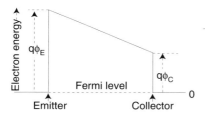

Figure 6.8   Energy levels in a short-circuited thermionic diode.

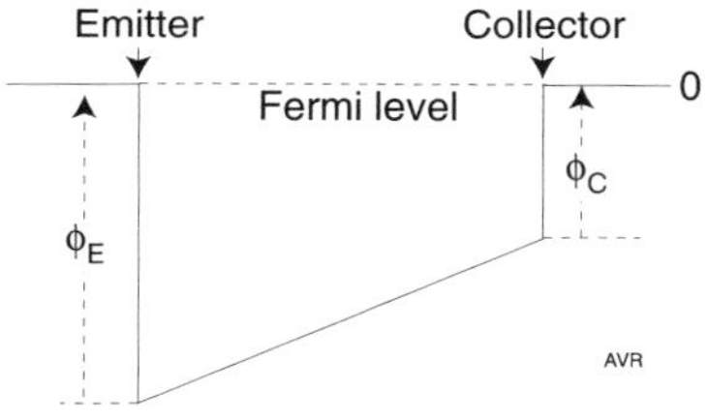

Figure 6.9   Voltages in a short-circuited thermionic diode.

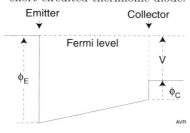

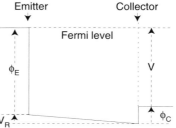

Figure 6.10   Voltages in a thermionic diode connected to a load. $(V + \phi_C < \phi_E)$

Figure 6.11   Voltages in a thermionic diode connected to a load. $(V + \phi_C > \phi_E)$

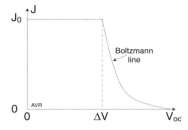

Figure 6.12   Current density versus voltage in a diode generator with neutralized space charge.

Under these circumstances, only electrons that, inside the emitter, have energy larger than the total barrier, $q(V + \phi_C)$, will reach the collector. The current density is still governed by Richardson's law, but the effective barrier is now larger than before. Electrons from the emitter now have to overcome not only the emission barrier, $q\phi_C$, but also the additional barrier, qV_R. Only electrons with more than $q(\phi_E + V_R) = q(\phi_C + V)$ joules can reach the collector:

$$J = AT^2 \exp\left[-\frac{q}{kT}(V + \phi_C)\right] = AT^2 \exp\left[-\frac{q}{kT}(\phi_E + V - \Delta V)\right]$$
$$= J_0 \exp\left[-\frac{q}{kT}(V - \Delta V)\right]. \tag{27}$$

6.4.2 V-J Characteristics

The J versus V characteristic of a thermionic generator is shown in Figure 6.12.

Consider a diode with a tungsten emitter ($\phi_E = 4.52$ eV) and a cesium plasma ($\phi_C = 1.81$ eV). Because of the cesium condensation on the collector, the latter operates with the cesium work function.

The break point in the J versus V curve should occur at $V = 4.52 - 1.81 = 2.71$ V. Experimentally, it occurs at 2.5 V owing to, at least in part, the thermoelectric voltage created by the temperature difference between the electrodes.

6.4.3 The Open-Circuit Voltage

Under the assumed Maxwellian energy distribution of the emitted electrons (which assigns nearly infinite velocity to some electrons), it would appear that the open-circuit voltage of a thermionic diode should be infinite because only an infinite retarding potential can stop *all* electrons. In fact, the open circuit voltage is not large because the feeble current of high energy electrons from the emitter is counterbalanced by small photoelectric and thermionic currents from the collector.

Under open-circuit conditions (ignoring photoelectric and thermoelectric effects),

$$J_E = J_C, \tag{28}$$

and

$$|V + \phi_C| > |\phi_E|. \tag{29}$$

$$A_E T_E^2 \exp\left[-\frac{q}{kT_E}(V + \phi_C)\right] = A_C T_C^2 \exp\left[-\frac{q}{kT_C}\phi_C\right], \tag{30}$$

$$\exp\left[-\frac{q}{kT_E}V - \frac{q}{k}\phi_C\left(\frac{1}{T_E} - \frac{1}{T_C}\right)\right] = \frac{A_C}{A_E}\left(\frac{T_C}{T_E}\right)^2, \tag{31}$$

$$\frac{q}{kT_E}V = -\frac{q}{k}\left(\frac{1}{T_E} - \frac{1}{T_C}\right)\phi_C - \ln\left[\frac{A_C}{A_E}\left(\frac{T_C}{T_E}\right)^2\right], \tag{32}$$

$$V = \phi_C\left(\frac{T_E}{T_C} - 1\right) - \frac{k}{q}T_E \ln\left[\frac{A_C}{A_E}\left(\frac{T_C}{T_E}\right)^2\right]. \tag{33}$$

Example:
A diode has an emitter made of tungsten and a collector made of thoriated tungsten. Operating temperatures are: $T_E = 2500$ K and $T_C = 500$ K. The open circuit voltage is

$$V = 2.63\left(\frac{2500}{500} - 1\right) - \frac{1.38 \times 10^{-23}}{1.6 \times 10^{-19}} \times 2500 \ln\left[\frac{30,000}{600,000}\left(\frac{500}{2500}\right)^2\right]$$

$$= 10.5 + 1.3 = 11.8 \text{ V}.$$

6.4.4 Maximum Power Output

The J versus V characteristics of an ideal thermionic diode with space charge neutralization can be divided into two regions:

Region I: $|V| \leq |\phi_E - \phi_C|$ where $J = J_0$;
Region II: $|V| > |\phi_E - \phi_C|$ where $J = J_0 \exp\left\{\frac{q}{kT}[V - (\phi_E - \phi_C)]\right\}$.

In Region I, the diode is saturated, while in Region II, its current depends exponentially on V. The plot of J versus V in this latter region is known as the **Boltzmann line** (see Figure 6.12).

In Region I, the power density, P_{out}, delivered to the load increases linearly with the output voltage (thus increasing with the load resistance, R_L). It reaches a maximum of $J_0(\phi_E - \phi_C)$ when $V = (\phi_E - \phi_C)$. If the load resistance is increased further, there is a small increase in V coupled with an exponential decrease in J so that the output power decreases. This intuitive conclusion can be reached mathematically by determining the maximum of the function (applicable to Region II):

$$P_{out} = VJ = VJ_0 \exp\left\{ \frac{q}{kT} [V - (\phi_E - \phi_C)] \right\}.$$ (34)

The maximum occurs for $V = kT/q$. Even at a temperature as high as 2500 K, kT/q is only 0.22 V, well below the break point $\phi_E - \phi_C$. This means that the maximum occurs at a voltage not valid in Region II. Thus,

$$P_{out_{max}} = J_0(\phi_E - \phi_C).$$ (35)

For the tungsten emitter, cesium plasma diode of the previous example, $\phi_E - \phi_C = 2.5$ V and, at 2500 K, $J_0 = 3000$ A m^{-2}. The maximum power output is 7500 W m^{-2}.

6.5   Losses in Vacuum Diodes with No Space Charge

6.5.1   Efficiency

If a thermionic generator had no losses of any kind and if the heat applied caused electrons to simply ooze out of the emitter, then all the heat input would be used to evaporate electrons. In reality, the evaporated electrons leave the emitter with some kinetic energy. Assuming for the moment that this kinetic energy is (unrealistically) zero, then the heat input to the generator will be simply $P_{in} = J\phi_E$ and the maximum power output would be $J_0(\phi_E - \phi_C)$. Consequently, the efficiency would be

$$\eta = \frac{P_{out}}{P_{in}} = \frac{J_0(\phi_E - \phi_C)}{J_0\phi_E} = 1 - \frac{\phi_C}{\phi_E}.$$ (36)

For the current example,

$$\eta = 1 - \frac{1.81}{4.52} = 0.60.$$ (37)

Nothing was said about the electrode temperatures, although they determine the limiting (Carnot) efficiency of the device. The implicit assumption is that $T_E \gg T_C$, otherwise the collector, itself, will emit a current that may not be negligible compared with that of the emitter.

In a real generator, the heat source must supply numerous losses in addition to the energy to evaporate electrons. These losses are related to:

1. heat radiation,
2. excess energy of the emitted electrons,
3. heat conduction,
4. lead resistance,

and, in plasma diodes,

5. heat convection,
6. ionization energy, and
7. internal resistance (called **plasma drop**).

6.5.2 Radiation Losses

The most serious loss mechanism is radiation from the hot emitter.

6.5.2.1 Radiation of Heat

If a *black body* is alone in space, it will radiate energy according to the **Stefan-Boltzmann** law:

$$P_r = \sigma T_E^4 \ \text{W/m}^2. \tag{38}$$

where σ is the **Stefan-Boltzmann** constant (5.67×10^{-8} W m^{-2} K^{-4}). However, real bodies do not exactly obey this law. A fudge factor called the **emissivity**, ϵ, must be used and the Stefan-Boltzmann law becomes

$$P_r = \sigma \epsilon T_E^4 \ \text{W/m}^2. \tag{39}$$

Heat emissivity compares the actual heat radiated from a given material at a certain temperature with that of a black body at the same temperature. It is always smaller than 1.

To complicate matters, it turns out that ϵ depends on the frequency of the radiation. However, one can use average data taken over a wide frequency band and define a **total emissivity**, which is then, of course, frequency independent but is valid only for the frequency band considered. The body being modeled with this frequency-independent emissivity is called a **gray body**. The total emissivity is temperature-dependent as can be seen from Figure 6.13. Emissivity data valid only for a narrow band of frequencies are referred to as **spectral emissivity**, ϵ_λ.

When the emitter is in the neighborhood of another object (the collector of a thermionic generator, for instance), the net emissivity is altered. We are particularly interested in the estimation of the radiation exchanged between a planar emitter and a planar collector that are in close enough proximity to allow us to treat them as parallel infinite planes. When radiation strikes a surface, it can be transmitted (if the material is translucent or transparent), reflected or absorbed. In the cases of interest here, the materials are opaque and no transmission occurs.

Reflection can be specular (mirror-like) or diffuse, or a combination of both. We will consider only diffuse reflection. The part that is not reflected must be absorbed and an **absorptivity coefficient**, α, is defined.

Based on equilibrium considerations, Kirchhoff's (thermodynamic) law concludes that the emissivity of a surface must equal its absorptivity—that is,

$$\epsilon = \alpha. \tag{40}$$

When the hot emitter radiates energy toward the collector, a fraction $\alpha_c = \epsilon_c$ is absorbed by the latter and a fraction, $1 - \epsilon_c$, is returned to the emitter. Here, again, part of the radiation is absorbed and part is returned to the collector. The radiation continues to bounce between the two plates altering the apparent emissivity of the emitter. The **effective emissivity**, ϵ_{eff}, takes into account the environment around the radiating material.

Clearly, if the radiator is surrounded by a perfect mirror, its effective emissivity is zero: all radiated energy is returned to it. The effective emissivity of the emitter (in terms of the emissivities of the emitter and the collector) can be calculated by adding up the energies absorbed by the collector in all these bounces. Problem 6.1 asks you to do the math. The result is

$$\epsilon_{eff_E} = \frac{1}{1/\epsilon_E + 1/\epsilon_C - 1}. \tag{41}$$

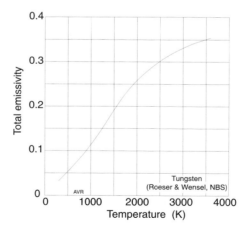

Figure 6.13   The total heat emissivity of tungsten as a function of temperature.

In the cases where either the emitter is alone in space or where it is surrounded by a black body (both situations corresponding to unity absorptivity and consequently, unit emissivity [$\epsilon_C = 1$]), the above equation reduces to

$$\epsilon_{eff_E} = \epsilon_E, \tag{42}$$

that is, the effective emissivity is simply the total emissivity of the emitter. Notice also that the effective emissivity of the collector is

$$\epsilon_{eff_C} = \frac{1}{1/\epsilon_C + 1/\epsilon_E - 1} = \epsilon_{eff_E}. \tag{43}$$

Thus, two plates in close proximity have the same effective emissivity.

Example

Consider a tungsten plate heated to 2500 K and having a total emissivity, $\epsilon_E = 0.33$. It is mounted near another metallic plate whose total emissivity is $\epsilon_C = 0.2$. By using the formula above, one calculates an effective emissivity, for both plates, of 0.14.

The emitter radiates (from the surface facing the collector) a heat power density of

$$P_r = \sigma \epsilon_{eff}(T_E^4 - T_C^4) \tag{44}$$

The collector is usually at a much lower temperature than the emitter, but it may nevertheless be quite hot. Assume that $T_C = 1800$ K, then the radiated power (using the calculated $\epsilon_{eff} = 0.14$) is 226,700 W/m^2.

When $T_C \ll T_E$, Equation 44 reduces to

$$P_r = \sigma \epsilon_{eff} T_E^4, \tag{45}$$

and the collector temperature can be ignored.

6.5.2.2 Efficiency with Radiation Losses Only

From an examination of Equation 36 or from the general idea of how thermionic generators work, one could conclude that, for a fixed collector work function, ϕ_C, the larger the emitter work function, ϕ_E, the larger the efficiency. However, introducing heat radiation losses into the equation, changes this picture. In fact, there is one given value of ϕ_E that maximizes the efficiency.

Let us return to the tungsten emitter at 2500 K. At this temperature, tungsten will emit a current of 3000 A/m². If the collector is cesium plated, the output voltage will be about 2.5 V and the output power density will be 7.5 kW/m². To emit the 3000 A/m², the tungsten emitter consumes a heat energy of $3000 \times 4.52 \approx 14$ kW. For simplicity's sake, we will assume, that the emitter radiates only toward the collector. The opposite side of the emitter faces a heat source that totally reflects any heat coming from the emitter. Then, as we saw in the example in the previous subsection, the emitter will radiate away into a cold collector a total of some 310 kW. This corresponds to the minuscule efficiency of $7.5/(310 + 14) = 0.023$. Considering that there are really many more losses, it would seem that there is little hope of building a useful thermionic generator. Not so!

Owing to the large ϕ of tungsten, significant emitted currents require a high temperature that translates into a large radiation loss (owing to the T^4 temperature dependence). The solution is to find an emitting material capable of producing large currents even at relatively low temperatures—that is, a material with lower ϕ_E. However, although this reduces the radiation losses, it will also reduce the power output which is proportional to $\phi_E - \phi_C$. Which effect dominates?

The output power density is

$$P_L = J_0(\phi_E - \phi_C), \tag{46}$$

while the heat input power must be (neglecting all loss mechanisms other than heat radiation) the sum of the energy, $J_0\phi_E$, needed to evaporate the electrons plus the radiation losses, $\sigma\epsilon_{eff}T_E^4$. The efficiency is, then,

$$\eta = \frac{J_0(\phi_E - \phi_C)}{J_0\phi_E + \sigma\epsilon_{eff}T_E^4} = \frac{\phi_E - \phi_C}{\phi_E + \dfrac{\sigma}{A_E}\epsilon_{eff}T_E^2 \exp\left(\dfrac{q}{kT_E}\phi_E\right)} \tag{47}$$

Here, we replaced J_0 by its Richardson's equation value.

We can now use the formula above to do some numerical experimentation. We select an arbitrary collector work function, ϕ_C (1.81 V, in this example), and an emitter temperature, T_E. We also chose a value for the emission constant, A_E, which we took as 600,000 A m⁻² K⁻².[†] We used $\epsilon_{eff} = 0.14$ as in the previous examples.

From Figure 6.14, one can see that, for maximum efficiency, an emitter at 2500 K should have a work function of about 3.3 V and if at 2000 K, the work function should be about 2.7 V,

---

[†] The emission constant depends on the material chosen for the emitter, as does our free variable, ϕ_E. Thus, strictly speaking, it is not correct to use a constant A_E when varying ϕ_E. But this is just an exercise.

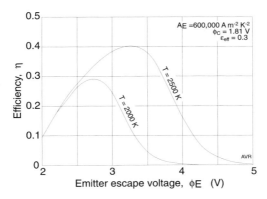

Figure 6.14   The efficiency of a thermionic generator at a given
emitter temperature peaks for a given emitter work function.

6.5.3 Excess Electron Energy

The tail end of the energy distribution of electrons in the emitter exhibits an exponential decay. To be emitted, the electrons need only an energy equal to $q\phi_E$. The excess appears as kinetic energy of the free electrons. As this energy must be supplied by the input heat source and because it is not available as electric output, it constitutes a loss. The average excess energy of the emitted electrons is between $2kT_E$ and $2.5kT_E$. Therefore, for each J_0 A/m^2, approximately $2J_0 kT_e/q$ watts/m^2 is wasted and is returned as heat when the electrons impact the collector.

Including this type of loss, the efficiency of the generator is

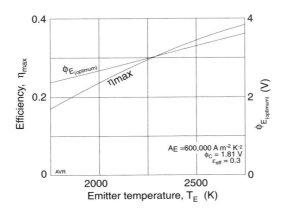

Figure 6.15   Dependence of efficiency and optimum work
function on emitter temperature.

$$\eta = \frac{\phi_E - \phi_C}{\phi_E + 2\dfrac{k}{q}T_E + \dfrac{\sigma}{A_E}\epsilon_{eff}T_E^2 \exp\left(\dfrac{q}{kT_E}\phi_E\right)}. \tag{48}$$

6.5.4 Heat Conduction

The structure that supports the emitter will, unavoidably, conduct away a certain amount of heat power, $\dot{Q}_E$. The heat source must supply this power. This means that the heat input must be increased by an amount $\dot{Q}_E$, and the efficiency is now,

$$\eta = \frac{\phi_E - \phi_C}{\phi_E + 2\dfrac{k}{q}T_E + \dfrac{\dot{Q}_E}{J_0} + \dfrac{\sigma}{A_E}\epsilon_{eff}T_E^2 \exp\left(\left(\dfrac{q}{kT_E}\phi_E\right)\right)}. \tag{49}$$

6.5.5 Lead Resistance

The lead resistance, R_{int}, will dissipate $I^2 R_{int} = J_0^2 S^2 R_{int}$ watts, reducing the available load power. S is the effective area of the emitter. The efficiency becomes.

$$\eta = \frac{\phi_E - \phi_C - J_0 S^2 R_{int}}{\phi_E + 2\dfrac{q}{q}T_E + \dfrac{\dot{Q}_E}{J_0} + \dfrac{\sigma}{A_E}\epsilon_{eff}T_E^2 \exp\left(\left(\dfrac{q}{kT_E}E\right)\right)}. \tag{50}$$

6.6 Real Vacuum Diodes

The short-circuit current of a real vacuum thermionic power generator is severely limited by the space charge created by the electrons in transit from emitter to collector. The V-J characteristic of a possible generator of this type is displayed in Figure 6.16, together with the corresponding ideal characteristic of a diode of similar construction but whose space charge has, magically, been eliminated.

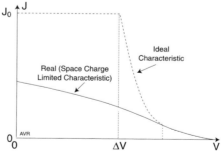

Figure 6.16   Space charge limitations degrade the output of a vacuum diode.

To alleviate the current limiting effect of negative space charges, the interelectrode spacing must be made extremely small. However, it is difficult to make devices with spacings below some 10 μm. Even then, space charge limits the current to such an extent that the output power density can't exceed some 20 kW/m^2 (2 W/cm^2), about one order of magnitude too small for practical applications. This explains why vacuum diodes have found no practical application. Space charge neutralization arrangements discussed in the next sections are the solution.

The correct analysis of the V-J characteristic of a vacuum diode is surprisingly complicated. It was first presented by Langmuir[†] in 1913 and is reproduced in Volume II of the Hatsopoulos book. We will skip this particular subject, especially because, as mentioned above, vacuum diodes appear to have no great future.

6.7   Vapor Diodes

Vacuum diodes with no space charge are essentially a contradiction in terms because the only practical way to eliminate space charges is to replace the vacuum by a low pressure ion gas. To create ions, one has to fill the diode with a low pressure neutral gas and then cause the gas to become partially ionized. The two ionization processes of interest here are **contact ionization** and **collision ionization**.

The cesium vapor used in most plasma diodes has five major effects on the performance of the diode:

1. Cesium ions can cancel the negative interelectrode space charge.
2. Cesium atoms condense on the electrodes changing their effective work function in a predictable way.
3. Cesium atoms and ions scatter the electrons, interfering with their smooth travel from emitter to collector.
4. Inelastic electron collisions with cesium atoms can ionize them, increasing the ion current.
5. The heat convected away by the cesium vapor increases the losses of the diode.

When the cesium vapor concentration is sufficiently small, only Effects 1 and 2 are significant and we have **low-pressure plasma diodes**; otherwise we have **high-73.2 pressure plasma diodes**. The latter are capable of high power densities (over 300 kW/m^2 or 30 W/cm^2).

[†] Irving Langmuir was the multitalented scientist who, working for General Electric, made vast contributions for the development of thermionics. Langmuir, besides being the introducer of the word "plasma" into the technical vocabulary, was also the developer of cloud seeding for producing rain. He received the 1932 Nobel Prize for Chemistry for "discoveries and inventions in surface chemistry."

Cesium vapor, from which ions will be produced, is obtained from the evaporation of the element stored in a reservoir whose temperature, T_r, can be adjusted so as to permit selection of a desired cesium vapor pressure, p_{cs}. The vapor pressure of cesium was tabulated by Stull and, fitting a curve to the data he gathered, one obtains an empirical expression relating T_r, to the pressure, p_{cs}, of the resulting vapor,

$$p_{cs} = \frac{32.6 \times 10^9}{\sqrt{T_r}} \exp\left(-\frac{8910}{T_r}\right) \qquad \text{Pa.} \qquad (51)$$

The vaporization of cesium, being a heat-driven phenomenon, should obey Boltzmann's law,

$$p = p_0 \exp\left(\frac{\Delta H_{vap}}{RT_r}\right), \qquad (52)$$

where $p_0 = 10^9$ pascals. (Razor)

The listed value of the enthalpy of vaporization of cesium varies considerably depending on the source consulted. The most frequently cited value is 67.74 MJ/kmole. However, using this value in the above equation leads to considerable departures from the experimentally measured values of Stull. Much closer agreement with the empirically determined data is obtained when one uses $\Delta H_{vap} = 73.2$ MJ/kmole, which corresponds to the 0.75 eV suggested by Razor in the paper mentioned before.

The value of p_{cs} determines:

1. The degree of cesium coverage of the electrodes—that is, the value of their respective ϕ.
2. The degree of the space charge neutralization, measured by the parameter, β.

We are going to show that $\beta = 1$ corresponds to exact neutralization, $\beta < 1$ corresponds to incomplete ionization—that is, to an **electron-rich** condition, and $\beta > 1$ corresponds to over-neutralization—that is, to an **ion-rich** condition.

6.7.1 Cesium Adsorption

Thermionic emission is a surface phenomenon, hence, it is not surprising that a partial deposition of cesium on the electrodes will alter their work function, ϕ, whose exact value will then depend on the **degree of coverage**, Θ.

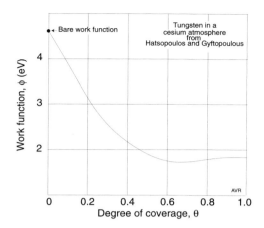

Figure 6.17   Effect of cesium adsorption on tungsten work function.

Figure 6.17 is an example of how the work function of tungsten (**bare work function** of 4.52 eV) is influenced by the degree of cesium coverage. It is interesting to observe that in the region near $\Theta = 0.6$, the work function of the cesiated tungsten electrode is lower than that of pure cesium, which is 1.81 eV. The minimum work function of cesiated tungsten is 1.68 eV, a value lower than that of any pure metal. The tungsten-cesium combination shows sort of "eutectic-like" behavior.[†]

The degree of coverage depends on both the temperature of the emitter and on the pressure of the cesium vapor. The higher the temperature, the smaller the coverage. Consequently, Θ tends to be higher on the cooler collector than on the hotter emitter.

Obviously, the degree of coverage is quite sensitive to the cesium vapor pressure, p_{cs}. Altering this pressure is the main method of controlling ϕ, thus "tuning" this quantity to its chosen design value. As we saw, p_{cs} can be adjusted in a simple manner by varying the temperature, T_r, of the liquid cesium reservoir.

The relationship between ϕ and Θ may be of academic interest, but it is not too useful for modeling the behavior of vapor diodes because there is no simple way to measure Θ. It is better to develop a formula for ϕ based on more easily measured parameters. We used the data reported by Houston et al. (collected by Taylor and Langmuir) to come up with a formula that yields the work function of a representative tungsten electrode as a function of the ratio of the electrode temperature, T, to the cesium reservoir temperature, T_r. The formula appears to be acceptably accurate for T/T_r ratios above 2 but fails to reproduce the minimum in ϕ that occurs

---

[†] The word "eutectic" normally applies to the alloy that has the lowest melting point of all alloys of similar constituents.

near this point. Nevertheless the formula is useful for general modeling work and for solving homework problems.

$$\phi = 9.499 - 9.9529T/T_r + 4.2425(T/T_r)^2$$
$$- 0.67592(T/T_r)^3 + 0.03709(T/T_r)^4. \tag{53}$$

In using this formula, one must be careful not to let ϕ exceed the bare metal value of 4.52 V. If the formula yields larger values (thus showing that its validity interval was exceeded), clamp the values of ϕ to 4.52 V, independently of the T/T_r ratio. Also clamp the value of ϕ to 1.81 V for $T/T_r < 1.9$.

The value of ϕ is a strong function of T/T_r, but also depends weakly on the value of T_r. Therefore, the formula above cannot be trusted for precision work on problems in which the latter temperature varies widely.

Symbology

A more complicated symbology is useful in the analysis of vapor diodes, where, in addition to electron currents, there are also important ion currents to consider.

In the case of low-pressure diodes, one has to deal with four different currents:

1. J_{e_E}, the electron current from emitter to collector,
2. J_{e_C}, the electron current from collector to emitter,
3. J_{i_E}, the ion current from emitter to collector resulting from contact ionization of the cesium gas at the emitter,
4. J_{i_C}, the ion current from collector to emitter resulting from contact ionization of the cesium gas at the collector.

Both electron and ion currents are thermionically emitted by the electrodes, the former according to Richardson's law discussed previously, and the latter according to a law we will derive in one of the coming sections. Only under saturation conditions will the whole emitted currents reach the opposite electrode. We shall distinguish the *emitted* or *saturation* current by appending a "0" to the appropriate symbol. For example, the ion saturation current generated by the emitter is $J_{i_{E_0}}$.

The load current, J_L is given by

$$J_L = J_{e_E} - J_{e_C} - J_{i_E} + J_{i_C}. \tag{54}$$

Currents from collector to emitter are called **back emission**.

6.7.2 Contact Ionization[†]

A glance at the periodic table of elements reveals that alkali metals (Column I) have ionization potentials lower than those of elements in the other columns. It also becomes apparent that the ionization potential of any element in a given column tends to be lower the higher its atomic number. Hence, one would expect that francium, occupying the southwest tip of the table, should have the lowest ionization potential on record. However, it is cesium, the next-door neighbor to the north, that holds this honor with a meager 3.89 eV. This is one of the reasons why cesium (pure or with some additives) is so popular for space charge neutralization.

Cesium vapor can undergo **contact ionization** by simply "touching" a hot metal surface provided the material of this surface has a work function larger than the ionization energy of the vapor. These conditions are satisfied by, among others, tungsten. Figure 6.18 illustrates what happens.

In a metal, there are many unoccupied levels above the Fermi level. According to classical mechanics, an electron in a gas atom can only leave the atom and transfer itself to one of the unoccupied levels in the neighboring metal if it has more energy than the ionization potential, ϕ_{ion}. In other words, the electron in cesium would have to first free itself from the parent atom. Quantum mechanics, however, predicts a finite probability that an electron, even with less than the ionization potential, will **tunnel** through and will migrate to the metal. The atom becomes "spontaneously" ionized, provided $\phi_{ion} < \phi_E$ and provided also that the atom is near enough to the metal. Remember that gas atoms in "contact" with a hot electrode are, in fact, a certain distance away from that electrode.

Do not confuse the ionization potential of cesium (3.89 eV) with its work function (1.81 eV). The former represents the energy necessary to remove the outer electron from the atom, while the latter represents the (smaller) energy to remove a conduction electron from the solid.

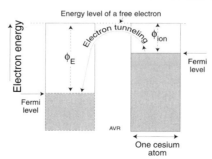

Figure 6.18   Energy levels in a gas near a metal.

[†] Also known as "thermal" or "surface" ionization.

6.7.3 *Thermionic Ion Emission*

We shall derive an expression that yields the magnitude of the thermionically emitted ion current resulting from the contact ionization of cesium gas discussed in the preceding section. Contact ionization captures an electron from a cesium atom liberating the corresponding ion.

In a one-dimensional gas, the flux, Φ, of molecules moving in either of the two possible directions is

$$\Phi = \frac{1}{2}nv. \tag{55}$$

The factor, $1/2$, results from half the molecules moving in one direction and the other half, in the opposite direction.

Since $\frac{1}{2}m_{cs}v^2 = \frac{1}{2}kT_{cs}$ (here, T_{cs} is the temperature of the cesium gas), the velocity is

$$v = \sqrt{\frac{kT_{cs}}{m_{cs}}}, \tag{56}$$

where m_{cs} is the mass of the cesium molecule (or atom, in this case). This results in a flux of

$$\Phi = \frac{n}{2}\sqrt{\frac{kT_{cs}}{m_{cs}}}. \tag{57}$$

From the perfect gas law, $n = p_{cs}/kT_{cs}$.

$$\Phi = p_{cs}\sqrt{\frac{1}{4m_{cs}kT_{cs}}}. \tag{58}$$

For a three-dimensional gas, the correct expression for the flux is

$$\Phi = p_{cs}\sqrt{\frac{1}{2\pi m_{cs}kT_{cs}}}. \tag{59}$$

The gas in the interelectrode space consists of electrons and two different species of larger particles: neutral atoms and ions (both with essentially the same mass, m_{cs}). Hence, the flux we calculated is the sum of the flux of ions, Φ_i, and the flux of neutrals, Φ_n. Thus, Equation 59 becomes

$$\Phi_i + \Phi_n = p_{cs}\sqrt{\frac{1}{2\pi m_{cs}kT_{cs}}}. \tag{60}$$

The ionization of neutral atoms is a heat activated process, and as such, should follow **Boltzmann's law**:

$$\frac{\Phi_i}{\Phi_n} \propto \exp\left[\frac{q}{kT}(\phi - \phi_{ion})\right]. \tag{61}$$

T is the temperature of the electrode, ϕ is its work function, and ϕ_{ion} is the ionization potential of cesium.

For cesium, the coefficient of proportionality is usually taken as $1/2$, hence

$$\frac{\Phi_i}{\Phi_n} = \frac{1}{2\exp\left[\frac{q}{kT}(\phi_{ion} - \phi)\right]}. \tag{62}$$

The expression above is known as the **Saha-Langmuir** equation. Ionization does not stop completely when $\phi < \phi_{ion}$. However, it decreases rapidly with decreasing ϕ.

We are interested not in the ratio of Φ_i to Φ_n, but rather in the ratio of Φ_i to the total flux of particles, $\Phi_i + \Phi_n$,

$$\frac{\Phi_i}{\Phi_i + \Phi_n} = \frac{1}{1 + \frac{\Phi_n}{\Phi_i}} = \frac{1}{1 + 2\exp\left[\frac{q}{kT}(\phi_{ion} - \phi)\right]}. \tag{63}$$

Combining Equation 63 with Equation 60,

$$\Phi_i = \frac{p_{cs}}{\sqrt{2\pi m_{cs}kT_{cs}}\left[1 + 2\exp\left(\frac{q(\phi_{ion} - \phi)}{kT}\right)\right]}, \tag{64}$$

$$J_{i_0} = \frac{qp_{cs}}{\sqrt{2\pi m_{cs}kT_{cs}}\left[1 + 2\exp\left(\frac{q(\phi_{ion} - \phi)}{kT}\right)\right]}, \tag{65}$$

This is the ion saturation current emitted by a hot electrode exposed to cesium vapor.

6.7.4 Space Charge Neutralization Conditions

Cesium vapor was introduced into the diode in the hope of neutralizing the negative space charge resulting from the streaming cloud of electrons in transit from emitter to collector. In the preceding subsection, we showed that the cesium gas can, indeed, give rise to ion currents in the device. Let us investigate what magnitude these currents must have in order to cancel the electron space charge.

The average thermal velocities of ions and of electrons are

$$v_i = \sqrt{\frac{kT_i}{m_i}} \tag{66}$$

and

$$v_e = \sqrt{\frac{kT_e}{m_e}} \tag{67}$$

The two current densities are (defining $T \equiv T_e = T_i$)

$$J_i = qn_iv_i = qn_i\sqrt{\frac{kT}{m_i}}, \qquad (68)$$

and

$$J_e = qn_ev_e = qn_e\sqrt{\frac{kT}{m_e}}, \qquad (69)$$

hence,

$$\frac{J_i}{J_e} = \frac{n_i}{n_e}\sqrt{\frac{m_e}{m_i}}. \qquad (70)$$

For exact space charge neutralization, the electron concentration must equal the ion concentration—that is, $n_e = n_i$.

$$J_i = \sqrt{\frac{m_e}{m_1}}J_e = \frac{J_e}{492}. \qquad (71)$$

Hence, the plasma will be **ion-rich** if

$$J_i > \frac{J_e}{492}. \qquad (72)$$

If the above inequality is not satisfied, the plasma is **electron-rich** and the negative space charge is not neutralized.

It is convenient to define an **ion-richness parameter** β:

$$\beta \equiv 492\frac{J_i}{J_e}. \qquad (73)$$

When $\beta < 1$, negative space charge is not completely neutralized (electron-rich regimen), and when $\beta > 1$, negative space charge is completely neutralized (ion-rich regimen).

6.7.5 V-J *Characteristics*

It was shown that the negative space charge caused by an electron flow, J_e, can be neutralized by the flow of an ion current, J_i, about 500 times smaller. This is, of course, the result of cesium ions being some 500 times more "sluggish" than electrons. The small ion current needed to neutralize the negative space charge has negligible influence on the output current of the device. In fact, one can ignore the influence of J_i on the output current provided $\beta < 10$—that is, provided $J_i < J_e/50$. However, the ion current is frequently much larger than that required for negative space charge neutralization and it has then to be taken into account when computing the

V-J characteristics of a diode. Most of the time, the back emission from the collector is negligible owing to the low collector temperature. Yet, this is not invariably true—sometimes the back emission current, J_{e_C}, has to be included. The same applies to the (rare) case when the ion emission current, J_{i_C}, from the collector is significant. In a general treatment of this subject, one should write the load current, J_L, as

$$J_L = J_{e_E} - J_{e_C} - J_{i_E} + J_{i_C}. \tag{74}$$

The values of these individual currents are

	$V < \Delta V$	$V > \Delta V$
J_{e_E}	$J_{e_{E_0}}$	$J_{e_{E_0}} \exp\left[-\frac{q}{kT_E}(V - \Delta V)\right]$
J_{e_C}	$J_{e_{C_0}} \exp\left[-\frac{q}{kT_C}(\Delta V - V)\right]$	$J_{e_{C_0}}$
J_{i_E}	$J_{i_{E_0}} \exp\left[-\frac{q}{kT_E}(\Delta V - V)\right]$	$J_{i_{E_0}}$
J_{i_C}	$J_{i_{C_0}}$	$J_{i_{C_0}} \exp\left[-\frac{q}{kT_C}(V - \Delta V)\right]$

The assumptions are

1. The interelectrode space charge is either 0 or positive—that is, $\beta \geq 1$.
2. If the interelectrode space charge is positive, it is not big enough to seriously limit the ion currents. This particular assumption is often violated and the ion currents in the region in which $V > \Delta V$ can be much smaller than that calculated with the above formulas.

To illustrate several operating conditions, we consider a tungsten electrode diode whose emitter and collector temperatures are, respectively, 1800 K and 700 K. Different cesium reservoir temperature are used to obtain selected values of βs. The plots computed for values of β of 1, 10, and 100 appear in Figures 6.19, 6.20, and 6.21.

In Figure 6.19, the cesium reservoir temperature was adjusted to 580 K to achieve an essentially exact cancellation of the space charge. The V-J characteristic is that of an ideal vacuum diode (with no space charge). Neither ion currents nor back emission are noticeable.

When β was adjusted to 10, the ion current became large enough to influence the output. At large load voltages this ion current is dominant and the output current reverses its direction. Finally, at $\beta = 100$, the calculated ion current is very significant. In reality, measured results would probably show a much smaller ion current because, in our calculations, we did not take into account the large positive space charge that develops, a space charge that will severely limit the ion current.

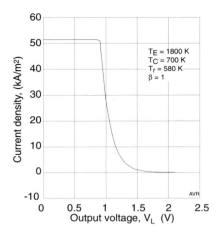

Figure 6.19   When $\beta = 1$, the only significant current is the electron current from the emitter.

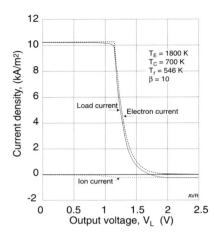

Figure 6.20   When $\beta = 10$, the ion current begins to affect the output, which is slightly negative at high output voltages.

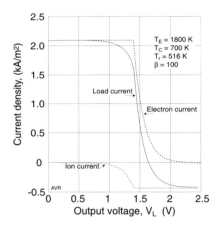

Figure 6.21   When $\beta = 100$, the ion current greatly affects the output.

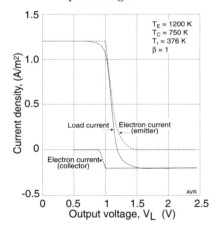

Figure 6.22   When the temperature of the emitter is not much higher than that of the collector, back emission becomes important.

By lowering the emitter temperature to 1200 K, while raising the collector temperature to 750 K, we create a situation in which the back emission becomes important (Figure 6.22). The collector work function under these conditions is 1.81 (owing to 100% cesium coverage) so that a reasonably large electron current is emitted while the hotter emitter is still operating at a relatively high work function of 2.82 V.

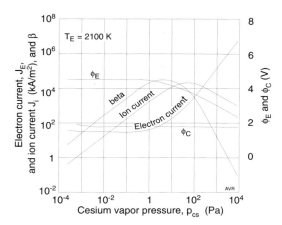

Figure 6.23   Behavior of the electron and the ion current as a function
of the cesium vapor pressure.

Those who studied the three V-J plots of Figures 6.20 to 6.22 may have
spotted an apparent paradox. It would seem that increasing the cesium
reservoir temperature (and consequently the cesium vapor pressure) should
increase β, because a larger amount of vapor is available to be ionized and
to contribute to the ion current. Yet, it was necessary to reduce T_r in order
to raise β. To understand what happens, please refer to Figure 6.24.

At the relatively high emitter temperature of 2100 K, there is little cesium condensation on that electrode until the cesium vapor pressure reaches
about 1 Pa. The emitter work function remains at the high bare metal value
of 4.52 V, and, consequently, the emitted electron current is relatively low
and unchanging. On the other hand, the ion current is, in this low vapor
pressure region, essentially proportional to the p_{cs}—in other words, it rises
as p_{cs} rises. β, being proportional to J_i/J_E, also rises. Once the vapor pressure exceeds 1 Pa, ϕ_E starts decreasing causing an exponential increase in
J_E. But as ϕ_E grows, the difference, $\phi_E - \phi_{ion}$, begins to fall, and so does
the rate of ionization of the gas. The ion current diminishes causing β to
come down again. So the dependence of β on p_{cs} is not monotonic. In this
example, β reaches a maximum at around $p_{cs} = 8$ Pa. Figures 6.20 to 6.22
correspond to a region in which β is in the descending part of its trajectory.

Plasma diodes operate in this region of decreasing β, where, even at
moderate temperatures, tungsten, being covered with cesium, emits current
densities in the 100 kA/m^2 range. This would lead to acceptable output
power densities except that the high-output power region tends to occur
at such high vapor pressures that even for small interelectrode spacings,
electron-cesium collisions become significant. As a rule of thumb, when
$p_{cs}d > 0.0033$ N/m, collisions cannot be disregarded and we leave the

domain of low-pressure diodes to enter that of high-pressure ones. In the preceding inequality, d, is the interelectrode spacing, in meters.

The shaded region in which $\beta < 1$ corresponds to the presence of negative space charge. In this region, the performance of the diode is intermediate between that of a vacuum device and the ideal device with no space charge. See Figure 6.16. The unshaded region (where $\beta > 1$ and $p_{cs}d < 0.0033$ N/m) is the ion-rich region in which a low-pressure diode exhibits ideal characteristics. Finally, the vertically elongated rectangular region on the right is a region in which the inequality $p_{cs}d < 0.0033$ N/m is violated. The width of this shaded area is for a diode with a small interelectrode spacing of 100 μm. In this area, we no longer have a low-pressure diode. With diodes of wider spacing the shaded area is broader and the output limitations are correspondingly more stringent. Although diodes operating in the unshaded region can have (very nearly) ideal characteristics, their output power density is disappointing—it is comparable to that of vacuum devices. We have come back to the situation in which, similar to what happened with vacuum devices, useful devices would require prohibitively small separations between emitter and collector.

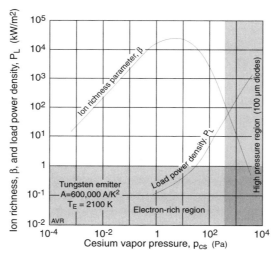

Figure 6.24   Dependence of beta and of the output power on the cesium vapor pressure.

6.8   High-Pressure Diodes

The difficulty with a low pressure diode is that it has two contradictory requirements: First, the emitter must have a *low* work function in order to produce the needed high current density. Second, it simultaneously must have a *high* work function in order to generate sufficient ionization to eliminate the negative space charge that would otherwise impede the flow of the desired current. A solution to this dilemma is the use of an ionization scheme that does not depend on a high emitter work function. Such a mechanism is collision ionization in which the very electrons whose charge is to be neutralized collide with a cesium atom with enough violence to ionize them. Although for each created ion, one additional electron is generated, there is a net gain as far as charge neutralization is concerned because the ion will linger in the interelectrode space about 500 times longer.

Inelastic collisions between the streaming electrons and the cesium atoms in the interelectrode space can, under some circumstances, cause the ionization of the cesium. This is an important process in the operation of high pressure vapor diodes. It is intuitive that the probability of an electron colliding with a gas atom is proportional to both the gas concentration (and hence, to the pressure, p_{cs}) and to the length of the path the electron must take—that is, to the separation, d, between the electrodes. It can be shown that when the $p_{cs}d$ product is smaller than 0.0033 N/m, the electrons suffer negligible collisions. Here, the cesium pressure is measured in pascals and the interelectrode distance, in meters.

In high-pressure diodes, collisions between electrons and the interelectrode gas are numerous, so substantial scattering occurs. This causes power losses (**plasma drop**) resulting in appreciably reduced output currents and output voltages. Nevertheless, the collisions may induce so much cesium vapor ionization that the population of carriers is substantially increased and large load current can result. Current densities of more than 500,000 A/m^2 are realizable.

When the disadvantages and advantages of high-pressure operation are weighed, the latter make a compelling case for using high-pressure diodes. Indeed, practical high-pressure diodes can realize a comparatively high output power density, far beyond that achievable with their low-pressure cousins. Even though load voltages tend to be smaller than those in low-pressure diodes, output power densities of some 30 W/cm^2 have been demonstrated.

Analysis of the performance of high-pressure diodes is very complicated owing to the interaction of many phenomena. Here, we will have to be satisfied with a simple explanation of what goes on.

In Figure 6.25a, $\beta \ll 1$ suggesting considerable negative space charge and, thus, small load currents. This is, indeed, the situation at high output voltage where the characteristics resemble those of a space charge limited device: lowering the voltage causes the current to increase modestly and to saturate at a low level. This behavior corresponds to the ledge in the figure.

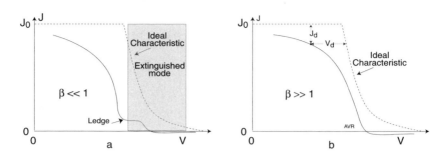

Figure 6.25   Characteristics of a high-pressure diode. Left, with very small beta; right, with very large beta.

Further reduction in voltage moves the diode into a regimen in which frequent collisions generate enough ions to drive the load current up sharply. In this regimen, ions are derived mainly from inelastic collisions (ions from the emitter being negligible), and the diode is said to be in the **ignited mode**, contrasting with the initial region called the **unignited** or **extinguished** mode in which ions from the emitter are dominant.

When $\beta \gg 1$ as depicted in Figure 6.25b, there is no perceptible extinguished mode—the ignited mode starts already at large load voltages.

It is this abundance of ions created by inelastic collision that leads to the large currents in high-pressure diodes and results in acceptable output power densities. But this operation comes at a cost: the real characteristics depart from the ideal by both a current and a voltage deficiency (J_d and V_d, in Figure 6.25b). V_d, called **plasma drop**, is made up of three components:

1. V_i, the voltage drop (about 0.4 V) associated with the ionization of cesium atoms by electron impact,
2. V_p, the voltage drop owing to the electron scatter by cesium atoms, and
3. V_s, a potential barrier that forms next to the electrodes.

The relatively large cesium vapor pressure conducts substantial heat from emitter to collector diverting some of the input energy. On the other hand, the heavy cesium plating that occurs in high-pressure diodes reduces the emitter work function and brings down the amount of input heat power required to evaporate the electrons from the emitter.

Although, as pointed out, the high-pressure diode is by far the most practical thermionic generator, it tends to be an expensive device owing in good part to the special materials that have to be used.

For a much more complete discussion of high pressure thermionic diodes, read the article by Ned Razor.

References

Iannazzo, S., A survey of the present status of vacuum microelectronics, *Solid-State Electronics*, Vol. 26, No. 3, 301–320, **1993** (Pergamon Press).

Hatsopoulos, G. N., and E. P. Gyftopoulos, *Thermionic Energy Conversion* (2 Volumes). The MIT Press, **1973**.

Houston, J. M., and P. K. Dederick, The electron emission of the Pt-group metals in Cs vapor, Report on the Thermionic Conversion Specialist Conference, San Diego, Calif., Oct. 25–27, (**1965**), 249–257.

Razor, N. S., Thermionic energy conversion plasmas, IEEE Trans. on Plasma Sci., 19, page 191, **1991**.

Shavit, A., and G. N. Hatsopoulos, Work function of polycrystalline rhenium, Proc IEEE (Inst. Elec. Electron. Eng.), (54): 777–781 (**1996**).

Stull, D.R., *Industrial and Engineering Chemistry*, 39, 517 (**1947**).

PROBLEMS

6.1   Consider two metallic plates with linear dimensions much larger than their separation from one another (so they can be treated as infinite planes). Prove that their effective emissivity on the sides that face one another is given by

$$\epsilon_{eff} = \frac{1}{1/\epsilon_1 + 1/\epsilon_2 - 1} \quad ,$$

where ϵ_1 and ϵ_2 are the total emissivity of the two plates when taken by themselves.

6.2   Two metallic plates are held parallel to one another, separated by a small spacing. There is neither electric nor thermal conduction between them. This "sandwich" is installed in outer space, at 1 AU from the sun. One of the plates (made of pure tungsten) is exposed to concentrated sunlight, while the other (made of thoriated tungsten) is in shadow.

The tungsten plate has its outer surface (the one that receives the sunlight) treated so that its thermal emissivity and its albedo are both 10%. The inner surface (the one that faces the thoriated tungsten plate) has an emissivity of 25%. This is the absolute emissivity and does not take into account the radiation reflected by the colder plate. The latter has 100% emissivity on each side.

There is no radiation in space other than that from the two plates and from the concentrated sunlight. Unconcentrated sunlight, at 1 AU from the sun, has a power density of 1350 W/m^{-2}.

The area of each plate is 1 m^2.

There is no space charge between the plates.

The emissivities are frequency independent.

1. What is the sunlight concentration ratio that causes the hotter plate to be 1000 K above the colder plate, under steady state conditions? What are the temperatures of the plates?

2. Now, an external electric load is connected between the plates and the concentration ratio is adjusted so that the tungsten plate heats up to 3100 K. The load draws the maximum available power. What is the temperature of the colder plate? What is the required concentration ratio? What is the electric power generated? Assume steady state and no space charge. Consider the system as an ideal plasma diode, but disregard any ion current.

6.3   A low-pressure cesium vapor diode operating under a no negative space charge condition has emitter and collector with a 2 cm^2 area. The collector has a work function of 1.81 V. The work function of the tungsten collector can be adjusted by varying the pressure, p_{cs}, of the cesium vapor.

6.38

The emission constant is unaltered by the cesium and is 600,000 A m^{-2} K^{-2}.

The temperatures of the emitter and of the collector are 2100 K and 1100 K, respectively. The effective emissivity of the emitter is 0.3 on the side facing the collector and 0 on the opposite side.

The only loss mechanisms are the cost of emitting electrons, excess energy of these electrons, and the radiation from the emitter toward the collector.

Which emitter work function maximizes the efficiency?

What is this efficiency?

6.4   Refer to the diode in Problem 6.3.

You want to operate with an emitter work function of 3.0 V.

1. Estimate the fraction, Θ, of the emitter area that has to be plated with cesium.
2. If the interelectrode spacing is large, collisions will be significant. What is the largest interelectrode spacing, d, that allows the diode to operate in the "no collision" regimen (i.e., as a low-pressure device)? Please comment.

6.5   In a low-pressure cesium vapor diode, the actual electron emitter current will be equal to the emitter saturation current only if there is no negative interelectrode space charge. Is there a negative space charge in a diode with an emitter work function of 3.3 eV, an emitter emission constant of 600,000 A m^{-2} K^{-2}, and an emitter temperature of 2350 K?

6.6   If the diode of Problem 6.5 is to operate as a low-pressure diode, what is the largest allowable interelectrode gap?

Assume that the cesium vapor pressure is 2000 Pa.

6.7   What is (approximately) the maximum power output density of the diode of Problem 6.5?

Assume that the diode operates as a low-pressure diode with no negative space chage. Assume also that there is no significant back emission.

6.8   Consider a cesium vapor thermionic generator with a 10 cm^2 emitter having an emission constant of 5×10^5 A m^{-2} K^{-2} and a work function of 3.0 eV. The collector is at 1000 K and, owing to the cesium condensation on it, has a work function of 1.8 eV. The effective emissivity of the emitter (operated at 2000 K) is 0.3 for the side that faces the collector. The other side does not radiate. Losses in the device are only through radiation and through excess kinetic energy of the emitted electrons. No conduction losses and no plasma drop. There is no negative space charge. Disregard any ion current.

What load resistance causes the device to furnish the highest possible output power?

Under the above conditions, what is the efficiency of the device?

What is the required input power?

6.9   Consider two solid plates: one made of tantalum and one of tungsten. Which of the two can be made to emit, in vacuum, the higher thermionic current density?

6.10   A thermionic generator has a tungsten emitter heated to 2500 K and a cesium coated collector at 1000 K. Electrodes measure 10×10 cm. Under all circumstances, the diode behaves as if it had an internal resistance of 10 mΩ. Disregard collector emission and thermoelectric effects. Assume there is no space charge. Calculate the power delivered to loads of 80 and 100 mΩ.

6.11   Assume the two materials being compared are at the same temperature. Which one emits thermionically, regardless of temperature, the larger current density? Chromium or tantalum?

6.12   A cube (side d meters) of a highly radioactive material floats isolated in intergalactic space. Essentially, no external radiation falls on it. Owing to its radioactivity, it generates 1200 kW per cubic m. Call this generation rate Γ.

The material has the following characteristics:

Thermionic emission constant: 200,000 A m^{-2} K^{-2},

Work function: 3.0 V,

Heat emissivity (at the temperature of interest): 0.85.

What is the temperature of the cube?

1. if $d = 1$ m,
2. if $d = 10$ cm.

6.13   A vacuum diode is built with both emitter and collector made of pure tungsten. These electrodes are spaced 1 μm apart. The "emitter" is heated to 2200 K and the "collector" to 1800 K.

1. Estimate the maximum power that the diode can deliver to a load.
2. Introduce cesium vapor. Explain what happens when the cesium vapor pressure is progressively increased. Plot the output power as a function of the cesium vapor pressure, p_{cs}. Disregard collision ionization and all ion currents.

6.14   A thermionic diode has parallel plate electrodes 2 by 2 cm in size and separated by 1 mm. Both emitter and collector are made of tantalum.

6.40

The emitter temperature is 2300 K and the collector is at 300 K.

If a vacuum is maintained between the plates, what is the current through the diode if a $V_{CE} = 100$ V are applied across it making the collector positive with respect to the emitter.

Repeat for $V_{CE} = 50V$.

6.15   A cube (edge length: 1 m) is placed in a circular orbit around the sun at 0.15 AU. This means that its distance to the sun is 15% of the sun-earth distance and that the cube is way inside the orbit of Mercury.

The orbit is stabilized so that a flat face always faces the sun.

The cube is made of a material that has 90% light absorptivity (i.e., has an albedo of 10%, hence it looks pretty black). Its heat emissivity is 30%.

The solar constant at the orbit of earth is 1360 W/m^2.

The material of the cube has a low work function for thermionic emission: $\phi = 1.03$ V. Its emission constant is 4000 A m^{-2} K^{-2}.

What is the equilibrium temperature of the cube?

6.16   In the text, we have presented an empirical equation relating the work function, ϕ, of tungsten to the ratio, T/T_r, of the electrode temperature to the cesium reservoir temperature. This assumes that, although ϕ is a function of T/T_r, it is not a function of T, itself. This is not quite true. ϕ is also a weak function of T. The formula value has an uncertainty of ± 0.1 V, for the $1500 < T < 2000$ K range.

What is the corresponding uncertainty factor in the calculated value of the emitted current when $T = 1500$ K? And when $T = 2000$ K?

6.17   What is the output power density of a cesium plasma diode with tungsten electrodes? Assume that there are no negative space charges and that collisions between electrons and cesium atoms are negligible.

Emitter temperature, $T_E = 2100$ K.
Collector temperature, $T_C = 1000$ K.
Cesium reservoir temperature, $T_r = 650$ K.
Cesium vapor temperature, $T_{cs} = T_r$.
Electrode emission constant, $A = 600,000$ A m^{-2}K^{-2}.
Interelectrode spacing, $d = 0.1$ mm.
Assume that the ion current has no effect on the load current. Nevertheless, it definitely has an effect of the space charge.
Check if your assumptions are valid. What is the nature of the space charge? Is it negative, zero, or positive? Can you neglect electron-cesium collisions?
Is the output power density that you calculated valid? If not valid do not recalculate.

Remember that there are limitations on the validity of the expression that yields ϕ as a function of T/T_r. Use the following boolean test:

IF $T/T_r < 2.2$ THEN $\phi = 1.81$ ELSE IF $\phi > 4.52$ THEN $\phi = 4.52$ ELSE $\phi = \phi$.

6.42

Chapter 7

AMTEC[†]

7.1 Operating Principle

AMTEC stands for **Alkali Metal Thermal Electric Converter** and is an example of a class of devices known as **Concentration-Differential Cells**. It's operation is conceptually simple. It takes advantage of the special properties of β' alumina, which is an excellent conductor of alkali metal ions but a bad conductor of electrons. In other words, it is an electrolyte. A slab of this material constitutes β' **alumina solid electrolyte** or **BASE**, a somewhat unfortunate acronym because it may suggest the "base" electrode of some semiconductor devices.

If the concentration of sodium ions (see Figure 7.1) on the upper part of the slab (anode) is larger than that at the lower (cathode), sodium ions will diffuse downward through the slab, accumulating in the lower interface between it and the porous electrode that serves as cathode. Such an accumulation of ions will create an electric field that drives an upward drift of ions. In equilibrium, the downward diffusion flux equals the upward drift flux.

What is the voltage developed by a concentration cell? Let N be the concentration of the migrating ion (Na$^+$, in the cell described above) and x be the distance into the BASE slab, counting from the top toward the bottom. The flux, ϕ_D, owing to the diffusion of ions is

$$\phi_D = -D\frac{dN}{dx}, \tag{1}$$

where D is the diffusion constant. The ion migration causes their concentration at the bottom of the slab to differ from that at the top. An electric field appears driving a return flux, ϕ_E,

$$\phi_E = N\mu E, \tag{2}$$

where μ is the mobility of the ions and $E = -dV/dx$ is the electric field.

In equilibrium, the two fluxes are of the same magnitude,

$$D\frac{dN}{dx} = \mu N\frac{dV}{dx}, \tag{3}$$

$$\frac{D}{\mu}\frac{dN}{N} = dV. \tag{4}$$

---

† Much of this chapter is based on the article by Cole (see References).

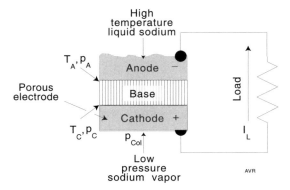

Figure 7.1   A concentration-differential cell.

Integrating from anode to cathode,

$$\frac{D}{\mu}(\ln N_A - \ln N_C) = V,\tag{5}$$

where N_A is the ion concentration at the anode (top) and N_C is that at the cathode (bottom). V is the potential across the cell.

According to Einstein's relation, $D/\mu = kT/q$. Thus,

$$V = \frac{kT}{q}\ln\frac{N_A}{N_C}.\tag{6}$$

In the usual case in which $T_A = T_C$, we have

$$V = V_{oc} = \frac{kT}{q}\ln\frac{p_A}{p_C},\tag{7}$$

because $p = NkT$.

Equation 7 gives the value of the *open-circuit* voltage of the AMTEC.

In the case of an ideal (reversible) AMTEC, the load voltage, V_L, is the same as V_{oc} because the internal resistance of the device is zero. Then, when μ kilomoles[†] of sodium ions (which are singly ionized and thus carry a positive charge equal, in magnitude, to that of an electron) pass through the BASE, the electric energy delivered to the load is

$$W_e = \mu N_0 q V_L = \mu N_0 q \frac{kT}{q}\ln\frac{p_A}{p_C} = \mu RT \ln\frac{p_A}{p_C},\tag{8}$$

which is the energy delivered by the isothermal expansion of μ kilomoles of gas from p_A to p_C.

[†] μ represents, now, the number of kilomoles, not the mobility as in the preceding page.

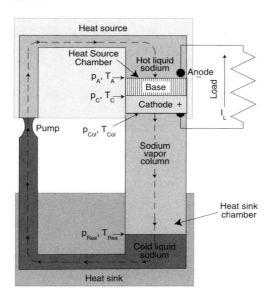

Figure 7.2   The configuration of an AMTEC cell.

An AMTEC can be built in the manner shown in Figure 7.2. A heat source raises the temperature and, consequently, the pressure of the sodium vapor in contact with the anode of the cell, while a heat sink causes the condensation of the metal into a liquid, which is circulated back to the heat source by means of a pump. Wires are attached to the electrodes in contact with the BASE.

p_A, the sodium pressure at the anode is the vapor pressure of sodium at the temperature, T_A, while p_{Res} is the vapor pressure of the liquid sodium in the heat-sink chamber reservoir at temperature T_{Res}.

The sodium in the heat-sink chamber can be either in the liquid or in the vapor state. In our examples we will consider the former case, the case of the **liquid-fed AMTEC**.

7.2   Vapor Pressure

From the description in the preceding subsection, it is clear that the values of the sodium vapor pressure in the hot and in the cold ends of the device play a decisive role in determining its performance.

The vapor pressure of sodium and of potassium is listed in Table 7.1. These elements have melting and boiling points of, respectively, 370.9 K and 1156 K for sodium and 336.8 K and 1047 K for potassium. Potassium has much less favorable characteristics than sodium. Consequently, we will concentrate on the latter.

Table 7.1
Vapor pressure of
Sodium and Potassium

Temperature (K)	Sodium	Potassium
	(Pa)	
400	0.0003	–
500	0.153	4
600	7	124
700	117	1235
800	1004	6962
900	5330	26832
1000	19799	78480
1100	60491	189460
1200	150368	–
1300	336309	–
1400	626219	–

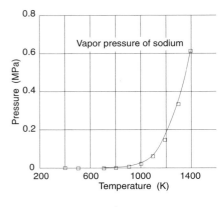

Figure 7.3   The vapor pressure of sodium.

The sodium vapor pressure can be estimated with an accuracy adequate for modeling AMTEC performance by using the formula below:

$$\ln p = -62.95 + 0.2279T - 2.9014 \times 10^{-4}T^2$$
$$+ 1.7563 \times 10^{-7}T^3 - 4.0624 \times 10^{-11}T^4. \tag{9}$$

The formula yields acceptable values of p for $400 \leq T \leq 1400$.

7.3 Pressure Drop in the Sodium Vapor Column

We saw that the pressure, p_A, on the top (anode) face of the BASE is the vapor pressure of the sodium vapor at the temperature, T_A, and that the pressure, p_{Res}, just above the surface of the liquid sodium pool near the heat sink, is the vapor pressure of sodium at the temperature, T_{Res}. However, the voltage across the BASE slab is determined by the pressure ratio, p_A/p_C, not p_A/p_{Res}. It is, therefore, necessary to relate p_C to p_{Res}.

If one assumes that the flow of the sodium vapor in the column between the bottom (cathode) of the BASE and the liquid sodium surface is sufficiently small not to cause, by itself, a pressure gradient and assuming a normal ideal gas behavior, then the column is isobaric—that is, $p_C = p_{Res}$. However, if the mean free path of the sodium atoms is large compared with the diameter of the column, then isobaric conditions do not hold and the column may sustain a pressure differential without a corresponding gas flow. Under such circumstances, $p_C \neq p_L$, and it becomes necessary to establish a relationship between these pressures.

7.4

Consider a gas in a horizontal pipe. If there is no net mass flow, then across any surface normal to the pipe, the flux, ϕ_1, from left to right, must equal the flow, ϕ_2, from right to left:

$$\phi_1 = \phi_2. \tag{10}$$

Here, we assumed that the flux is uniform over the cross section of the pipe. We will **not** impose the constraint, $p_1 = p_2$, which would, of course, lead to isobaric conditions.

Let n_1 and v_1 be, respectively, the concentration and the mean thermal velocity of the molecules that cross the surface coming from the left, and n_2 and v_2 the corresponding quantities for the flux from the right.

$$n_1 v_1 = n_2 v_2, \tag{11}$$

or

$$\frac{n_1}{n_2} = \frac{v_2}{v_1} = \sqrt{\frac{T_2 m_1}{T_1 m_2}}, \tag{12}$$

because

$$mv^2 = kT. \tag{13}$$

Thus,

$$\frac{n_1}{n_2} = \sqrt{\frac{T_2}{T_1}}. \tag{14}$$

Since $p = nkT$, we get

$$\frac{p_1}{p_2} = \sqrt{\frac{T_1}{T_2}}. \tag{15}$$

This means that, throughout the column of gas, when the mean free path, ℓ, is much larger that the dimensions of the column, the pressure in the gas obeys the rule,

$$\frac{p}{\sqrt{T}} = constant, \tag{16}$$

provided there is no gas flow in the column—in other words, provided no current is withdrawn from the AMTEC.

These circumstances are known as **Knudsen's conditions**.[†]

For example, consider a situation in which $T_A = 1300$ K and $T_L = 400$ K. The corresponding vapor pressures are 332,000 Pa and 3×10^{-4} Pa. This would lead to an open-circuit voltage of

$$V_{oc} = 1300 \frac{k}{q} \ln \frac{332,000}{3 \times 10^{-4}} = 2.33 \text{ V}, \tag{17}$$

[†] Martin Knudsen, Danish physicist 1871–1949. Knudsen number is the ratio of the mean free path length of a molecule to some characteristic length in the system.

provided isobaric conditions did prevail because then $p_C = p_L = 7$ Pa.[†]

However, if Knudsen conditions prevail, then

$$\frac{p_C}{\sqrt{T_C}} = \frac{p_L}{\sqrt{T_L}}, \tag{18}$$

and

$$p_C = p_L \sqrt{\frac{T_C}{T_L}} = 3 \times 10^{-4} \sqrt{\frac{1300}{400}} = 540 \times 10^{-6} \text{ Pa}, \tag{19}$$

and the open-circuit voltage is

$$V_{oc} = 1300 \frac{k}{q} \ln \frac{332,000}{540 \times 10^{-6} \times 100} = 2.27 \text{ V}, \tag{20}$$

just a tiny bit lower than in the isobaric case.

It is of interest to determine which is the regimen of the pressure behavior in the sodium column. For this, we need to know the dimensions of the column and the mean free path of the sodium ions.

7.4   Mean Free Path of Sodium Ions

The mean free path, ℓ, of a gas molecule can be estimated from

$$\ell = \frac{1}{4nA}, \tag{21}$$

where n is the gas concentration and A is the cross-sectional area of the molecule which, for sodium can be taken as 110×10^{-21} m^2. Thus,

$$\ell = \frac{2.28 \times 10^{18}}{n}. \tag{22}$$

From the perfect gas law, $p = nkT$, or $n = \frac{p}{kT}$, hence,

$$\ell = 2.28 \times 10^{18} k \frac{T}{p} = 31 \times 10^{-6} \frac{T}{p}. \tag{23}$$

[†] The temperature span in this example is quite optimistic. Advanced Modular Power Systems, Inc., manufacturer of AMTEC cells, for example, quotes heat source temperatures from 870 K to 1120 K and heat sink temperatures from 370 K to 670 K.

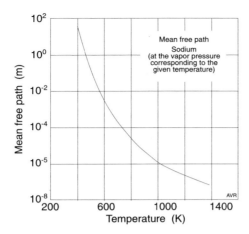

Figure 7.4   The mean free path of sodium molecules as a function
of temperature. The gas is assumed to be at its vapor pressure.

If the gas is at its saturation pressure, then for each value of T, the
value of pressure can be determined from Table 7.1 and a plot of ℓ as a
function of T can be constructed as shown in Figure 7.4.

At sufficiently low temperatures, the mean free path becomes surprisingly large because the vapor pressure is so low and may very well exceed
the dimensions of the sodium vapor column in the device. Under these
circumstances, the Knudsen conditions prevail.

7.5   *V-I* Characteristics of an AMTEC[†]

Under open-circuit conditions, there is no net flow of sodium ions
through the BASE and through the cathode slab of the device. However,
if there is an external current through a load connected to the terminals
of the AMTEC, then ions must flow through the base and recombine with
electrons in the cathode. A net flux, Φ, of sodium atoms is established in
the vapor column between the cathode and the cold liquid sodium. To drive
this flux, a vapor pressure differential, Δp must appear between the top of
the BASE and the bottom of the cathode slab. The relationship between
the flux and the pressure differential is

$$\Phi = \frac{\Delta p}{\sqrt{2\pi mkT_A}}, \tag{24}$$

where m is the mass of a sodium ion (22.99 daltons or 38.18×10^{-27} kg).
This equation, known as the **Langmuir Assumption**, was derived in Sub-

---

[†] With thanks to John Hovell.

section 6.7.3 of Chapter 6. We are making the assumption that $T_A = T_C$ (i.e., that the BASE is at uniform temperature).

The current density is $J = q\Phi$, hence,

$$\Delta p = \frac{\sqrt{2\pi m k T_A}}{q} J = \chi J, \tag{25}$$

where

$$\chi \equiv \frac{\sqrt{2\pi m k T_A}}{q}. \tag{26}$$

The sodium vapor pressure at the top of the sodium vapor column is now,

$$p_{col} = p_{col_0} + \chi J = p_{Res}\sqrt{\frac{T_A}{T_{Res}}} + \chi J, \tag{27}$$

and, extending Equation 7.7,

$$V = \frac{kT_A}{q} \ln \frac{p_A}{p_{Res}\sqrt{\dfrac{T_A}{T_{Res}}} + \chi J} = \frac{kT_A}{q}\left[\ln p_A - \ln\left(p_{Res}\sqrt{\frac{T_A}{T_{Res}}} + \chi J\right)\right]$$

$$= \frac{kT_A}{q}\left\{\ln p_A - \ln\left[\chi\left(\frac{p_{Res}}{\chi}\sqrt{\frac{T_A}{T_{Res}}}\right) + J\right]\right\}$$

$$= \frac{kT_A}{q}\left\{\ln p_A - \ln\chi - \ln\left(\frac{p_{Res}}{\chi}\sqrt{\frac{T_A}{T_{Res}}} + J\right)\right\}. \tag{28}$$

Defining

$$J_\vartheta \equiv \frac{p_{Res}}{\chi}\sqrt{\frac{T_A}{T_{Res}}}, \tag{29}$$

$$V = \frac{kT_A}{q}\left[\ln p_A - \ln\chi - \ln(J_\vartheta + J)\right] = \frac{kT_A}{q}\left[\ln \frac{p_A}{\chi} - \ln(J_\vartheta + J)\right]. \tag{30}$$

Defining

$$V_p \equiv \frac{kT_A}{q}\ln\frac{p_A}{\chi}, \tag{31}$$

$$V = V_p - \frac{kT_A}{q}\ln(J_\vartheta + J). \tag{32}$$

When $J = 0$, $V = V_{oc}$,

$$V_{oc} = V_p - \frac{kT_A}{q}\ln(J_\vartheta). \tag{33}$$

Using Equation 29, we find that Equation 33 is exactly the same as Equation 7.

7.8

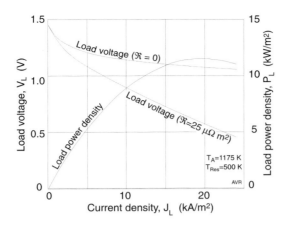

Figure 7.5   Calculated characteristics of an AMTEC.

In the preceding derivation we assumed that there is no resistance to the flow of ions in the BASE. This ideal situation is not achievable and a voltage drop, $J\Re$, appears across that component. $\Re = AR$ is the **specific resistivity** of the cell. Here A is the cross-sectional area of the BASE and R is its resistance. The load voltage becomes,

$$V = V_p - \frac{kT_A}{q} \ln\left(J_\vartheta + J\right) - J\Re. \tag{34}$$

The above equation can be used for modeling the behavior of an AMTEC. Cole states that under normal operating conditions, the calculated values are within 50 mV from the experimentally determined ones.

Figure 7.5 displays the V-I characteristics of a sodium AMTEC operating between 1175 K and 500 K. The values were calculated using Equation 34. One curve corresponds to the situation in which there is no BASE resistance, and the other, to a more realistic situation in which $\Re = 25 \ \mu\Omega m^2$. The load power is also plotted in the figure. It peaks at 11.6 kW/m^2. Both the load current density and the load power density are quite high—a characteristic of AMTECs.

7.6   Efficiency

The efficiency of an AMTEC is (see Cole)

$$\eta = \frac{\text{Electric power delivered to the load}}{\text{Total heat input power}}. \tag{35}$$

Heat is used up by the AMTEC by the following four processes:

1. The liquid sodium has to have its temperature raised from T_{Res} to T_A. This amounts to a heat power of

$$\dot{Q}_1 = c(T_A - T_{Res})\Phi = c(T_A - T_{Res})\frac{J}{qN_0}$$
$$= 311.4 \times 10^{-6}(T_A - T_{Res})J, \tag{36}$$

where $c = 30$ kJ K^{-1} kmole^{-1} is the specific heat of sodium, and Φ is the sodium flux, which is, of course, proportional to the current density through the AMTEC.

2. The heat power needed to vaporize the liquid sodium. This amounts to a heat power of

$$\dot{Q}_2 = \overline{\Delta h_{vap}}\Phi = \dot{Q}_2 = \overline{\Delta h_{vap}}\frac{J}{qN_0} = 0.924J, \tag{37}$$

where $\overline{\Delta h_{vap}} = 89$ MJ K^{-1} kmole^{-1} is the heat of vaporization of sodium.

3. The heat power needed to drive the output electric power,

$$\dot{Q}_3 = VJ. \tag{38}$$

4. The sum of all parasitic heat losses, $\dot{Q}_4 \equiv \dot{Q}_{loss}$. The main contributors to these parasitic losses are heat conduction through the output leads and heat radiation from the cathode.

Equation 35 becomes

$$\eta = \frac{JV}{J\left[V + \frac{c(T_A - T_{Res})}{qN_0} + \frac{\overline{\Delta h_{vap}}}{qN_0}\right] + \dot{Q}_{loss}}$$
$$= \frac{VJ}{J\left[V + 311.4 \times 10^{-6}(T_A - T_{Res}) + 0.924\right] + \dot{Q}_{loss}}. \tag{39}$$

In the hypothetical case in which there are no parasitic losses ($Q_{loss} = 0$), the AMTEC would operate at its maximum efficiency which Cole calls the **electrode efficiency**. It is,

$$\eta_{electrode} = \frac{1}{1 + \left[311.4 \times 10^{-6}(T_A - T_{Res}) + 0.924\right]/V}. \tag{40}$$

The efficiency, for given values of T_A and T_{Res}, depends, of course, on the operating point. Figure 7.6 shows how the electrode efficiency of a sodium AMTEC operating between 1175 and 423 K varies with the load current, and, consequently, with output power. Although the electrode efficiency exceeds 60% at low power outputs, at the maximum power of 6.64 kW/m^{-2} ($I_L = 11.5$ kA/m^{-2}), the efficiency is down to 33.3%.

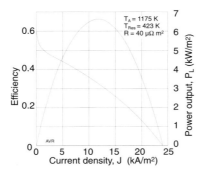

Figure 7.6   Electrode efficiency of an AMTEC.

Parasitic losses considerably reduce the real efficiency of the device. For instance, consider only radiation losses from the cathode (neglecting heat conduction losses through the output leads). If both the cathode and the surface of the liquid sodium pool were to act as black body radiators, the radiation loss from the cathode would be (read about radiation losses in Chapter 6),

$$P_r = \sigma(T_A^4 - T_{Res}^4) \approx \sigma T_A^4$$
$$= 5.67 \times 10^{-8} \times 1175^4 = 108{,}000 \ \mathrm{W/m^2}. \tag{41}$$

With such large radiation losses, the efficiency of the AMTEC in our example would barely exceed 5% at maximum power. Fortunately, the surface of the liquid sodium in the heat-sink chamber is far from a black body (which would absorb 100% of the radiation falling on it). In fact, the liquid sodium will do almost the opposite: it reflects more than 98% of the infrared radiation it receives. If the reflected radiation is reabsorbed by the cathode (even if the latter were a black body radiator), the net losses would be 50 times smaller than that calculated above. This would result in an efficiency of 30% for the AMTEC in the example. For such a reduction of radiation losses to take place, it is necessary to have the reradiation from the liquid sodium focused back onto the cathode. This may not be easy to arrange.

Conduction losses are far from negligible. The output leads connect a very hot BASE to the cool world outside. These leads must have high electric conductance which implies high heat conductance (see the Wiedemann-Frank-Lorenz law in Chapter 5). Low conductance leads reduce heat conduction losses but increase the I^2R losses in the device. One way to reduce conduction losses is to operate a number of AMTEC cells in series so that for all but the first and the last cell, there is no temperature differential across the leads.

7.7 Thermodynamics of an AMTEC[†]

We will examine the thermodynamics of a liquid-fed sodium AMTEC cell by following the states of the working fluid through a complete cycle. Please refer to the p-V and to the T-S diagrams in Figure 7.7. The fluid sodium will be in one of the several regions of the cell described in Figure 7.2:

1. Heat-sink chamber
2. Pump
3. Heat-source chamber
4. Base
5. Cathode
6. Sodium column

In this example, the device uses 0.001 kmole ($\mu = 0.001$ kmoles) of sodium (about 0.032 kg) and operates between the 1200 K and 600 K. We will start at State 1 when the sodium vapor is in the heat-sink chamber and is in contact with the heat sink. Its temperature is 600 K and its vapor pressure is 7 Pa. The vapor occupies a volume,

$$V = \frac{\mu R T}{p} = \frac{1 \times 10^{-3} \times 8314 \times 600}{7} = 713 \ \text{m}^3. \tag{42}$$

State "1" (sodium vapor, in the heat-sink chamber).
$V = 713 \ \text{m}^3$,
$T = 600 \ \text{K}$,
$p = 7 \ \text{Pa}$,
$S = 150$ aeu (arbitrary entropy units).

In Step 1 → 2, the sodium vapor condenses isothermally and isobarically. When fully condensed, the 0.001 kilomoles of liquid occupy $23.3 \times 10^{-6} \ \text{m}^3$ (assuming a liquid sodium density of 970 kg/m^3).

State "2" (liquid sodium, in the heat-sink chamber).
$V = 23.3 \times 10^{-6} \ \text{m}^3$,
$T = 600 \ \text{K}$,
$p = 7 \ \text{Pa}$,
$S = 32$ aeu.

[†] For more detailed information on this topic, please refer to the article by Vining et al.

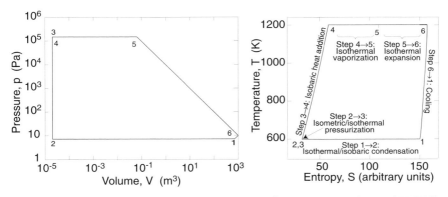

Figure 7.7   Pressure/volume and temperature/entropy graph for an AMTEC.

In Step 2 → 3, a pump pressurizes the liquid sodium to a pressure equal to the vapor pressure at 1200 K. In this example, it will be 150,400 Pa.[†] The pressurization is carried out at constant temperature, so points 2 and 3 cannot be distinguished in the temperature/entropy plot, but are quite apart in the pressure/volume plot.

State "3" (liquid sodium moving through the pump).
$V = 23.3 \times 10^{-6}$ m^3,
$T = 600$ K,
$p = 150{,}400$ Pa.
$S = 32$ aeu.

In Step 3 → 4, the liquid sodium is now in the heat-source chamber and temperature to rises to 1200 K. The process is isobaric and essentially isometric because of the small expansion of the liquid sodium. For this reason, points 3 and 4 cannot be distinguished in the pressure/volume plot but, owing to the large increase in temperature and a modest increase in entropy, they are widely apart in the temperature/entropy plot.

State "4" (liquid sodium, in the heat-source chamber).
$V = 23.3 \times 10^{-6}$ m^3,
$T = 1200$ K,
$p = 150{,}400$ Pa,
$S = 60$ aeu.

[†]   In practice, a certain overpressurization is used to make sure that the sodium remains in the liquid state when heated to 1200 K. A depressurization phase is then used to bring the pressure to the desired value.

In Step 4 → 5, the liquid sodium vaporizes isothermally and isobarically.

State "5" (sodium vapor, in the heat-source chamber).
$V = 0.0663$ m^3,
$T = 1200$ K,
$p = 150{,}400$ Pa,
$S = 80$ aeu.

In Step 5 → 6, the sodium vapor expands isothermally until the pressure is slightly above that of the initial State "1." Say, $p = 10$ Pa.

State "6"
$V = 997.7$ m^3,
$T = 1200$ K,
$p = 10$ Pa,
$S = 157$ aeu.

Finally, in Step 6 → 1, the sodium vapor cools to 600 K, and is returned to the initial State "1."

State "1" (sodium vapor, in the heat-sink chamber).
$V = 713$ m^3,
$T = 600$ K,
$p = 7$ Pa,
$S = 150$ aeu.

References

Cole, Terry, Thermoelectric energy conversion with solid electrolytes, *Science* **221** 4614, p. 915 2 Sept. **1983**.

Vining, C. B., R. M. Williams, M. L. Underwood, M. A. Ryan, and J. W. Suitor, Reversible thermodynamic cycle for AMTEC power conversion, *J. Electrochem. Soc.*, 140, 10, Oct. **1993**.

Chapter 8
Radio-Noise Generators

A subcommittee of the Committee on Government Operations of the House of Representatives of the Ninety-Fourth Congress of the United States heard, on June 11, 1976, testimony on "Converting Solar Energy into Electricity: A Major Breakthrough." Joseph C. Yater, the inventor, read a prepared statement. Although the implementation of the the idea is improbable, we will discuss it here as an interesting intellectual exercise.

The idea is fundamentally simple: It is well known that a resistor generates electric noise owing to the random motion of electrons. The available noise power is proportional to the temperature but is independent of the value of the resistance. If two resistors are connected in parallel and maintained at different temperatures, there is a net flow of electric noise power from the hotter to the colder. This energy can be converted into direct current and used for any desired purposed. The system described converts heat into electricity directly.

There is also heat transfer by convection, conduction, and radiation. The crucial question is how much of the input heat is lost by these parasitic processes compared with what is transformed into electricity.

By taking appropriate precautions, one can, at least conceptually, eliminate both convection and conduction, but not radiation.

Radiation losses are proportional to the surface area of the heated part while the generated noise power is independent of this area. By reducing the dimensions of the device, it is possible to reduce radiation losses without diminishing the useful power output. Can one build a device small enough to achieve acceptable efficiencies?

The electric output from a heated resistor is in the form of "white" noise—in other words, noise whose power density (power per unit frequency interval) is independent of frequency, up to a given upper limiting frequency or **upper cutoff** frequency, f_U.

The **available noise power** generated by a resistor is

$$P = kTB, \tag{1}$$

where k is Boltzmann's constant, B is the bandwidth under which the noise is observed, and T is the temperature. Available power is the power a generator can deliver to a matched load. The bandwidth is

$$B = f_U - f_L \approx f_U, \tag{2}$$

because, in general, $f_U >> f_L$.

When two resistors of equal resistance are interconnected, the net flow of radio-noise power from the hotter to the colder is

$$P = k \left(f_{UH} T_H - f_{UC} T_C \right). \tag{3}$$

The subscripts H and C designate quantities associated with the hotter and the colder resistors, respectively. Notice the assumption that the bandwidth, f_{UH}, for generation of power by the hotter resistor is different from that for the colder, f_{UC}.

It must be remembered that P is the total power transferred, not the power density—the dimensions of the device play no role.

The upper cutoff frequency is a parameter of utmost importance. It is determined by the mean collision frequency of the electrons in the material of the resistors. The mean thermal velocity of the electrons is

$$v = \sqrt{\frac{k}{m}} T^{1/2}. \tag{4}$$

The mass to be used in the formula is the effective mass of the electrons in the conductor. If one takes this mass as 10% of that of the free electron, then $v \approx 12{,}000 T^{1/2}$.

For $T = 700$ K, $v \approx 3.3 \times 10^5$ m/s. The mean free path, ℓ, of the electron can be estimated as, roughly, one order of magnitude larger than the lattice constant of the material which is, typically, 5×10^{-10} m. Take $\ell = 10^{-8}$ m. The collision frequency is then

$$f_U = \frac{v}{\ell} \approx 1.2 \times 10^{12} T^{1/2}. \tag{5}$$

Again, for $T = 700$ K, $f_U = 3 \times 10^{13}$ Hz (30 THz). The red end of the visible spectrum occurs at 400 THz. Therefore, f_U is in the infrared.

The available power from any resistor is

$$P_{avail} = 16 \times 10^{-12} T^{3/2}. \tag{6}$$

For a resistor at 700 K, this power amounts to 3×10^{-7} W. The power delivered to the load is

$$P_L = P_{avail,\ H} - P_{avail,\ C}. \tag{7}$$

In our example, $P_L = 2 \times 10^{-7}$ W. This small amount of power must be compared with the losses incurred when a resistor is heated to 700 K.

8.2

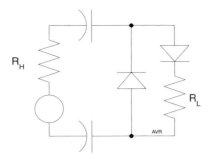

Figure 8.1   Half-wave rectifier.

If there is a direct connection between the hot and the cold resistors, heat will be conducted through this path and will cause excessive losses. Fortunately, it is possible to transfer the noise from the hot to the cold resistor through a vacuum capacitor having essentially zero heat conductance. Several circuits can be used. Figure 8.1 shows a simple half-wave rectifier. The diode that shunts the load provides a dc path otherwise blocked by the capacitors. The circuit in Figure 8.2 is a full-wave rectifier and consists of two of the preceding rectifiers connected back-to-back.

Figure 8.3 shows a perspective of a possible realization of the full-wave rectifier. It can be seen that in the arrangement, the major radiative losses occur across the capacitor plates—one plate is at T_H, while the opposing one is at T_C.

Such losses are

$$P_R = A\sigma\epsilon(T_H^4 - T_C^4). \tag{8}$$

A is the total area of the three capacitors (one side only), σ is the Stefan-Boltzmann radiation constant, and ϵ is the effective thermal emissivity (see Chapter 6).

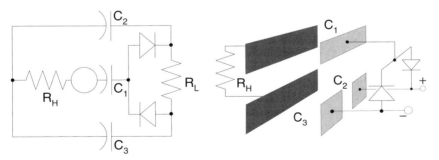

Figure 8.2   Full-wave rectifier.   **Figure 8.3**   A possible geometry
                                                              for the radio-noise converter.

To minimize P_R at a given temperature, it is necessary to reduce both A and ϵ. The latter depends on the nature of the material, on surface roughness, on the surroundings and on the temperature. At low temperatures, typical values of ϵ are given in Table 8.1.

Gold seems to be a good candidate for capacitor plates in the radio-noise converter because it combines low emissivity with good stability at high temperatures.

Let us assume that, to assure good efficiency, radiation losses are to be limited to 10% of the useful power transferred to the load. The latter is

$$P_L = 1.38 \times 10^{-23} \times 1.2 \times 10^{12}(T_H^{3/2} - T_C^{3/2}). \tag{9}$$

For $T_H = 700$ K and $T_C = 350$ K, $P_L = 2 \times 10^{-7}$ W. Thus, we must limit P_R to, say, 2×10^{-8} W. Then the total area of capacitors must not exceed

$$A = \frac{P_R}{\sigma\epsilon(T_H^4 - T_C^4)} \approx 10^{-10} \text{ m}^2, \tag{10}$$

assuming gold-plated capacitors. This amounts to 3.3×10^{-11} m^2 per capacitor, a square of about 6×10^{-6} m to the side.

If R_H occupies the same area as one capacitor, the whole device will have an area of 1.3×10^{-10} m^2, leading to a device density of 7.5×10^9 m^{-2} and a power density of about 1500 W m^{-2}.

What is the limit on power densities? The smaller the device, the larger the ratio between useful power and radiation losses. We must investigate what is the smallest size of a resistor that still produces white noise.

The mean free path of the electrons was estimated at 10^{-8} m. A resistor with 10 times this linear dimension will satisfy the statistical requirements for random noise. Thus, the resistor should have a minimum area of $10^{-7} \times 10^{-7} = 10^{-14}$ m^2, which is much less than the area used in the previous example.

Table 8.1
Heat Emissivity of Some Materials

Lamp black	0.95
Some white paints	0.95
Oxidized copper	0.60
Copper	0.15
Nickel	0.12
Silver	0.02
Gold	0.015

If, again, we assume that the area of the capacitors is equal to that of the resistor, the device will occupy a total area of 4×10^{-14} m^2 and the power density will be 5 MW m^{-2}.

The capacitance of a capacitor with area, A_c, is

$$C = \epsilon_0 \kappa \frac{A_c}{d}, \tag{11}$$

where ϵ_0 is the permittivity of free space (8.9×10^{-12} F/m), d is the separation between the plates and κ is the dielectric constant ($\kappa = 1$, because there is no dielectric, the capacitor being in vacuum). With an area of 10^{-14} m^2, $C = 8.9 \times 10^{-26}/d$ farads. If the device can be built with a separation of 10^{-7} m between the plates, the capacitance will be about 10^{-18} F. The lower cutoff frequency is

$$f_L = \frac{1}{2\pi C(R_H + R_L)}. \tag{12}$$

Since for maximum power transfer, the load must match the generator, $R_L = R_H$. To have the lowest possible f_L, we want the largest possible R_H. High resistivity material should be used. Possibly, some semiconductor would be appropriate. Let us, however, try a metal—gold—as an example. Assume that the resistor is a ribbon of gold 20 atoms thick and 10^{-7} by 10^{-7} m in area.

The resistivity of gold at room temperature is 2.44×10^{-8} Ωm and its temperature coefficient[†] is 3.4×10^{-11} Ωm/K. Thus, at 700 K, the resistivity is 3.8×10^{-8} Ωm. The ribbon would have a resistance of 0.4 Ω, leading to a lower cutoff frequency of 2×10^{17} Hz, which is much higher than the upper cutoff frequency. The device as planned is patently too small. If the resistor could be made of lightly doped silicon, the lower cutoff frequency would be brought down by a factor of, say, 10^5, leading to $f_L = 2 \times 10^{12}$. Such a cutoff frequency is acceptable because it would result in a bandwidth of $3 \times 10^{15} - 2 \times 10^{12} \approx 3 \times 10^{15}$.

References

Yater, Joseph C., Power conversion of energy fluctuations, *Phys. Rev. A, 10*, No. 4, 1361, Oct. **1974**.

Yater, Joseph C., Rebuttal to "Comments on Power conversion of energy fluctuation," *Phys. Rev. A, 20*, 623, Aug. **1979**.

[†] Temperature coefficient of resistivity is the ratio of the change in resistance to the corresponding change in temperature.

Part II

The World of Hydrogen

Chapter 9

Fuel Cells

9.1 Introduction

It is said that the 19th was the century of mechanical engineering, the 20th, that of electronics, and the 21st, that of biology. In fact, the 20th century could just as well be known as the century of the mechanical heat engine. Counting cars, trucks, and buses, the United States alone, built, from 1900 to 1999, slightly more than 600 million vehicles. If one adds the rest of the world's automotive production, and includes lawn mowers, motorcycles, motorboats, railroad locomotives, airplanes, and heavy construction machinery, the production of internal combustion engines in the 20th century probably reached the 2 billion mark!

Mechanical heat engines use the heat released by the reaction of a chemical substance (fuel) with oxygen (usually from from air). The heat is then upgraded to mechanical energy by means of rather complicated machinery. This scheme is inherently inefficient and cumbersome. It is the final outcome of our millenarian struggle to control and use fire. Converting chemical energy directly into electricity is more straightforward, especially in view of the electric nature of the chemical bond that holds atoms in a molecule. Devices that convert chemical energy directly into electricity are called **electrochemical** cells.

Flashlight batteries, automobile batteries, and fuel cells are examples of electrochemical cells.

Because electrochemical cells transform chemical energy directly into electricity without requiring an intermediate degradation into heat, they are not limited by the Carnot efficiency.

The words "cell" and "battery" are, in modern parlance, interchangeable. Cell suggests one single unit (although "fuel cell" most frequently consists of a number of series-connected units). Battery suggests a number of units, but a single 1.5 V flashlight cell is commonly called a battery.

If the battery is not worth preserving after its first discharge, it is an **expendable** (also called **primary**) battery. If the device is reusable after discharge, it may fall into one of two categories:

Rechargeable (also called **secondary**) devices, in which the activity is restored by means of an electric **charging** current, as is the case of automobile batteries.

Refuelable devices (**fuel cells**), which deliver a sustained output because their consumables are replenished. To facilitate such replenishment, these consumables are usually fluids, although some fuel cells use solid consumables as is the case of **zinc-air cells**, described later in this chapter.

$$\text{Electrochemical Cells} \begin{cases} \text{Expendable} \\ \text{Nonexpendable} \begin{cases} \text{Rechargeable} \\ \text{Refuelable} \end{cases} \end{cases}$$

Although fuel cells date back to 1839 when Sir William Groves demonstrated his "gaseous voltaic battery," until recently they remained in their technological infancy.

NASA revived fuel cell research: both *Gemini* and *Apollo* used fuel cells, and so does the space shuttle. The most important applications of fuel cells in the near future are as power sources for buses and automobiles, as central utility power plants, as dispersed (including residential) power suppliers, and as power sources for cell phones and other small electronic devices.

9.2   Electrochemical Cells

The purpose of electrochemical cells is to provide a flow of electrons in an external circuit where useful work can be done. To this end, the cells must consist of a source and a sink of electrons.

The reactions used in electrochemical cells are called **reduction-oxidation (redox)** reactions, because the buzz word for releasing electrons is **oxidation** and for capturing electrons is **reduction**.

Numerous old scientific terms are confusing or at least not self-explanatory. The terms **reduction** and **oxidation** require explanation.

The word *oxygen* stems from *oxús* = acid or sharp, and means generator of acids, a name that appears in de Morveau and Lavoisier's "Nomenclature Chimique" in 1787, when chemists were under the wrong impression that oxygen was an essential element in acids. Actually, it is hydrogen that is essential. When an acid is dissolved in water, some of its hydrogen atoms lose their electron—the water becomes *acid*; the hydrogen is *oxidized*. By extension any reaction that involves the loss of electrons is called **oxidation**. The reverse reaction—gaining electrons—is called **reduction**.

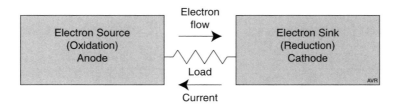

Figure 9.1   An electrochemical cell must consist of a source and a sink of electrons.

In electrochemical cells, the full reaction is broken down into two **half-cell reactions** or **half-reactions** that occur in physically separate regions of the device. These regions are interconnected by an **electrolyte** that conducts ions but not electrons. The latter, having been released by the oxidizing half-reaction, can move to the reduction side only via an external circuit. This establishes the external current that is the purpose of the cell. The *conventional* direction of this external current is from the reduction to the oxidizing side—the current exits the device from the reduction side which thus becomes the **cathode** of the cell, and enters the device at the oxidizing side which becomes the **anode**. As in any source of electricity, the cathode is the positive electrode and the anode the negative one, exactly the opposite of what happens in sinks of electricity (**loads**). See the Introduction to Chapter 6 for a discussion of the words "anode" and "cathode."

As an example of an electrochemical cell, consider a membrane capable of acting as an electrolyte. Put hydrogen in contact with one side of this membrane. At ambient conditions, most of the gas will be in the form of H_2 molecules; however, a small amount will **dissociate**:

$$H_2 \rightarrow 2H, \tag{1}$$

and some of the resulting H will **oxidize** (ionize)—that is, lose an electron:

$$H \rightarrow H^+ + e^-. \tag{2}$$

Since the membrane does not conduct electrons, these will remain on its surface while the ions will **diffuse** across it and arrive at the other side. Because the ions carry a positive charge, the hydrogen side becomes negative owing to the excess electrons that remain on it and the opposite side becomes positive owing to the positive ions that arrived there. The resulting electric field causes some of the ions to drift back to the hydrogen side. A **dynamic equilibrium** is established when the diffusion exactly equals the returning drift. It is easy to calculate the potential developed (Chapter 7, Section 7.1).

Now sprinkle a conducting powder on both sides of the membrane so as to create two porous electron-conducting layers, i.e., two **electrodes**. Interconnect the electrodes externally through a load resistance, R_L. Ions cannot flow through this external connection, but electrons can and, when they do, they flow from the hydrogen side where they are abundant to the opposite side establishing an electric current as indicated in Figure 9.2. The reaction of interest that occurs at the hydrogen electrode is

<u>Anode reaction:</u> $2 H_2 \rightarrow 4 H^+ + 4 e^-.$ \hfill (3)

The difficulty with this picture is that it contradicts the first law of thermodynamics in that it causes an $I^2 R_L$ amount of heat to be generated in the load, while, at the cathode, the incoming electrons will combine with the H^+ that diffused through the membrane regenerating the hydrogen atom, H, and, eventually recreating the H_2 gas used as "fuel." We would generate heat without using any fuel.

The external circuit creates a *path* for the electrons, but cannot by itself force a current to circulate, just as a pipe with one end dipped into a lake cannot cause water to flow up inside it. For the water to flow, the open end of the pipe must be lower than the level of the lake. Similarly, to have an external current, it is necessary to lower the (thermodynamic) potential on the cathode side. This can conveniently be done by introducing oxygen so that, combined with the electrons and the H^+, water is formed:

<u>Cathode reaction:</u> $4\ e^- + 4\ H^+ + O_2 \rightarrow 2\ H_2O.$ (4)

This reaction is strongly exothermal—that is, it releases energy (although, in this case, not mostly as heat, as in case of the combustion of hydrogen, but mainly as electricity). This is, of course the energy that powers the fuel cell.

The electrochemical cell just described is shown in Figure 9.2.

Under STP conditions, the degree of hydrogen dissociation at the anode is small. It can be increased somewhat by altering physical conditions (for example, increasing the temperature,). Remember the Le Chatelier's principle. It also can be increased by the action of catalysts.

The overall cell reaction is:

$$2\ H_2\ (g) + O_2\ (g) \rightarrow 2\ H_2O\ (g).\qquad(5)$$

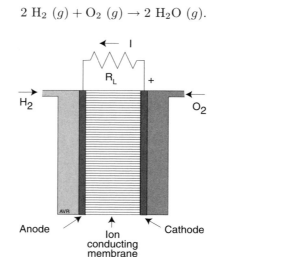

Figure 9.2   The simplest electrochemical cell.

The electrochemical cell invented by Alessandro Volta (1745–1827) in 1800 was the first device capable of generating sustained electrical currents. It consisted of zinc disks separated from silver (or copper) disks by a sheet of paper soaked in salt. A battery or pile was formed by stacking the cells so that there was a direct contact between the copper electrode of one with the zinc electrode of the next—the cells were all in series.

A "Volta" cell can be made by dipping a zinc and a copper electrode into a dilute (say, 10%) sulfuric acid solution. The zinc will oxidize:

$$Zn \rightarrow Zn^{++} + 2\,e^-, \tag{6}$$

providing electrons for the external current. Zinc ions are soluble in water. The sulfuric acid, being a strong acid,[†] will mostly dissociate into ions:

$$H_2SO_4 \rightarrow 2\,H^+ + SO_4^{--}. \tag{7}$$

The zinc ions combine with the sulfate ions forming zinc sulfate. The protons, in the form of hydronium, $H^+(H_2O)_x$, migrate through the electrolyte to the copper where they are reduced to hydrogen (by combining with the electrons arriving via the external circuit) and evolve as gas bubbles. This type of cell is of little practical use because soon the copper electrode is covered with adhering hydrogen bubbles that severely limit the current delivered. The so called "dry cells" use some scheme to avoid the formation of the insulating gas layer at the cathode. The chemical used to absorb the hydrogen is called a **depolarizer**. One of the consumables is in general a metal that can be easily oxidized, zinc being a common choice. Note that the copper in Volta's cell remains unchanged, not undergoing any chemical reaction.

Inexpensive batteries were, until recently, of the **Leclanché** type in which the anode is made of zinc and the cathode of a graphite rod surrounded by pulverized manganese dioxide mixed with carbon (to increase the conductivity). The MnO_2 combines with the liberated hydrogen and keeps it from coating the cathode. The electrolyte is ammonium chloride. More modern batteries use an alkaline electrolyte (**alkaline batteries**).

Perfectly pure zinc is consumed only when a current is drawn. The presence of impurities causes corrosion of the electrode even when the cell is inactive (the impurities form numerous microscopic electrochemical cells within the mass of the metal). To insure a long shelf life the zinc is alloyed with mercury (**amalgamated**). Leclanché cells have been mostly replaced by alkaline ones.

[†] The strength of an acid is a measure of the degree of its dissociation when in aqueous solution. Hydrochloric acid dissociates completely into H^+ and Cl^-; it is a very strong acid. Sulfuric acid is weaker, but is still a strong acid. Surprisingly, hydrofluoric acid, in spite of its corrosiveness, is a weak acid: when in water solution at room temperature, the concentration of H^+ is less than 3% of the concentration of neutral HF molecules.

9.3   Fuel Cell Classification

As in many technical areas, there is here a proliferation of acronyms capable of driving the uninitiated to distraction. We will use a few:

AFC	Alkaline fuel cell
DMFC	Direct methanol fuel cell
MCFC	Molten carbonate fuel cell
PAFC	Phosphoric acid fuel cell
SAFC	Solid acid fuel cell
SOFC	Solid oxide fuel cell (ceramic)
SPFC	Solid polymer fuel cell

Just as in the beginning of the 20th century at least three different technologies—steam, electric, and internal combustion—were competing for the automotive market, now, in the beginning of the 21st century, the different fuel cell technologies, listed above, are vying for dominance. Although AFC, MCFC, and PAFC are still in commercial production, it would appear that only SPFC and SOFC have a really good chance of emerging victorious.

Fuel cells can be classified according to several criteria:

9.3.1   Temperature of Operation

The ideal open-circuit voltage of a fuel cell depends on the nature of the fuel used and, to a small extent, goes down as the temperature increases. However, the maximum deliverable current density rises quickly with increasing temperature because it is related to the rapidity of the chemical reaction–that is, on the chemical kinetics which are, of course, also a function of the type of fuel and can be improved by catalysts. The higher the temperature, the larger the current the cell can deliver. On the other hand, high temperatures may be incompatible with the electrolyte or with other materials in the cell and tend to limit life time.

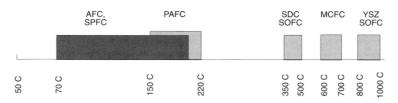

Figure 9.3   The operating temperatures of the different fuel cell types fall into relatively narrow ranges. Notice the two different types of SOFC: SDC = samaria-doped ceria, YSZ = yttria-stabilized zirconia.

Large power plants that work continuously may profit from high operating temperature: the kinetics are favorable, catalysts are either unnecessary or, when necessary, are immune to CO poisoning,[†] and exhaust gases can be used for cogeneration increasing the overall efficiency. Plants that operate at somewhat lower temperatures can produce hot water for general use, an advantage in the case of district or residential generators. For intermittent use, especially in in automobiles, it is important to use fuel cells that operate at temperatures low enough to permit short start up times. However, low temperatures bring the disadvantage of higher vulnerability to carbon monoxide poisoning, of high catalyst loading requirements and of the need for more complicated cooling systems.

At present, the more common SPFC operating at slightly below 100 C are too cool for optimal performance, while SOFC are too hot especially for use in cars. Indeed, at the low temperature of present day SPFC, problems of catalysis and sensitivity to CO poisoning are important,while, in SOFC, the advantages of high temperature mentioned before, are counterbalanced by a number of disadvantages listed in Subsection 9.5.4. For this reason, SPFC research strives for high temperature plastics such as the polybenzimidazole membranes developed by Celanese AG that operate at 200 C. while SOFC research is seeking lower temperature ceramics, such as doped lanthanum gallate (LSGM) and samarium-doped ceria (SDC).

9.3.2   State of the Electrolyte

Most early fuel cells used liquid electrolytes which can leak, require liquid level and concentration management, and may be corrosive. Modern cells use either ceramics for high temperatures or plastics for low temperature. Some second generation utility type fuel cells use molten carbonates.

9.3.3   Type of Fuel

At least in a laboratory, the simplest fuel to use in a fuel cell is hydrogen. However, this gas is difficult to store (see Chapter 11). Especially for automotive use, efforts are being made to use use easily storable substances from which hydrogen can be extracted as needed (see Chapter 10). The extraction process frequently consists of a **steam reforming** step in which a carbon-bearing feed stock is transformed into a **syngas** or **reformate** consisting mainly of H_2 and CO, followed, when necessary, by a **shift reaction** in which the CO in the reformate is made to react with additional water and is transformed into H_2 and CO_2.

[†] CO poisoning of the catalyst may be a serious problem in low temperature cells operating with fossil fuel-derived hydrogen.

Higher temperature fuel cells such as MCFCs and SOFCs can use CO as a fuel so that the H_2–CO reformate can be fed directly to the cells.

Methanol is used in some fuel fells, especially in small ones for use in portable electronic equipment. High reactivity fuels were used when the technology was quite immature. They included such combinations as hydrazine (NH_2NH_2) with hydrogen peroxide (H_2O_2) as oxidant. These fuels are, however, corrosive and expensive.

9.3.4 Chemical Nature of the Electrolyte

Electrolytes can be alkaline, acid, molten carbonates, or ceramic.

Alkaline electrolytes, such as potassium hydroxide (KOH) are commonly used in **electrolyzers** (the dual of fuel cells—instead of generating electricity from fuels, they produce fuels from electricity). Alkalis are avoided in most fuel cells because the presence of CO_2 in air (used as oxidant) causes the formation of insoluble carbonates that spoil the electrolyte. For special applications where pure oxygen is available, KOH fuel cells offer the advantage of high efficiency. This is the case of fuel cells used in space.

Acids tend to be more corrosive than alkalis but relatively weak acids perform well in fuel cells. Phosphoric acid, in particular, was popular.[†] It tolerates carbon dioxide. For good performance it must operate at temperatures between 150 and 220 C and, to keep the liquid from boiling away, a certain degree of pressurization is needed. At lower temperatures, the conductivity of the solution is too small and at higher temperatures, there are problems with the stability of the materials. Solid acids (Subsection 9.5.7) have been proposed as electrolytes for fuel cells. They may contribute to the solution of the vexing methanol cross-over problem (Subsection 9.5.6).

Most ceramic electrolytes, as for instance yttria-stabilized zirconia (YSZ) and SDC, are anion conductors (conductors of negative ions such as O^{--}). However, cation conductors have been proposed.

Solid polymers act, in general as proton conductors—that is, as acids although, as in the case of ceramics, cation conductors have been investigated.

9.4 Fuel Cell Reactions

The chemical reaction in a fuel cell depends on both the type of fuel and the nature of the electrolyte. Some of the most common combinations are listed below and are displayed in the illustrations that follow this subsection.

[†] Phosphoric acid is a benign acid as attested by its daily consumption by the millions of Coca-Cola drinkers.

9.4.1 Alkaline Electrolytes

Hydrogen-oxygen fuel cells with alkaline electrolytes (generally, KOH) use OH^- as the current-carrying ion. Because the ion contains oxygen, water is formed at the anode.

The KOH in the electrolyte dissociates:

$$KOH \rightleftharpoons K^+ + OH^-. \tag{8}$$

Neutral hydrogen at the anode combines with the hydroxyl ion to form water, releasing the electrons that circulate through the external load:

$$2\,H_2 + 4\,OH^- \rightarrow 4\,H_2O + 4\,e^-. \tag{9}$$

At the cathode, the electrons regenerate the hydroxyl ion:

$$4\,e^- + O_2 + 2\,H_2O \rightarrow 4\,OH^-. \tag{10}$$

The KOH is, of course, not consumed. The overall reaction is:

$$2\,H_2 + O_2 \rightarrow 2\,H_2O. \tag{11}$$

9.4.2 Acid Electrolytes

When the electrolyte is acid, H^+ ions are available. These can come from the ionization of the hydrogen (as in the SPFC cells) or from the dissociation of the acid in the electrolyte. Take phosphoric acid:

$$H_3PO_4 \rightleftharpoons 3\,H^+ + PO_4^{---}. \tag{12}$$

In either case, the H^+ ion is replenished by the anode reaction:

$$2\,H_2 \rightarrow 4\,H^+ + 4\,e^-. \tag{13}$$

At the cathode, the H^+ is reduced in the presence of O_2 by the electrons that circulate through the load, forming water:

$$4\,H^+ + 4\,e^- + O_2 \rightarrow 2\,H_2O. \tag{14}$$

The overall reaction is the same as in the previous case. Water is formed at the cathode, and the active ion is hydronium.

9.4.3 Molten Carbonate Electrolytes

Molten carbonate electrolytes are not bothered by carbon oxides. They operate at relatively high temperatures, hence under more favorable kinetics.

When fueled by hydrogen, the reactions are:

$$H_2 + CO_3^{--} \rightarrow H_2O + CO_2 + 2 \ e^-, \tag{15}$$

$$\frac{1}{2}O_2 + CO_2 + 2 \ e^- \rightarrow CO_3^{--}, \tag{16}$$

When the fuel is CO, the anode reaction is:

$$CO + CO_3^{--} \rightarrow 2 \ CO_2 + 2 \ e^-, \tag{17}$$

while the cathode reaction is the same as in the hydrogen case.

9.4.4 Ceramic Electrolytes

Ceramic electrolytes usually conduct negative ions. At the cathode, oxygen is ionized by capturing the electrons that constitute the load current:

$$O_2 + 4 \ e^- \rightarrow 2 \ O^{--}. \tag{18}$$

At the anode, the fuel combines with the O^{--} ions that drifted through the electrolyte, freeing electrons. The fuel may be hydrogen,

$$2 \ H_2 + 2 \ O^{--} \rightarrow 2 \ H_2O + 4 \ e^-, \tag{19}$$

or carbon monoxide,

$$2 \ CO + 2O^{--} \rightarrow 2 \ CO_2 + 4 \ e^-. \tag{20}$$

9.4.5 Methanol Fuel Cells

The anode reaction is

$$CH_3OH + H_2O \rightarrow CO_2 + 6 \ H^+ + 6 \ e^-. \tag{21}$$

The cathode reaction is

$$6 \ H^+ + 6 \ e^- + \frac{3}{2}O_2 \rightarrow 3 \ H_2O. \tag{22}$$

Thus, the overall reaction is

$$CH_3OH + \frac{3}{2}O_2 \rightarrow CO_2 + 2 \ H_2O. \tag{23}$$

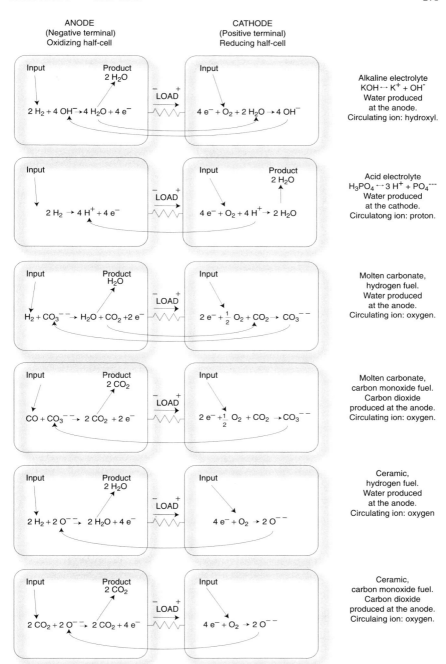

ANODE
(Negative terminal)
Oxidizing half-cell

CATHODE
(Positive terminal)
Reducing half-cell

Input   Product
2 H₂O

LOAD

$2 H_2 + 4 OH^- \rightarrow 4 H_2O + 4 e^-$

Input

$4 e^- + O_2 + 2 H_2O \rightarrow 4 OH^-$

Alkaline electrolyte
$KOH \cdots K^+ + OH^-$
Water produced
at the anode.
Circulating ion: hydroxyl.

Input

LOAD

$2 H_2 \rightarrow 4 H^+ + 4 e^-$

Input   Product
2 H₂O

$4 e^- + O_2 + 4 H^+ \rightarrow 2 H_2O$

Acid electrolyte
$H_3PO_4 \cdots 3 H^+ + PO_4^{---}$
Water produced
at the cathode.
Circulatong ion: proton.

Input   Product
H₂O

LOAD

$H_2 + CO_3^{--} \rightarrow H_2O + CO_2 + 2 e^-$

Input

$2 e^- + \frac{1}{2} O_2 + CO_2 \rightarrow CO_3^{--}$

Molten carbonate,
hydrogen fuel.
Water produced
at the anode.
Circulating ion: oxygen.

Input   Product
2 CO₂

LOAD

$CO + CO_3^{--} \rightarrow 2 CO_2 + 2 e^-$

Input

$2 e^- + \frac{1}{2} O_2 + CO_2 \rightarrow CO_3^{--}$

Molten carbonate,
carbon monoxide fuel.
Carbon dioxide
produced at the anode.
Circulating ion: oxygen.

Input   Product
2 H₂O

LOAD

$2 H_2 + 2 O^{--} \rightarrow 2 H_2O + 4 e^-$

Input

$4 e^- + O_2 \rightarrow 2 O^{--}$

Ceramic,
hydrogen fuel.
Water produced
at the anode.
Circulating ion: oxygen

Input   Product
2 CO₂

LOAD

$2 CO_2 + 2 O^{--} \rightarrow 2 CO_2 + 4 e^-$

Input

$4 e^- + O_2 \rightarrow 2 O^{--}$

Ceramic,
carbon monoxide fuel.
Carbon dioxide
produced at the anode.
Circulaing ion: oxygen.

9.11

9.5   Typical Fuel Cell Configurations

9.5.1   Demonstration Fuel Cell (KOH)

It is relatively simple to build a small demonstration fuel cell. The design is self-evident from Figure 9.4.

The six holes near the rim of the two Lucite covers allow the passage of screws that hold the system together. Diameter of the device is 8 cm.

Fuel sources are two toy balloons, one containing oxygen and the other, hydrogen. Excess gas is vented into a beaker with water.

It can be seen that the cell is symmetrical—there is no structural difference between the anode and the cathode side. The electrodes are made of platinum mesh. Prior to using the cell, platinum black is sprinkled over the mesh to enhance the catalytic action.

Open-circuit voltage is about 1 V and the cell will deliver some 200 mA at 0.7 V, a power of 140 mW.

The electrolyte consists of a disc of filter paper soaked in KOH solution.

The above design is quite primitive and dates from the times when it was difficult to obtain ion exchange membranes. Much better demonstration fuel cells can currently be obtained from a number of vendors. Ask Google for "Fuel cell demonstration kits." Some of the available kits are well designed and easy to operate. They all require distilled water, but that should be no problem. Particularly instructive are kits that incorporate a photovoltaic converter, an electrolyzer and a simple hydrogen storage system, in addition to the fuel cell. They demonstrate that hydrogen can be produced with absolutely no CO_2 emission. The problem is, of course, that it is not yet economically feasible to do so in large scale.

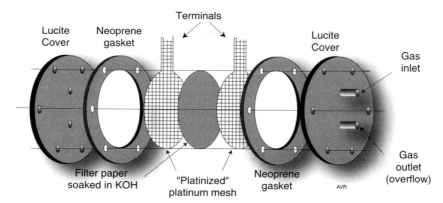

Figure 9.4   Exploded view of a demonstration fuel cell.

9.5.2 *Phosphoric Acid Fuel Cells (PAFC)*

9.5.2.1 A Fuel Cell Battery (Engelhard)

Since fuel cells are low-voltage, high current devices, they must be connected in series. The simplest way of doing so is to use a **bipolar** configuration: a given electrode acts as anode for one cell and (on the other face) as cathode for the next.

The Engelhard phosphoric acid cell (obsolete) is of the air-cooled bipolar type. Its construction can be seen in Figure 9.5.

The **bipolar electrode** was made of either aluminum or carbon. Gold was plated on aluminum for protection and flashed on carbon to reduce contact resistance.

The plate was 3 mm thick and had grooves or channels machined on both faces. The oxidant channels ran perpendicular to those for hydrogen on the opposite face. The plate was rectangular with its smaller dimension along the air-flow direction to minimize pressure drop. The channels, in addition to leading in the gases, served also to increase the active surface of the electrodes.

The electrolyte soaked a "cell laminate" held in place by a rubber gasket. It can be seen that this type of construction does not differ fundamentally from that used in the demonstration cell described in the preceding subsection.

The cell operated at 125 C. The oxidant was air, which entered at ambient temperature and left hot and moist carrying away excess heat and water (formed by the oxidation of the fuel in the cell). The air flow was adjusted to provide sufficient oxygen and to assure proper heat and water removal.

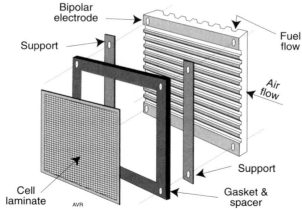

Figure 9.5   One element of the Engelhard PAFC.

Apparently, no catalysts were used. This, combined with the relatively low operating temperature (phosphoric acid cells frequently operate at temperatures above 150 C), results in somewhat adverse kinetics and, consequently, in low voltage efficiency.

Engelhard marketed cells of this type in a 750 W battery fed by hydrogen obtained from the cracking of ammonia. Ammonia is a hydride of nitrogen and is a convenient way of storing hydrogen. The system consisted of an ammonia source (a pressure cylinder), a dissociator or **cracker**, a scrubber, the fuel cells, and ancillary pumping and control mechanisms.

The fuel cells themselves operated at roughly 45% efficiency (optimistic?) referred to the higher heating value of ammonia, but the system needed to divert fuel to heat the dissociator that worked at 850 C thus reducing the overall efficiency at rated power to about 30% and to much less at lower power levels.

9.5.2.2 First-Generation Fuel Cell Power Plant

One early effort to adapt fuel cells for dispersed utility-operated power plant use was made by United Technologies Corporation. It was a 4.8 MW module fueled by natural gas, to be installed in Manhattan.

The UTC fuel cells used phosphoric acid and delivered 0.64 V per cell. At the operating temperature of 150 to 190 C, current densities of 1600 to 3100 A m^{-2} were possible. The life of the cells was expected to be some 40,000 hours (around 5 years).

The UTC project in New York suffered such lengthy delays in the approval process by the city government that when finally it was cleared for operation in 1984, the equipment had deteriorated to the point that it was uninteresting to put it in operation. The first commercial demonstration of this fuel cell application took place in Japan where a 4.5 MW unit started operation in 1983. The same company (Tokyo Electric Power Company) now operates an 11 MW PAFC facility built by Toshiba using US build cells (International Fuel Cell Company a subsidiary of UTC).

ONSI, a subsidiary of the International Fuel Cell Co. manufactured a 200-kW PAFC power plant, known as PC25. Since 1991 until the beginning of 2001, worldwide sales exceeded 220 plants. They delivered both electricity (at 37% efficiency) and heat. The latter could be in the form of 260 kW of hot water (at 60 C) or 130 kW of hot water plus another 130 kW of heat at 120 C. They were fueled by natural gas or, in some cases, by biogas from anaerobic digesters. By 2004, some PC25s could be bought in the surplus market.

Modern PAFCs use platinum catalysts and are vulnerable to CO in the fuel stream (see SPFC, further on). Fortunately, the tolerance of Pt to CO increases with operating temperature: at 175 C, CO concentration can be as high as 1%, and, at 190 C, as high as 2%. This contrasts with the more stringent requirements of the cool SPFCs, which, at 80 C require hydrogen with less than 10 ppm of CO unless special procedures or special catalysts are employed.[†] On the other hand, whereas the SPFCs have a very stable electrolyte, the phosphoric acid in PAFCs can be degraded by ammonia or sulfur whose concentrations in the fuel stream must be kept low.

9.5.3 Molten Carbonate Fuel Cells (MCFC)

In 1988 the American Public Power Association (APPA) together with the Electric Power Research Institute (EPRI) promoted an international competition to design fuel cells tailored to urban needs. The winning project was a 2-MW MCFC developed by Energy Research Corporation (ERC), now renamed Fuel Cell Energy, Inc. These "second generation" cells were evaluated and in 1996 appeared to have come close to their design performance. The plant achieved the excellent measured efficiency of 43.6% when delivering 1.93 MW ac to the grid.

The cells were assembled in bipolar stacks (see the Engelhard battery) bracketed by nickel-coated stainless steel end plates and separated by bipolar plates made of the same material. The cells themselves were a sandwich of an anode, an electrolyte matrix, and a cathode. The nickel-ceramic anode and the nickel oxide cathode were porous. The reactive gases were fed in through the side opposite to the one in contact with the electrolyte which was a mixture of lithium and potassium carbonates held in lithium aluminate matrix.

Operating temperature was between 600 and 700 C, high enough not to require expensive platinum based catalysts and to permit internal reforming of natural gas. The exhaust gases could efficiently drive a steam turbine in co-generation schemes. Life was limited by the slow dissolution of the cathode and the carbonate poisoning of the reforming catalyst.

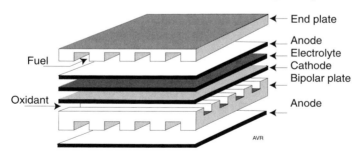

Figure 9.6   Exploded view of a MCFC unit.

[†] Roughly, the tolerance, λ (in ppm), of platinum catalysts to CO as a function of the temperature, T (in kelvins), is given by $\lambda = 255 \times 10^{-12} \exp(T/14.5)$.

Based on the experience gained from operating the Santa Clara plant and from other development work, Fuel Cell Energy, Inc. was (2004) offering 2 MW plants for commercial sale.

9.5.4 Ceramic Fuel Cells (SOFC)

One of the most critical components in any type of fuel cell is the electrolyte. SOFC are no exception. The ceramics used as electrolytes must have a much higher conductivity for ions than for electrons, a requirement that is not too difficult to achieve; however, high ionic conductivity usually requires operation at elevated temperatures. Figure 9.7 illustrates the strong influence of the operating temperature on the achievable power density in a fuel cell using zirconia (YSZ) electrolyte. The performance of the cell is limited by the relatively low conductance of the latter. The data in the figure are from Global Thermoelectric Inc. See Ghosh, D., et al. No information on the thickness of the electrolyte was given. There is an obvious temptation to operate SOFC at temperatures lower than the current 1100 K:

1. High temperatures require the use of expensive alloys.
2. Temperature cycling introduces mechanical stresses.
3. The electrodes (but not the electrolyte) must be porous, however high temperatures promote their sintering causing them to become progressively more impermeable to fuel and air.
4. High temperatures promote diffusion of constituent of the electrodes into the electrolyte.

Items 2 though 4 reduce the lifetime of the fuel cell.

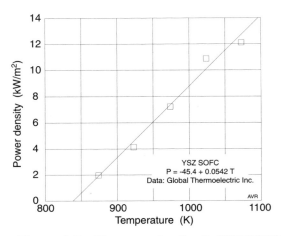

Figure 9.7   The power density of a YSZ SOFC rises sharply with operating temperature.

The conductance of a ceramic electrolyte depends on three factors:

1. The operating temperature discussed above.
2. The thickness of the electrolyte.
3. The nature of the electrolyte, which, in most current cells is yttria-stabilized zirconia (YSZ), typically $(ZrO_2)_{0.9}(Y_2O_3)_{0.1}$.

In order to achieve acceptable conductances, the electrolyte must be quite thin. Global Thermoelectrics has demonstrated 5 μm-thick electrolytes (Gosh et al.) but current commercial cells use thicknesses almost one order of magnitude larger. The reason for this is that the electrolyte must be impermeable to gas, a condition difficult to satisfy when very thin layers are used—porosity will be high and small pin holes are apt to occur. Layer densities of some 98% of the theoretical density are desirable. Additionally, thin electrolytes are exceedingly fragile and must be supported by either the anode or the cathode (i.e., they must be thin compact ceramic layers deposited on one of the two electrodes).

Lower temperature operation can be achieved by employing ceramic materials other than the popular YSZ. For example, SDC is a ceramic electrolyte that, at any given temperature, has much higher ionic conductivity than YSZ. See Sub-subsection 9.5.4.2. Ceramics that conduct protons instead of negative ions have been demonstrated in laboratory. It is expected that, at 700 C, these ceramics, working with the same power density as zirconia at 1000 C, will deliver some 10% higher efficiency. A promising "low temperature" ceramic proton conductor is $BaCeO_3$. See Rocky Goldstein (EPRI, Advanced Fossil Power System Business Unit). Any ceramic used as electrolyte must not react chemically with the materials with which it is in contact.

The most popular cathode material for YSZ fuel cells is strontium-doped lanthanum manganite, LSM ($La_xSr_{1-x}MnO_3$), which is compatible with the electrolyte. The strontium doping increases the electronic conductivity of the material. It is mainly an electron conductor and, consequently, a chemical reaction is required in which the negative ion conducted by the electrolyte must be transformed into electrons conducted by the cathode. This reaction must occur at a triple-point where cathode, electrolyte, and oxygen come together. The triple-point area can be increased, as shown by Yoon et al., by coating the pores of the cathode with macroporous YSZ. The cathode material must exhibit high porosity (say, 50%) and must be stable in an oxidizing atmosphere. Porosity is achieved by incorporating a pore forming substance such as starch in the powder mixture that will form the ceramic.

The fuel electrode must have good catalytic action if internal fuel reforming is used.

The anode, being in a reducing atmosphere, could be made of porous nickel were it not for the high coefficient of thermal expansion of the metal. To correct this, the nickel is dispersed in a matrix of yttria-stabilized zirconia forming a cermet.[†] YSZ powder is mixed with NiO_2 and sintered together in an environment that reduces the nickel oxide to dispersed metallic nickel leaving behind a porous ceramic. The fuel electrode must have good catalytic action if internal fuel reforming is used. If the temperature is much smaller than some 700 C, internal reformation may become impossible or difficult.

All three layers, anode, electrolyte, and cathode, must have matching thermal expansion to avoid delamination and must not react chemically with one another or diffuse into the neighbor.

To form a stack, interconnections between the different cells are required. They must have good mechanical properties and good electric conductivity; they must have appropriate thermal expansion coefficients and good resistance to corrosion and chemical compatibility with the rest of the cells. Above all, they must be of low cost.

Iron alloys, rendered corrosion resistant by the addition of chromium, satisfy most of the above requirements. As in all stainless steels, their rust resistance depends on the formation of a thin scale of chromia on the surface that effectively seals the bulk of the material from the corroding atmosphere. Unfortunately, chromia, at the high operating temperature of the fuel cells forms volatile CrO_3 and $CrO_2(OH)_2$, which contaminate the cathode severely degrading the performance. CrO_3 is the dominant chromium vapor in cells using oxygen or very dry air as oxidant, while $CrO_2(OH)_2$ (chromium oxyhydroxide) is dominant when moist air is used. Chromium vaporization can be reduced by operating at low temperature and using dry air. The latter may not be economically attractive.

Power densities as high as 12 kW/m^2 have been demonstrated in the laboratory (Pham, A. Q., et al.). These levels are below the 20 kW/m^2 reached by modern SPFC (Ballard) and refer to individual cells, not to the whole stack. SOFC in or near production show a more modest power density of some 3 kW/m^2 (Siemens Westinghouse). This may not be a major disadvantage in stationary applications, but may be more of a problem in automotive use where compactness is desirable.

SOFC have high efficiencies (over 50%) especially if used in hybrid (cogeneration) arrangements when overall efficiencies approaching 60% have been achieved. Since no corrosives are used (such as phosphoric acid in PAFC or molten carbonates in MCFC), long life is possible.

[†] Cermets are combinations of ceramic materials with metals.

Chromium

Chromium forms compounds in which it can assume at least three different oxidation states,[†] therefore it has a number of different oxides: chromous oxide, CrO, also called chromium monoxide, oxidation state II, chromic oxide, Cr_2O_3, also called chromium sesquioxide or **chromia**, oxidation state III, and chromium trioxide, CrO_3, oxidation state VI.

The significance, as far as the use of iron-chromium alloys in fuel cells is concerned, is that, whereas chromia is highly refractory (melting point 2435 C), chromium trioxide is quite volatile, having a melting point of only 196 C. The conditions inside of a fuel cell lead to the transformation of stable chromium oxides into volatile compounds capable of contaminating the cathode of the cell.

[†] Oxidation state is the difference in the number of electrons of a neutral atom and that of the corresponding ion in an ionic compound.

Continuous operation for over 35,000 hours has been demonstrated. See Bessette et al. SOFC have unlimited shelf life. Many of the ceramics used in fuel cells have a perovskite structure. See the box on the next page.

In negative ion solid electrolyte fuel cells, the air electrode (cathode) reaction is that of Equation 21 and the fuel electrode reaction (anode) is either that of Equation 22 or 23 or both, depending on the fuel used.

SOFC are usually either planar or tubular. The latter, as exemplified by the Siemens Westinghouse cells, have the great advantage of doing away with seals that plague planar cells.

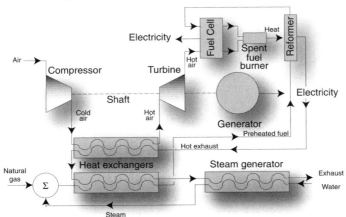

Figure 9.8   An arrangement for a SOFC/microturbine combined cycle plant (Fuller et al.).

Perovskite

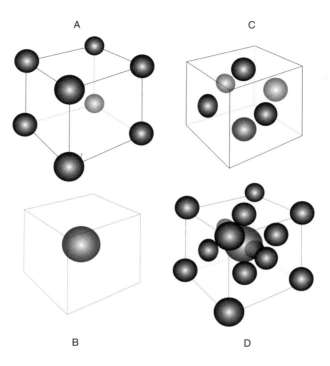

Perovskite is the mineral $CaTiO_3$. By extension, the word designates any substance having the empirical formula ABC_3 (where, C usually corresponds to oxygen) which crystallizes in the perovskite arrangement.

Consider a cubic crystalline unit cell having 8 atoms, A, at the eight vertices (see "A" in the figure). When these unit cells are stacked together, each of the atoms is shared among 8 different, adjacent cells. In other words, there is, on average, exactly 1 atom A per cell.

Now refer to "B" in the figure. There is 1 single B atom per unit cell.

Finally, refer to "C." C atoms are centered on the face of the cube. There are 6 faces, but each atom is shared by 2 adjacent unit cells. Thus, on average, there are 3 C atoms per cell. In "D," the three arrangements are put together yielding a unit cell with the empirical formula ABC_3.

Perovskites are important in SOFC electrolytes and in high temperature superconductors.

9.5.4.1   High Temperature Ceramic Fuel Cells

Much work in the area of tubular cells has been done by Siemens West-inghouse. This company had in 2001 a 100 kWe plant in the Netherlands that operated unattended, generating power into the grid. It also had a 220 kWe hybrid (operating in conjunction with a micro gas turbine) at the Southern California Edison Co., in Los Angeles, CA. A 1 MWe plant with 58% efficiency was being installed in Ft. Meade, Maryland, for the Environmental Protection Agency while another plant of equal power was planned for Marbach, Germany.

When starting up and when stopping, SOFC are subjected to large am-plitude thermal cycling, which creates serious problems with seals. Siemens Westinghouse has come up with a clever seal-free configuration using tubu-lar cells. Each individual cell consists of a triple-layered ceramic tube, as depicted in Figure 9.9. The inner layer is the air electrode (cathode), the middle layer, the electrolyte, and the outer, the fuel electrode (anode). Manufacture of the fuel cell starts with an air electrode made of lanthanum manganite ($La_{0.9}Sr_{0.1}$). The YSZ electrolyte is built up on the cathode tube and the anode is deposited on top of that. Connections to the anode are made directly to the outer electrode. To reach the cathode, an inter-connecting strip is placed longitudinally on the tube and penetrates to the inner layer (see Figure 9.9). This interconnecting strip must be stable at both the oxidizing environment of the air electrode and the reducing one at the fuel electrode. It must also be impermeable to gases. These re-quirements are met by lanthanum chromite. To enhance conductivity, the material is doped with Ca, Mg, Sr, or other metals of low valence.

Figure 9.10 shows how bundles of tubes can be stacked in series-parallel connections forming modules. Nickel felt made of nickel fibers sintered together provides mechanically compliant interconnections. Notice that all these interconnections take place in a chemically reducing environment.

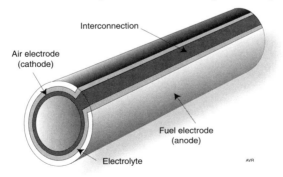

Figure 9.9   Tubular concentric electrodes of the
Siemens Westinghouse SOFC.

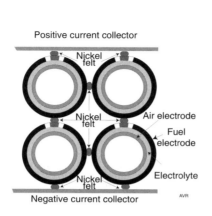

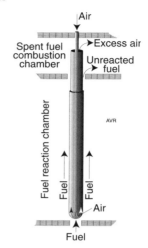

Figure 9.10   Tubular cells can easily be stacked in series-parallel connections.

Figure 9.11   Tubular cells are stacked in seal-free modules.

The modules consist of two chambers as indicated in Figure 9.11. The lower and longer **fuel reaction chamber** receives the fuel from openings at the bottom. The fuel flows up and most of it is used up by the anode.

Unreacted fuel, some 15% of the total, exits into the **spent fuel combustion chamber** where it is mixed with the excess air that comes out from the top of each tube and burns producing heat used in preheating the input air and increasing the temperature (up to 900 C) of the exhaust gases in order to drive a bottoming cycle generator such as a turbine.

If an adequate amount of water vapor is mixed with fossil fuel gases, automatic reformation will take place thanks to the catalytic action of the nickel in the anode. (See Subsection on modern hydrogen production in Chapter 10.)

$$CH_4 + H_2O \rightarrow CO + 3H_2.$$

"Standard" Siemens Westinghouse cells are 22 mm diameter tubes of varying lengths, usually 1.5 m. Better performance is obtained by the use of of alternate geometries.

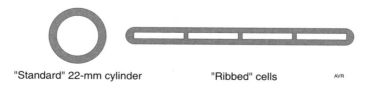

"Standard" 22-mm cylinder   "Ribbed" cells

Figure 9.12   Cross-section of cylindrical and flattened "ribbed" cells.

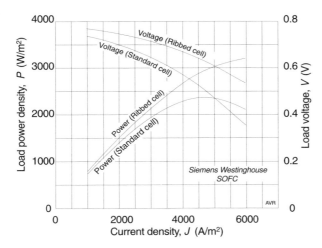

Figure 9.13   "Ribbed" cells show substantially better
performance than cylindrical ones.

Flattening the cells into the "ribbed" configuration not only improves
the stacking factor (so that more units fit into a given volume) but also
reduces production cost. These flattened cells have incorporated improvements that substantially increase cell efficiency. See the comparison of the
performances of cylindrical and "ribbed" cells shown in Figure 9.13.

9.5.4.2 Low Temperature Ceramic Fuel Cells

It would be useful to operate SOFC at temperatures lower than those
currently used with YSZ. Among other things the life of the stacks would
presumably be longer. It would also be simpler if one could build a **single
chamber** fuel cell.

In the usual fuel cell, fuel and oxidizer are fed separately to the device,
each gas to its own electrode. In a single chamber cell, the fuel is mixed with
air in proportions which are too rich to allow an explosion. The mixture
is fed simultaneously to both electrodes, one of which reacts preferentially
with the fuel, the other with the oxygen.

In the cell described by Hibino, the SDC electrolyte was a ceramic disc
ground to 0.15 mm thickness. The anode was a layer of nickel-enriched
SDC while the cathode consisted of $Sm_{0.5}Sr_{0.5}CoO_3$—samarium strontium
cobalt oxide.

The ethane (18%)/air mixture in contact with the nickel-rich anode is
reformed into a hydrogen/carbon monoxide mixture according to

$$C_2H_6 + O_2 \rightarrow 3\,H_2 + 2\,CO \tag{24}$$

The two gases in the reformate are oxidized:

9.23

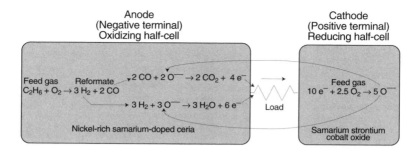

Figure 9.14   Cell reactions of the SDC low temperature SOFC.

$$3 \text{ H}_2 + 3 \text{ O}^{--} \rightarrow 3 \text{ H}_2\text{O} + 6 \text{ e}^-, \tag{25}$$
$$2 \text{ CO} + 2 \text{ O}^{--} \rightarrow 2 \text{ CO}_2 + 4 \text{ e}^-. \tag{26}$$

Thus, each ethane molecule yields 10 electrons that circulate in the load. The 5 oxygen ions come from the cathode (by moving through the electrolyte). They are produced by combining the 10 electrons with 2.5 oxygen molecules from the fuel/air mixture:

$$10 \text{ e}^- + 2.5 \text{ O}_2 \rightarrow 5 \text{ O}^{--} \tag{27}$$

The single chamber configuration simplifies the construction and makes the cell more resistant to both thermal and mechanical shocks.

The performance of some of the small experimental cells prepared by Hibino is quite promising: over 4 kW/m^2 at 500 C and over 1 kW/m^2 at 350 C. Compare this with the 20 kW/m^2 of the much more developed Ballard SPFC.

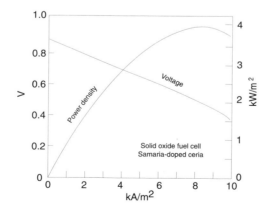

Figure 9.15   Characteristics of an experimental SDC fuel cell developed by Hibino.

The $v-i$ characteristics of one of the experimental SDC cells, with 0.15 mm electrode thickness, is shown in Fig 9.15.

9.5.5 Solid-Polymer Electrolyte Fuel Cells

This is conceptually the simplest type of fuel cell and, potentially, the easiest to manufacture. As the name suggests, the electrolyte is a solid membrane made of an ion-conducting polymer. The ion can be positive (a cation), usually a proton, or negative (an anion) usually an OH^-.

SPFCs are safer than liquid or molten electrolyte cells because they use noncorrosive (and of course, nonspillable) electrolytes. They can stand fairly large differential pressures between the fuel and the oxidant sides, thus simplifying the management of these gases.

A short history of the development of SPFC cells can be found in an article by David S. Watkins.

SPFCs were pioneered by GE, starting in 1959. In 1982, the technology was transferred to United Technology Corporation/Hamilton Standard. Little additional progress was made until Ballard Power Systems[†] took up further development and pushed it, with the collaboration of Daimler Benz, to the production stage. Prior to Ballard, fuel cell cars were predicted to come on the market by 2020. Now, the general expectation is for fuel cell cars in general use by 2010!

The progress made by the original GE effort is illustrated by the the growth of power densities from the initial 50 W/m^2 in 1959 to over 8 kW/m^2 in 1982, a 160-fold improvement.

Many factors contributed to such progress, including better membranes (early devices used sulfonated polystyrene, later ones used Nafion), thinner membranes (from 250 μm to 123 μm), higher operating temperatures (from 25 C to 150 C), and better catalysts.

SPFC are fast emerging as the preferred solution for automotive use and as a competitor for fixed power plants. The extremely fast advances in this type of cell, a sign of a young and immature technology, is attested to by the exponential improvement in power density of Ballard's cells illustrated in Figure 9.16. Ballard cells exceed 20 kW per square meter of active surface (compare with 8 kW/m^2 for the best GE cells). The power-to-mass ratio (gravimetric power density) has also improved substantially—it exceeds 0.7 kW/kg, approaching the values for a modern aircraft engine.[††] Thus, power-to-volume and power-to-mass ratios of a modern SPFC are already well into the range of acceptability in automotive applications. What needs to be improved—substantially—is the power-to-cost ratio.

[†] For Ballard's history, read Tom Koppel.

[††]To be fair, one must compare the aircraft engine weight with the **sum** of the fuel cell stack weight plus that of the electric motor.

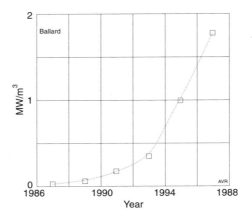

Figure 9.16   The power density of Ballard fuel cells has risen exponentially since 1986. Now, more than 2 MW can be generated by 1 m^3 of cells (2 kW/liter).

Cost can be reduced in many areas. Most obvious is the need to create a market large enough to permit mass production. Cheaper membranes and cheaper catalysts are a must. Daimler forecasts prices of \$20 to \$30 per kW for the fuel cell stack in automobiles when the technology is mature.

9.5.5.1 Cell Construction

The construction of Ballard fuel cells illustrates the manner in which SPFCs are put together. The heart of a cell is the **membrane/electrode assembly** (MEA), a triple-layer structure consisting of two thin porous sheet electrodes (frequently carbon cloth) separated by the ion exchange membrane electrolyte.

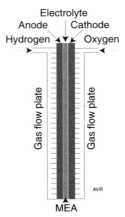

Figure 9.17   Structure of a SPFC.

A small amount of catalyst is applied to the side of the electrodes in contact with the electrolyte. The three layers are bonded together by heating the sandwich under pressure to a point where the electrolyte becomes soft. The MEA measure less than a millimeter in thickness.

Contacts to the electrodes are made through gas flow plates (metal or, more commonly, graphite) on whose surface grooves have been formed to distribute the fuel and the oxidant uniformly over the whole active area of the cell. An example of the complicated pattern of grooves in the gas flow plates appears in Figure 9.18.

The plates have a number of holes near the corners constituting a manifold through which fluids flow. All but the two end plates are bipolar— one face connects to the anode, the other, to the cathode of the cells. The plate depicted shows its cathode side to which oxidant is fed and from which oxygen-depleted oxidant and product water are removed. The anode side of the flow plate is fed through the smaller hydrogen holes. This is because air is delivered in larger quantities than what would be demanded by stoichiometry—the excess air is used to cool the cells and remove the product water.

The design of the groove pattern can greatly influence the performance of the cell. The grooves should not trap any of the reaction water thereby "drowning" areas of the cell and must distribute the reactants uniformly. The bipolar configuration facilitates stacking the cells so that they are electrically in series.

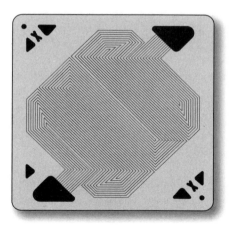

Figure 9.18   Grooves are formed on the surface of the gas flow plates to distribute fuel and oxidant to the cell electrodes. In this cathode face of the plate, oxidant enters and exits through the large, roughly triangular holes at the upper right-hand and lower left-hand corners of the plate. The remaining holes are for hydrogen and water (Ballard).

9.27

9.5.5.2 Membrane

Though most membranes are of the proton-exchange type, anion-exchange ones have been proposed claiming a number of advantages. Among these are their immunity to CO_2 fouling, simplified water management requirements, and the possibility of using less expensive non-noble metal catalysts.

Membranes that act as electrolytes must be excellent ionic conductors, but must not conduct electrons. They must be reasonably impermeable to gases and sufficiently strong to withstand substantial pressure differences.

Many such membranes consist of a Teflon-like backbone to which side chains containing a sulfonic (SO_3) group are attached. These acid groups are immobile and cannot be diluted or leached out.

The thinner the membrane, the smaller its resistance to the flow of protons. Typically, membranes 50 to 180 μm thick are used.[†] If they are too thin, gas crossover occurs and the performance of the cell is degraded, as witnessed by a reduction in the V_{oc} of the cell.

For a membrane to be a good proton conductor, its SO_3 concentration must be sufficiently high. The ratio between the mass of dry polymer (in kg) and the number of kilomoles of SO_3 sites is called the **equivalent weight** (EW). Smaller EWs lead to higher ion conductivity. For membranes with the same backbone, the shorter the side chains, the smaller the EW. Thus Nafion having a side chain,

$$-O - CF_2 - CF_3CF - O - CF_2 - CF_2 - SO_3H,$$

has a larger EW than the Dow membrane with its shorter chain,

$$-O - CF_2 - CF_2 - SO_3H.$$

In fact, Nafion has a typical EW of 1100 kg per kmoles whereas the Dow membrane has an EW of some 800.

The proton conductivity of a membrane also depends on it being thoroughly hydrated—a dry membrane loses most of its conductivity and, for this reason, careful water management is essential (see below).

Membranes can be surprisingly costly. In the 80s, Nafion sold at \$800 per m^2. One reason is that this material was created to last 100,000 hours of operation in a chlorine producing plant. On the other hand, an automobile is not expected to survive much more than 300,000 km, which, even at a very modest speed of 30 km/h, represents only some 10,000 hours of operation. Nafion is, as far as automotive fuel cells are concerned, "overdesigned" by one order of magnitude. Ballard set out to develop its own proprietary membranes which contains much less fluorine and is much cheaper in spite of being even more efficient than Nafion and the Dow membrane. Surprisingly the life of the membrane exceeds 15,000 hours, more than enough for an automotive fuel cell.

[†] The paper commonly used in copying machines is roughly 100 μm thick.

9.5.5.3 Catalysts

Because of their low operating temperature, the kinetics of SPFC are unacceptable unless special catalysts are used. Catalysts typically use expensive noble metals and greatly contribute to the cost of the cells. Great effort has been made to lower the amount of material used. In earlier cells, platinum loading of about 4 mg/cm^2 added more than \$500 to the cost of each square meter of active fuel cell area. Fortunately, techniques that permit a reduction of a factor of 10 in the platinum loading are now practical and the Los Alamos National Laboratory has had success with even less catalyst. The British firm Johnson and Matthey, producers of catalytic converters for conventional (IC) cars, is working with Ballard to reduce the cost of catalysts in SPFCs. The quest for reduced platinum loading is partially based on the fact that in a catalyst it is only the surface that is active. By reducing the size of the grains, the surface to volume ratio is increased—that is, less platinum is needed for a given catalyst area.

Another approach is to use small grains of graphite upon which a thin layer of platinum has been deposited. It is also possible to reduce the amount of platinum by alloying it with ruthenium which is cheaper, acts as a catalyst, and is more CO-tolerant. Indeed, one problem with some catalysts is their vulnerability to the presence of certain contaminants—especially CO—in the hydrogen stream. If the cell is to be fed with hydrogen extracted from a carbon-bearing "carrier" such as methane or methanol, then the reformate will invariably contain some carbon monoxide—perhaps as much as 1% to 2%. Much lower concentrations are required for proper operation of the cells. It may be costly to reduce the CO content below 100 ppm.

One possibility is to force the reformate through a palladium filter which will eliminate most of the contaminant. The Los Alamos National Laboratory has built such a filter based on a thin tantalum sheet plated with palladium on both sides. Idatech, of Bend, OR, has developed fuel processors capable of producing very pure hydrogen thanks to the use of palladium filters (see Chapter 10). Other types of catalysts are being considered and it may even be possible to use enzymes for this purpose.

Hydrogen containing small amounts of CO (say 100 ppm) will drastically impair the performance of low temperature fuel cells with pure Pt catalysts as shown in Figure 9.19. However, if a small amount of O_2 is added to the fuel, it will selectively combine with the carbon monoxide and almost complete recovery of the cell performance is achieved.

The CO vulnerability of Pt catalysts diminishes with increasing temperature. At 80 C, CO concentration must be kept below 10 ppm. At 90 C, where Ballard cells operate, higher concentrations are acceptable. Alloy catalysts (Pt+Ru, for instance) are less vulnerable to CO. Ballard has shown that a reformate containing as much as 2% CO can be used if it is passed over a platinum-on-alumina catalyst and Pt-Ru is used in the anode.

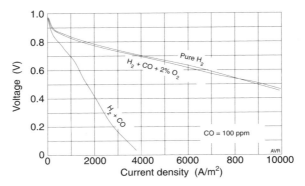

Figure 9.19   When a SPFC using pure Pt catalysts is fed hydrogen contaminated by 100 ppm of CO, its performance suffers drastically. However, if a small amount of oxygen is added to the fuel stream, it will selectively combine with the CO and the performance is then essentially indistinguishable from that with pure hydrogen.

There is a possibility of avoiding altogether the use of expensive platinum based catalysts. The anode reaction consists of splitting the hydrogen molecule into two protons and two electrons (see Equation 3), a feat performed by many anaerobic bacteria (Chapter 13) that use enzymes called **hydrogenases**. Synthetic hydrogenases have been created by Robert Hembre of the University of Nebraska, Lincoln, using compounds containing ruthenium and iron. Ruthenium being substantially cheaper than platinum may have a future as a catalyst in low temperature fuel cells.

If the fuel cell catalysts become poisoned, all is not lost—a short exposure to pure H_2 will completely rejuvenate the cell.

CO is not the only contaminant to avoid. SPFCs have extremely low tolerance for sulfur compounds (H_2S, for instance) and for ammonia (NH_3).

9.5.5.4 Water Management

Present-day ion-exchange membranes are an acid electrolyte. Therefore, water forms at the cathode where it would collect and "drown" the electrode—that is, it would impede the penetration of the oxygen to the active surface. To reduce the tendency of the water to collect in the pores and interstices of the MEA, a hydrophobic material is applied (typically a Teflon emulsion).

Water is removed by the flow of oxidant which is generally supplied well in excess of the stoichiometric requirement. Usually the amount of air circulated is double of that needed to furnish the correct amount of oxygen.

Although the anode reaction (Equation 3) indicates the formation of protons, the ion that actually migrates through the electrolyte is hydronium —a hydrated proton. Thus, water is consumed at the anode because it is electrically "pumped" to the cathode, tending to desiccate the membrane.

This drying is compensated, in part, by water from the cathode that diffuses back through the membrane driven by the concentration gradient. Nevertheless, there is a tendency to dehydrate the proton-exchange membrane which tends to wrinkle and have its proton conductivity drop catastrophically. To avoid desiccation, a partial pressure of at least 50 kPa of H_2O must be maintained in both the fuel and oxidant streams. If the feed pressure is 100 kPa (1 atmos), the partial pressure of H_2 would be some 50 kPa and the output voltage would fall correspondingly. For this reason, the fuel cell is usually pressurized to 300 kPa (3 atmos) leading to a hydrogen partial pressure of 250 kPa, compatible with an efficient fuel cell operation.

Careful water management is necessary to avoid either "drowning" or "desiccation" of the membrane.

9.5.6 Direct Methanol Fuel Cells

As discussed in the preceding subsection, one attractive alternative for carrying hydrogen on board of a fuel-cell vehicle is to use a hydrogen carrier, a substance from which hydrogen can be extracted as needed by some chemical reforming processes. Methanol is a leading candidate as a hydrogen carrier for vehicular applications. Nevertheless, the need for separate fuel processing equipment in the vehicle adds to the cost and complexity of the system. An obvious solution is to develop fuel cells that can use methanol directly without the need of pre-processing.

The Jet Propulsion Laboratory has done considerable work on the development of direct methanol fuel cells (DMFC) (see Halpert et al.).

The fuel used is a low concentration (3%) methanol in water solution while air is the oxidant.

The anode reaction is

$$CH_3OH + H_2O \rightarrow CO_2 + 6H^+ + 6e^-. \tag{28}$$

The cathode reaction is

$$6H^+ + 6e^- + \frac{3}{2}O_2 \rightarrow 3H_2O. \tag{29}$$

Thus, the overall reaction is

$$CH_3OH + \frac{3}{2}O_2 \rightarrow CO_2 + 2H_2O. \tag{30}$$

Halpert (1997) reports a cell producing 5.0 MJ of energy per liter of methanol consumed when operated at 90 C and 2.4 atmospheres of air pressure. This can be interpreted as an efficiency of 28% if referred to the higher heat of combustion of the alcohol or 32% if referred to the lower.

The JPL DMFC is a solid-polymer fuel cell using a Nafion 117 membrane with a fairly high loading of noble-metal catalyst (2 to 3 mg of Pt-Ru per cm^2). The combination of Nafion with high catalyst loading leads to high costs. The search for better anode catalysts can be greatly accelerated by the technique described by Reddington (1998) that permits massively parallel testing of large number of catalyst samples. See also *Service,* 1998.

One advantage of DMFCs is that since the anode is in direct contact with water, problems of membrane dehydration are circumvented and the constantly flowing liquid simplifies heat removal. Pollution problems are greatly alleviated because only carbon dioxide and water are produced whereas carbon monoxide is generated when methanol is reformed into hydrogen for use in other types of cells.

"Methanol crossover" (i.e., the transport of methanol from anode to cathode through the Nafion membrane) severely reduces the cell efficiency. In the JPL cell mentioned above, crossover consumed 20% of the fuel. Efforts are being made to develop new low cost membranes less subject to this difficulty. It appears that the figure of 10% has already been achieved. Lower crossover rates will permit operation with more concentrated methanol mixtures yielding better efficiencies. JPL feels confident that efficiencies above 40% can be attained. Crossover reduces the efficiency of the cell, not only because some fuel is diverted, but also because the methanol that reaches the cathode is prone to undergo the same reaction (Equation 28) that normally takes place at the anode, generating a counter-voltage and, thus, reducing the voltage delivered to the load. International Fuel Cells Corporation has a patent for a catalyst that promotes the reaction of Equation 29, but not that of Equation 28.[†]

If the methanol at the cathode is not consumed, perhaps it can be recovered. This can be done by condensing the methanol-water vapor mixture exhausted from the cathode.

At present, direct methanol fuel cells are being considered for transportation applications and for powering portable electronic equipment. For the latter, "air breathing"—that is, delivery of air to the cathode by diffusion only (no pumps)—is important.

Table 9.1
Thermodynamic Data for Methanol Fuel Cells at RTP

Methanol	Water	ΔH° MJ/kmole	ΔG° MJ/kmole
liquid	gas	−638.5	−685.3
gas	gas	−676.5	−689.6
liquid	liquid	−726.5	−702.4
gas	liquid	−764.5	−706.7

Data from Dohle et al.

[†] Chu et al. at the U.S. Army Research Laboratory have developed an iron-cobalt tetraphenyl porphyrin catalyst that can promote the oxygen reduction to water (Eq. 29) but will not catalyze methanol oxidation (Eq. 28). Consequently, it has possibilities as a cathode catalyst in DMFC.

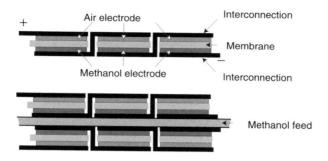

Figure 9.20   Bipolar configuration reduces the series resistance between individual cells, an advantage when high currents are involved. Flat-packs can be extremely thin making it easy to incorporate them into small-sized equipment. In the flat-pack configuration (developed by the Jet Propulsion Laboratory, see Narayanan et al.) cell interconnections pierce the membrane as shown. Two flat-packs deployed back-to-back permit the sharing of the methanol feed.

Instead of employing the conventional "bipolar" configuration, direct methanol fuel cells designed for the powering of cell phones, laptops, and other small portable equipment may employ the "flat-pack" design shown in Figure 9.20 (top) or the "twin-pack" design (bottom). This lends to the batteries (or "stacks") a shape that is more compatible with the equipment with which they are used.

9.5.7 Solid Acid Fuel Cells

Phosphoric acid and most solid polymer fuel cells are examples of cells that use hydrated acid electrolytes. Solid acids, on the other hand, can exhibit anhydrous proton conductivity and may have certain advantages over the popular SPFC. See Boysen et al.

H_2SO_4 is a common acid. However it is liquid at temperatures one would expect fuel cells to operate. On the other hand, replacing one of the hydrogens by a cesium atom leads to $CsHSO_4$ an acid (known as cesium hydrogen sulfate) with a high melting point. Above 414 K, this acid becomes a good proton conductor (commonly referred to as a **superprotonic** or **superionic** conductor) and has been proposed as a fuel cell electrolyte by groups at the California Institute of Technology and the University of Washington, among others.

Since these solid acids are soluble in water and since water is produced by the fuel cell reaction, it is essential to operate the cells at temperatures high enough to assure that any water that comes in contact with the electrolyte is in vapor form. This means that operating temperatures should be above some 150 C.

A major difficulty with $CsHSO_4$ (and with the corresponding selenate) is that they react with hydrogen forming, in the sulfate case, hydrogen sulfide. Greater stability is expected from phosphate-based compounds. CsH_2PO_4 fuel cells, for instance, were demonstrated at 235 C (see Boysen).

The advantage of SAFC over SPFC is that, operating at higher temperatures, they are less susceptible to CO poisoning and require less catalysts. In addition, since the water is always in vapor form, they do not need the careful water management required by the polymer cells.

Perhaps the most promising aspect of SAFC is that the electrolyte is not permeable to methanol thus eliminating the serious methanol crossover problem of polymer cells. See the subsection on direct methanol fuel cells. The power densities of solid acid methanol fuel cells (Boysen) are within a factor of 5 of the density of the most advanced DMFC (2004). The electrolyte thickness of these cells was 260 μm. Future cells with substantially thinner electrolytes may exhibit considerable improvement in their performance. However, one important difficulty with this type of cell is the fragility of the electrolyte.

9.5.8 Rechargeable Fuel Cells
Nickel Metal Hydride Battery (NiMH)

The growingly popular NiMH battery is useful in, among other applications, electrical vehicles. Although not strictly a fuel cell, it deserves mention here. It is a secondary battery (i.e., it is rechargeable) that derives its energy from the change in the Ni oxidation state from +3 to +2.

A metal alloy anode is immersed in a 30% (by weight) solution of KOH in water. This strong alkali dissociates almost completely into K^+ and OH^- ions. Hydrogen supplied to this electrode will combine with the hydroxyl ion forming water and liberating an electron, which becomes available to circulate through an external load constituting the useful output of the cell:

$$H + OH^- \rightarrow H_2O + e^-. \tag{31}$$

The cathode is made of nickel oxyhydroxide (NiOOH) which, upon receiving an electron, is reduced to nickel hydroxide liberating a hydroxyl ion:

$$NiOOH + H_2O + e^- \rightarrow Ni(OH)_2 + OH^-. \tag{32}$$

Notice that the average amount of water and of hydroxyl ions is constant—at the anode water is consumed at exactly the rate at which it is produced at the cathode. This contrast with NiCd batteries in which, during the discharge, water is consumed at both electrodes. The OH^- concentration remains constant during the discharge because this ion is produced at the anode and consumed at the cathode.

The materials consumed, hydrogen and nickel oxyhydroxide, are regenerated during the charge when an external power supply forces electrons into the metal electrode. This causes the water to be electrolyzed into H^+ and OH^-. At the other electrode, the nickel hydroxide is oxidized into nickel oxyhydroxide consuming hydroxyl ions. The metallic alloy on the hydrogen side absorbs this gas storing it in the form of hydride (see Chapter 11), so that no gas evolves. The hydrogen becomes available during the discharge. Nominal voltage delivered by each element of a NiMH battery is 1.2 V, some 60% of that of a lead-acid cell. When first introduced, prototypes of this battery exceeded most of the requirements of the US Advanced Battery Consortium (USABC),[†] as shown in Table 9.2 taken from an article by Ovshinsky et al. The larger the energy density of the battery, the larger the range of the vehicle. The GM EV1 (the first car designed from the ground up as an electric vehicle) used lead-acid batteries storing 49 MJ of energy (roughly the energy of one kg of gasoline) and had a range of 140 km. Equipped with NiMH batteries of the same mass, a more modern version of the vehicle had a range of over 250 km. Solectria, a manufacturer of electric vehicles, has built a demonstration car[††] with a composite body and equipped with a 32.5 kWh (117 MJ), 210-cell NiMH Ovonic battery and has covered a 601.25 km range (through rural roads and city streets) on a single charge. This was, at the end of 1996, a new record for EV. Ovonic Battery Co. of Troy, Mich., is the developer of MiMH batteries.

Table 9.2
USABC Performance Goals Compared with
Prototype NiMH Battery Performance

Property	USABC	NiMH
Mass energy density (MJ kg^{-1})	0.29 (0.36 desired)	0.29
Volume energy density (MJ m^{-3})	486	775
Mass power density (kW kg^{-1})	0.15 (>0.2 desired)	175
Volume power density (kW m^{-3})	250	470
Cycle life (cycles)	600	1000
Life (years)	5	10
Environmental operating temp.	−30 to 65 C	−30 to 60 C
Recharge time	<6 hours	15 min (60%) <1 hour (100%)
Self-discharge	<15% in 48 hours	<10% in 48 hours

Energy density values were measured with 50-Ah prototype cells, discharged at a rate that exhausted the cells in 3 hours. The mass power density was determined by discharging the battery in 30 seconds to 20% of its capacity. In determining the cycle life, the battery was repeatedly discharged to 20%.

[†] USABC, under the Department of Energy, promotes the development of batteries for electrical vehicles (EVs).

[††] The "Sunrise" carries two passengers and has a curb weight of 1000 kg.

As reported by the Argonne National Laboratories, the NiMH battery compares favorably with other candidates for EV application:

Table 9.3
Peak Power Delivered by Different Batteries
(at 25% depth of discharge)

Battery type	Peak power W/kg
NiMH	235
NiCd	200
NaS	145
NiFe	120
Pb-Acid	100
LiS	100
ZnBr	65

Material requirements of the NiMH battery are complicated. The energy storage capacity of the battery depends on the amount of hydrogen that can be absorbed by the metal alloy electrode. It is necessary that the hydride formation be easily reversible (see Chapter 11) and this is determined by the enthalpy of formation of the hydride that must fall in the 25 to 50 MJ/kmole range. If the enthalpy of formation is too small, hydrogen will fail to react with the alloy and gas will evolve. If too large, the electrode will be oxidized. Surface properties of the metal alloy are critical in determining the catalytic activity, electric conductivity, and the porosity and area of the surface.

The crucial question of cost of the battery will depend on many factors, especially on the demand for these devices. Nevertheless, at the present stage of development, the NiMH battery looks like a promising candidate as a storage system for EVs.

Small NiMH batteries are also popular for use in camcorders, cellular phones, and other small devices. In these applications, NiMH batteries face a serious competitor—lithium ion batteries.

9.5.9 Metallic Fuel Cells—Zinc-Air Fuel Cells

From the practical point of view, one of the advantages of **refuelable** cells, over **rechargeable** ones is that refueling tends to be much faster than recharging (minutes versus hours). However, hydrogen fuel cells suffer from the difficulty of transporting and storing the gas. The use of certain metals as fuel may lead to simple and safe transportability and large volumetric energy density. Metallic cells are an exception to the rule of using fluids as fuel. Two of the metals considered for this purpose are aluminum and zinc. Aluminum, however, is corroded by the caustic electrolyte even when no current is being generated. Aluminum-Power, Inc. has developed a type of cell in which the electrolyte is pumped away when the cells is not in use. This apparently leads to a cumbersome system. It appears that it is easier to use zinc. Metallic Power, Inc. has created a promising zinc-air fuel cell. Small zinc pellets (about 1 mm in diameter) are one of the reactants, the other being oxygen, in general just plain air. The electrolyte in the cells is a concentrated KOH solution.

At the anode (the negative electrode) the reaction is

$$Zn + 4OH^- \rightarrow Zn(OH)_4^{--} + 2 \ e^-. \tag{33}$$

The ion on the right-hand side of the equation is a zincate ion which consists of a positively charged Zn^{++} surrounded by four hydroxyls (OH^-). The electrons, after circulating through the load, recombine with the oxygen at the cathode and with water regenerating half of the hydroxyl ions used at the anode,

$$2 \ e^- + \tfrac{1}{2}O_2 + H_2O \rightarrow 2OH^-. \tag{34}$$

The electrolyte containing the zincate ion is pumped to an **Electrolyte Managing unit**—an integral part of the fuel cell—where the reaction,

$$Zn(OH)_4^{--} \rightarrow ZnO + H_2O + 2OH^-, \tag{35}$$

not only supplies the water consumed at the cathode and the other half of the needed hydroxyls, but also generates zinc oxide which precipitates out and can be removed from the cell for recovery of the zinc.

The overall fuel cell reaction is

$$Zn + \frac{1}{2}O_2 \rightarrow ZnO. \tag{36}$$

The cell is refueled by loading it's hopper with zinc pellets. As the zinc is consumed, the hopper automatically drops additional pellets into the reaction area. The zinc oxide is transferred to a stand-alone **Zinc Regeneration/Refueling System** in which the the oxide is electrochemically reduced to metal which is then pelletized for delivery to the fuel cell.

9.6 Fuel Cell Applications

Fuel cells are in a stage of rapid development and are on the verge of achieving maturity. As time goes on and its economic potential becomes well documented, this technology will occupy a growing number of niches.

At present, much of the effort is being concentrated in two distinct areas of application: stationary power plants and automotive power plants.

9.6.1 Stationary Power Plants

Stationary power plants of various types include central utility-operated power plants of large capacity (say, up to 1 GW), dispersed utility-operated power plants (perhaps in the tens of MW sizes) and on-site electrical generators (some 10 to 100 kW). For these applications, fuel cells present the following advantages over conventional heat engines:

1. Absence of noise.
2. Little pollution.
3. Ease of expansion (owing to modular construction).
4. Susceptibility to mass production (again, owing to modularity).
5. Possibility of dispersion of power plants. Owing to the low pollution and low noise, plants can be operated even in residential areas, thus economizing transmission lines.
6. Possibility of using reject heat for ambient heating because fuel cells can be near populated areas where there is a demand for hot water.
7. Possibility of cogeneration, using the high temperature exhaust gases in some types of plants.
8. Fast response to demand changes.
9. Good efficiency at a fraction of rated power.
10. Extremely good overload characteristics.
11. Small mass/power ratio, in some types of plants.
12. Small volume/power ratio, in some types of plants.
13. Great reliability (potentially).
14. Low maintenance cost (potentially).

Owing to the modular construction of many fuel cells, plant capacity can be easily expanded as demand grows. Capital investment can be progressive, considerably lessening the financial burden. Clearly, not all of the above advantages can be realized simultaneously. Cogeneration can only be achieved with high temperature fuel cells such as MCFCs and SOFCs, especially the latter. When cogeneration is used, the low noise advantage may be lost.

9.6.2 Automotive Power Plants

Imagine a hypothetical country having a modern industrial base but, for some unexplained reason, lacking completely any automotive industry. Imagine also that someone wants to build a small number of automobile engines of the type that drives present-day cars and that in the United States, Germany, or Japan can be bought for roughly $4000 a piece. Since specialized mass production machines do not exist, the engines have to be built one by one and their cost would be certainly more than an order of magnitude higher than that of the imported model.

Exactly the same situation prevails, all over the world, regarding fuel-cell power plants for cars. Current technology is already quite adequate as far as efficiency, life-time, power-to-mass ratio, power-to-volume ratio, and so forth are concerned. Further improvements will be made, but the existing technology is acceptable. However, the cost of manufacturing is not. It will only come down when a sufficient number of units can be sold; a sufficient number will only be sold when the cost comes down. This vicious circle is hard to break. One must remember that the low cost of automobiles results from millions upon millions being sold. The retooling for a new model runs to 1 or 2 billion dollars and can only be justified by large sales. Once the vicious circle above is broken, a second, very attractive, area of application of fuel cells will be in transportation.

For small and medium sized (and perhaps even for large) vehicles, the compact SPFCs are nearly ideal. Among the many advantages that can be claimed for them, one must count the extremely low pollution, the high efficiency (guaranteeing fuel economy), and the high power density (permitting compact designs). In addition, their low operating temperature permits rapid startups.

The expected long life time of SPFC stacks may lead to an unusual situation. A cell with a plausible 50,000 hour life would, even at only 40 km/h average speed, drive a car some 2 million kilometers. The fuel cell would outlive by far the automobile body. It would, then, make sense to re-use the cell in a new car. Different vehicle models could be designed to operate with one of a small number of standardized fuel cell types. This would have the adverse effect of reducing the market for fuel cells.

Hydrogen is the usual fuel for SPFC. The gas can be used as such (compressed or liquefied as discussed in Chapter 11) or in the form of a hydrogen "carrier" such as methanol (Chapter 10) or a metallic hydride (Chapter 11). Hydrogen carriers can be derived from fossil fuels (Chapter 10) or from biomass (Chapter 13). Fuel-cell cars may need energy storage devices (batteries, flywheels, or ultra-capacitors) to provide startup power, and to accumulate the energy recovered from dynamic braking. This stored energy can be used to supplement the fuel cell during fast accelerations. Hence, the needed fuel-cell power might be closer to the average power required by the vehicle than to the peak power. This would cut down the size and cost of the cell stack. Under all circumstances, the total electrically stored energy would be only a small fraction of that of a purely electric car.

Ballard Power Systems of Vancouver, BC, is one of the manufacturers closest to bringing fuel cells for vehicular use into regular production. In 1999, their fuel cell busses were already in revenue service in Vancouver and in Chicago. These vehicles, fueled by compressed hydrogen, have a range of 560 km, with excellent acceleration and a top speed of over 95 km/h. They are driven by a 275-hp motor and are totally nonpolluting.

Daimler-Benz, who has a major financial interest in Ballard, produces a fuel-cell van, the "Necar II," that is being tested by various users in Germany. The company plans to soon have production fuel cell cars for sale. Following this example, Ford, Toyota, Mazda, and General motors have made similar promises. Honda delivered the first few fuel cell FCX vehicles in 2003. Manufacturers have been accumulating experience with cars that may help the transition away from purely internal combustion engines. Extremely efficient cars like the General Motors EV1 (whose production stopped in 1999) can travel over 180 km on 14 kWh of electricity. This energy can be obtained from less than 1 kg of hydrogen stored in less than 100 kg of hydride. Honda also produced a limited number of electric cars—the EV Plus. Toyota has sold thousands of hybrid (electric-IC) cars—the Prius.

9.6.3 Other Applications

The first practical application of fuel cells was in space, where reliability far outweigh cost. The cells work with the hydrogen and the oxygen already available for other uses in the spacecraft and provide valuable drinking water as an output. GE SPFCs started the trend; at the moment AFCs are more popular—they were used in the Apollo program and currently supply energy for the space shuttle and for the International Space Station.

Small submersibles as well as full scale submarines benefit from the clean operation of fuel cells. The German navy developed 400-kW fuel cells for their submarines. This is half the power of their standard Diesel engines. The fuel cells will probably be used in a hybrid combination with the Diesels. Some military applications benefit from the low heat signature and absence of noise. Once fuel cells become even more compact and light, they may allow the operation of "cold" aircraft invisible to heat seeking missiles. Fuel cells are important in the propulsion of other naval vessels. Very compact power plants can be built in combination with superconducting motors. The U.S. Department of Defense is actively pursuing this development. Presumably a number of other nations are doing the same.

Micro fuel cells may be used to trickle portable electronics. In this application, micro fuel cells may have to face stiff competition from **nuclear batteries** using β^- emitters such as tritium or nickel-63. Such batteries take advantage of the enormous energy density of nuclear materials, (thousands of times more than chemical batteries). One nanotechnological implementation consists of a metallic target that, collecting the emitted electrons, becomes negatively charged. The resulting electrostatic attraction causes a thin silicon cantilever to bend until the target touches the source and discharges. The cantilever snaps back and the process repeats. The energy of the vibrating cantilever is transformed into electrical energy by piezoelectric converters. Efficiencies of 4% have been demonstrated, but Lal and Blanchard expect 20% with more advanced configurations.

9.7   The Thermodynamics of Fuel Cells

Some students will profit from re-reading the first few pages of Chapter 2 to reacquaint themselves with such concepts as internal energy, enthalpy, and entropy. We will repeat here the convention (Appendix to Chapter 2) for representing thermodynamic quantities.

Symbology
G, free energy,
H, enthalpy
Q, heat,
S, entropy, and
U, internal energy:
    1   Capital letters will indicate the quantity associated with an arbitrary amount of matter or energy.
    2   Lowercase letters indicate the quantity per unit. A subscript may be used to indicate the species being considered. For example, the free energy per kilomole of H_2 will be represented by $\overline{g}_{H_2}$.

    g = free energy per kilogram.

    $\overline{g}$ = free energy per kilomole.

    g^* = free energy per kilogram, at 1 atmosphere pressure.

    $\overline{g}^*$ = free energy per kilomole, at 1 atmosphere pressure.

    $\overline{g}_f$ = free energy of formation per kilomole.

    $\overline{g}_f^\circ$ = free energy of formation per kilomole, at 298 K, 1 atmosphere,
        i.e., at RTP, (Standard Free Energy of Formation).

9.7.1   Heat of Combustion

Let us return to reaction

$$2H_2 + O_2 \rightarrow 2H_2O \tag{5}$$

(in Section 9.2) in which hydrogen and oxygen combine to form water. In rearranging 4 hydrogen atoms and 2 oxygen atoms to form 2 molecules of water, some energy is left over. The force that binds atoms into molecules is electric in nature. Nevertheless, when H_2 reacts directly with O_2, only heat results because electrons and ions collide and their energy is randomized.

Assume that a measured amount, μ, of hydrogen is introduced into a constant pressure calorimeter containing sufficient oxygen and is made to react. A certain amount of heat, Q, is released. The ratio, $Q/\mu = \overline{h}_{comb}$, is the **heat of combustion** of hydrogen in an atmosphere of oxygen (joules/kmole). The exact amount of heat released depends on the temperature of both the reactants and the product and on the state of the latter. If the water produced is liquid, the heat of combustion is larger than when water is in the form of gas because the heat released includes the heat of condensation of water. As explained in Chapter 3, all fuels containing hydrogen have two different heats of combustion: the **higher heat**

of combustion in case of liquid water and the **lower heat of combustion** in case of water vapor.

Since the reaction occurs at constant pressure, the heat released equals the change in enthalpy, ΔH, of the system (see *enthalpy* in Chapter 2).

$$Q = \Delta H. \tag{37}$$

Q is taken as positive if it is added to the system as happens in a heat engine. In **exothermic** reactions, $Q < 0$ and so is ΔH.

Notice the sign convention used here. To conform with Chapter 2, energies introduced into a system, $\sum W_{in}$, are taken as positive; energies rejected by the system, $\sum W_{out}$, are negative. Consequently the energy balance reads

$$\sum W_{in} + \sum W_{out} = 0. \tag{38}$$

Enthalpy, as any other form of energy, has no absolute value; only changes can be measured. We can, therefore, arbitrarily assume any value for the enthalpy of the reactants. By convention, *at 298.15 K, the enthalpy of all elements in their natural state in this planet is taken as zero.* This means that, at 298.15 K, the enthalpies of H_2 and O_2 are zero, but those of H and O are not.

The difference between the enthalpy of a product and those of its elements is the **enthalpy of formation**, ΔHf, of the product. If both reactants and products are at the **reference temperature and pressure (RTP)**,[†] then the enthalpy of formation is called the **standard** enthalpy of formation, $\Delta Hf°$.

From the calorimeter experiment, for water (expressing the enthalpies in a per kilomole basis),

$$H_2O \text{ (g)}: \quad \overline{h}^{\circ}_{f_g} = -241.8 \text{ MJ/kmol,} \tag{39}$$

$$H_2O \text{ (}\ell\text{)}: \quad \overline{h}^{\circ}_{f_\ell} = -285.9 \text{ MJ/kmol.} \tag{40}$$

The subscripts "g" and "ℓ" indicate the state of the product water. The difference of -44.1 MJ/kmol between the enthalpies of formation of liquid and gaseous water is the **latent heat of condensation**, $\overline{h}_{con}$, of water. Clearly, $\overline{h}_{con} = -\overline{h}_{vap}$, where $\overline{h}_{vap}$ is the **latent heat of vaporization**.

[†] **Standard temperature and pressure** (STP), frequently used by chemists, corresponds to 1 atmosphere and 0 C (273.15 K); the *CRC Handbook of Chemistry and Physics*, however, lists the standard thermodynamic properties at 1 atmosphere and 298.15 K, which we will call RTP.

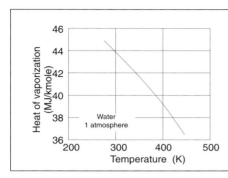

For coarse estimates, the latent heat of vaporization can be taken as constant. However, it is, in fact, temperature dependent, falling somewhat with rising temperature as can be seen in the figure.

9.7.2 Free Energy

If the reaction under consideration occurs not in a calorimeter but in an ideal fuel cell, part (but not all) of the energy will be released in the form of electricity. It is important to investigate how much of the ΔH of the reaction will be converted into electric energy and to understand why some of the energy must take the form of heat even in the ideal case.

Let V_{rev} be the voltage produced by the cell. Each kilomole of water (or of any other substance) contains N_0 molecules, where $N_0 = 6.022 \times 10^{26}$ is **Avogadro's number**.[†] From Equation 4, Section 9.2, describing the cathode reaction of the cell,

$$4\,e^- + 4\,H^+ + O_2 \rightarrow 2\,H_2O \tag{4}$$

we see that for each molecule of water, 2 electrons (charge q) circulate in the load. The energy delivered to the load is the product of the charge, $2qN_0$, times the voltage, V_{rev}. More generally, the electric energy produced per kilomole of product by a reversible fuel cell is

$$W_e = n_e q N_0 V_{rev} \equiv n_e F V_{rev}, \tag{41}$$

where

n_e = number of kmoles of electrons released per kilomole of products;

q = charge of the electron in coulombs;

N_0 = Avogadro's number in MKS;

$F \equiv qN_0 \equiv$ **faraday** $= 1.602 \times 10^{-19} \times 6.022 \times 10^{26}$

$= 96.47 \times 10^6$ Coulombs/kmole = charge in one kmole of electrons;

V_{rev} = voltage.

[†] Observe that, owing to our use of kilomoles instead of moles, Avogadro's number is three orders of magnitude larger than the value usually listed.

Consider a hypothetical experiment in which the open-circuit voltage of a reversible fuel cell is accurately measured. The voltage at RTP would be 1.185 V. The voltage delivered by a reversible fuel cell is called the **reversible voltage** and is designated by V_{rev}, as was done above.[†] Thus, the electric energy produced by a reversible fuel cell of this type is

$$|W_e| = 2 \times 96.47 \times 10^6 \times 1.185 = 228.6 \text{ MJ/kmole.} \tag{42}$$

The electric energy delivered to a load by a reversible cell is called the **free energy change** owing to the reaction (designated ΔG). Usually, if there is a single product, ΔG is given per kilomole of product and is represented by $\overline{g}_f^\circ$ (if at RTP). Again, since the cell *delivers* electric energy, the free energy change is negative, conforming to the convention for the sign of the enthalpy change of the reaction. As in the case of enthalpies, by convention, at RTP, *the free energy of all elements in their natural state in this planet is taken as zero.*

In most cases (but not always—see Problem 9.3), $|\Delta G| < |\Delta H|$. Thus the energy from the reaction usually exceeds the electric energy delivered to a load even in an ideal reversible cell. The excess energy, $\Delta H - \Delta G$ must appear as heat. Consider the entropies involved. Each substance has a certain entropy which depends on its state. Entropies are tabulated in, among others, the *Handbook of Chemistry and Physics, CRC Press*. In the reaction we are examining, the absolute entropies, at RTP, are:

$$H_2 \text{ (g)}: \qquad \overline{s}^\circ = 130.6 \text{ kJ K}^{-1}\text{kmol}^{-1}$$
$$O_2 \text{ (g)}: \qquad \overline{s}^\circ = 205.0 \text{ kJ K}^{-1}\text{kmol}^{-1}$$
$$H_2O \text{ (g)}: \qquad \overline{s}^\circ = 188.7 \text{ kJ K}^{-1}\text{kmol}^{-1}$$

When 1 kilomole of water is formed, 1 kilomole of H_2 and 0.5 kilomoles of O_2 disappear and so do the corresponding entropies: a total of $130.6 + 205.0/2 = 233.1$ kJ/K disappear. This is, in part, compensated by the appearance of 188.7 kJ/K corresponding to the entropy of the water formed. The matter balance leads to an entropy change of $188.7 - 233.1 = -44.4$ kJ/K. In a closed system, the entropy cannot decrease (Second Law of Thermodynamics), at best—under reversible conditions—its change is zero. Consequently, an amount of entropy, $\Delta S = Q/T$ must appear as heat:

$$Q = T\Delta S = 298 \times (-44.4 \times 10^3) = -13.2 \text{ MJ/kmol.} \tag{43}$$

This amount of heat must come from the enthalpy change of the reaction leaving $-241.8 - (-13.2) = -228.6$ MJ/kmol as electricity.

[†] Owing to irreversibilities in practical cells, such a measurement cannot be carried out accurately. However, the reversible voltage can be estimated by connecting a voltage generator to the cell and observing its voltage-versus-current characteristic as the applied voltage is varied. If this voltage is sufficiently high, current will be driven into the cell and if it is low, current will be delivered by the cell to the generator. The characteristic should be symmetrical around the reversible voltage.

Chemical energy can be thought of as consisting of two parts: an entropy free part, called **free energy**, that can entirely be converted to electricity and a part that must appear as heat. The free energy,[†] G, is the enthalpy, H, minus the energy, TS, that must appear as heat:

$$G = H - TS \tag{44}$$

and

$$\Delta G = \Delta H - \Delta(TS). \tag{45}$$

In isothermic cases (as in the example above),

$$\Delta G = \Delta H - T\Delta S. \tag{46}$$

The electric energy, W_e, delivered by the reversible fuel cell is ΔG:

$$\Delta G = -n_e q N_0 N |V_{rev}|. \tag{47}$$

where N is the number of kilomoles of water produced. Note that since ΔG is removed from the cell it must be < 0. Per kilomole of water,

$$\overline{g}_f = -n_e q N_0 |V_{rev}|. \tag{48}$$

Table 9.4.
Some Thermodynamic Values

	$\Delta \overline{h} f^{\circ}$ (MJ/kmole)	$\Delta \overline{g} f^{\circ}$ (MJ/kmole)	$\overline{s}^{\circ}$ (kJ K^{-1} kmole^{-1})
H_2O (g)	-241.8	-228.6	188.7
H_2O (l)	-285.9	-237.2	70.0
H_2 (g)	0	0	130.6
O_2 (g)	0	0	205.0

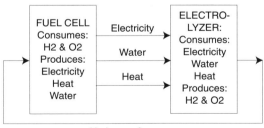

Hydrogen & oxygen

Figure 9.21   An ideal fuel cell and its dual, the ideal electrolyzer, act as a reversible system. The inputs of one are precisely the same as the output of the other. For this to be true, the input to the electrolyzer must be water vapor.

[†] The idea of free energy was first proposed by Josiah Willard Gibbs (1839–1903), hence the symbol "G."

The dual of a fuel cell is an electrolyzer (Chapter 10). A fuel cell may use hydrogen and oxygen, generating electricity and producing water and heat. The electrolyzer consumes water and electricity, producing hydrogen and oxygen. In the ideal case, the electrolyzer absorbs heat from the environment, acting as a heat pump. If there is insufficient heat flow from the environment to the electrolyzer, the later will cool down.

For a given amount of gas handled, the electric energy generated by the reversible fuel cell is precisely the same as that required by the reversible electrolyzer and the heat produced by the reversible fuel cell is precisely the same as that absorbed by the electrolyzer. This amount of heat is reversible.

Clearly the reversibility is destroyed if the system has losses. A lossy (read, practical) fuel cell generates more heat than $T\Delta S$, while a lossy electrolyzer will generate heat which may (and frequently does) exceed the thermodynamically absorbed heat. Nevertheless, some realizable electrolyzers may operate with sufficient efficiency to actually cool down during operation.

9.7.3 Efficiency of Reversible Fuel Cells

The efficiency of a fuel cell is the ratio of the electric energy generated to the enthalpy change of the chemical reaction involved. In reversible cells,

$$\eta_{rev} = \frac{\Delta G}{\Delta H}. \tag{49}$$

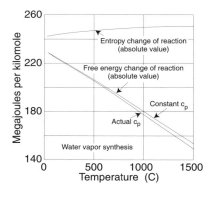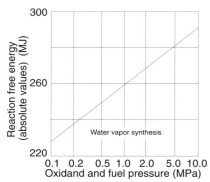

Figure 9.22 (Left) At constant pressure (0.1 MPa), increasing equally the temperature of reactants and product increases slightly the absolute value of the enthalpy change of the reaction, but reduces substantially the corresponding free energy. (Right) At constant temperature (298 K), increasing equally the pressure of reactants and product increases substantially the absolute value of the free energy change of the reaction.

It is of interest to examine how this efficiency depends on the temperature and the pressure of reactants and products, or, in other words, how the enthalpy and free energy changes of the reaction depend on such variables.

Figure 9.22 (left) depicts the manner in which ΔH and ΔG vary with temperature and (right) how ΔG varies with pressure. The data assume that both reactants, H_2 and, O_2 and the product, H_2O, are at identical pressures and identical temperatures.

Increasing the temperature while maintaining the pressure constant has a small effect on ΔH, but reduces appreciably the available free energy (the electric energy generated). Thus, for reversible cells, the higher the temperature the lower the efficiency. It must be noted that exactly the opposite happens with practical cells. The reason is that the improvement in the cell kinetics with temperature more than offsets the free energy loss. This point will be discussed in more detail farther on.

We will now derive the dependence of the enthalpy and the free energy changes of a reaction on temperature and pressure.

9.7.4 Effects of Pressure and Temperature on the Enthalpy and Free Energy Changes of a Reaction

9.7.4.1 Enthalpy Dependence on Temperature

Let ΔH be the enthalpy change owing to a chemical reaction.

By definition, ΔH is equal to the sum of the enthalpies of the products of a reaction minus the sum of the enthalpies of all reactants. Thus, $\Delta H =$ *enthalpy of products − enthalpy of reactants*. When there is a single product, then the enthalpy change of the reaction can be expressed per kilomole of product by dividing ΔH by the number of kilomoles of product created. An equivalent definition holds for the free energies.

In mathematical shorthand,

$$\Delta H = \sum n_{P_i} H_{P_i} - \sum n_{R_i} H_{R_i}. \tag{50}$$

Here,

n_{P_i} = number of kilomoles of the ith product,
n_{R_i} = number of kilomoles of the ith reactant,
H_{P_i} = enthalpy of the ith product,
H_{R_i} = enthalpy of the ith reactant.

For the water synthesis reaction,

$$\Delta H = 1 \times H_{H_2O} - 1 \times H_{H_2} - \frac{1}{2} \times H_{O_2}. \tag{51}$$

where H_{H_2O} is the enthalpy of water, and so on.

Of course, at RTP, for water vapor, $H_{H_2O} = -241.8$ MJ/kilomole, and H_{H_2} and H_{O_2} are both zero by choice.[†] Hence, $\Delta H = -241.8$ MJ/kilomole.

If the value of the enthalpy, H_0 of a given substance (whether product or reactant), is known at a given temperature, T_0, then at a different temperature, T, the enthalpy is

$$H = H_0 + \int_{T_0}^{T} c_p dT, \tag{52}$$

where c_p is the specific heat at constant pressure.

c_p is somewhat temperature dependent and its value for each substance can be found either from tables or from mathematical regressions derived from such tables. For rough estimates, c_p, can be taken as constant. In this latter case,

$$H = H_0 + c_p \Delta T, \tag{52a}$$

where $\Delta T \equiv T - T_0$.

Replacing the various enthalpies in Equation 50 by their values as expressed in Equation 52,

$$\Delta H = \sum n_{P_i} H_{0_{P_i}} + \sum n_{P_i} \int_{T_{0_{P_i}}}^{T_{P_i}} c_{p_{P_i}} dT - \sum n_{R_i} H_{0_{R_i}} - \sum n_{R_i} \int_{T_{0_{R_i}}}^{T_{R_i}} c_{p_{R_i}} dT$$

$$= \Delta H_0 + \sum n_{P_i} \int_{T_{0_{P_i}}}^{T_{P_i}} c_{p_{P_i}} dT - \sum n_{R_i} \int_{T_{0_{R_i}}}^{T_{R_i}} c_{p_{R_i}} dT. \tag{53}$$

For a simple estimate of the changes in ΔH, one can use the constant c_p formula:

$$\Delta H = \Delta H_0 + \sum n_{P_i} c_{p_{P_i}} (T_{P_i} - T_{0_{P_i}}) - \sum n_{R_i} c_{p_{R_i}} (T_{R_i} - T_{0_{R_i}}). \tag{53a}$$

Example 1

The standard enthalpy of formation of water vapor is -241.8 MJ/kmoles. What is the enthalpy of formation when both reactants and product are at 500 K?

We need to know the specific heats at constant pressure of H_2, O_2, and $H_2O(g)$. To obtain an accurate answer, one must use Equation 53 together with tabulated values of the specific heat as a function of temperature. An approximate answer can be obtained from Equation 53a using constant values of the specific heats. We shall do the latter.

(continued)

[†] Remember that, by convention, the enthalpy of all elements in their natural state under normal conditions on earth is taken as zero.

(continued)

We saw in Chapter 2 that, if one can guess the number of degrees of freedom, ν, of a molecule, one can estimate the specific heat by using the formula

$$c_p = R \left(1 + \frac{\nu}{2} \right).$$

The advantage of this procedure is that it is easier to remember ν than c_p. For diatomic gases, ν can be taken as 5, yielding a $c_p = 29.1$ kJ kmole^{-1}K^{-1}, and for water vapor, ν can be taken as 7, yielding $c_p = 37.4$ kJ kmole^{-1}K^{-1}.

Using Equation 53a (in this problem, all ΔT are the same: 500 K -298 K),

$$\Delta H = -241.8 + (500 - 298) \left[0.0374 - (0.0291 + \frac{1}{2} \times 0.0291) \right]$$

$$= -243.1 \ \text{MJ per kmole of water vapor.}$$

Since the enthalpy changes only little with temperature, these approximate results are close to the correct value of -243.7 MJ/kmole obtained through the use of Equation 53.

9.7.4.2 Enthalpy Dependence on Pressure

In Section 2.12, we defined enthalpy as the sum of the internal energy, U, with the pressure-volume work, pV:

$$H \equiv U + pV. \tag{54}$$

The internal energy of the gas is the energy stored in its molecules. Such storage can take the form of excitation, ionization, etc. However, in this chapter, we will limit ourselves to the internal energy stored as kinetic energy of the molecules, a quantity measured by the temperature of the gas.

Changing the pressure at constant temperature does not change the average energy of the molecules, and since the mass of gas (the number of molecules) is constant, the total internal energy remains unaltered—that is, U remains constant when p is changed provided T is unchanged.

We have $pV = \mu RT$. At constant temperature, pV must also be constant.

Thus, neither U nor pV change when the pressure is altered isothermally. Consequently, H does not depend on pressure provided the temperature and the mass of the gas is unaltered.

9.7.4.3 Free Energy Dependence on Temperature

The free energy is

$$G = H - TS. \tag{55}$$

The behavior of H as a function of temperature was discussed previously. We must now investigate the behavior of the entropy, S.

From Chapter 2, we have the relationship (for isobaric processes and per kilomole of gas),

$$dS = c_p \frac{dT}{T}, \tag{56}$$

which integrates to

$$S = S_0 + \int_{T_0}^{T} c_p \frac{dT}{T}. \tag{57}$$

The change in free energy when the temperature is changed under constant pressure is

$$G - G_0 = H - H_0 - (TS - T_0 S_0)$$
$$= \int_{T_0}^{T} c_p dT - \left(T S_0 + T \int_{T_0}^{T} c_p \frac{dT}{T} - T_0 S_0 \right)$$
$$= \int_{T_0}^{T} c_p dT - T \int_{T_0}^{T} c_p \frac{dT}{T} - S_0 \Delta T. \tag{58}$$

Given a table of values for c_p as a function of T, the change, $G - G_0$, in the free energy can be numerically calculated from Equation 58 and the ΔG of reaction can be obtained from

$$\Delta G = \sum n_{P_i} G_{P_i} - \sum n_{R_i} G_{R_i}, \tag{59}$$

an equation equivalent to Equation 50 for ΔH.

For the case when c_p is assumed constant, Equation 58 reduces to

$$G - G_0 = (c_p - S_0)\Delta T - T c_p \ln \frac{T}{T_0}. \tag{58a}$$

Example 2

Estimate the free energy of the $H_2(g) + \frac{1}{2}O_2(g) \rightarrow H_2O\ (g)$ reaction at standard pressure and 500 K, using constant c_p.

The necessary values are

	Entropy kJ K^{-1}kmole^{-1} (at RTP)	Specific Heat kJ K^{-1}kmole^{-1}
H_2 (g)	130.6	29.1
O_2 (g)	205.0	29.1
H_2O (g)	188.7	37.4

(continued)

(continued)

Since the product and all reactants are at the same temperature, $\Delta T = 500 - 298 = 202$ for all gases. Let us calculate individually their change in free energy remembering that elements, in their natural state at RTP, have zero free energy.

From Equation 58a,

$$G_{H_2} = 0 + (29.1 \times 10^3 - 130.6 \times 10^3) \times 202$$
$$- 500 \times 29.1 \times 10^3 \times \ln \frac{500}{298} = -28.03 \times 10^6 \text{ J/kmole,}$$

$$G_{O_2} = 0 + (29.1 \times 10^3 - 205.0 \times 10^3) \times 202$$
$$- 500 \times 29.1 \times 10^3 \times \ln \frac{500}{298} = -43.06 \times 10^6 \text{ J/kmole,}$$

$$G_{H_2O} = -228.6 \times 10^6 + (37.4 \times 10^3 - 188.7 \times 10^3) \times 202$$
$$- 500 \times 37.4 \times 10^3 \times \ln \frac{500}{298} = -268.8 \times 10^6 \text{ J/kmole.}$$

Thus, the ΔG of the reaction is

$$\Delta G = G_{H_2O} - G_{H_2} - \frac{1}{2} G_{O_2}$$
$$= -268.8 - (-28.0) - \frac{1}{2}(-43.1) = -219.3 \text{ MJ/kmole.}$$

This estimate came fortuitously close to the correct value of -219.4 MJ/kmole. If we had calculated ΔG at, say, 2000 K, we could be making a substantial error.

Nevertheless, the assumption of constant c_p leads to estimates that indicate the general manner in which the free energy depends on temperature.

More accurate values can be obtained from numerical integration of Equation 59. The values of c_p for H_2, O_2, and H_2O can be read from experimentally determined tables (reproduced below from Haberman and John).

Table 9.5

Specific Heats at Constant Pressure and Gammas

T (C)	T (K)	H_2 c_p (kJ/K) per kmole	H_2 γ	O_2 c_p (kJ/K) per kmole	O_2 γ	H_2O c_p (kJ/K) per kmole	H_2O γ
-50	223.18	27.620	1.426	29.152	1.399	33.318	1.333
0	273.18	28.380	1.410	29.280	1.397	33.336	1.332
25	298.18	28.560	1.406	29.392	1.406	33.489	1.330
50	323.18	28.740	1.402	29.504	1.403	33.642	1.328
100	373.18	28.920	1.399	29.888	1.386	34.020	1.323
150	423.18	28.980	1.398	30.336	1.378	34.434	1.318
200	473.18	29.020	1.397	30.816	1.369	34.902	1.312
226.8	500	29.031	1.397	31.090	1.365	35.172	1.309
250	523.18	29.040	1.397	31.328	1.361	35.406	1.307
300	573.18	29.080	1.396	31.840	1.354	35.946	1.301
350	623.18	29.120	1.395	32.320	1.346	36.522	1.295
400	673.18	29.180	1.394	32.768	1.340	37.098	1.288
450	723.18	29.240	1.393	33.184	1.334	37.710	1.283
500	773.18	29.340	1.391	33.536	1.329	38.322	1.277
550	823.18	29.440	1.389	33.888	1.325	38.952	1.271
600	873.18	29.560	1.387	34.208	1.321	39.564	1.266
650	923.18	29.720	1.384	34.496	1.318	40.194	1.261
700	973.18	29.880	1.381	34.752	1.315	40.788	1.256
750	1023.2	30.040	1.378	34.976	1.312	41.382	1.251
800	1073.2	30.240	1.375	35.200	1.309	41.958	1.247
850	1123.2	30.420	1.372	35.392	1.307	42.642	1.242
900	1173.2	30.640	1.369	35.584	1.305	43.326	1.237
950	1223.2	30.840	1.365	35.744	1.303	43.920	1.233
1000	1273.2	31.060	1.362	35.904	1.301	44.514	1.229
1050	1323.2	31.280	1.358	36.064	1.300	45.072	1.226
1100	1373.2	31.500	1.355	36.224	1.298	45.630	1.223
1150	1423.2	31.720	1.351	36.352	1.296	46.170	1.219
1200	1473.2	31.940	1.348	36.480	1.295	46.674	1.216
1250	1523.2	32.140	1.345	36.608	1.294	47.178	1.214
1300	1573.2	32.360	1.342	36.736	1.292	47.664	1.211
1350	1623.2	32.560	1.339	36.864	1.291	48.114	1.209
1400	1673.2	32.760	1.337	36.992	1.290	48.564	1.206
1450	1723.2	32.960	1.344	37.120	1.289	48.978	1.204
1500	1773.2	33.160	1.331	37.248	1.287	49.392	1.202

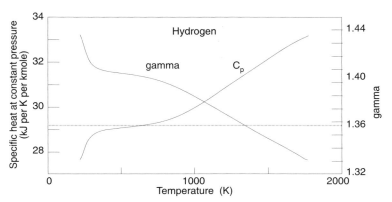

Figure 9.23   Specific heat and gamma of hydrogen.

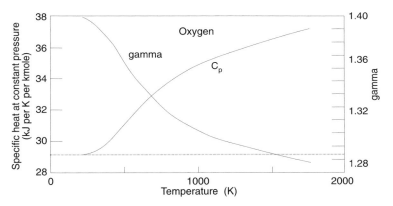

Figure 9.24   Specific heat and gamma of oxygen.

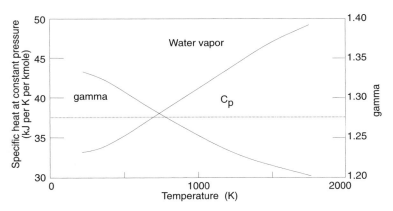

Figure 9.25   Specific heat and gamma of water vapor.

The data of Table 9.5 are displayed in the plots of Figures 9.23 through 9.25. They show that the specific heats at constant pressure, c_p, and the γ of the three gases of interest, far from being temperature independent as suggested by simple theory (see Chapter 2), do vary substantially.

The value of c_p derived from a guess of the number of degrees of freedom are indicated by the horizontal dotted line in each figure. For hydrogen and oxygen, two diatomic molecules, a reasonable number of degrees of freedom would be $\nu = 5$ which leads to $c_p = 29.1$ kJ K^{-1} per kilomole. This matches the actual value for hydrogen at temperatures between 350 K and 600 K. At higher temperatures, the c_p of this gas seems to be heading towards the 37.4 kJ K^{-1} per kilomole that correspond to 7 degrees of freedom. For a somewhat more detailed discussion of these changing degrees of freedom refer to Chapter 2.

Water, with its presumed 7 degrees of freedom, should have $c_p = 37.4$ kJ K^{-1} per kilomole. This is actually the correct value at some 700 K. However, the c_p of water, like that of most gases, varies fairly rapidly with temperature.

We fitted a 5th order polynomial to the data in Table 9.5, so that the values can be calculated as a function of T with reasonable accuracy:

$$c_p = a + bT + cT^2 + dT^3 + eT^4 + fT^5 \tag{60}$$

where the constants a through f are given in Table 9.6.

It should be noticed that these regressions must be used only in the 220 K $< T <$ 1800 K interval. Outside this range, the errors become, in some cases, unacceptably large.

Table 9.6

Coefficients of the Polynomial Used to Calculate the Specific Heats at Constant Pressure 100,000 Pa (1 atmosphere)

Gas	a	b	c	d	e	f
H$_2$	22.737	37.693E-3	$-$85.085E-6	89.807E-9	$-$42.908E-12	7.6821E-15
O$_2$	30.737	$-$19.954E-3	72.554E-6	$-$80.005E-9	38.443E-12	$-$6.8611E-15
H$_2$O	32.262	1.2532E-3	11.285E-6	$-$3.7103E-9	—	—

9.7.4.4 Free Energy Dependence on Pressure

The free energy, G, is defined as

$$G = H - TS. \tag{61}$$

We saw that the enthalpy does not change when the pressure is altered isothermally. Thus, isothermally, pressure can only alter the free energy through its effect on the entropy, S:

$$\Delta G = -T\Delta S \tag{62}$$

When an amount, Q, of heat is added to a system at constant temperature, the entropy increases by Q/T:

$$\Delta S = \frac{Q}{T}. \tag{63}$$

From the first law of thermodynamics,

$$\Delta U = Q - W \tag{64}$$

But we saw previously that in an isothermal compression, $\Delta U = 0$, hence, $Q = W$. However, the work done per kilomole of gas isothermally compressed (see Section 2.10) is

$$W = RT \ln \frac{p_1}{p_0}. \tag{65}$$

Thus,

$$-\Delta G = T\Delta S = Q = RT \ln \frac{p_1}{p_0}. \tag{66}$$

Consequently, the energy of isothermal compression of a gas is entirely free energy. This is an important effect. It is possible to change the efficiency (and the voltage) of a fuel cell by changing the pressure of products and reactants.

Example 3

A reversible fuel cell, when fed hydrogen and oxygen at RTP, delivers a voltage of 1.185 V. Calculate the voltage delivered by the same cell if air (at RTP) replaces the oxygen.

Air contains roughly 20% of oxygen. Thus the partial pressure of this gas is 0.2 atmosphere, a 5:1 decompression relative to the pure oxygen case.

The energy of isothermal decompression is

$$W_{decompr.} = \frac{1}{2}RT \ln \frac{1}{2} = \frac{1}{2}8314 \times 298 \times \ln 0.2 = -2 \times 10^6 \ \text{J/kmole}.$$

The factor, $\frac{1}{2}$, results from the stoichiometric proportion of one half kilomole of oxygen per kilomole of water. This energy must be *subtracted* from the ΔG of the reaction.

$$\Delta G = -228.6 - (-2) = -226.6 \ \text{MJ/kmole}.$$

The voltage is now

$$V = \frac{|\Delta G|}{n_e q N_0} = \frac{226.6}{2 \times 1.60 \times 10^{-19} \times 6.02 \times 10^{26}} = 1.174 \ \text{V}.$$

9.7.4.5   Voltage Dependence on Temperature

We have derived expressions that show how the free energy depends on both pressure and temperature. This was done under the assumption that the specific heats of the different substances involved in the reactions are independent of temperature. We will now derive the temperature dependence of the voltage of an ideal fuel cell in a rigorous manner. We will, however, limit ourselves to the constant pressure case.

By definition,

$$G = H - TS \tag{67}$$

and

$$H = U + pV, \tag{68}$$

hence

$$G = U + pV - TS, \tag{69}$$

from which

$$dG = dU + pdV + V\,dp - T\,dS - S\,dT. \tag{70}$$

From the combined laws of thermodynamics,

$$dU = T\,dS - pdV, \tag{71}$$

hence

$$dG = V\,dp - S\,dT. \tag{72}$$

G is a function of the independent variables p and T. Thus, formally,

$$dG = \left(\frac{\partial G}{\partial p}\right)_T dp + \left(\frac{\partial G}{\partial T}\right)_p dT. \tag{73}$$

Comparing the last two equations, one can see that

$$\left(\frac{\partial G}{\partial T}\right)_p = -S. \tag{74}$$

From here on, V represents voltage, not volume, as before.

$$\sum n_{P_i} n_e q N_0 V = -\Delta G = -\left(\sum n_{P_i} G_{P_i} - \sum n_{R_i} G_{R_i}\right), \tag{75}$$

$$\sum n_{P_i} n_e q N_0 \left(\frac{\partial V}{\partial T}\right)_p = -\left(\sum n_{P_i} \frac{\partial G_{P_i}}{\partial T} - \sum n_{R_i} \frac{\partial G_{R_i}}{\partial T}\right)$$

$$= \sum n_{P_i} S_{P_i} - \sum n_{R_i} S_{R_i} = \frac{\sum n_{P_i} T\, S_{P_i} - \sum n_{R_i} T\, S_{R_i}}{T}. \tag{76}$$

But $TS = H - G$, hence,

$$\sum n_{P_i} n_e q N_0 \left(\frac{\partial V}{\partial T} \right)_p$$
$$= \frac{\sum n_{P_i} H_{P_i} - \sum n_{R_i} H_{R_i} - \sum n_{P_i} G_{P_i} + \sum n_{R_i} G_{R_i}}{T}. \quad (77)$$

Therefore,

$$\left(\frac{\partial V}{\partial T} \right)_p = \frac{V + \Delta H / (\sum n_{P_i} n_e N_0 q)}{T}, \quad (78)$$

where $\Delta H / (\sum n_{P_i} n_e N_0 q)$ is the voltage the cell would have if all the enthalpy-change of the reaction were transformed into electrical energy. Let us call this the **enthalpy voltage**.

For an H_2/O_2 fuel cell producing water vapor,

$$\frac{\Delta H}{\sum n_{P_i} n_e N_0 q} = \frac{-241.8 \times 10^6}{2 \times 6.022 \times 10^{26} \times 1.6 \times 10^{-19}} = -1.225 \text{ V},$$

where we set $\sum n_{P_i} = 1$ because the value of ΔH used is the one for a single kilomole of water.

$$\left(\frac{\partial V}{\partial T} \right)_p = \frac{1.185 - 1.255}{298} = -2.3 \times 10^{-4} \text{ V/K}.$$

9.8 Performance of Real Fuel Cells

In examining the performance of real fuel cells, we must inquire

1. What current can the cell deliver?
2. What is the efficiency of the cell?
3. What are the current/voltage characteristics?
4. What is the heat balance?
5. How can the excess heat be removed?

9.8.1 Current Delivered by a Fuel Cell

If $\dot{N}$ is the rate (in kilomoles/sec) at which the product is generated (water, in case of hydrogen/oxygen cells) and n_e is the number of electrons per molecule of product (2, for hydrogen/oxygen cells), then the rate at which electrons are delivered by the cell to the load is $n_e \dot{N}$ or, expressing this in kilomoles of electrons per second, $n_e N_0 \dot{N}$. Consequently, the current is

$$I = q n_e N_0 \dot{N}. \quad (79)$$

One defines a **current efficiency** as the ratio of the actual load current, I_L, to the theoretical current calculated above. In many cases, one can safely assume 100% current efficiency.

9.8.2 Efficiency of Practical Fuel Cells

It was shown that the theoretical efficiency of a reversible fuel cell is

$$\eta_{rev} = \frac{\Delta G}{\Delta H}.$$

One has the choice of using for ΔH either the higher or the lower heat of combustion of the reactants, and, in stating a given efficiency, reference should be made to which was chosen. As to ΔG, one should use the one appropriate for the formation of water vapor (if, indeed, water is involved).

The efficiency of practical fuel cells is defined as the ratio of the electric power, P_L, delivered to the load to the heat power that would be generated by combining the reactants in a calorimeter, under the same temperature and pressure used in the cell,

$$\eta_{practical} = \frac{P_L}{P_{in}} = \frac{I_L V_L}{\Delta \overline{h} \dot{N}} = \frac{q n_e N_0 V_L}{\Delta \overline{h}} \tag{80}$$

Again, one has a choice of which $\Delta \overline{h}$ to use. Assuming 100% current efficiency, for hydrogen/oxygen fuel cell at RTP referred to the lower heat of combustion, the efficiency is, at RPP,

$$\eta_{pract_{RTP}} = 0.798 V_L. \tag{81}$$

Practical fuel cells have lower efficiency than ideal ones owing to:

1. Not all reactants used up take part in the desired reaction—some may simply escape, others may take part in undesired side-reactions. Sometimes, part of the fuel is consumed to operate ancillary devices such as heating catalytic crackers, and so on, or it may be "after burned" to raise the exhaust gas temperature in co-generation arrangements,
2. Not all the current produced will go through the load—some may leak through parallel paths (a minor loss of current) or be used to drive ancillary equipment such as compressors.
3. The voltage, V_L, the cell delivers to a load is smaller than, V_{rev}, the **reversible voltage** (the theoretical voltage associated with the change in the free energy).
   A number of factors contribute to such voltage loss:

   3.1 Unavoidably, fuel cells have internal resistance to the flow of electrons and, in the electrolyte, to the flow of ions.

3.2 The rate at which the chemical reactions take place—the **kinetics** of the chemical reaction—limits the rate at which electrons are liberated—that is, limits the current produced.

3.3 Unwanted reactions generate voltages that oppose the normal potential of the cell. Some fuel may leak to the oxidizer side (**fuel crossover**), a worrisome problem with direct methanol fuel cells.

4. The effective electrode area may become reduced because

4.1 Excess water may "drown" the electrodes disturbing the "triple point" contact between reactants, electrolyte, and electrodes.

4.2 Insufficient moisture may dry out the solid electrolyte membrane of SPFCs further increasing the resistance to ion flow.

Incomplete use of fuel (Effect 1), diversion of some of the generated current (Effect 2), and loss of effective area (Effect 4) are problems dealt mostly by system design whereas voltage loss (Effect 3) is an inherent property of the individual cell itself. We will discuss this voltage loss by examining the voltage–current (or voltage–current density) characteristics of the cell.

9.8.3 Characteristics of Fuel Cells

Reversible fuel cells will deliver to a load a voltage, $V_L = V_{rev}$, which is independent of the current generated. Their *V-I* characteristic is a horizontal line. Such ideal cells require reaction kinetics fast enough to supply electrons at the rate demanded by the current drawn. Clearly, reversible cells cannot be realized. In practical fuel cells, two major deviations from the ideal are observed:

1. The open-circuit voltage, V_{oc}, is smaller than V_{rev}.
2. The load voltage, V_L, decreases as the load current, I_L, increases.

Frequently, the *V-I* characteristics approaches a straight line with some curvature at low currents as, for example, those of the Figure 9.26 (left), while other cells exhibit characteristics in which the linearity can be seen only in a limited region (right).

The representative *V-I* characteristic of Figure 9.26 (left) can be mathematically described by

$$V_L = V_{oc} - I_L R_{int} - V_{act}, \tag{82}$$

where $I_L R_{int}$ is the voltage drop owing to the internal resistance, R_{int}, and V_{act}—which we shall call the **activation voltage**—is a voltage that has exactly the correct dependence on I_L needed to reproduce the observed values of V_L. Having good data of V_L as a function of I_L should permit the determination of the value of R_{int} and of the functional dependence of V_{act} on I_L. However, some additional theoretical considerations should help in this task.

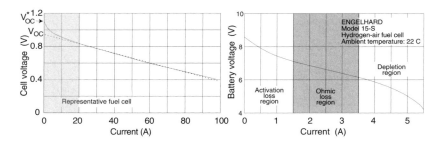

Figure 9.26   The typical modern fuel cell tends to have V-I characteristic consisting of a long stretch of apparently linear relationship between current and voltage with a small curvature at the low current end (left, above). The small Engelhard liquid-electrolyte demonstration cell had a limited region in which voltage decreases linearly with current. At both lower and higher currents the characteristic exhibited marked curvature (right, above).

As discussed later on, there are reasons to believe that V_{act} should have a logarithmic dependence on I_L. At reasonably large currents,

$$V_{act} = V_2 \ln \frac{I_L}{I_0}, \tag{83}$$

where V_2 and I_0 are parameters derived from the measured characteristics.

The slope of the V-I characteristic is not an exact measure of the internal resistance. The actual slope is (except at very low currents)

$$-\frac{dV_L}{dI_L} = R_{int} + \frac{V_2}{I_L} \equiv R_{app}. \tag{84}$$

As defined above, R_{app} is a function of I_L. However, in many cases we expect to operate entirely in a region of the characteristics in which there is an apparent straight-line relationship between V and I, i.e., $V_2/I_L \ll R_{int}$. This allows the modeling of the cell as a voltage generator in series with an internal resistance, R_{int}, as suggested by the diagram in Figure 9.27.

The two circuits of Figure 9.27 are entirely equivalent. The one on the left is obvious; in the one on the right the open-circuit voltage, V_{oc}, is represented by two opposing voltage generators, V_{rev} and $V_{rev} - V_{oc}$. Such representation facilitates the calculation of internal heat generation later in this chapter. V_{oc} is, of course, the intercept of the V-I (or V-J) line with the ordinate axis.

From the circuit model,

$$V_L = V_{oc} - I_L R_{int} = V_{oc} - JAR_{int}, \tag{85}$$

Not uncommonly, voltages are plotted versus current densities, $J = I/A$, A being the effective surface area of the electrodes:

$$V_L = V_{oc} - JAR_{int} = V_{oc} - \Re J \tag{86}$$

where $\Re$, the **specific resistance** of the cell, has dimensions of ohms $\times$ m^2.

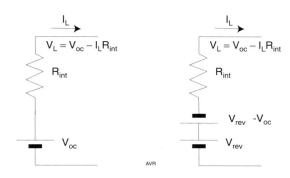

Figure 9.27   Circuit model for a fuel cell with straight line
V-I characteristics.

For the particular Ballard cells whose characteristics appear in Figure 9.28, the linear regressions derived from the measured data lead to:

at 50 C, $V_L = 0.912 - 54.4 \times 10^{-6} J$,
and at 70 C, $V_L = 0.913 - 49.3 \times 10^{-6} J$,

where J is the current density in amperes per m^2.

Since the active area of this particular cell is $A = 0.0232$ m^2, the equations above can be written in terms of the current, I where $I = J \times A$,

at 50 C, $V_L = 0.912 - 2.34 \times 10^{-3} I$,
and at 70 C, $V_L = 0.913 - 2.12 \times 10^{-3} I$.

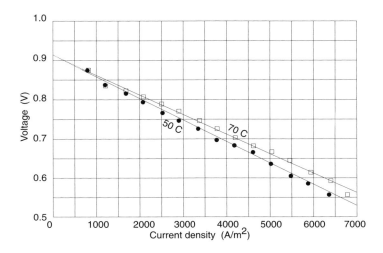

Figure 9.28   Characteristics of a Ballard SPFC cell.

The open-circuit voltages are

at 50 C,   0.912 V   or   77.4% of V_{rev} which is 1.178 V,

and at 70 C,   0.913 V   or   78.0% of V_{rev} which is 1.171 V.

The open-circuit voltage is only slightly influenced by the temperature. As explained before, V_{rev} became a bit smaller with the increase in temperature, while V_{oc} actually became marginally larger owing to improved kinetics.

The internal resistance was affected by the temperature in a more substantial way. It fell from 2.34 mΩ to 2.12 mΩ (more than 9%) with the 20 K increase in temperature.

9.8.3.1 Scaling Fuel Cells

Some of the data on fuel cell performance are presented in the form of V-J rather than V-I characteristics. Such practice permits the scaling of the cells—estimating the performance of a larger cell based on the data from a smaller one of the same type.

Consider a fuel cell with an active area A_0 and having a V-I characteristic

$$V_L = V_{oc} - R_{int_0} I. \tag{87}$$

Since

$$J = \frac{I}{A_0}, \tag{88}$$

the equation can be written

$$V_L = V_{oc} - R_{int_0} A_0 J = V_{oc} - \Re J, \tag{89}$$

where, remember, $\Re$ is the **specific resistance** of the fuel cell.

If now another fuel cell is built with exactly the same configuration and the same materials but with a different active area, A, then plausibly its internal resistance will be

$$R_{int} = R_{int_0} \frac{A_0}{A}, \tag{90}$$

because $R = \rho A/L$ (assuming that the thickness of the cell does not change).

V_{oc} does not depend on the area and will, therefore, remain the same. Thus, the load voltage will be

$$V_L = V_{oc} - R_{int_0} \frac{A_0}{A} I = V_{oc} - \Re J. \tag{91}$$

In other words, the specific resistance of the larger cell is the same as that of the smaller and both have the same V-J characteristics although they may have quite different V-I characteristics.

In practice, scaling up a particular fuel cell is not necessarily simple. Problems with heat removal and water management arise when dimensions are changed. The same occurs when cells are operated together in "stacks" to raise the overall voltage, as it is almost invariable the case.

9.8.4 *More Complete* V-I *Characteristics of Fuel Cells*

In the previous subsection, it was suggested that the activation voltage, V_{act}, is logarithmically dependent on the current. Here, we will first infer this dependence from empirical observations and then present a relatively simple argument as theoretical support. We will use real data from an experimental KHO fuel cell. These data were published in a book titled *Fuel Cells* by Will Mitchell, Jr. We simply scaled a graph that appeared in page 153 of the book. This introduced additional noise in the data but still permitted the carrying out of the necessary computer experimentation. The device is designated "New Cell" and refers to an old (1960) alkaline (KOH), high pressure hydrogen-oxygen fuel cell described (in the Mitchell book) by Adams et al. It operates at 200 C and, to keep the electrolyte from boiling away, had to be pressurized to 42 atmospheres. Owing to its high operating temperature this experimental cell is reasonably efficient as witnessed by the large open-circuit voltage.

We re-scaled the published data obtaining a tabulation of V_L as a function of I. The values are re-plotted in Figure 9.29.

Our goal is to develop a mathematical formula that, given a certain load current, yields the correct load voltage. This can be accomplished by writing,

$$V_L = V_{OC} - R_{int}I - V_{act}, \tag{92}$$

where V_{act} is a function of I constructed in such a way as to yield the correct V_L. This definition of V_{act} is not unique: it depends on our choice of R_{int}. To resolve this uncertainty, we must impose the additional requirement that V_{act} be a relatively straightforward function of I and, if possible, find a physical mechanism that justifies such a behavior. A mechanism does indeed exist and it requires, as we are going to show farther on, that V_{act} have a logarithmic dependence on I. This would lead to

$$V_L = V_{OC} - R_{int}I - V_2 \ln \frac{I}{I_0}, \tag{93}$$

where V_2 and I_0 are appropriate constants.

The procedure to find the correct expression for V_{act} and, incidentally, the correct value of R_{int} is as follows:

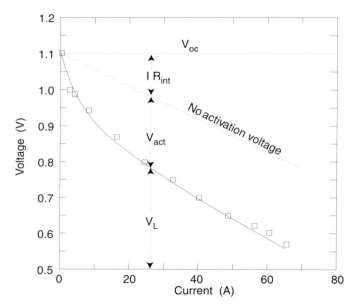

Figure 9.29   *V-I* characteristics of a high pressure hydrogen-oxygen KOH fuel cell of 1960.

Choose an arbitrary (but not implausible) value for R_{int}. If the only loss mechanism in the cell were this internal resistance, then the *V-I* characteristic would simply be the straight line marked "No activation voltage" in the figure. Owing, however, to the existence of an activation voltage drop, the true load voltage is given by Equation 93. Solving for V_{act},

$$V_{act} = V_{OC} - R_{int}I - V_L. \tag{94}$$

For each selected value of I, our data provide a value for V_L, and, since we know V_{OC} and assumed a value for R_{int}, we can calculate the corresponding V_{act}. Thus a table of V_{act} versus $\ln I$ can be constructed and these data can be plotted as seen, for example, in the curve marked $R = 0.002\ \Omega$ in Figure 9.30.

Observe that the curve in question is far from straight—that is, the relationship between V_{act} and $\ln I$ is not linear, as hoped. Repeating the procedure for $R = 0.006\ \Omega$ still results in a non-straight graph but with a reversed concavity. A little experimentation reveals that using $R_{int} = 0.0046\ \Omega$ yields an acceptably linear relationship between V_{act} and $\ln I$. One can then write a first degree regression between these two variables,

$$V_{act} = V_1 + V_2 \ln I \tag{95}$$

This empirical expression is called the **Tafel Equation**.

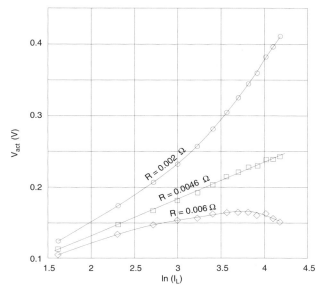

Figure 9.30   Only a specific value of R_{int} will
yield a linear relationship between V_{act} and $\ln I$.

For the current example, the Tafel Equation is
$$V_{act} = 0.0277 + 0.0521 \ln I. \tag{96}$$

This can also be written as
$$V_{act} = V_2 \ln \frac{I}{I_0} = 0.0521 \ln \frac{I}{0.588,} \tag{97}$$

because
$$I_0 = \exp\left(-\frac{V_1}{V_2}\right) = \exp\left(-\frac{0.0277}{0.0521}\right) = 0.588. \tag{98}$$

The characteristics of the fuel cell are
$$V_L = 1.111 - 0.0046I - 0.0521 \ln \frac{I}{0.588}. \tag{99}$$

The open-circuit voltage is 1.111.

The equation above fits well the measured data, with one major exception. At low currents, the predicted V_L exceeds the observed value. At 0.1 A, the predicted value is 1.20 V which is larger than the observed V_{OC} and when $I = 0$, the equation predicts a value of $V_L = \infty$ which is patently absurd.

It is obvious that our regression for calculating V_{act} is not valid when small load currents are considered. In fact, we discarded the value $I_L = 0$, when we made our calculations. Can we modify Equation 96 or 97 so as to extend our model into the low current range?

Equation 97 can be inverted,

$$I = I_0 \exp\left(\frac{V_{act}}{V_2}\right). \tag{100}$$

The preceding expression reminds us of Boltzmann's law which states: "the probability of finding molecules in a given spatial arrangement varies exponentially with the negative of the potential energy of the arrangement, divided by kT."

The potential energy of the electron in the presence of a voltage, V_{act}, is qV_{act}. Equation 100 can then be written as

$$I = I_0 \exp\left(\alpha \frac{qV_{act}}{kT}\right) \tag{101}$$

Here, we introduced the arbitrary factor, α, to adjust the magnitude of the argument of the exponential. Clearly, if Equation 101 is the same as Equation 100, then

$$\frac{V_{act}}{V_2} = \alpha \frac{qV_{act}}{kT}, \tag{102}$$

and, for the present example in which $T = 473$ K,

$$\alpha = \frac{kT}{qV_2} = 0.783. \tag{103}$$

Observe that after all the manipulations, Equation 101 is simply a mathematical representation of most of the experimental data. However, it fails badly in representing the obvious condition that V_{act} must be zero when $I = 0$. On the other hand, Equation 104 will also fit the data provided β is sufficiently large (because then the second term is essentially zero). However, this same equation does also fit the condition, $V_{act} = 0$ for $I = 0$.

$$I = I_0 \exp\left(\alpha \frac{qV_{act}}{kT}\right) - I_0 \exp\left(-\beta \frac{qV_{act}}{kT}\right). \tag{104}$$

Later on we are going to show that there are theoretical reasons for assuming that $\alpha + \beta = 1$, or $\beta = 1 - \alpha$. If so, Equation 104 becomes

$$I = I_0 \exp\left(\alpha \frac{qV_{act}}{kT}\right) - I_0 \exp\left((\alpha - 1)\frac{qV_{act}}{kT}\right), \tag{105}$$

and, for the current example,

$$\begin{aligned} I &= 0.588 \ \exp\left(0.783 \times 24.5 V_{act}\right) - 0.588 \ \exp\left(-0.217 \times 24.5 V_{act}\right) \\ &= 0.588 \left[\exp(19.2 V_{act}) - \exp(-5.32 V_{act})\right], \end{aligned} \tag{106}$$

where 24.5 is the value of q/kT when the temperature is 473 K.

The first term of the preceding expression is, of course, Equation 101. The second term is equal to the first when $V_{act} = 0$ forcing $I = 0$ under such conditions. As V_{act} grows, the second term quickly decreases in magnitude becoming negligible for even small values of the activation voltage so that Equations 101 and 106 yield then the same numerical result.

Thus, Equation 105 can be made to represent with acceptable accuracy the characteristics of a fuel cell. It is, nevertheless totally empirical. There is a simple, albeit somewhat handwaving, model that justifies its use.

When two dissimilar materials, at uniform temperature, are placed in contact with one another, a contact potential develops. The most familiar case (at least for the electrical engineer) is the potential that appears across a p-n junction.[†] In a single semiconductor crystal consisting of an n-region and a p-region, free electrons, more abundant in the n-side, diffuse toward the p-side, whereas holes from the p-side migrate to the n-side.

Were these particles uncharged, the diffusion process would only stop when the concentration became uniform across the crystal. This does not occur because a compensating drift current causes carriers to flow back in a direction opposite to that of the diffusion current. The drift current is driven by a **contact potential** created by the following mechanism: Being charged, the migrating electrons not only transport negative charges to the p-side but, also, leave uncompensated positively charged donors in the n-side. The holes also contribute to the accumulation of positive charges in the n-side and, the uncompensated acceptors, to the accumulation of negative charges in the p-side. Thus, the n-side becomes positive and the p-side negative.

In an open-circuited junction in equilibrium, there is no net current. This is the result of the precise cancellation of the diffusion current by the drift current. Although these **exchange currents** add to exactly zero,[††] they are not individually zero; their magnitude is surprisingly large. In silicon, under normal conditions, they may be of the order of 1 million A cm^{-2}. The above is a good example of the dynamic equilibria that occur frequently in nature—a zero net effect is the result of the precise cancellation of two large effects.

One peculiarity of the contact potential is that it cannot be measured directly because around any closed loop the potentials cancel each other.

The behavior of the electric potential across a p-n junction is easy to predict (see Figure 9.31).

In the metal-electrolyte junction, the situation is more complicated because, at this interface, there is a change of carriers—in the metal electrode, the current is transported by electrons, while in the electrolyte, it

---

[†] See Section 14.6.

[††] There are minute statistical fluctuations that give rise to the radio noise discussed in Chapter 8.

is transported by ions, either positive or negative (**cations** or **anions**, respectively). Thus, the flow of current from electrode to electrolyte or vice versa always involves a chemical reaction.

Consider an inert metal electrode in simultaneous contact with an electrolyte and with adsorbed hydrogen atoms some of which will ionize spontaneously.

The H^+ goes into solution and the electrons stay in the metal. As a consequence, the solution becomes more positive than the metal and some of the dissolved ions are attracted back to the electrode. Just as in the p-n case, two exchange currents are established:

1. A current carried by the ions that leave the metal and diffuse into the electrolyte under the influence of the gradient of ion concentration near the metal.
2. A current carried by the ions from the electrolyte that drift back to the metal under the influence of the electric field.

When there is no external electric connection between metal and electrolyte, the two currents must, under steady state conditions, be equal and opposite so that their sum is zero. If the diffusion current were larger than the return drift current, the ion concentration in the electrolyte would grow causing the latter to become even more positive. This would reduce the potential barrier for the returning ions increasing the drift current until it matched the diffusion current.

However, the hydrogen ionization reaction is not straightforward. Intermediate high-energy species are formed near the metal and the potential distribution acquires a hump as seen in Figure 9.32. This is the usual situation in chemical reactions. Catalysts are used to reduce the height of the hump. Notice that the difference in potential between the bulk of the electrolyte and the metal is independent of the size or the presence of the hump. To diffuse from metal to electrolyte, ions must have the energy, qV_A, of the hump, or more. Owing to their Maxwellian energy distribution, the fraction of ions with more than an energy, qV_A, is proportional to $\exp(-qV_a/kT)$.[†] The diffusion current, i_{f_0}, must be of the form

$$i_{f_0} = I_f \exp\left(-\frac{qV_A}{kT}\right). \tag{107}$$

The return current consists of ions from the electrolyte that have energy larger than qV_B. Consequently, the current must be of the form

$$i_{r_0} = I_r \exp\left(-\frac{qV_B}{kT}\right). \tag{108}$$

The total current, i_0, is of course

$$i_0 = i_{f_0} + i_{r_0} = 0. \tag{109}$$

[†] This is a situation like that of thermionic emission. See Section 6.2.

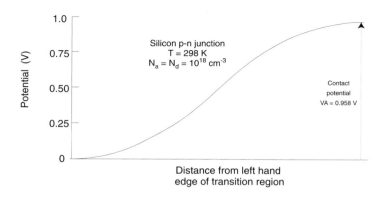

Figure 9.31   The potential varies with distance from the edge of a
p-n junction in a simple and predictable manner.

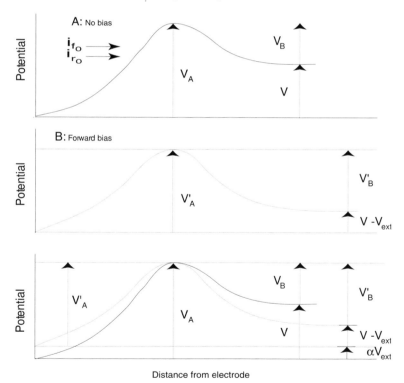

Distance from electrode

Figure 9.32   Potential versus distance from electrode at a metal-
electrolyte interface.   A—Unbiased, B—Biased, C—Biased and
unbiased superposed.

The subscript "0" indicates that these currents are in absence of bias. Notice the positive direction of i_{r_0} as defined in Figure 9.32A. This means that $i_{r_0} < 0$ because it represents H^+ ions moving *towards* the electrode.

Now, assume that a voltage is applied to the system so that the potential between the electrode and the solution is reduced by an amount, V_{ext}. In other words, V_{ext} forward-biases the metal-electrolyte junction (see Figure 9.32B). The magnitude of the potential barrier for ions diffusing from the electrode is now

$$V_A' = V_A - \alpha V_{ext}, \tag{110}$$

where $\alpha < 1$ is a fraction of the applied potential. Its value depends on the particular circumstances.

The barrier for ions from the electrolyte returning to the electrode is now

$$V_B' = V_B + (1 - \alpha)V_{ext}. \tag{111}$$

This can be shown by observing (Figure 9.32B) that

$$V_A' = V_A - \alpha V_{ext} = V - V_{ext} + V_B', \tag{112}$$

and (Figure 9.32A) that

$$V_A = V + V_B. \tag{113}$$

Thus, the forward bias reduces the barrier to the diffusion current and enhances that of the return drift current. These currents no longer cancel one another and a net current circulates: The two currents are now

$$i_f = I_f \exp\left[-\frac{q(V_A - \alpha\,V_{ext})}{kT}\right] = i_{f_0} \exp\left[\alpha\frac{q\,V_{ext}}{kT}\right], \tag{114}$$

$$i_r = I_r \exp\left[-q\frac{V_B + (1 - \alpha)\,V_{ext}}{kT}\right] = i_{r_0} \exp\left[-(1 - \alpha)\frac{q\,V_{ext}}{kT}\right]. \tag{115}$$

The total current that circulates in the external circuit is

$$i = i_f + i_r = i_{f_0} \exp\left[\alpha\frac{q\,V_{ext}}{kT}\right] + i_{r_0} \exp\left[-(1 - \alpha)\frac{q\,V_{ext}}{kT}\right]. \tag{116}$$

If we let $i_{f_0} = I_0$ and $i_{r_0} = -I_0$ (remember that $i_{f_0} = -i_{r_0}$), then the expression in Equation 116 becomes the same as that in Equation 109, derived empirically. The lower the potential barrier, V_A, the larger the exchange current, I_0. This current is, thus, a measure of the catalytic activity that controls the barrier.

In the derivation of Equation 109, V_{act} is the total activation voltage drop of the cell—that is, the sum of the voltage drop between anode and electrolyte plus the drop between electrolyte and cathode. In Equation 116, V_{ext}, is the drop between a single electrode and the electrolyte.

9.8.5 Heat Dissipation by Fuel Cells

Departures from reversibility constitute losses that appear as heat. Under thermal equilibrium, the rate of heat rejection is equal to the difference between the total power available from the reaction and the electric power delivered to the load. The power available from the reaction is $|\Delta\overline{h}|\dot{N}$ where $\Delta\overline{h}$ is the enthalpy change owing to the reaction per kilomole of product and $\dot{N}$ is the rate at which the product is being generated. If P_{heat} is the heat rejected, and P_L is the electric power in the load,

$$P_{heat} = |\Delta\overline{h}|\dot{N} - P_L. \tag{117}$$

In practical cells there are several mechanisms for heat generation:

1. The thermodynamic heat power, $P_{thermodynamic} = T|\Delta\overline{s}|\dot{N}$. This heat is rejected (rarely, absorbed) even by reversible cells.
2. The heat resulting from the difference between the reversible, voltage, V_{rev}, and the open-circuit voltage, V_{oc}. It is $P_{oc} = (V_{rev} - V_{oc})I$ watts.
3. The heat dissipated in the internal resistance, R_{int}, of the cell. This amounts to $P_{Joule} = I^2 R_{int}$ watts.
4. The heat owing to other departures, V_{extra}, of the cell voltage from the simple $V_L = V_{oc} - IR_{int}$ behavior. These departures may be due to the activation voltage drop, V_{act}, or due to the voltage drop resulting from electrolyte depletion, as mentioned. This amounts to $P_{extra} = V_{extra}I$.
5. The heat of condensation, $P_{cond} = |\Delta\overline{h}_{cond}|\dot{N}$, of the product water. If $\Delta\overline{h}$ of the reaction (Equation 117) is the value for formation of water vapor and the product water is removed as vapor, or if $\Delta\overline{h}$ is for the formation of liquid water, and the water is removed as liquid, then P_{con} must be taken as zero.

Example 4

Consider the Ballard fuel cell of Figure 9.28. What is the maximum power that can transferred to a load and what heat is generated? What is the efficiency of the cell? Use the V-I characteristics for 70 C, but, to simplify the problem, assume that the operating conditions are at RTP. The product water is removed from the cell in vapor form. The V-J characteristic of the cells is

$$V_L = 0.913 - 49.3 \times 10^{-6}J, \tag{118}$$

consequently the power output is,

$$P_L = V_L J = 0.913J - 49.3 \times 10^{-6}J^2 \ \text{ W m}^{-2}. \tag{119}$$

(continues)

(continued)

This is the power the cell delivers to a load per square meter of active electrode surface. The cell delivers maximum power when

$$\frac{dP}{dJ} = 0.913 - 98.6 \times 10^{-6}J = 0, \tag{120}$$

or

$$J = 9260 \text{ A/m}^2. \tag{121}$$

At this current, the cell would deliver 4230 W/m^2.
With 100% current efficiency the rate of water synthesis is

$$\dot{N} = \frac{J}{qn_e N_0} = \frac{9260}{1.60 \times 10^{-19} \times 2 \times 6.02 \times 10^{26}}$$

$$= 48 \times 10^{-6} \text{ kilomoles } (H_2O)s^{-1}m^{-2}. \tag{122}$$

Hence, the energy input to the cell is

$$P_{in} = \Delta\overline{h}\dot{N} = 242 \times 10^6 \times 48 \times 10^{-6} = 11,600 \text{ W/m}^2. \tag{123}$$

Of these, 4230 W/m^2 appear as electric energy in the load, so that $11,600 - 4230 = 7370$ W/m^2 of heat are generated.

Notice that the Joule losses inside the cell amount to

$$P_{Joule} = R_{int}J^2 = 49.3 \times 10^{-6} \times 9260^2 = 4220 \text{ W/m}^2. \tag{124}$$

This is, of course, equal to the power delivered to the load because, for maximum power transfer the load resistance must equal the internal resistance of the generator.

The thermodynamic heat is

$$P_{therm} = T|\Delta\overline{s}|\dot{N} = 298 \times 44.4 \times 10^3 \times 48 \times 10^{-6} = 635 \text{ W/m}^2. \tag{125}$$

The losses owing to V_{oc} being different from V_{rev} amount to

$$P_{oc} = (V_{rev} - V_{oc})I = (1.185 - 0.913) \times 9260 = 2519 \text{ W/m}^2. \tag{126}$$

Necessarily,

$$P_{in} = P_{therm} + P_{oc} + P_{Joule} + P_L \tag{127}$$

The efficiency of this cell is

$$\eta = \frac{P_L}{P_{in}} = \frac{4230}{11,600} = 0.365. \tag{128}$$

In the above example, the current delivered by the cell (9260 A/m^2) exceeds the maximum current (7000 A/m^2) in the Ballard data. This probably means that the cell cannot deliver all this power—that is, it cannot dissipate the 7400 W/m^2 of heat it generates.

The efficiency of the cell, if operated at lower output levels, will be substantially larger than that in the extreme example above. In fact, at $J = 7000$, the efficiency would be about 45%.

9.8.5.1 Heat Removal from Fuel Cells

The operating temperature of a fuel cell depends on the type of cell. In any cell, if the temperature is to remain unchanged, an amount of heat power, P_{remov}, equal to the generated heat power, P_{heat} must be removed from the device.

This heat removal can be accomplished passively or by using special heat exchange schemes. Heat can, for example, be removed by using a rate of air flow in excess of that which is required to satisfy the oxygen demand of the cell. This same stream of air may also be useful in the removal of the reaction water.

References

Adams, A. M., F. T. Bacon, and R. G. H. Watson, *The high pressure hydrogen-oxygen cell*, Chapter 4 of Fuel Cells, ed. Will Mitchell, Jr. Academic Press, **1963**.

Bessette, N. F. and J. F. Pierre, Status of Siemens Westinghouse tubular solid oxide fuel cell technology and development program, *Fuel Cells—Powering the 21st Century*, Oct. 30 to Nov. 2, **2000**, Portland, Oregon.

Boysen, Dane A., Tetsuya Uda, Calum R. I. Chisholm, and Sossina M. Haile, *High-performance solid acid fuel cells through humidity stabilization*, Science **303**, *68* 2 January 2004.

Chu, Deryn, R. Jiang, and C. Walker, Methanol tolerant catalyst for direct methanol fuel cell applications, *Fuel Cells—Powering the 21st Century*, Oct. 30 to Nov. 2, **2000**, Portland, Oregon.

Dodelet, J. P., M. C. Denis, P. Gouérec, D. Guay, and R. Scholz, CO tolerant anode catalysts for fuel cells made by high energy ball-milling, *Fuel Cells—Powering the 21st Century, Fuel Cell Seminar*, October 30–November 2, **2000**, Portland, OR.

Dohl, H., S. von Adrian, J. Divisek, B. Höhlein, and J. Meusinger, Development and process analysis of direct methanol fuel cell systems, *Fuel Cells—Powering the 21st Century*, October 30–November 2, **2000**, Portland, Oregon.

Forbes, C. A., and J. F. Pierre, The solid fuel-cell future, *IEEE Spectrum*, Oct. **1993**, p 40.

Goldstein, Rocky, *Proton conductors for solid electrolyte fuel cells*, Exploratory Research Letter, Electric Power Research Institute (EPRI), Palo Alto, CA.

Ghosh, D., M. E. Pastula, R. Boersma, D. Prediger, M. Perry, A. Horvath, and J. Devitt, Development of low temperature SOSF systems for remote power and home cogen applications, page 511, *Fuel Cells—Powering the 21st Century, Fuel Cell Seminar*, October 30–November 2, **2000**, Portland, OR.

Haberman, William L., and James E. A. John, *Engineering thermodynamics with heat transfer*, Allyn and Bacon, **1989**.

Hibino, Takashi, Atsuko Hashimoto, Takao Inoue, Jun-ichi Tokuno, Shin-ichiro Yoshida, and Mitsuru Sano, A low-operating-temperature solid oxide fuel cell in hydrocarbon-air mixtures, Science **288**, 2031 16 June **2000**.

Koppel, Tom, *Powering the Future (The Ballard fuel cell and the race to change the world)*, John Wiley, **1999**.

Lal, Amit, and James Blanchard, The daintiest dynamos, *IEEE Spectrum* September **2004**.

Mitchell, Jr., Will, *Fuel Cells*, Academic Press, **1963**.

Narayanan, S. T., T. I. Valdez and F. Clara, Design and development of miniature direct methanol fuel cell sources for cellular phone applications. *Fuel Cells—Powering the 21st Century*, Oct. 30 to Nov. 2 **2000**, Portland, Oregon.

Ovshinsky, S. R., M. A Fetcenko and J. Ross, A nickel metal hydride battery for electrical vehicles, *Science* **260**, 176, April 9 **1993**.

Pham, A. Q., B. Chung, J. Haslam, J. DiCarlo, and R. S. Glass, Solid oxide fuel cell development at Lawrence Livermore National Laboratory, *Fuel Cells—Powering the 21st Century, p. 787*, Oct. 30 to Nov. 2, **2000**, Portland, Oregon.

Reddington, E. et.al., Combinatorial Electrochemistry: a highly parallel, optical screening method for discovery of better electrocatalysts, *Science*, 280, p. 1735, June 12, **1998**.

Service, R. F., The fast way to a better fuel cell, *Science*, 280, p. 1690, June 12, **1998**.

Watkins, David S., Research, development and demonstration of solid polymer fuel cell systems, *Fuel cell systems*, Leo J. M. J. Blomen and Michael N. Nugerwa, editors, Plenum Press, New York, **1993**.

Yoon, S. P., S. W. Nam, T.-H. Lim, I.-H. Oh, H. Y. Ha, and S.-A. Hong, Enhancement of cathode performance by sol-gel coating of yttria-stabilized zirconia, *Fuel Cells—Powering the 21st Century, Fuel Cell Organizing Committee*, page 611–614, Oct 30.—Nov. 2, **2000**, Portland Oregon.

Further reading

European Fuel Cell R&D Review, Sep. **1994**, *Argonne National Laboratory*.

Fuel Cells, A Handbook (Revision 3), Jan. **1994**, *USDE, Office of Fossil Energy*.

Halpert, Gerald, Sekharipuram R. Narayanan, Thomas Valdez, William Chun, Harvey Frank, Andrew Kindler, and Subbarao Surampudi (Jet Propulsion Laboratory), and Jack Kosek, Cecelia Cropley, and Anthony LaConti (Giner Inc), Progress with the direct methanol liquid-feed fuel cell system, *IECEC*, **1997**.

Prater, Keith B., Polymer electrolyte fuel cells: a review of recent developments, *J. Power Sources*, 51, p. 129, **1994**.

Reynolds, W. C., Thermodynamic properties in SI, *Department of Mechanical Engineering*, Stanford University, **1979**.

Technology Development Goals for Automotive Fuel Cell Power Systems, Jul. **1995**, *Argonne National Laboratory*.

Williams, Robert H., The clean machine, *Technology Review*, Apr. **1994**.

PROBLEMS

9.1   Every substance is endowed with a certain amount of energy and a certain amount of entropy. While the latter is well defined, the former—the energy—has no absolute value; only changes in energy can be measured. For this reason (entirely by convention), the enthalpy of formation of elements in their natural state is taken as zero.

Consider aluminum and oxygen. In their natural states, their standard enthalpy of formation (i.e., the energy of formation at RTP) is, as we said, zero. Every kilogram of aluminum has (at RTP) an entropy of 1.05 kJ/K, whereas every kilogram of oxygen has an entropy of 6.41 kJ/K.

Aluminum burns fiercely forming an oxide (Al_2O_3) and releasing energy. The standard enthalpy of formation of the oxide is -1.67 GJ/kmole. The entropy of the oxide is 51.0 kJ/K per kilomole.

According to the second law of thermodynamics, the entropy of a closed system suffering any transformation cannot diminish. It can, at best, remain unchanged as the result of the transformation or else, it must increase. If you add up the entropies of the aluminum and of the oxygen, you will discover that the sum exceeds the entropy of the oxide formed. This means that when aluminum combines with oxygen, only part of the chemical energy can be transformed into electricity, while the rest must appear as the heat associated with the missing entropy. That part that can (ideally) be converted into electricity is called the **free energy**.

Calculate the free energy of the aluminum/oxygen reaction.

9.2   Daniel cells used to be popular last century as a source of electricity, especially in telegraph systems. These cells consisted of a container divided into two compartments by a membrane permeable to ions. In one compartment, a zinc electrode was dipped in a zinc sulfate solution and, in the other, a copper electrode in a copper sulfate solution.

The zinc oxidizes (i.e., loses electrons):

$$Zn \rightarrow Zn^{++} + 2\ e^-$$

The zinc is eroded, going into the solution in the form of ions.

At the other electrode, the copper is reduced (i.e., the copper ions in the sulfate accept electrons and the copper from the sulfate plates out onto the copper electrode):

$$Cu^{++} + 2\ e^- \rightarrow Cu$$

The cell will deliver current until it runs out of either zinc or sulfate, whichever is less.

Assume the cell has 95 g of zinc and 450 ml of a 0.1 M $CuSO_4$ solution. M stands for molarity: moles of solute per liter of solution. How long can this cell deliver a current of 2 A to a load?

9.76

9.3 A fuel cell has the following reactions:
ANODE: $C + 2\ O^{--} \rightarrow CO_2 + 4\ e^-$,   (1)
CATHODE: $4\ e^- + O_2 \rightarrow 2\ O^{--}$.   (2)

Changes of enthalpy and free energy (RTP), per kmole of CO_2, are:
$$\Delta \bar{h}_f^\circ = -393.5\ \text{MJ},$$
$$\Delta \bar{g}_f^\circ = -394.5\ \text{MJ}.$$

What is the overall reaction? What is the ideal emf (i.e., V_{rev})? What is the difference in entropy between reactants and products?

Assume that the internal resistance of the cell is 1 milliohm. Otherwise, the cell behaves as an ideal voltage source. How much carbon is needed to deliver 1 MWh of electricity to the load in minimum possible time? What is the load resistance under such conditions?

9.4 The enthalpies and the free energies of formation (at RTP) of each species of interest in this problem are:

	$\Delta \bar{h}_f^\circ$	$\Delta \bar{g}_f^\circ$
	(MJ/kmole)	
CH_3OH (g)	−201.2	−161.9
CH_3OH (l)	−238.6	−166.2
O_2 (g)	0	0
CO_2 (g)	−393.5	−394.4
H_2O (g)	−241.8	−228.6
H_2O (l)	−285.9	−237.2

Owing to the reaction, the changes in enthalpy and in free energy are:

Methanol	Water	$\Delta \bar{h}^\circ$ MJ/kmole	$\Delta \bar{g}^\circ$ MJ/kmole
liquid	gas	−638.5	−685.3
gas	gas	−676.5	−689.6
liquid	liquid	−726.5	−702.4
gas	liquid	−764.5	−706.7

Data from Dohle et al.

Consider methanol, a fuel that has been proposed for both internal combustion (IC) engines and for fuel cells. Methanol can be derived from fossil fuels and also from biomass. Being liquid at RTP conditions, it is a convenient fuel for automobiles. It has reasonable reactivity in fuel cells.

In IC engines, methanol is first evaporated and then burned. The engine exhausts water vapor. In fuel cells, the methanol reacts in liquid form but the product water is in vapor form.

1. What heat do you get by burning 1 kg of methanol in an IC engine?
2. How much electric energy will an ideal fuel cell (using methanol and air) produce per kg of fuel?
3. How much heat does the cell reject?

9.77

4. A practical OTTO cycle engine has an efficiency of, say, 20%, while a practical methanol fuel cell may have an efficiency of 60% (this is the efficiency of the practical cell compared with that of the ideal cell). If a methanol fueled IC car has a highway performance of 10 km per liter, what is the performance of the fuel cell car assuming that all the other characteristics of the car are identical?

5. If you drive 2000 km per month and a gallon of methanol costs $1.20, how much do you save in fuel per year when you use the fuel cell version compared with the IC version? Can you think of other savings besides that in fuel?

6. You get a ten year loan with yearly repayments of principal plus interest of 18% of the initial amount borrowed. By how much can the initial cost of the fuel-cell car exceed that of the IC car for you to break even? Assume that after 10 years the car is totally depreciated.

7. What is the open circuit voltage of an ideal methanol fuel cell at RTP? To answer this question, you need to make an intelligent guess about the number of electrons freed per molecule of methanol.

In the above questions, assume 100% current efficiency and 100% efficiency of the electric motor.

9.5   You want to build a hydrogen manometer based on the dependence of the output voltage on the pressure of the reactants. Take an H_2/O_2 fuel cell at 298 K. Assume that it produces water vapor and and acts as an ideal cell. The oxygen pressure is maintained at a constant 0.1 MPa while the hydrogen pressure, p_{H_2}, is the quantity to be measured.

1. What is the output voltage when p_{H_2} is 0.1 MPa?
2. What is the output voltage when p_{H_2} is 1 MPa?
3. Develop an expression showing the rate of change of voltage with p_{H_2}. What is this rate of change when p_{H_2} is 0.1 MPa?
4. The output voltage of the cell is sensitive to temperature. Assume that a ±10% uncertainty in pressure measurement can be tolerated (when the pressure is around 1 MPa). In other words, assume that when a voltage corresponding to 1 MPa and 298 K is read, the actual pressure is 0.9 MPa because the temperature of the gases is no longer 298 K. What is the maximum tolerable temperature variation?

9.6   A certain gas, at 10^5 Pa, has a specific heat given by the expression

$$c_p = a + bT + cT^2 \tag{1}$$

for 298 K < T < 2000 K.

a = 27.7 kJ K^{-1} kmole^{-1},
b = 0.8×10^{-3} kJ K^{-2} kmole^{-1},
c = 10^{-6} kJ K^{-3} kmole^{-1}.

9.78

At 298 K, the enthalpy of the gas is 0 and its entropy is 130.0 kJ K^{-1} kmole^{-1}. What are H, G, and S of the gas (per kilomole) at $T = 1000$ K and $p = 10^5$ Pa? Please calculate with 4 significant figures.

9.7   A fuel cell has the reactions:

ANODE: $2 A_2 = 4 A^{++} + 8 e^-$.   (1)

CATHODE: $4 A^{++} + B_2 + 8 e^- = 2 A_2 B$ (2)

All data are at RTP.

The overall reaction, $2 A_2 + B_2 = 2 A_2 B$, releases 300 MJ per kmole of $A_2 B$ in a calorimeter. The entropies of the different substances are:

A_2:   200 kJ K^{-1} kmole^{-1},

B_2:   400 kJ K^{-1} kmole^{-1},

$A_2 B$:   150 kJ K^{-1} kmole^{-1}.

A_2 and B_2 are gases, whereas $A_2 B$ is liquid.

What is the voltage of an ideal fuel cell using this reaction at RTP? Estimate the voltage at standard pressure and 50 C.

How much heat does the ideal fuel cell produce per kilomole of $A_2 B$, at RTP? What is the voltage of the cell if the gases are delivered to it at 100 MPa? The operating temperature is 25 C. If the internal resistance of the cell (operating at RTP) is 0.001 Ω, what is the maximum power the cell can deliver to a load? What is the fuel consumption rate of the cell under these circumstances? What is the efficiency of the cell?

9.8   Owing to its ceramic electrolyte, a fuel cell can operate at 827 C. Pure oxygen is used as oxidizer. The gases are fed to the cell at a pressure of 1 atmosphere. Use the data below:

	$\Delta \overline{h}^\circ_f$ MJ per kmol	$\Delta \overline{g}^\circ_f$ MJ per kmol	γ	$\overline{s}^\circ$ kJ K^{-1} per kmol
$CO(g)$	-110.54	-137.28	1.363	197.5
$CO_2(g)$	-393.51	-394.38	1.207	213.7
$O_2(g)$	0	0	1.341	205.0

The values of γ are those appropriate for the the 25 C to 827 C interval.

We want to examine the influence of temperature on the performance of an ideal fuel cell.

1. Calculate the reversible voltage and the efficiency of the above (ideal) cell at both 25 C and 827 C.
2. As expected (if you did the calculations correctly), you will have found that both V_{rev} and η are larger at the lower temperature. Yet, the cell is operated at 827 C where its ideal performance is not as good. Why? Explain in some detail.

9.9   A fuel cell was tested in the laboratory and yielded the following:

Open-circuit voltage	0.600 V
Internal resistance	0.01 Ω
Voltage (I = 1 A)	0.490 V
Voltage (I = 10 A)	0.331 V

Thermodynamic data indicate that V_{rev} of a fuel cell is 0.952 V and that the enthalpy change of the reaction is 1.26 times the reaction free energy. Two electrons circulate in the load per molecule of product.
  1. What is the power the cell delivers to a load when the current is 5 A?
  2. What is the heat power dissipated internally when $I_L = 5$ A?
    Assume the Tafel equation is valid. The internal resistance given above is the slope of the straight line portion of the v-i characteristic of the cell.

9.10   A hydrogen-oxygen fuel cell, operating at RTP, has the following v-i characteristics:
$$V_L = 0.8 - 0.0001 I_L.$$

Assume 100% current efficiency.

  1. What is the hydrogen consumption rate (in mg/s) when the cell delivers 1 kW to a load?
  2. What is the heat generated by the cell? Liquid water is produced.

9.11   A fuel cell is prismatic in shape and measures $d \times 2d \times 2d$ (where $d = 33$ cm). It is fed with H_2 and O_2 which are admitted at 25 C and 1 atmosphere. Product water is exhausted at 1 atmosphere and 110 C.
    The inside of the cell is evenly at 110 C. The outside is maintained at 50 C by immersing it totally in running water admitted at 20 C and exhausted at 45 C. Liquid water has a heat capacity of 4 MJ per K per m^3. Assume that the temperature gradient across the walls is uniform. The walls are made of 10-mm thick stainless steel with a heat conductivity of 70 W per m per K. The only energy input to the cell comes from the fuel gases admitted. Heat is removed by both the coolant water and the product water that is exhausted from the cell at a rate of $\dot{N}$ kmoles/s.
    The load voltage is $V_L = 0.9 - R_{int} I$ volts. $R_{int} = 10^{-7} \Omega$.

  1. What heat energy is removed every second by the coolant water?
  2. What is the flow rate of the coolant water?
  3. Express the heat removal rate by the product water in terms of $\dot{N}$.
  4. What is the input power in terms of $\dot{N}$?
    What is the power delivered to the load in terms of $\dot{N}$?
  5. Write an equation for thermal equilibrium of the cell using your results from above.
  6. What is the value of I that satisfies the above equation?

9.80

To simplify the solution, assume that the fuel cell reaction proceeds at RTP and liquid water is produced at 25 C. This water is then heated up so that the exhaust is at the 110 C prescribed.

9.12   A fuel cell has a cooling system that allows accurate measurement of the heat dissipated and precisely controls the operating temperature which is kept, under all circumstances, at 298 K.

When a current of 500 A is generated, the cooling system removes 350 W of heat, while when the current is raised to 2000 A, the amount of heat removed is 2000 W.

Estimate the heat removed when the current is 1000 A. The input gases are at RTP and liquid water is created in the process.

Assume a linear dependence of the load voltage on the load current.

9.13   An ideal fuel cell operates at 1000 K. Two reactant gases (not necessarily H_2 and O_2) are fed in at 1 atmosphere. The reversible voltage is 1.00 V. What will the voltage be if gas A is fed in at 100 atmospheres and gas B at 200 atmospheres, both still at 1000 K? The reaction is

$$2A + 3B \rightarrow A_2B_3.$$

The product is liquid (in spite of the high temperature). A total of 6 electrons circulate in the load for each molecule of product.[†]

9.14   A fuel cell employs the reaction below,

$$A + B \rightarrow AB.$$

At STP, the relevant thermodynamic data are

	$\Delta \overline{h}_f^{\,\circ}$ [MJ/kmole]	$\overline{s}$ [kJ/(K kmole)]
A(g)	0	100
B(g)	0	150
AB(g)	−200	200

What is the reversible voltage of the above fuel cell? For each molecule of AB, 2 electrons circulate in the load.

9.15   A fuel cell battery is to be used in a satellite. It must deliver a 2 kW at 24 V for 1 week. The mass of the cell must be the smallest possible.

The fuel cell manufacturer has a design with the following characteristics:

Open-circuit voltage: 1.10 V,

[†] The very large difference in pressure between the two reactant gases would require a strong diaphragm or electrolyte. This suggests a ceramic electrolyte and hence the high operating temperature. Clearly, the realizability of this fuel cell is highly doubtful.

Internal resistivity: 92×10^{-6} ohm m^2,
Cell mass: 15 kg per m^2 of active electrode area.
There is a linear relationship between V_L and I_L.

How many cells must be connected in series?
What is the total mass of all fuel cells in the battery?

9.16   The open circuit voltage of a hydrogen-oxygen fuel cell operating at RTP is 0.96 V and its internal resistance is 1.2 mΩ. The activation voltage drop is given by

$$V_{act} = 0.031 + 0.015 \ln I,$$

where I is in amperes.

From thermodynamic data, the reversible voltage, V_{rev}, is known to be 1.20 V.

200 of the above cells are connected in series forming a battery that feeds a resistive load rated at 2.5 kW at 100 V.

What is the actual load voltage?
How much heat does the battery generate internally?

9.17   A fuel cell battery is fed by H_2 and O_2 (both at 300 K) and produces water vapor that promptly condenses inside the cell.

ΔT is the difference, in kelvins, between the temperature of the active area of the cell and 300 K, which is supposed to be the temperature of the cooling liquid and of the environment.

Two main mechanisms remove heat from the cells:

1. Some heat is conducted away at a rate of 40 W per m^2 of active electrode surface for each kelvin of ΔT. This, of course, implies some cooling system whose exact nature is irrelevant as far as this problem is concerned.
2. To simplify the solution, assume that water vapor is synthesized in the cell at the temperature of the incoming gases (300 K) and then immediately condenses at this temperature and then heats up by an amount ΔT to reach the operating temperature of the cell. The water is then removed, carrying with it a certain amount of heat and, thus, cooling the cell. If the temperature of the product water exceeds 100 C, assume that the cell is pressurized so that water does not boil. However, assume that all the reactions actually occur at RTP, i.e., use thermodynamic data for RTP.
   The V-J characteristic of the cell is

$$V = 1.05 - 95.8 \times 10^{-6} J,$$

where J is the current density in A/m^2.

The current efficiency is 100%.

Although the cell will operate at conditions that differ from RTP, use RTP thermodynamic data to simplify the problem.

1. The battery is not going to be operated at full power because it probably will exceed the maximum allowable temperature. Nevertheless, calculate what the equilibrium temperature would be if full power operation were attempted.

2. In fact, the battery will operate at a much lower power. It must deliver 20 kW to a load at 12 V. It consist of several cells connected in series. The mass of each cell is 15 kg for each m^2 of active electrode area.

The battery must deliver this power for a week. The total mass (fuel plus battery) must be minimized. Ignore the mass of the fuel tanks.

How many cells must be employed?

What is the total mass?

How many kg of H_2 and how many of O_2 are needed?

What is the operating temperature of the cell?

9.18   Fill in the answers as follows:

If output voltage rises, mark "R," if it falls, mark "F," if there is no effect, mark "N."

	IDEAL FUEL CELL	PRACTICAL FUEL CELL
Higher temperature		
Higher reactant pressure		
Higher product pressure		

9.19   A fuel cell, generating water vapor, has a straight line V-I characteristic:

$$V_L = V_0 - R_{int}I$$

Both V_0 and R_{int} are temperature dependent and are given by the expressions below, over the temperature range of interest.

$$V_0 = \beta_0(1 + \alpha_V T)V_{rev},$$

$$R_{int} = (1 + \alpha_R T)R_{int_0}.$$

The coefficients are:

1. $\beta_0 = 0.677$,
2. $\alpha_V = 443.5 \times 10^{-6}$ per K,
3. $R_{int_0} = 0.00145$ Ω,
4. $\alpha_R = -1.867 \times 10^{-3}$ per K.

What are the efficiencies of the fuel cell at 298 K and at 500 K when feeding a 1 milliohm load?

9.20   A fuel cell battery is to be used aboard the space station. The bus voltage (the voltage the battery has to deliver under full load) is 24 V when delivering 30 kW. Since the space craft has a hydrogen and oxygen supply, the battery will use these gases which are delivered to it at 1 atmosphere and 298 K. A manufacturer-submitted sample cell was tested in the laboratory with the following results:

When no current is drawn from the cell, its voltage is 1.085 V. When delivering 2000 A, the voltage is 0.752 V. A straight line relationship was found to exist between V and I. The cell masses 75 kg and, when taken apart, it was found that the active electrode area is 1.5 m^2. It is clear that if the battery is to deliver 30 kW under 24 V, it must generate a current of 1250 A. Since all the cells are in series, this is also the current through each cell.

The sample cell operates with a current density of $1250/1.5 = 833.3$ A/m^2. If the manufacturer constructs a cell, in all aspects identical to the sample, except with different active electrode area, S, the new cell must still deliver the 1250 A but under different current density and, consequently, under different cell load voltage. Since the battery load voltage must still be 24 V, the battery will contain a different number, N, of cells.

Assume that the mass of the new cell is proportional to the active area of the electrodes. The total mass (mass, M_B, of the battery plus mass, M_F, of the fuel, H$_2$ and O$_2$) is to be minimized for a 30-day long mission during which the battery delivers a steady 30 kW at 24 V. Ignore the mass of the fuel tanks. The current efficiency is 100%.

Calculate this minimum total mass. How many cells are needed in series?

In the cell above, assume that water is synthesized as vapor at the temperature of the incoming gases (298 K) and promptly condenses into a liquid and then heats up to T_{op}, the operating temperature of the device. The product water is continuously removed from the cell at this latter temperature. In addition, a cooling system also removes heat. It does this at a rate of 6 W per degree of temperature difference ($T_{op} - 298$ K) for each square meter of active electrode surface.

What is T_{op} when the battery delivers 30 kW as in the first part of this problem?

9.84

9.21   A hydrogen-oxygen fuel cell has the following characteristics when both reactants are supplied at the pressure of 1 atm:

$$V_{oc} = 0.75 + 0.0005T \text{ V},$$

$$R_{int} = 0.007 - 0.000015T \text{ } \Omega.$$

Estimate, roughly, the power this fuel cell delivers to a 5 milliohm load. In any fuel cell, heat may be removed by

1. circulation of a coolant,
2. excess reactants that leave the cell at a temperature higher than the input temperature,
3. products that leave the cell at a temperature higher than that at which they were synthesized.

To simplify this problem assume that the contribution of mechanism a is always 30 times that of mechanism c and that of mechanism b is negligible. Assume that the reactants are fed in at 298.2 K and the product water is created as a vapor at this temperature and then heated to T_{op} by the heat rejected by the cell.

Pure hydrogen and pure oxygen are supplied at 1 atm.

1. What is the temperature of the cell when temperature equilibrium has been reached?
2. What is the load current and the power delivered to the load under the above conditions?

9.22

R_L Ohms	I Amps
0.05	14.98
0.10	8.23
0.15	5.71
0.20	4.37
0.25	3.54

The reactions in a fuel cell are:

$$\text{Anode}: \quad A \rightarrow A^{++} + 2e^-$$
$$\text{Cathode}: \quad A^{++} + 2e^- + B \rightarrow AB$$

The gases, A and B, are fictitious (and so are their properties). The atomic mass of A is 16 daltons and that of B is 18 daltons. Both A and B behave, over the temperature range of interest, as if they had 5 degrees of freedom, while the product, AB, as if it had 7.

The fuel cell was tested in a laboratory by observing the current delivered as the load resistance was altered.

The results are displayed in the table.

1. What are the open circuit voltage and the internal resistance of the cell?

2. Careful calorimetric observations show that when the fuel cell is delivering 10.0 A to a load, the heat dissipated internally is 3.40 W. From this information, determine the ΔH of the reaction.

3. Notice that P_{heat} depends on $(V_{rev} - V_{oc})I$. This would suggest that it is possible to determine V_{rec} from the knowledge of P_{heat}. Demonstrate that it is not possible to do so, i.e., that for a fixed V_{oc} and R_{int}, P_{heat} is not sensitive to the value of V_{rev}.

4. In order to estimate the ΔG of the reaction, an external voltage was applied to the fuel cell so as to cause it to act as an electrolyzer. When this external voltage was 1.271 V, the electrolyzer produced A at a rate of 2.985 g/hour. Making plausible assumptions, estimate the ΔG of the reaction.

9.23   A hydrogen/oxygen fuel cell has the V-J characteristic (at RTP):

$$V_L = 0.98 - 10^{-3}J.$$

The active area of its electrodes is 0.444 m^2. The water is exhausted from the cell in gaseous form.

1. What is the rate of heat production when the cell is connected to a load of

   1.1   Open circuit?
   1.2   Short circuit?
   1.3   A load that maximizes the power output?

2. What are the efficiencies of the cell under the 3 conditions above?
3. What is the efficiency of the cell if it delivers half of its maximum power? Use the more efficient solution.
4. Assume that V_{oc} is a constant fraction of V_{rev}. Thus, under all circumstances $V_{oc} = (0.98/1.185) \times V_{rev} = 0.827V_{rev}$.
   What is the V-I characteristic of the cell when fed air at 1 atmos and 25 C instead of oxygen?
5. To simplify this problem assume that the only way to remove heat from the cell is via the exhaust stream which consists of water vapor and excess input gases. The input gases (hydrogen and air) are at 298 K. Assume that the water is produced at 298 K and then heated by the fuel cell to the exhaust temperature, T.

   What is the value of T when the cell, fed by the minimum amount of air that satisfies the oxygen requirement of the device, produces the electric output of Item 3 (half its maximum power). Although the oxidizer is air, not pure oxygen, use, for simplicity, the V-I characteristic for pure oxygen as given in Item 1.
6. If you made no mistake, you have found that the temperature calculated in the preceding item is too high. Much more vigorous cooling

will be necessary. This can be accomplished by injecting much more air than is required by the stoichiometry. Assume that the temperature raise should not exceed 80 K. How much must the flow of air be compared with that required in Item 5?

9.24   The EV1 was an exceedingly well-designed automobile, having excellent aerodynamics and, all over, low losses. With an energy supply of 14 kWh, it had range of over 100 km. It's 100 kW motor allowed very good acceleration making it a "sexy" machine. The problem was that, no matter how good a battery it used, it took a long time to recharge it. If instead of a battery, it had used fuel cells, then *refueling* would take only minutes versus hours for *recharging*.

Imagine that you want to replace the NiMH batteries by a fuel cell battery which, of course, must supply 100 kW of power. The V-I characteristic of the available hydrogen/oxygen fuel cell operating at RTP is

$$V_l = 1.1 - 550 \times 10^{-6} I.$$

The maximum internal heat dissipation capability is 300 W. Product water exits the cell in vapor form.

The fuels cells deliver their energy to a power conditioning unit (inverter) that changes its dc input into ac power. The efficiency of this unit can be taken as 100%.

1. What is the input voltage of the power conditioning unit, in other words, what is the voltage that the fuel cell battery (at 100 kW) must deliver assuming the the smallest possible number of individual cells are used.

2. The 100 kW are needed only for acceleration. For cruising at 110 km/h, only 20 kW are required.[†] How many kg of hydrogen are needed to provide a range of 800 km to the car (using 20 kW)?

3. If the hydrogen is stored at 500 atmospheres, how much volume does it occupy at 298 K?

9.25   A single chamber low-operating-temperature solid oxide fuel cell somewhat similar to the one described by Hibino et al., when operated at a current density of 6000 A/m², delivers a load voltage that depends on the thickness of the electrolyte in the manner indicated in the table below:

Electrolyte thickness mm	Load Voltage V
0.15	0.616
0.35	0.328
0.50	0.112

[†] Just a wild guess!

The cell has essentially straight V-I characteristics. Its specific resistance, $\Re$, (see Subsection 9.8.3.1) consistis of two components, $\Re_1 + \Re_2$, where $\Re_1$ is the resistance of the electrolyte and $\Re_2$ represents all other resistances of the cell. The open-circuit voltage is 0.892 V.

1. If it were possible to use a vanishingly thin electrolyte, what maximum power would the cell be able to deliver?
2. What would be the corresponding load resistance if the cell has an effective electrode area of 10 by 10 cm?
3. Compare the power output of the cell with that for the cell with 0.15 mm-thick electrolyte.

9.26   Solid oxide fuel cells manufactured by Siemens Westinghouse have a very pronounced curvature in their V-J characteristics. One class of cells using "ribbed" units behaves according to
$$V_L = 0.781 - 1.607 \times 10^{-6}J - 6.607 \times 10^{-9}J^2,$$
where J is the current density in A/m^2 and V_L is the load voltage in V.

1. What is the open-circuit voltage of the cell?
2. What is the voltage of the cell when delivering maximum power?

9.27   The V-I characteristics of a given fuel cell (measured with incredible precision) are tabulated as shown. See plot. The measurements were made at RTP. Water leaves the cell as a gas.

1. Calculate the efficiency of the cell when 10 A are being delivered.
2. Calculate the rate of heat generated by the cell when $I_L = 10$ A.
3. Visually, the characteristics appear as a straight line for sufficiently large current. This suggests that, in the relatively large current region, one can use the equation
$$V_L = V_{oc} - R_{app}I,$$
where R_{app} is the apparent internal resistance of the cell as inferred from the straight line. Estimate the value of R_{app} using
    3.1 the region $30 \leq I \leq 41$ A.
    3.2 the region $10 \leq I \leq 41$ A.
4. For each of the values of R_{app} above, determine the magnitude of the various sources of heat (Joule effect, etc.) when the cell delivers 10 A to the load. Clearly, because you used a straight-line approximation, the activation voltage does not contribute to the heat calculation.
5. Now, determine accurately the value of the internal resistance, R_{int}, i.e., include the activation voltage in the V-I characteristics.
6. Write equations describing the manner in which the load voltage depends on the load current. Check the values obtained against the tabulated data. Do this for $I_L = 40$ A and for $I_L = 0.5$ A.
7. Explain why your equation overestimates V_L at small currents.

9.88

Load Current (A)	Load Voltage (V)	Load Current (A)	Load Voltage (V)
0.00	0.90000		
0.5	0.84919		
1.00	0.83113	21.00	0.72622
2.00	0.81144	22.00	0.72390
3.00	0.79922	23.00	0.72163
4.00	0.79018	24.00	0.71942
5.00	0.78291	25.00	0.71725
6.00	0.77677	26.00	0.71514
7.00	0.77142	27.00	0.71306
8.00	0.76664	28.00	0.71103
9.00	0.76230	29.00	0.70902
10.00	0.75831	30.00	0.70706
11.00	0.75461	31.00	0.70512
12.00	0.75114	32.00	0.70322
13.00	0.74786	33.00	0.70134
14.00	0.74476	34.00	0.69949
15.00	0.74180	35.00	0.69766
16.00	0.73896	36.00	0.69586
17.00	0.73624	37.00	0.69408
18.00	0.73361	38.00	0.69232
19.00	0.73107	39.00	0.69058
20.00	0.72861	40.00	0.68886
		41.00	0.68715

9.28 Although low voltage automotive batteries have been standardized at 12 V, no such standards have been agreed on for automotive traction batteries. Some hybrid cars use 275 V motors and 275 V batteries (some use 550 V motors powered by 275 V batteries.)

Consider a fuel cell battery rated 100 kW at 275 V. It uses pure hydrogen and pure oxygen, both at 1 atmosphere pressure. The battery, consisting of 350 cells, operates at 390 K. To simplify the problem, assume 298 K thermodynamics. Assume a linear V vs I characteristic for the fuel cells.

1. What is the hydrogen consumption (in kg of H_2 per hour) when the battery delivers 100 kW?

2. The retarding force on a car can be represented by a power series in U (the velocity of the car):
$$F = a_0 + a_1 U + a_2 U^2. \tag{1}$$

$a_1 U$ represents mostly the force associated with the deformation of the tires. $a_2 U^2$ is the aerodynamic retarding force and is
$$a_2 = \frac{1}{2}\rho C_D A U^2 \tag{2}$$

where $a_0 = 0$, $\rho = 1.29$ kg/m^3 is the air density, $C_D = 0.2$ is the drag coefficient, and $A = 2$ m^2, is the frontal area of the vehicle.

3. When delivering 50 kW, the battery voltage is 295 V. When cruising at a constant, moderate speed of 80 km/hr the car uses only 15 kW. What is the range of the car under such conditions if the hydrogen tank can store 4 kg of the gas? This assumes flat, horizontal roads and no wind.

4. How slow must the car drive to do 1000 km on 4 kg of H_2?

9.89

9.29   To test a fuel cell in a laboratory, an ac voltage generator (peak-to-peak voltage $v_{pp} = 0.001$ V) was connected in series with the load and an ac ammeter (peak-to-peak current i_{pp}) was used to measure the load current fluctuations caused by the varying V_L. The frequency used was low enough to cause any reactive component in the measurement to be negligible. The following measurements were obtained:

| I_L | V_L | i_{pp} |
A	V	A
5.34	0.956	0.186
10.67	0.936	0.366

It was observed that there was a 180° phase relationship between v_{pp} and i_{pp}—that is, that increasing the voltage, actually reduced the current. Calculate the true internal resistance of the cell.

9.30   Hydrogen-oxygen fuel cell. Although the temperature of the cell will vary throughout its operation, use thermodynamic data for RTP so as not to complicate the computation.

Each cell is 3 mm thick and has a total area of 10 by 10 cm.

The density of each cell is equal to twice the density of water, and the specific heat capacity of the cell is 10% of the specific heat of water. This means that it takes 24 J of heat to raise the cell temperature by 1 kelvin.

Under all circumstances, the product water is removed in vapor form. The highest allowable operating temperature of the cell is 450 K.

Although heat is removed from the cell by several different mechanisms, the net effect is that the rate of heat removal is proportional to $T - 300$: In fact the heat removal rate, $\dot{Q}_{rem} = 0.3(T - 300)$ W.

Laboratory tests reveal that when the load current is 2 A, the load voltage is 0.950 V, and when the load current is 20 A, the load voltage falls to 0.850 V.

1. Write an equation relating V_L to I_L, assuming a linear relationship between these variables.
2. What is the maximum power the cell can deliver?
3. Show that this maximum power cannot be delivered continuously because it would cause the cell temperature to exceed the maximum allowable operating temperature.
4. What is the maximum power that the cell can deliver continuously to a load?
5. Although the cell cannot deliver maximum power continuously, it can do so for a short time if it starts out cold—that is, if its initial temperature is 300 K. It will generate more heat than it can shed and its temperature will rise. How long can the cell (initially at 300 K) deliver maximum power to the load without exceeding 450 K?

9.90

Chapter 10
Hydrogen Production

10.1 Generalities

In mid-2004, if you looked up "Hydrogen use" in Google, you would get about 1,910,000 entries. This, better than any other statistics, attests to the enormous interest in this gas. For the production of ammonia alone, there has been a 46-fold raise in hydrogen utilization between 1946 and 2003, when over 19 million tons of the gas were produced for this purpose alone. The impending massive introduction of fuel cells into the economy will cause the demand for hydrogen to rise much more rapidly in the near future. Techniques for both bulk production and for local generation (especially in vehicles) will be perfected by established industries and by a host of start-ups all trying profit from the new market. An interesting study of hydrogen production methods can be find in the paper by Brinkman.

It is important to start out with the clear understanding that, although extremely abundant, hydrogen, unlike fossil fuels, is not a source of energy. Much of the existing hydrogen is in the form of water—hydrogen ash—and considerable energy is required to extract the desired element. Hydrogen is, at best, an excellent vector of energy. It holds great promise as:

1. Fuel for land and sea vehicles especially when used in high efficiency fuel cells.
2. Fuel for large air- and spacecraft owing to its high energy-to-weight ratio when in cryogenic form.
3. Industrial and domestic fuel for generation of heat and electricity.
4. A means for transporting large quantities of energy over long distances.

The advantages of hydrogen include:

1. Low pollution
   Hydrogen burns cleanly producing only water. It is true that, depending on the flame temperature when burned in air, small amounts of nitrogen oxides may also be generated. However, pollution may be associated with some hydrogen production processes.
2. Controllability
   At ambient temperatures, hydrogen reacts extremely slowly with oxygen. Catalysts permit adjusting the reaction speed over a large range from very low temperature flames to intense ones.
3. Safety

Hydrogen's reputation as a dangerous gas stems mostly from the spectacular 1937 explosion of the Hindenburg in Lakehurst, NJ, when 36 people were killed. Yet, a good case can be made that the explosion actually proved how *safe* the gas is. Indeed, the Hindenburg carried 200,000 cubic meters of hydrogen, equivalent to 2.5×10^{12} joules of energy. An energetically equivalent amount of gasoline would correspond to over 80 cubic meters which could form a pool of fiery liquid covering the area of some 15 football fields.[†]

3.1 Being the lightest of all gases, it quickly rises and disperses, while liquid fuels form pools that spread the fire.

3.2 The smallness of the hydrogen molecule causes this gas to leak easily through small cracks and holes making it difficult to accumulate in explosive concentrations.

3.3 Owing to its low density, a given volume of hydrogen contains little energy and thus represents a much smaller hazard than natural gas or gasoline (the vapor of the latter contains 20 times the energy of H_2 on a same volume basis).

3.4 At 1 atmosphere, the auto-ignition temperature for hydrogen is about 580 C whereas that for gasoline is as low as 260 C. The likelihood of accidentally starting a fire is much higher with the latter fuel.

3.5 Hydrogen/air mixtures with less than 4.1% fuel (in volume) will not catch fire while the flammability limit for gasoline is 1%.

3.6 A pure hydrogen flame radiates little energy allowing firemen to approach much more closely the site of a fire.

3.7 Hydrogen is totally non-toxic and can be breathed in in high concentration (of course, it can asphyxiate you and can also cause you to explode if hydrogen filled lungs are accidentally ignited).

Accumulation of hydrogen in high points of equipment or buildings can be prevented by installing catalysts that cause the (relatively) slow oxidation of the gas and its conversion to water. Odorants such as mercaptans can be added to the hydrogen to alert humans of any escaping gas.

The numerous processes for the production of hydrogen include:

1. chemical,
2. electrolytic,
3. thermolytic,
4. photolytic, and
5. biological.

[†] This assumes a perfectly flat cement surface. If the spill is over a flat compacted earth surface, the area will be something like 20% of that on a non-absorbing surface.

10.2 Chemical Production of Hydrogen

10.2.1 Historical

On September 19, 1783, in the presence of Louis XVI, a hot-air balloon built by the Montgolfier brothers rose up carrying a duck, a rooster, and a sheep. This marked the first recorded aeronautical accident: the sheep kicked the rooster, breaking its wing. Otherwise, the experiment was a success and encouraged humans to try an air trip by themselves. On November 21 of the same year, De Rozier and the marquis d'Arlandes stayed aloft for 25 minutes and rose to 150 m altitude. It is noteworthy that no sheep was taken along and that the Montgolfiers, prudently, also stayed on the ground.

Hydrogen balloons were introduced surprisingly early—on December 1, 1783, only ten days after the first manned hot-air flight, Jaques Charles and one of the Robert brothers[†] made their ascent. The test was a little too successful; after a short trip, Robert disembarked to salute the onlookers, causing the lightened balloon to rise quickly to 2700 m carrying the flustered Charles, who, eventually, opened a valve dumping some hydrogen and returning safely to the ground. Charles is the physicist celebrated in "Charles's Law" (1787), which states that the volume of a fixed amount of gas at constant pressure is proportional to the temperature. This is, of course, a part of the ideal gas law.

It is interesting to observe that the more primitive technology of hot-air balloons has survived and greatly grown in popularity, whereas no one uses hydrogen for this purpose any more.

Early on, hydrogen was produced by passing steam over red-hot iron filings. The iron combines with oxygen in the water, liberating hydrogen:[††]

$$3Fe + 4H_2O \rightleftharpoons Fe_3O_4 + 4H_2. \tag{1}$$

The gas was then washed by bubbling it through water.

The iron oxide formed is the ferroso-ferric oxide in which iron appears in two different states of oxidation—"ferrous" (iron II) usually forming pale green salts, and "ferric" (iron III) usually forming yellow-orange-brownish salts. The mineral magnetite is naturally occurring Fe_3O_4, a compound that is also used in ferrites employed in some electronic devices.

After 1850, hydrogen for balloons was frequently produced from the reaction of iron with sulfuric acid. The high price of the latter led to costly hydrogen.

[†] Marie-Noel Robert. Strangely, he and his brother, Anne-Jean, had feminine first names.

[††] The iron/water reaction is now being considered as a way of storing hydrogen. See Chapter 11.

10.2.2 Modern Production

Small amounts of hydrogen are produced even now by making aluminum chips react with caustic soda (NaOH). This is sometimes the source of the gas used in meteorological balloons. However, the bulk of the hydrogen produced in the world is made from fossil fuels. Oil, naphtha, and natural gas are still the main materials used. Owing to their growing scarcity, some effort is being made to use the more abundant coal.

Hydrogen production can fall into one of several categories among which one can list:

1. Production of hydrogen in modest amounts for food industry and other small consumers. Frequently, electrolytic processes discussed later on are employed because they yield purer gas.
2. Production of hydrogen in massive amounts at stationary plants as for instance in the production of ammonia.
3. Production of hydrogen in small amounts by compact on-board plants for use in fuel cell vehicles. This last application is only now being developed and promises to become of significant economic interest.
4. Production of hydrogen for use in compact residential or local electricity (and hot water) generation.

Hydrocarbons and alcohols, among other substances, can yield hydrogen when submitted to **partial oxidation, steam reforming**, or **thermal decomposition**. These processes lead to a mixture of CO and H_2 called **syngas**.

When any of the above reactions is used in a fuel processor to feed fuel cells with pure hydrogen, the efficiency, η, can be defined as

$$\eta \equiv \frac{\text{Lower heat value of the hydrogen delivered to the fuel cell}}{\text{Higher heat value of feed stock} + \text{higher heat value of fuel used for heat}}.$$

10.2.2.1 Partial Oxidation

Partial oxidation can be carried out noncatalytically (POX) or catalytically (ATR or **autothermal reaction**).

Partial oxidation is preferred when the raw material is a heavier fraction of petroleum, while steam reforming is more convenient for lighter ones. However, small fuel processors for automotive use based on partial oxidation of methanol are being seriously considered.

In the partial oxidation process, air is used as oxidant and this results in nitrogen being mixed with the hydrogen produced, reducing the partial pressure of the latter and, consequently, lowering the fuel cell output.

Partial oxidation is accomplished by reacting a fuel with a restricted amount of oxygen:

$$C_nH_m + \frac{n}{2}O_2 \rightarrow n\ CO + \frac{m}{2}H_2 \tag{2}$$

Thus, for the case of methane,

$$CH_4 + \frac{1}{2}O_2 \rightarrow CO + 2H_2. \tag{3}$$

These reactions take advantage of oxygen having a greater affinity for carbon than for hydrogen.

10.2.2.2 Steam Reforming

In steam reforming, the fuel reacts with water that adds its hydrogen to that from the fuel and does not introduce any nitrogen into the reformate. This contrasts with the partial oxidation process. Steam reforming of a generalized hydrocarbon proceeds according to:

$$C_nH_m + n\ H_2O \rightarrow n\ CO + \left(\frac{m+2n}{2}\right)H_2 \tag{4}$$

This reaction is also known as the **carbon-steam** reaction.

As an example consider carbon itself (let $m = 0$ to cancel out the hydrogen in the hydrocarbon). Note that in this case, all the hydrogen comes from the water, and the fuel contributes only energy.

$$C + H_2O \rightarrow CO + H_2 \tag{5}$$

Consider also methane,

$$CH_4 + H_2O \rightarrow CO + 3H_2 \tag{6}$$

10.2.2.3 Thermal Decomposition

Thermal decomposition of alcohols can be exemplified by the methanol and ethanol reactions indicated below:

$$CH_3OH \rightarrow CO + 2H_2 \tag{7}$$

and

$$C_2H_5OH \rightarrow CO + H_2 + CH_4. \tag{8}$$

All the hydrogen comes from the fuel used.

10.5

10.2.2.4 Syngas

Syngas, the mixture of CO and H_2 that results from all the reactions discussed so far, can be used directly as fuel. It can even be directly used in molten carbonate and ceramic fuel cells, but, owing to the presence of the carbon monoxide, it is totally incompatible with low temperature fuel cells such as SPFCs.

Syngas has been used as domestic and industrial fuel, but its low energy per unit volume makes it unattractive if it has to be pumped to a distant consumer. For such application, the gas can be **enriched** by transforming it into methane. See Equation 14. This is the basis of many coal gasification processes. Observe that the preceding syngas is dangerously poisonous owing to the carbon monoxide it contains.

An important use of syngas is as a feedstock for the production of an amazing number of chemicals. Many of these have an H/C ratio substantially larger than that of syngas. For this reason, and for its use in low temperature fuel cells, a hydrogen enriching step may be needed. This is known as a **shift** reaction.

10.2.2.5 Shift Reaction

The shift reaction promotes the combination of carbon monoxide with water. The result is carbon dioxide and more hydrogen:

$$CO + H_2O \rightarrow CO_2 + H_2 \tag{9}$$

By using the shift reaction, it is possible to adjust the H/C ratio of syngas over a wide range of values. For fuel cells, the shift reaction is used to (nearly) eliminate all the CO.

As an example, consider the production of hydrogen from natural gas (methane):

$$CH_4 + H_2O \rightarrow CO + 3\ H_2 \tag{10}$$

followed by

$$CO + H_2O \rightarrow CO_2 + H_2 \tag{11}$$

with the overall result:

$$CH_4 + 2\ H_2O \rightarrow CO_2 + 4\ H_2 \tag{12}$$

Notice that the heat of combustion of methane into liquid water is 890 MJ/kmole, while the combustion of 4 kilomoles of hydrogen, again into liquid water, yields $4 \times 286 = 1144$ MJ. Thus, in the above reaction, the products have more energy than the reactants—the reaction is endothermic. The extra energy comes from the heat necessary for the reaction to proceed.

In fixed installations, this heat comes, usually, from the combustion of hydrocarbons:

$$C_nH_m + (n + \frac{m}{4})O_2 \rightarrow n\ CO_2 + \frac{1}{2}mH_2O \tag{13}$$

In the more compact automotive and residential uses, the heat may conveniently come from the combustion of part of the hydrogen in the reformate. See the example further on.

10.2.2.6 Methanation

The transformation of syngas into methane, part of the process of transforming any fossil fuel into the (usually) more valuable "natural gas," is called **methanation**. Besides being of great industrial importance, methanation is of interest to us in this text because it provides a technique for eliminating most of the CO impurity from the stream of hydrogen produced from carbon bearing fuels. The methanation reaction is

$$CO + 3\ H_2 \rightarrow H_2O + CH_4 \tag{14}$$

This is the reverse of steam reforming of methane, Equation 6. Incidentally, carbon dioxide can also be transformed into methane:

$$CO_2 + 4H_2 \rightarrow CH_4 + 2H_2O. \tag{15}$$

10.2.2.7 Methanol

In addition to being a valuable fuel and chemical, methanol is an important intermediate in the production of many other chemicals. This is particularly true because it is the only substance that can be produced singly and with good efficiency from syngas. Efficient production of other substances leads to mixtures that can be difficult to separate.

Methanol may become the fuel of choice for fuel cell cars. It can be produced from syngas:

$$CO + 2\ H_2 \rightarrow CH_3OH \tag{16}$$

This reaction, discovered in 1902 by Sabatier and Sedersens, is the base of the **Fischer-Tropsch** process that attained such fame in Germany during World War II, generating liquid fuels from coal.

The plant that produces methanol bears a strong resemblance to an ammonia plant. The difference lies in the type of syngas and in the catalysts used in the reactor.

It should be pointed out that the end product of syngas based products is controlled by adjusting temperatures and pressures and by the choice of catalysts.

Methanol can also be produced directly from biomass such as wood.

10.7

10.2.2.8 Syncrude

Syncrude is the term that describes the liquid products resulting from coal liquefaction. Liquefaction is a more efficient process of converting coal than gasification. It requires little water and can use all sorts of coal including bituminous coals that tend to cake when submitted to gasification. There are essentially four syncrude processes:

1. The Fischer-Tropsch process similar to that used for production of methanol. Here, however, selectivity is not desired: instead of pure methanol, the process yields a complex mixture of hydrocarbons.
2. **Pyrolysis**, the destructive distillation of coal in the absence of air, results in gases, liquids, and solids (char). Coal is flash heated because prolonged heating will cause the liquid fraction to crack forming gases.
3. Direct hydrogenation of coal.
4, Solvent extraction of liquids. The solvents used are produced in a preliminary direct hydrogenation step.

10.2.3 Hydrogen Purification

Hydrogen derived from electrolysis (see later) comes close to being acceptably pure when leaving the electrolyzer. On the other hand, when hydrogen is derived from fossil fuels, it is accompanied by many impurities including massive amounts of CO_2, objectionable traces of CO and, in some processes, large amounts of nitrogen. In addition, the feed stock itself may contain undesirable components, such as sulfur, which must be removed prior to processing.

10.2.3.1 Desulfurization

If the feedstock is in gaseous form, sulfur can be removed by spraying it with a calcium-based (limestone, for instance) slurry. The SO_2 in the gas reacts with the slurry producing sulfites or sulfates which are then removed.

Catalysts containing molybdenum disulfide with low concentrations of cobalt or nickel can be used to convert sulfur-bearing molecules in heavy crude into H_2S gas.

A number of other desulfurization processes exist.

10.2.3.2 CO_2 Removal

Syngas as well as biogas (see Chapter 13) contains large percentages of carbon dioxide which, at best, acts as a dilutant. In the presence of water, a destructive acid is formed that can damage equipment and pipelines. Removal of CO_2 is a technique central to sequestering schemes aimed at reducing CO_2 emissions into the atmosphere.

The removal of the CO_2 can be accomplished by a number of processes including

1. Chemical methods that use, for example, calcium hydroxide to absorb the carbon dioxide, forming calcium carbonate. In a later step, the carbonate is regenerated to hydroxide.

2. Physical processes called **temperature swing adsorption, TSA**, that take advantage of the solubility variation of CO_2 with temperature. The solvents may be water, methanol, or one of the three ethanol amines (mono- or MEA, di- or DEA, and tri- or TEA).

3. Currently, the most popular technique is **pressure swing adsorption, PSA**, that employs the ability of certain substances such as some zeolites to selectively adsorb carbon dioxide (or some other substances) at high pressure and then desorb them when the pressure is lowered. In an effort to design better adsorbers, organic molecules are being investigated. See Atwood et al.

4. Partial removal of CO_2 can be achieved by the use of special membranes (cellulose acetate, for example) that display a higher permeability to carbon dioxide than for other molecules. CO_2 is a relatively large molecule, hence the selectivity can not be based on pore size. In other words, the membranes do not act as filters—they are nonporous. Carbon dioxide dissolves into the membrane, diffuses through it, and emerges on the other side. Some of the useful gas is lost in the process: 85% can be recovered if 3% CO_2 is tolerated in the exhaust, and 90–92% is recovered if 8% CO_2 in the exhaust is acceptable.

10.2.3.3 CO Removal and Hydrogen Extraction

Hydrogen extraction (with removal of most of the CO) can be achieved by means of metallic membranes that allow the passage of H_2 but not of other gases. Again, although known as "filters" they are nonporous and depend on the dissociation of H_2 into H which then forms a hydride (see Chapter 11) that diffuses rapidly through the sheet and reconstitutes the molecular hydrogen on the other side. This is a two step mechanism: the dissociation and the hydridization and diffusion. Tantalum allows the second step to proceed efficiently but is a poor catalyst for the dissociation. Palladium performs both steps very well but is expensive. One possible solution is the use of tantalum with a very thin palladium plating. Since hydrogen tends to embrittle the palladium, the latter is alloyed with gold, silver, or copper (typically 60% Pa and 40% Cu).

Sizable flow rates through the membrane without excessive pressure differentials demand small thicknesses. This economizes palladium but can result in very fragile sheets which may have imperfections in the form of minute pinholes. Unwanted gases seeping through such pinholes destroy the selectivity of the "filter." Deposition of very thin but very uniform layers of palladium on top of a high porosity substrate may solve the problem. The degree of undesirable porosity of the membrane can be inferred from a measurement of a helium flux through it. Helium, unlike hydrogen, can only cross the membrane by passing through the pinholes.

Data from Idatech (see Figure 10.1) show the relationship between the thickness of the membrane and the flux rate of hydrogen—that is, the number of kilowatts of hydrogen that flow each second through each square meter of membrane (kg/s being translated into watts using the lower heat value of hydrogen). The data were measured with the membrane at 400 C, and a pressure differential of 6.6 atmos. The difference in permeability of smooth compared to etched membranes is obvious. Etched membranes are rough and thus present a larger surface area than smooth ones thereby improving the surface reaction with hydrogen. This effect is more noticeable with thin membranes because, in the thicker ones, the rate of permeation is controlled mostly by the gas diffusion through the bulk of the material.

Palladium is an expensive metal. A 1 kW fuel cell operating at 60% efficiency requires a hydrogen input of around 1700 W. A 17 μm thick membrane will produce, at a Δp of 6.7 atmospheres, hydrogen at a rate of about 170 kW/m^2, hence a total effective membrane area of about 0.01 m^2 or 100 cm^2 is required. If the total area of the palladium sheet is 1.5 times larger than this and if the average thickness of the sheet is 20 μm, then the total amount of metal is 0.3 cm^3. Palladium constitutes 60% of the alloy or 0.18 cm^3. This corresponds to about 2.2 grams. Owing to the use of palladium in automobile catalytic converters, the price of the metal has risen considerably in the last decade. At the current \$25/gram, the cost of palladium in a 1 kW fuel cell would be \$55, not a negligible amount.

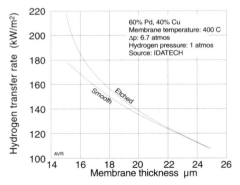

Figure 10.1   Rate of hydrogen transport through a palladium-alloy membrane. Thin etched membranes transport hydrogen better than the smooth ones owing to the increased surface area.

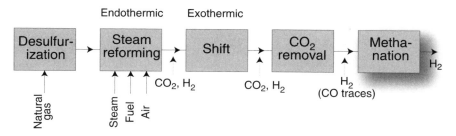

Figure 10.2   Typical hydrogen production sequence.

If it proves possible to reduce the palladium thickness to, say, 1 or 2 μm, the cost of the material would fall substantially. Also, the designer could opt for a lower Δp across the membrane.

10.2.4 Hydrogen Production Plants

Large scale hydrogen production starting from fossil fuels (frequently, for the production of ammonia) is a mature technology. A typical setup is shown in Figure 10.2. The first step in the process is frequently the desulfurization of the feedstock because sulfur tends to poison the catalysts required in some of the subsequent steps. Next, syngas is produced by steam reforming. This is, as pointed out previously, an endothermic reaction that requires heat input.

The shift reaction that eliminates most of the CO releases substantially less heat than that needed do drive the reforming (or decomposition) reaction. The shift reaction, being exothermic, profits from operation at low temperatures owing to equilibrium considerations. However, this influences unfavorably the kinetics of the reaction. Consequently, attention has been given to the development of good catalysts. Early catalysts based on nickel, cobalt, or iron oxide required temperatures above 700 K. Modern copper based catalysts allow operation at temperatures as low as 520 K. After the shift reaction, the gas contains large amounts of carbon dioxide mixed with hydrogen. This requires a CO_2 removal step. The final step in the hydrogen production sequence is the elimination of residual CO which, otherwise, would alter the catalyst in the ammonia synthesis process.

10.2.4.1 Compact Fuel Processors

Compact fuel processors for use in automobiles or for residential applications are vigorously being developed at the moment. Before discussing one such miniature plant, we will present an example of calculations required to specify the device. For this, we will need some thermodynamic data which, for convenience, we summarize in Table 10.1.

Table 10.1
Thermodynamic Data

Substance		Formula	Heat of Combust. (Higher) MJ/kmol	Heat of Combust. (Lower) MJ/kmol	$\overline{h}^*_{vap}$ kJ/kmol	$\overline{h_f}^{\circ}$ MJ/kmol	$\overline{g_f}^{\circ}$ MJ/kmol	$\overline{s}^{\circ}$ kJ K^{-1} /kmol
Carbon		C	-393.52			0	0	5.74
C dioxide		CO_2				-393.52	-394.36	213.80
C monoxide		CO	-282.99			-110.53	-137.15	197.65
Ethanol	(g)	C_2H_5OH	-1409.30	-1277	42.34	-235.31	-168.57	282.59
Ethanol	(l)	C_2H_5OH				-277.69	-174.89	160.70
Hydrogen		H				218.00	203.29	114.72
Hydrogen		H_2	-285.84	-241.80		0	0	130.68
Hydroxyl		OH				39.46	34.28	183.7
Methane		CH_4	-890.36	-802.16		-74.85	-50.79	186.16
Methanol	(g)	CH_3OH	-764.54	-676.34	37.9	-200.67	-162.00	239.70
Methanol	(l)	CH_3OH				-238.66	-166.36	126.80
Oxygen		O				249.19	231.77	161.06
Oxygen		O_2				0	0	205.04
Water	(g)	H_2O			44.1	-241.82	-228.59	188.83
Water	(l)	H_2O				-285.83	-237.18	69.92

Example

Consider a methanol-to-hydrogen converter appropriate for vehicular or for residential use. Such a converter or **fuel processor** might employ the direct methanol decomposition reaction,

$$CH_3OH \rightarrow CO + 2H_2. \tag{17}$$

If the methanol were evaporated and burned, it would yield (see table above) 676.34 MJ per kilomole of water vapor being produced. If the alcohol is first decomposed, then the resulting 2 kilomoles of hydrogen would yield $2 \times 241.82 = 483.64$ MJ, while the carbon monoxide would yield an additional 282.99 MJ, for a total of 766.63 MJ. In other words, the products of the decomposition yield, upon combustion, more energy than the original fuel—the decomposition is endothermic and requires a heat input of $766.63 - 676.34 = 90.29$ MJ per kilomole of methanol.

It is important to get rid of much of the CO (which tends to poison the catalyst in low temperature fuel cells) and extract some additional hydrogen. This can be accomplished by employing a water gas shift reaction,

(continued)

(continued)
$$CO + H_2O \rightarrow CO_2 + H_2. \tag{18}$$

When burned, CO releases 282.99 MJ/kmol, while the product hydrogen releases 241.8. The reaction is, thus, exothermic providing $282.99 - 241.8 = 41.19$ MJ of heat per kilomole of CO.

The overall reaction is

$$CH_3OH + H_2O \rightarrow CO_2 + 3H_2, \tag{19}$$

which, of course, is endothermic requiring $90.29 - 41.19 = 49.00$ MJ per kilomole of methanol.

For each kilomole of methanol, 1 kmole of water is required. The system works by taking in the correct alcohol/water mixture, evaporating it and heating it to the reaction temperature, which may be between 200 and 600 celsius. The energy budget is

$$\Delta H_{TOTAL} = \Delta H_{VAP} + \Delta H_{C_P} + \Delta H_{REACT} + \Delta H_{LOSS}. \tag{20}$$

ΔH_{VAP} is the amount of energy required to vaporize the fuel/water mixture. For the example under consideration,

$$\begin{aligned} \Delta H_{VAP} &= \Delta H_{VAP_{METHANOL}} + \Delta H_{VAP_{WATER}} \\ &= (37.9 \times 10^6 + 44.1 \times 10^6) \\ &= 82.0 \times 10^6 \text{ J/kmole of methanol}. \end{aligned} \tag{21}$$

ΔH_{C_P} is the energy required to raise the temperature of the vaporized fuel/water to the operating temperature. Since the specific heat of most substances varies considerably with temperature, and since the exact operating temperature in this example is unspecified, we will use representative values of c_p: 37 kJ K^{-1} for water and 39 kJ K^{-1} for methanol.

$$\begin{aligned} H_{c_p} &= c_{p_{METHANOL}} \Delta T + c_{p_{WATER}} \Delta T \\ &= 37 \times 10^3 \Delta T + 39 \times 10^3 \Delta T = 76 \times 10^3 \Delta T. \end{aligned} \tag{22}$$

If, somewhat arbitrarily, we take $\Delta T = 250$ K—that is, we chose a reaction temperature of $298 + 250 = 548$ K or 275 C, then the energy to vaporize the water/fuel mixture is

$$\Delta H_{c_p} = 19 \times 10^6 \text{ J/kmole of methanol}. \tag{23}$$

ΔH_{REACT} was calculated at the beginning of this section. It amounts to 49 MJ per kilomole of methanol.

Finally, ΔH_{LOSS} can be minimized by good insulation. We will consider it to be negligible in this example.

(continued)

(continued)

$\Delta H_{TOTAL} = 82 + 19 + 49 = 150$ MJ/kmole of methanol. This amount of energy can be obtained by burning $150/242 = 0.62$ kilomoles of hydrogen. Hence, of the 3 kmoles of hydrogen produced, about 20% must be diverted to the burner and about 80% is available at the output. The fuel processor converts 676 MJ of methanol into 575 MJ of hydrogen, a ratio of 85%.

If energy were not expended in vaporization and heating the fuel/water mixture and if there were no losses, then the amount of heat required would be solely that needed to drive the decomposition reaction: 49 MJ/kilomole of methanol. This requires 0.2 kmoles of hydrogen and the useful hydrogen output would be 2.8 kmoles or 676 MJ. In other words, the fuel processor would have, as expected, 100% efficiency.

The output gas stream is at a temperature much higher than that of the input fuel/water mixture. It appears that some of the output heat could be used to preheat the input and, thus, increase the overall efficiency of the conversion.

Idatech (of Bend, OR) has developed a series of processors that work on the lines of the above example.[†] They can handle different fuels such as methanol, methane, and others. Refer to Figure 10.3. An equimolar pressurized methanol/water mixture, or equivalent feedstock, is forced into the **reforming region** having first been heated and vaporized in a coil-type heat exchanger exposed to the high temperature of the **combustion region**. In the reforming region, the methanol is converted to H_2 and CO. The latter is water gas shifted into CO_2 and more hydrogen. However, considerable amounts of impurities are left. To separate the hydrogen from these impurities, part of the gas is allowed to pass through a palladium filter that is selectively permeable to H_2. A small part of the hydrogen produced in the reforming region (mixed with the rest of reformate leftover gases), instead of going through the palladium membrane, is forced out into the combustion region where it burns, combining with air introduced into the apparatus by means of a blower.[††] A spark igniter, not shown in the figure, starts the combustion. This hydrogen (part of the non-purified reformate) provides the energy necessary to drive the reforming reaction. No external source of heat is needed, except during start up. At start up, a small electric heater raises the temperature of the reforming region to initiate the reaction. Start up times as short as 3 minutes have been demonstrated.

[†] US Patents 5,861,132, 5,997,594, and 6,152,995.

[††]When operating with a fuel cell, the air for the fuel processor may be supplied by the fuel cell cathode exhaust stream.

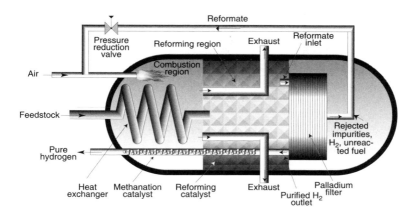

Figure 10.3   One configuration of the Idatech fuel processor.

Even after being filtered through the palladium, a small amount of CO and CO_2 is still present in the hydrogen. Before leaving the equipment, the hydrogen flows through an outlet tube inside of which an appropriate catalyst causes these impurities to be transformed into methane, a gas that is innocuous as far as the catalysts used in fuel cells are concerned. High purity hydrogen, containing less than 1 ppm of CO and less than 5 ppm of CO_2 is produced by this fuel processor.

In some fuel processor configurations, the membrane is formed into a tube, while in others it is left in planar shape. The latter solution, adopted by Idatech, leads to a very compact filter: the membrane consists of a stack of elements interconnected by manifolds so that all elementary filters are in parallel. The stack, see for example Figure 10.3, has one reformate inlet, one outlet for the purified hydrogen and one for the rejected impurities plus some hydrogen and some unreacted fuel.

Preparation of the palladium-alloy membrane by Idatech starts with a sheet of the material, typically 50 micrometers thick, which is held in place by a rectangular frame. A center region of the sheet is etched while the borders are left untouched and retain the original thickness. The etch area is defined by a piece of filter paper attached to the surface. The etchant, generally aqua regia,[†] saturates the paper, which distributes the chemical evenly over the selected area.

[†] Aqua regia ("royal water") is a mixture of 25% nitric acid and 75% hydrochloric acid. It has the property of dissolving gold, although neither acid will do this by itself.

10.3   Electrolytic Hydrogen

10.3.1   Introduction

Production of hydrogen by electrolysis is a relatively old art that has found industrial application in the food industry and in other activities which need only a moderate amount of the gas. Hydrogen produced by electrolysis has the advantage of being easily purified, whereas that produced from fossil fuels tends to contain several hard to remove contaminants.

To a small extent, electrolytic hydrogen has been used for the synthesis of ammonia but the low cost of oil prevalent up to 1972 and again in the 1980s, made this technology unattractive. However, the growing cost and uncertainty in the supply of fossil fuels may popularize electrolytic ammonia in countries where there is abundant hydroelectric power.

By far the most important electrolytic ammonia plant is one operated by Norsk Hydro in Glenfjord (Norway). This company had a long tradition of using electricity for the production of nitrogen fertilizers even before the invention of the Harber-Bosch ammonia process. Early in this century, Norsk Hydro used an electric arc operating in air to cause atmospheric nitrogen and oxygen to combine forming nitrogen oxides, to be converted into nitrates in an additional step. This is, of course, a simple but inefficient process of "fixing" atmospheric nitrogen.

In the 1920s, the plant was modified to produce ammonia by direct combination of electrolytic hydrogen with nitrogen extracted from the air. The Glenfjord plant also produced heavy water. At its peak, the plant used 380 MW of electricity and produced 1300 tons of ammonia and 85 kg of D_2O per day. At \$1/g, the heavy water sales amounted to over \$30 million per year. The current price of heavy water is about \$0.10/g and, for this reason, the Glenfjord operation became less profitable and the production was reduced to 600 ton/day of ammonia.

Another large electrolytic ammonia plant was operated by Cominco in Trail, B.C. (Canada). Production was about 200 ton/day. The operation was discontinued when, after the oil embargo, it became more profitable to export hydroelectric energy to the USA and switch to the abundant (in Canada) natural gas for ammonia synthesis.

Originally, the Trail plant used four different unipolar electrolyzers designed respectively by Fauser, Pechkranz, Knowles, and Stuart, each manufacturer accounting for 25% of the total capacity. The plant was later rebuilt and all electrolyzers were replaced by a Cominco design known as Trail-Cells.

An ammonia plant with a capacity of roughly 400 ton/day was erected in Aswan (Egypt) using, originally, Demag electrolyzers which were replaced by Brown Bovery equipment. A plant with 60 ton/day, capacity using Lurgi electrolyzers was installed in Cuzco (Peru).

Ammonia plants are not the only consumers of electrolytic hydrogen. In fact the vast majority of such plants use hydrogen from fossil fuels rather than from electrolysis. Electrolytic hydrogen is preferred by food and pharmaceutical industries owing to the ease with which high purity gas can be obtained—99.999% purity is not uncommon.

For certain applications an electrolyzer/fuel cell combination constitutes an excellent way to store energy.

One important coming application of electrolyzers is in **hydrogen gas stations** for refueling of fuel cell vehicles.

10.3.2 Electrolyzer Configurations

10.3.2.1 Liquid Electrolyte Electrolyzers

Up to relatively recently, almost all water electrolyzers used liquid electrolytes. Owing to its low conductivity, pure water cannot be used as an electrolyte: it is necessary to add a substance that increases the conductivity. Although acids satisfy this requirement (and are used in some electrolyzers) they are frequently avoided because of corrosion problems. Alkaline electrolytes are better; they almost invariably use potassium hydroxide in concentrations of 25% to 30% (about 6 M). Potassium is preferred to sodium owing to its higher conductivity. Care must be taken to avoid too much contact between air and the solution because the CO_2 in the former can combine with the hydroxide producing carbonate.

Smaller units tend to be of the **tank** type (unipolar), larger units use the **filter press** (bipolar) configuration.

Tank electrolyzers consist of single cells in individual tanks in which many plates can be connected in parallel.

Bipolar electrolyzers are arranged so that all but the two end electrodes act, simultaneously, as anode for one cell and cathode for the next. Thus, the cells are automatically connected in series. In such a configuration, it is convenient to have many relatively small cells in series instead of a few large cells as in the tank case. Consequently, bipolar electrolyzers operate with higher voltages than unipolar ones, a circumstance that reduces the cost of the power supply because, for the same power delivered, high voltage devices cost less than high current ones. In addition, higher voltages correspond to lower currents and demand substantially cheaper connecting bars.

To prevent the mixing of the gases generated, the cell is divided into two compartments separated by a diaphragm that, although impermeable to gases, allows free passage to ions. Asbestos is used as a diaphragm in aqueous electrolytes is. It is inexpensive but is attacked by the electrolyte when the temperature exceeds 200 C.

KOH electrolyzers become more efficient when operated at higher gas pressure because this causes the gas bubbles (whose presence reduces the effective area of the electrode to which they stick) to become smaller. Pressure is generally raised by throttling the gas outlets. Lurgi produces a KOH electrolyzer that operates at 3 MPa (30 atmospheres).

KOH cells produce wet hydrogen containing fine droplets of potassium hydroxide. Steps must be taken to remove the latter and to dry the gas. This type of cell requires electrolyte concentration and level controls.

Electrodes must be chosen to fulfill the following requirements:

1. corrosion resistance,
2. catalytic action,
3. large surface.

In KOH electrolytes, the cathode (H_2 electrode) can be made of iron because catalysis problems are minor and iron is cheap and stable in alkaline solutions. The anode is usually nickel—in general, nickel-plated iron. The plating is made spongy to increase the effective surface. KOH cells present the disadvantage of being very bulky.

10.3.2.2 Solid Polymer Electrolyte Electrolyzers

Solid polymer electrolyzers are dramatically more compact than KOH ones and do not require electrolyte controls. They can operate at much higher current densities than liquid electrolyte devices. They have the disadvantage of requiring de-ionized water and of producing wet gases. Ion exchange membranes are close to ideal electrolytes. Their advantages include:

1. The electrolyte can be made extremely thin leading not only to great compactness but also to reduced series resistance owing to the short path between electrodes. Electrolyte thicknesses as small as 0.1 mm are used.
2. No diaphragm is needed: the ion exchange membrane allows the motion of ions but not of gases.
3. The electrolyte cannot move. It has constant composition and no electrolyte concentration controls are needed.
4. There are no corrosives in the cells nor in the gases produced.
5. Notwithstanding its thinness, the membrane can be strong enough to allow large pressure differentials between the H_2 and the O_2 sides. Differentials as large as 3 MPa (30 atmospheres) are permitted in some cells.
6. Large current densities are possible.
7. Extremely long life (20 years ?) seems possible without maintenance.

10.3.2.3 Ceramic Electrolyte Electrolyzers

In Chapter 9, we discussed ceramic fuel cells using the anion conductor yttria-stabilized zirconia as electrolyte. Cation conducting ceramics have been proposed by Garzon et al. for use in electrolyzers. They work at temperatures between 450 and 800 C, use steam as feed stock, but produce pure, dry hydrogen. High temperature improves the kinetics of the reaction. $SrCe_{0.95}Yb_{0.05}O_{2.975}$ is an example of the material used as electrolyte.

10.3.3 Efficiency of Electrolyzers

Electrolyzers are the dual of fuel cells—much of their theory can be learned from studying Chapter 9 of this book. Before we can compare the performance of different electrolyzers, we have to say a few words about the efficiency of such devices.

We saw that the efficiency of an ideal fuel cell is the ratio of ΔG, the free energy change of the reaction, to ΔH, the enthalpy change. For a given ΔH, an amount, ΔG, of electricity is generated and an amount, $\Delta H - \Delta G$, of heat has to be *released*.

An ideal fuel cell, being perfectly reversible, can operate as an electrolyzer: when ΔG units of energy are supplied, gases capable of releasing ΔH units of energy on recombining are formed and a quantity of heat $\Delta H - \Delta G$ is *absorbed* by the electrolyzer from the environment. This means that the ideal water electrolyzer acts as a heat pump. Thus, the efficiency of an ideal electrolyzer is

$$\eta = \frac{\Delta H}{\Delta G} \tag{24}$$

which is larger than unity.

Consider a water electrolyzer operating at RTP. The change in enthalpy for the reaction is 285.9 MJ/kmole, while the free energy change is 237.2 MJ/kmole (all per kilomole of water).[†] Thus the ideal efficiency of a water electrolyzer, with reactant and products at RTP, is

$$\eta = \frac{285.9}{237.2} = 1.205. \tag{25}$$

If we divide top and bottom of Equation 25 by qn_eN_0, we obtain quantities with the dimension of voltage and we can express the efficiency as a voltage ratio,

$$\eta = \frac{V_H}{V}, \tag{26}$$

[†] In Chapter 9, we used 228.6 MJ/kmole as the free energy change while here, in Chapter 10, we are using a larger value—237.2 MJ/kmole. The reason is that fuel cells synthesize water vapor (even though it may immediately condense into a liquid), while electrolyzers, usually, dissociate liquid water. However, when working with steam electrolyzers, the value to use is that of Chapter 9.

where V_H is the hypothetical voltage an ideal fuel cell would generate if it could convert all the enthalpy change of the reaction into electricity. At RTP,

$$\eta = \frac{1.484}{1.231} = 1.205 \tag{27}$$

and the efficiency of any electrolyzer at RTP is

$$\eta = \frac{1.484}{V}, \tag{28}$$

where V is the required operating voltage.

This simple relation between required voltage and efficiency leads manufacturers to specify their equipment in terms of operating voltage or else in terms of **over-voltage** (the difference between the operating voltage and 1.484 V). If the operating voltage is less than 1.484 V, the electrolyzer absorbs heat and tends to work below ambient temperature. If the operating voltage is higher than 1.484 V, the electrolyzer must shed heat and tends to operate at above ambient temperature.

The amount of heat exchanged with the environment is

$$\dot{Q} = (V - 1.484)I. \tag{29}$$

If $\dot{Q} < 0$, the electrolyzer operates endothermically.

Clearly, the operating voltage of an electrolyzer depends on the current forced through the device. In a simple model, the device can be represented by a voltage generator of magnitude V_{oc} in series with an internal resistance accounting for the losses. The V-I characteristic is then a straight line as indicated in Figure 10.4.

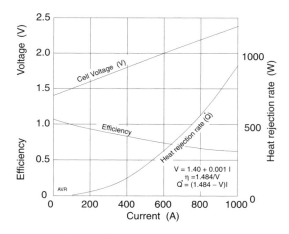

Figure 10.4   Voltage, efficiency, and heat rejection as a function of current in a typical electrolyzer.

The cell in the figure is assumed to have a characteristic given by

$$V = 1.40 + 0.001 \; I. \tag{30}$$

In this case, $V_{oc} = 1.40$ V is the voltage obtained by extrapolation as the current forced into the electrolyzer tends toward zero. The efficiency was calculated from Equation 28 and the heat exchange from Equation 29. Owing to the scale used, one cannot see from the figure that the values of $\dot{Q}$ are negative when the current is less than 80 A. Typical existing commercial electrolyzers (KOH) operate with voltages around 2 V or 74% efficiency.

Solid polymer electrolyzers (SPE) perform substantially better than KOH ones. Figure 10.5 shows how the efficiency of some GE SPE cells depends on current density. The lines represent different stages of development. The rapid progress made between 1967 and 1974 was mainly due to improved catalysts, especially the anodic one (oxygen electrode). Notice that the 1974 model will operate endothermically at low current densities.

The recommended operating point of any electrolyzer is determined by economics. At low current densities, the efficiency is high—that is, a large mass of hydrogen is produced for a given amount of energy used. However, operation at low current densities results in little hydrogen produced per dollar invested in the electrolyzer. On the other hand, at high current densities, although plenty of gas is produced per unit investment cost, little is produced per unit energy used.

There is an optimum current density that results in a minimum cost for the hydrogen produced. This depends on the cost of electricity, the interest rate, etc. See Problem 10.1.

A manufacturer's claim that his unit operates at, say, 80% efficiency is not sufficient information. Even primitive electrolyzers will operate at these efficiency levels if the current density is kept low enough. Good electrolyzers will operate economically at high current densities.

A typical, good KOH electrolyzer should operate at some 4 kA/m^2 while solid polymer devices will operate at the same efficiency with 20 kA/m^2.

The hydrogen production rate, $\dot{N}$, is strictly proportional to the current, I, forced through the electrolyzer.

$$\dot{N} = \frac{I}{2qN_0} \quad \text{kmoles/sec.} \tag{31}$$

The rate does not depend on the efficiency. However, the voltage necessary to drive a given current through the device is inversely proportional to the efficiency.

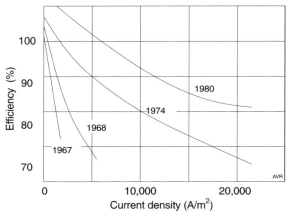

Figure 10.5   Efficiency versus current density of various GE (SPE) cells showing progress in the development.

10.3.4   Concentration-Differential Electrolyzers

In a normal electrolyzer, the minimum amount of electric energy required is equal to the free energy change, ΔG, necessary to separate water into its constituent molecules. In concentration-differential electrolyzers, part of this minimum energy can come from nonelectric sources.

Consider, for instance, the device shown in Figure 10.6. It is a common ion-exchange membrane electrolyzer except that the anode is bathed in a concentrated KOH solution while the cathode is in distilled water. As indicated in the figure, the KOH dissociates into K^+ and OH^-. The membrane, permeable to positive ions but not to negative ions nor to electrons, allows the migration of K^+ to the cathode where it reacts with water, regenerates KOH, and forms H^+ ions. The hydroxyl ion at the anode has its negative charge removed by the external power supply and decays into water, oxygen and electrons. The oxygen is one of the outputs of the system. The electrons, pumped by the power supply to the cathode, recombine with the protons to form hydrogen—the other output of the system.

Owing to the migration of the K^+ ions, the anode becomes more negative than the cathode so that the electrolyzer acts as a battery in series with the external power supply. To accomplish the electrolysis, the external source must provide only a voltage equal to that required by normal electrolyzers *minus* the internal "built-in" electrolyzer voltage.

As we are going to see, the built-in voltage is

$$V = 2.3\frac{kT}{q}(pH_A - pH_C), \tag{32}$$

where pH_A and pH_C are, respectively, the pH at the anode and at the cathode.

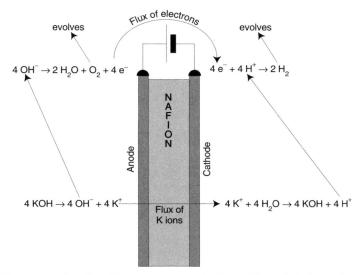

Figure 10.6   Reactions in a concentration-differential electrolyzer.

With concentrated KOH in the anode side and distilled water in the cathode side, we have $pH_A = 14$ and $pH_C = 7$ which leads to a built-in voltage of 0.42 V.

Thus, an ideal concentration-differential electrolyzer, at RTP, requires a voltage of $1.23 - 0.42 = 0.81$ volts and would have an efficiency of $1.48/0.81 = 1.83$. Even smaller voltages are required if, instead of distilled water, an acid solution is used to bathe the cathode.

Clearly, as the operation of the cell progresses, the KOH solution in the anode is depleted and that at the cathode is increased. To operate on a continuous basis, it is necessary to maintain the concentration disequilibrium. This can be done by washing the cathode continuously with distilled water and reconcentrating the weak KOH solution that results. This reconcentrated solution is then fed to the anode.

The energy necessary to perform the regeneration is low grade thermal energy of far less utility than the electric energy saved.

What is the voltage developed by a concentration cell? In Chapter 7, we derived an expression for the open-circuit voltage across a concentration cell:

$$V = \frac{kT}{q} \ln \frac{N_A}{N_C}. \tag{33}$$

Since $\ln N = 2.3 \log N$ and, by definition, $-\log N \equiv pH$,

$$|V| = 2.3 \frac{kT}{q} (pH_A - pH_C). \tag{34}$$

10.3.5 Electrolytic Hydrogen Compression

For many applications such as ammonia production or delivery of hydrogen to fuel cell cars, hydrogen must be available at high pressure. Thus, frequently a hydrogen plant must include a compressor.

An electrolyzer can produce gases at pressures above that of the environment provided its outlets are throttled, as is the case of the Lurgi, Teledyne, and GE equipment. This, however, limits the maximum pressure to values well below those necessary in some applications.

Alternatively, the pressure of the environment itself can be raised by housing the electrolyzer in a pressure vessel. For economic reasons, one needs compact equipment otherwise the cost of the pressure vessel becomes prohibitive.

From Table 10.2, it can be seen that only solid polymer electrolyzers (SPE)—and perhaps some of the proposed positive ion ceramic electrolyzers—lend themselves to such arrangement. The specific volume of the solid polymer electrolyzer is almost two orders of magnitude smaller than that of pressurized KOH machines.

The energy cost of introducing liquid water into the pressure vessel is minimal owing to the negligible volume of the liquid. There is, however, an energy cost in compressing "electrolytically" the product gases. Theoretically, per kilomole of hydrogen produced, an electrolyzer delivering oxygen at pressure p_{ox} and hydrogen at pressure p_h requires an excess energy, $RT_0 \ln(p_{ox}^{1/2} p_h p_o^{-3/2})$ compared with one delivering the same mass of gases at a pressure p_0, provided all gases involved are at the temperature T_0. This is, of course, the energy required to compress the gases isothermally.

In practice, the compression energy is somewhat higher owing to the decrease in the efficiency of the ion-exchange membrane electrolyzer with increasing pressure (as a result of the partial permeability of the membrane).

The great advantage of electrolytic compression is the simplicity and economy in maintenance since there are no moving parts in the system.

Table 10.2

Specific Volume of Various Electrolyzers

Manufacturer	Specific volume (m^3/MW)
NORSK-HYDRO	45
LURGI or TELEDYNE	20
GENERAL ELECTRIC (SPE)	0.3

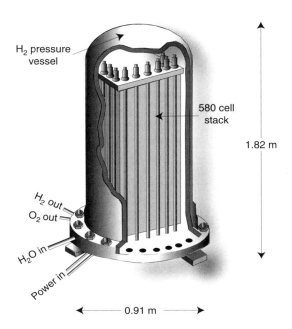

Figure 10.7   Owing to their compactness, SPE electrolyzers can be stacked in a pressure vessel to deliver gases at high pressure. Proposed GE 5 MW unit.

10.4 Thermolytic Hydrogen

10.4.1 Direct Dissociation of Water

It is well known that water vapor, at high temperatures, will dissociate into hydrogen and oxygen. The former could, in principle, be separated out by using an appropriate device such as the palladium filter discussed in the subsection on CO removal. Although, at a first glance, this scheme may seem attractive, its implementation is difficult.

Consider an experiment that consists of introducing 1 kilomole of water vapor (and nothing else) into a cylindrical container equipped with a piston weighed down so as to maintain a constant pressure of 1 atmosphere. The water is then heated to 3000 K. This pressure and this temperature have been chosen arbitrarily to serve as an example.

If all the molecules in the container remained as H_2O, one could look up their free energy by consulting an appropriate table. However, at least some of the water vapor will dissociate into its components, hydrogen and oxygen, according to

$$H_2O \leftrightarrow H_2 + \frac{1}{2}O_2, \tag{35}$$

and the mixture, containing, H_2O, H_2, and O_2, will have a different free energy.

If the dissociation were complete, the gases in the container would consist of 1 kilomole of hydrogen and 0.5 kilomoles of oxygen. The free energy of this gas (still at 1 atmosphere and 3000 K) will be found to be larger than that of the pure H_2O vapor. Notice that the 1 kilomole of H_2O has been transformed into 1 kilomole of H_2 plus 0.5 kilomoles of O_2—the total number of kilomoles is now, $N_{TOTAL} = 1.5$. Thus, the partial pressure of the hydrogen would be $1/1.5$ atmospheres and that of the oxygen must be $0.5/1.5$ atmospheres.

At any realistic temperature, water will only partially dissociate. Let F be the fraction of dissociated water. If so, the number of kilomoles of undissociated water (from the 1 kmole of water introduced into the container) is $(1 - F)$. The number of kilomoles of H_2 is F, and that of O_2 is $F/2$. The mixture consists of

$$(1 - F)H_2O + FH_2 + \frac{1}{2}FO_2. \tag{36}$$

The total number of kilomoles in the gas mixture is

$$N_{TOTAL} = (1 - F) + F + \frac{1}{2}F = 1 + \frac{1}{2}F. \tag{37}$$

The partial pressure of the three species in a mixture whose total pressure is p atmospheres is

$$p_{H_2O} = \frac{1 - F}{1 + F/2}p = \frac{1 - F}{1 + F/2}, \tag{38}$$

$$p_{H_2} = \frac{F}{1 + F/2}p = \frac{F}{1 + F/2}, \tag{39}$$

$$p_{O_2} = \frac{F/2}{1 + F/2}p = \frac{F/2}{1 + F/2}. \tag{40}$$

We took $p = 1$ atmosphere according to the conditions we selected for this example.

The free energy of a species varies with pressure as

$$\bar{g}_i = \bar{g}_i^* + RT \ln p_i. \tag{41}$$

where $\bar{g}_i^*$ is the free energy of species i *per kilomole at 1 atmosphere pressure*. See "Free Energy Dependence on Pressure" in Chapter 9.

The free energy of the gases in the mixture being considered is

$$G_{mix} = (1 - F)\left[\bar{g}^*_{H_2O} + RT \ln\left(\frac{1 - F}{1 + F/2}\right)\right] + F\left[\bar{g}^*_{H_2} + RT \ln\left(\frac{F}{1 + F/2}\right)\right]$$

$$+ F/2\left[\bar{g}^*_{O_2} + RT \ln\left(\frac{F/2}{1 + F/2}\right)\right]$$

$$= (1 - F)\bar{g}^*_{H_2O} + F\bar{g}^*_{H_2} + \frac{1}{2}F\bar{g}^*_{O_2} +$$

$$RT\left[(1 - F)\ln\left(\frac{1 - F}{1 + F/2}\right) + F\ln\left(\frac{F}{1 + F/2}\right) + F/2\ln\left(\frac{F/2}{1 + F/2}\right)\right]$$

$$(42)$$

Equation 42 is plotted in Figure 10.8 that shows that the free energy of the water-hydrogen-oxygen mixture at 3000 K and 1 atmosphere has a minimum when 14.8% of the water is dissociated. At this point, the reverse reaction $H_2 + \frac{1}{2}O_2 \rightarrow H_2O$ occurs at exactly the same rate as the forward reaction, $H_2O \rightarrow H_2 + \frac{1}{2}O_2$: equilibrium is established.

To find the point of equilibrium, we must seek the value of F that minimizes G_{mix}:

$$\frac{dG_{mix}}{dF} = -\bar{g}^*_{H_2O} + \bar{g}^*_{H_2} + \frac{1}{2}\bar{g}^*_{O_2}$$

$$+ RT\left[(1 - F)\frac{-3}{2(1 - F)(1 + F/2)} + \frac{1}{1 + F/2} + \frac{1}{2}\frac{1}{1 + F/2}\right.$$

$$\left. - \ln\left(\frac{1 - F}{1 + F/2}\right) + \ln\left(\frac{F}{1 + F/2}\right) + \frac{1}{2}\ln\left(\frac{F/2}{1 + F/2}\right)\right]$$

$$= -\bar{g}^*_{H_2O} + \bar{g}^*_{H_2} + \frac{1}{2}\bar{g}^*_{O_2}$$

$$+ RT\left[-\ln\left(\frac{1 - F}{1 + F/2}\right) + \ln\left(\frac{F}{1 + F/2}\right) + \frac{1}{2}\ln\left(\frac{F/2}{1 + F/2}\right)\right] = 0 \quad (43)$$

because

$$(1 - F)\frac{-3}{2(1 - F)(1 + F/2)} + \frac{1}{1 + F/2} + \frac{1}{2}\frac{1}{1 + F/2} = 0. \quad (44)$$

The arguments of the natural logarithms are the different gas pressures. Hence,

$$\frac{dG_{mix}}{dF} = -\bar{g}^*_{H_2O} + \bar{g}^*_{H_2} + \frac{1}{2}\bar{g}^*_{O_2} + RT\left[-\ln p_{H_2O} + \ln p_{H_2} + \frac{1}{2}\ln p_{O_2}\right] = 0,$$

$$(45)$$

$$\frac{dG_{mix}}{dF} = -\bar{g}^*_{H_2O} + \bar{g}^*_{H_2} + \frac{1}{2}\bar{g}^*_{O_2} + RT \ln\left(\frac{p_{H_2}p_{O_2}^{\frac{1}{2}}}{p_{H_2O}}\right) = 0. \quad (46)$$

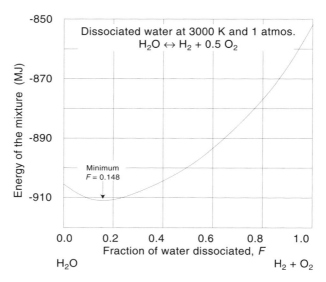

Figure 10.8   Free energy of a water-hydrogen-oxygen mixture as a function of the fraction, F, of dissociation.

Define an **equilibrium constant**, K_p,

$$K_p \equiv \frac{p_{H_2} p_{O_2}^{\frac{1}{2}}}{p_{H_2O}}, \qquad (47)$$

then

$$K_p = \exp\left(\frac{\overline{g}_{H_2O}^* - \overline{g}_{H_2}^* - \frac{1}{2}\overline{g}_{O_2}^*}{RT}\right). \qquad (48)$$

K_p depends on the temperature and on the *exact way the chemical equation was written*. The K_p for the $H_2O \rightleftharpoons H_2 + \frac{1}{2}O_2$ reaction is not the same as that for the $2H_2O \rightleftharpoons 2H_2 + O_2$ reaction. *However, K_p does not depend on pressure.* Indeed, referring to Equation 48, we see that the different $\overline{g}_i^*$ are values at 1 atmosphere and do not change if the actual pressure is changed. Furthermore, if an inert gas, say, argon, is mixed with the water vapor, it does not alter the value of K_p because its $\overline{g}^*$ is zero.[†]

The relationship between the equilibrium constant, K_p, and the fraction, F, of water that has been dissociated can be determined by expressing the partial pressures as a function of F as was done in Equations 38, 39, and 40 which are for the particular case in which the total pressure of the mixture is 1 atmosphere. For the more general case, when the mixture is at some arbitrary pressure, p, the partial pressures become

---

[†] Nitrogen would behave differently because, at the high temperatures employed, it would react with oxygen forming nitrogen oxides.

$$p_{H_2O} = \frac{1 - F}{1 + F/2}p, \tag{49}$$

$$p_{H_2} = \frac{F}{1 + F/2}p, \tag{50}$$

$$p_{O_2} = \frac{F/2}{1 + F/2}p. \tag{51}$$

Inserting these partial pressures into in (Equation 47), one obtains,

$$K_p = \frac{F^{\frac{3}{2}}}{\sqrt{2 - 3F + F^3}}p^{1/2}. \tag{52}$$

Since, as was pointed out, K_p is pressure independent, F must depend on pressure. For instance, we saw that for $T = 3000$ K and $p = 1$ atmosphere, 14.8% of the water dissociates ($F = 0.148$). However, if the pressure is increased to 100 atmospheres, the amount of dissociation falls to 3.4%. This is, of course, predicted by Le Chatelier's rules. In the reaction

$$H_2O \rightleftharpoons H_2 + \frac{1}{2}O_2,$$

an increase in pressure drives the equilibrium to the left because there are more kilomoles in the right-hand side than in the left-hand side.

It can be seen from the information developed above that the direct thermal dissociation of water would require impractically high temperatures. Figure 10.9 shows that at the temperature at which palladium melts (1825 K), only an insignificant fraction of the water is dissociated when the pressure is 1 atmosphere. This means that the partial pressure of the produced hydrogen is much too small to be useful.

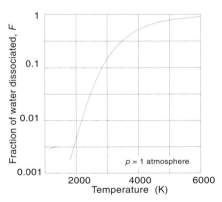

Figure 10.9   Water dissociation as a function of temperature. Pressure: 1 atmosphere.

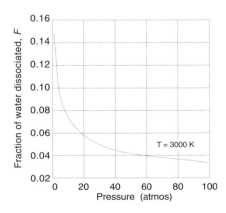

Figure 10.10   Water dissociation as a function of pressure. Temperature: 3000 K.

Increasing the pressure of the water does not help the situation because the degree of dissociation falls dramatically (Figure 10.10).

The definition of equilibrium constant can be extended to more complicated reactions. Thus, for instance, for the reaction

$$\mu_A A \ \mu_B B \rightleftharpoons \mu_C C \ \mu_D D, \tag{53}$$

the equilibrium constant is

$$K_p = \frac{p_A^{\mu_A} p_B^{\mu_B}}{p_C^{\mu_C} p_D^{\mu_D}}. \tag{54}$$

The equilibrium constant can then be related to the equilibrium composition by

$$K_p = \frac{N_A^{\mu_A} N_B^{\mu_B}}{N_C^{\mu_C} N_D^{\mu_D}} \left(\frac{p}{N_{TOTAL}} \right)^{\Delta\mu}, \tag{55}$$

where $\Delta\mu \equiv \mu_C + \mu_D - \mu_A - \mu_B$, and the N_i are for equilibrium composition, not for the theoretical reaction (Equation 53).

The value of the equilibrium constant, K_p, for different reactions can be found in thermodynamic tables. Figure 10.11 plots K_p versus $1/T$ for the $H_2O \leftrightarrow H_2 + \frac{1}{2}O_2$ reaction.

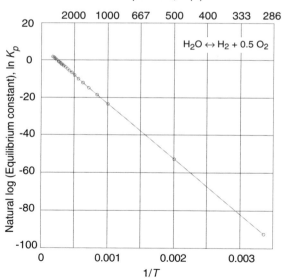

Figure 10.11   Natural logarithm of the equilibrium constant for the dissociation of water versus inverse temperature.

10.30

A linear regression of the data displayed in Figure 10.11 yields

$$K_p = 847.3 \exp - \frac{246 \times 10^6}{RT}. \tag{56}$$

-246 MJ/kmole is the energy of formation of water average over the 0 to 3000 K temperature interval. This is another example of the Boltzmann Equation.

10.4.2 Chemical Dissociation of Water

To circumvent the difficulties encountered with the direct dissociation of water, several chemical reactions have been proposed. In all of them, the intermediate products are regenerated so that, at least theoretically, there is no consumption of materials other than water itself. The temperatures required for these reactions must be sufficiently low to permit practical implementation of the process. In particular, it is desirable that the temperatures not exceed 1100 K to make the process compatible with nuclear fission reactors.

One proposed reaction chain is indicated below. The highest temperature step operates at 730 C. Overall efficiency is 50%. The major disadvantage is the use of corrosive hydrobromic acid.

$$Hg + 2\ HBr \rightarrow HgBr_2 + H_2 \tag{57}$$
$$HgBr_2 + Ca(OH)_2 \rightarrow CaBr_2 + HgO + H_2O \tag{58}$$
$$HgO \rightarrow Hg + \frac{1}{2}O_2 \tag{59}$$
$$CaBr_2 + 2\ H_2O \rightarrow Ca(OH)_2 + 2\ HBr \tag{60}$$

The overall reaction is:
$$H_2O \rightarrow H_2 + \frac{1}{2}O_2 \tag{61}$$

Another, more complicated, chain of reactions for the thermolytic production of hydrogen is:

Reaction 1

$$Cr_2O_3(s) + 4\ Ba(OH)_2(\ell) \xrightarrow{600\ C} 2\ Ba_2CrO_4(s) + 3\ H_2O(g) + H_2\ (g) \tag{62}$$

Reaction 2

$$2\ BaCrO_4(s) + Ba(OH)_2(\ell) \xrightarrow{850\ C} Ba_3(CrO_4)_2(s) + H_2O(g) + \frac{1}{2}O_2\ (g) \tag{63}$$

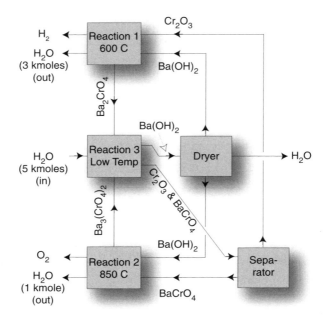

Figure 10.12   Block diagram of the barium chromate cycle.

Notice that, since hydrogen evolves from Reaction 1 and oxygen from Reaction 2 which occur in different parts of the equipment, it is easy to separate these gases. Input reactants are Cr_2O_3 and $BaCrO_4$. Products (besides H_2 and O_2) are Ba_2CrO_4 and $Ba_3(CrO_4)_2$. To recover the reactants, the last two chromates are made to react with water at low temperature:

Reaction 3

$$2\ Ba_2CrO_4(s) + Ba_3(CrO_4)_2 + 5\ H_2O$$
$$\rightarrow Cr_2O_3(s) + 2\ BaCrO_4(s) + 5\ Ba(OH)_2(d) \qquad (64)$$

As pointed out in Chapter 1, the heavy metal nuclear reactors currently under consideration as a means of reviving the popularity of fission reactors may be used as a source of heat to drive thermochemical hydrogen production processes. The high temperature, low pressure mode of operation of these reactors would facilitate the design of the necessary chemical machinery.

10.5 Photolytic Hydrogen

10.5.1 Generalities

> The technology for using solar light energy to produce hydrogen is well established. One certainly can produce this gas through entirely nonpolluting processes by using photovoltaic converters whose output drives electrolyzers. The main effort here is to develop processes that can accomplish this transformation more economically.

Water can be decomposed (and synthesized) according to

$$2H_2O \rightleftharpoons 2H_2 + O_2 \qquad (65)$$

Although simple looking, the reaction is the result of a chain of events that leads through the formation of several intermediate substances and can follow different paths depending on the circumstances and the catalysts.

The structural formula of water, H-O-H, suggests that the first step in the dissociation must be the breaking of the H-O bond so that H and OH (or the corresponding ions) are formed. Next, the OH is dissociated into O and another H, then these atoms coalesce into diatomic molecules. For the case in which no ionized species take part in the reactions, the different energies involved are given in Table 10.3.

Table 10.3
Dissociation Energies

	eV/ molecule	MJ/ kmole
$H_2O \rightarrow H + OH$	5.15	496.2
$OH \rightarrow H + O$	4.40	423.9
$H + H \rightarrow H_2$	−4.48	−431.7
$O + O \rightarrow O_2$	−5.12	−493.3
$H_2O \rightarrow H_2 + \frac{1}{2}O_2$	2.51	241.8

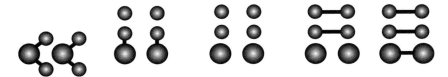

$$2\ H_2O \longrightarrow 2\ H + 2\ OH \longrightarrow 4\ H + 2\ O \longrightarrow 2\ H_2 + 2\ O \longrightarrow 2\ H_2 + O_2$$

$$\uparrow \qquad\qquad \uparrow \qquad\qquad \uparrow \qquad\qquad \uparrow$$

$$2 \times 5.15\ \text{eV} \quad + \quad 2 \times 4.40\ \text{eV} \quad - \quad 2 \times 4.48\ \text{eV} \quad - \quad 5.12\ \text{eV}$$

Figure 10.13   One possible path in the decomposition of water.

Figure 10.13 indicates schematically the progress of the reaction. The exact sequence of events may vary.

Although there is only a 2.51 eV energy difference between a molecule of water and its elements, to dissociate it directly, it is necessary to overcome a 5.15 eV barrier. Thus, photons or phonons with less than 5.15 eV cannot initiate the direct water dissociation reaction.

For direct lysis the input energy must be $5.15 + 4.40 = 9.55$ eV per molecule of water. The result is a hydrogen/oxygen mixture with 2.51 eV more energy than the initial water. The efficiency of this "fuel" production is $2.51/9.55 = 0.263$. The rest of the input energy will appear as heat as the result of the **back reaction** of the **activated intermediate** species.

10.5.2 Solar Photolysis

Figure 10.14 shows the cumulative energy distribution in the solar spectrum. It can be seen that about 22% of the energy is associated with photons having more than 2.51 eV, the energy necessary to dissociate water. However, as was discussed in the preceding subsection, without catalysts, photons with more than 5.15 eV are necessary to initiate the reaction. Thus, practical photolytic hydrogen production depends on the development of appropriate catalysts.

A catalyst, X, may, for instance, be oxidized in the presence of water under the influence of light:

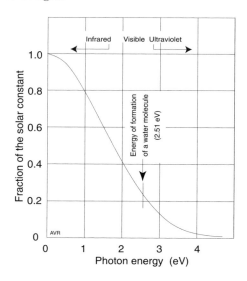

Figure 10.14   Cumulative energy distribution in the solar spectrum.

$$X + H_2O \rightarrow X^+ + H + OH^- \tag{66}$$

$$X^+ + \frac{1}{2}H_2O \rightarrow X + H^+ + \frac{1}{4}O_2 \tag{67}$$

$$H^+ + OH^- \rightarrow H_2O \tag{68}$$

The threshold energy for this reaction is 3.8 eV, somewhat less than the 5.15 eV necessary in the absence of catalysts. The reaction would use only 3% of the solar energy and, of that, a great deal would be lost in the $H^+ + OH^- \rightarrow H_2O$ back reaction.

Actually, for a ground-based system, the available energy is considerably less than 3% because the higher frequencies of the solar spectrum are selectively absorbed by the atmosphere.

A somewhat more favorable reaction is one in which the catalyst, Y, is reduced:

$$Y + H_2O \rightarrow Y^- + H^+ + OH \tag{69}$$

$$Y^- + H_2O \rightarrow Y + OH^- + \frac{1}{2}H_2 \tag{70}$$

$$OH^- + H^+ \rightarrow H_2O \tag{71}$$

$$\frac{1}{2}OH + \frac{1}{2}OH \rightarrow \frac{1}{2}H_2O + \frac{1}{4}O_2 \tag{72}$$

A third reaction avoids intermediate species that will back-react.

$$Z + H_2O \rightarrow ZO + H_2 \tag{73}$$

$$ZO \rightarrow Z + \frac{1}{2}O \tag{74}$$

The threshold for this reaction is 2.9 eV and it could use 13% of the solar energy. However, its realizability is uncertain.

10.6 Photobiologic Hydrogen Production

The majority of living organisms must respire—must consume oxygen and release carbon dioxide to fuel their anabolism. This is true of plants. These, however, under the influence of light, also perform photosynthesis, the opposite to respiration: they fix atmospheric carbon dioxide and release oxygen. Thus, when exposed to light, plants tend to be net oxygen producers and, in the dark, net oxygen consumers.[†]

[†] Some plants will shut off photosynthesis when exposed to too much light, and will, then, only respire. See "photorespiration" in Chapter 13.

In principle, some plants could generate the oxygen needed for their respiration by extracting it from water and releasing hydrogen. This occurs, for instance, with certain algae. Since, in darkness, the plant is almost dormant, the amount of hydrogen released is small. An enzyme—hydrogenase—promotes such release. Hydrogenase is inhibited by the presence of oxygen and, thus, does not work when photosynthesis is active.

Sulfur deprivation reversibly inactivates photosynthesis allowing the production of hydrogen in larger amounts when the plant is exposed to light.

Melis et al. have demonstrated a photobiological hydrogen production system using the alga Chlamydomonas reinhardtii. The processes proceed in two stages:

1. Algae are grown normally and build up their store of carbon compounds.
2. Sulfur is withdrawn from the system and the algae, still exposed to light, release hydrogen, consuming some of the accumulated carbon.

After Stage 2 has depleted much of the carbon, Stage 1 is re-initiated and the plant is again "refattened."

At present, it appears that substantially more research is required not only to perfect the parameters of the system but also to develop more efficient strains or algae.

The overall sunlight-to-hydrogen gas efficiency will be modest because photosynthesis is less than 8% efficient (see Chapter 13) and light is needed not only for photosynthesis but also for the hydrogen release. Nevertheless, what really counts from the practical point of view is the cost of the gas produced and not the efficiency of the system. It remains to be seen if the proposed Melis system will be economically attractive.

References

Atwood, Jerry L., Leonard J. Barbour, and Agoston Jerga, A new type of material for the recovery of hydrogen from gas mixtures, *Angewandte Chemie* , 2948–2950, **(43), 2004.**

Brinkman, Greg., *Economics and environmental effects of hydrogen production methods* http://www.puaf.umd.edu/faculty/papers/fetter/students/Brinkman.pdf.

Garzon, Fernando H., R. Mukundan, and Eric L. Brosha, Enabling science for advanced ceramic membrane electrolyzers, Proc. 2002 U. S. DOE Hydrogen Program Review, NREL/CP-610-32405, **2002.**

Melis, Anastasios, Liping Zhang, Marc Forestier, Maria L. Ghirardi, and Michael Seibert, Sustained photobiological hydrogen gas production upon reversible inactivation of oxygen evolution in green the alga Chlamydomonas reinhardtii., *Plant Physiol.* **122**, 127–136, Jan **2000.**

Pham, Ai-Quoc, High efficiency steam electrolyzer, Proc. of the 2000 US DOE Hydrogen Program Review, NREL/CP-610-32405.

Problems

10.1   Let C_D be the price of an electrolyzer installation expressed in dollars per kW of hydrogen produced when a given current density, J_0, is employed. The number of output kW is that heat power that would be generated if the produced hydrogen were burned. There is no compelling reason to operate the electrolyzer at the nominal (J_0) current density. It can be operated at any other current density, J, within the permissible range.

The cost of the hydrogen produced, C_H, is

$$C_H = C_{INV} + C_{OP} + C_E \qquad\qquad \$/(\text{kg of H}_2)$$

where

C_{INV} = part of the cost attributable to the cost of investment in the plant,

C_{OP} = the maintenance and operation cost (assume this to be zero), and

C_E = the cost of the electric energy which is a function of c_e, the price of electricity.

Assume that the efficiency of the electrolyzer depends linearly on the current density:

$$\eta = a + bJ.$$

Let Θ be the **plant factor**—that is, the fraction of the total time during which the plant is actually operating.

Let R be the yearly cost of the capital—that is, the interest, taxes, insurance, etc., expressed as a fraction of the borrowed capital.

Develop an expression that yields the value of J that minimizes the cost of the hydrogen produced. This must be a function of C_D, J_0, R, Θ, c_e, a, and b.

What is this optimum value of J for the case in which

C_D = 100 \$/kW,
J_0 = 10,000 A/m^2,
R = 0.2 per year,
Θ = 1,
c_e = 10 \$/MWh,
a = 0.74,
b = -6×10^{-6} m^2/A.

10.2   An electrolytic cell having 100% current efficiency, operates at a voltage of 1.9 V when the current is 20 kA. Its operating temperature is 86 C. The cell happens to be completely heat insulated: heat can only be removed by the flow of reactants and of products and by water flowing through a cooling system built into the cell.

10.38

Both feed and cooling water enter the system at 25 C. The cooling water leaves at 80 C. The gases leave at 85 C. Assume that the enthalpy change owing to the reaction is independent of temperature (285.9 MJ per kilomole of water).

1. What is the hydrogen production rate in kg/hour?
2. What is the feed water consumption rate?
3. What is the flow rate of the cooling water?
4. A consultant was called in and he improved catalysis in the anode reducing the operating voltage (at 20 kA) to 1.475 V. The cooling water can now be shut off. What is the new operating temperature?

10.3   Consider an ideal water electrolyzer installed at sea level and an ideal hydrogen/oxygen fuel cell installed 1000 m higher up.

The oxygen and the hydrogen produced electrolytically rise through a pipe and feed the fuel cell. Water is produced by the latter and flows down another pipe turning a turbine that drives a electric generator. The turbine/generator combination has 100% efficiency.

Here is a simple argument:

Since both electrolyzer and fuel cell are ideal (reversible), the electric energy generated by the fuel cell should be exactly the energy needed by the electrolyzer. But, there is some energy, W, generated by the flowing product water and this constitutes a net "profit."

Clearly,

1. we have finally invented a perpetual motion machine, or
2. we are extracting energy from some non-obvious source, or
3. there is a flaw in the argument.

If you believe (a), you are in trouble. It must be either (b) or (c). If it is (b), describe the hidden source and demonstrate that it delivers precisely W units of energy. If (c), explain what is wrong with the argument and demonstrate that W plus the energy delivered by the fuel cell is exactly equal to the energy required by the electrolyzer. Assume that the temperature of the electrolyzer is equal to that of the fuel cell.

10.4   Hydrogen, to be produced electrolytically, is needed at 40 MPa and 25 C. Two possibilities must be considered by the engineers who are designing the plant:

1. Produce hydrogen at 0.1 MPa and then compress it mechanically to 40 MPa.
2. Pressurize the electrolyzer to 37 MPa and remove the hydrogen at 40 MPa. This assumes that the internal pressure difference between the oxygen and the hydrogen side is 3 MPa and that the oxygen is produced at the environmental pressure of the electrolyzer.

When operating at 0.1 MPa, the electrolyzer efficiency is 85%. However, when pressurized, its efficiency falls to 80%.

The mechanical compressor has 65% efficiency.

How much energy is needed to produce 1 ton of hydrogen in each case? Calculate the energy required to supply the feed water (only the energy necessary to force the water into the pressurized environment).

The high pressure oxygen can be expanded to 0.1 MPa through a turbine. How much energy can you recover if you have a 100% efficient turbo-generator? What is the temperature of the exhaust oxygen assuming an adiabatic (completely insulated) turbo-generator? Assume the oxygen does not condense.

10.5   An electrolyzer consists of 100 cells, each with 1 m^2 of effective area. The efficiency of each cell is given by

$$\eta = 1.205 + bJ,$$

where J is the current density in A/m^2.

When $J = 1000$ A/m^2, the voltage required by each cell is 1.310 V.

The electrolyzer is installed in a cubical room (3 m edge) whose walls, floor, and ceiling conduct heat at a rate of 50 W/m^2 per kelvin of temperature difference. No other source or sink of heat is in the room. Outside temperature is 30 C.

When a current of 1000 A is forced through the cells,

1. How many kg of H$_2$ are produced per day?
2. How many kg of O$_2$ are produced per day?
3. How many kg of H$_2$O are consumed per day?
4. What is the equilibrium temperature of the room?
5. What current causes the room to reach the lowest possible temperature?

10.6   A water electrolyzer operating at RTP requires 1.83 V to produce 1 metric ton of hydrogen per day when operating continuously.

1. What current does it draw?
2. How many m^3 of water does it use per day?
3. How many MJ of heat does it either reject or absorb per day (state which)?

10.7   We want to estimate the performance of a system for the production of ammonia in geographically dispersed plants. Input energy is to be electrical.

The agricultural plot to be served by each ammonia plant is to be 20 by 20 km in effective cultivated area.

Nitrogen fertilization is to be intensive: 40 kg of ammonia per hectare per year (1 ha = 10,000 m^2).

The electrolyzers will operate at a current density that results in 85% overall practical efficiency (at RTP). Current efficiency is 100%.

1. Assuming that the plants operate 24 hours per day, how many tons (1000 kg) must each plant produce per day?
2. How much power does each electrolyzer demand?
3. What is the voltage of each electrolyzer cell?
4. What is the current through each electrolyzer cell?
5. Assume that instead of delivering the gases at 1 atmosphere, the electrolyzer is required to deliver them at 400 atmospheres. Clearly, additional electric power will be needed. Assume that this additional power is exactly the power for isothermal compression of the gases. How much is this additional power?
6. In the traditional ammonia process, 80% of the cost of production is the cost of energy. The price of ammonia in the international market is $200/ton. Thus, $40/ton cover all the non-energy related costs. How much can you afford to pay for 1 kWh and still make a 10% profit? Remember that the $40/ton remains unchanged.

10.8   A hydrogen producer has a battery of 100 cells connected in series. A v-i test shows that a current of 35.6 A must be driven into any of the cells to achieve a 1.482 V potential drop across it. As the current is decreased, the voltage also decreases and, as the current approaches 0, the voltage approaches 1.376 V.

A production rate of 1 liter (0.001 m^3) of H$_2$ per second at RTP is required. What voltage must be applied to the battery?

Model each cell as an ideal electrolyzer in series with an opposing voltage generator and a resistance of constant value (independent of the current).

10.9   Consider an ideal electrolyzer in series with a 0.01 Ω resistance. The gas outlets are connected to small closed vessels of equal volume so that when the device operates, gas pressure builds up. Initially, both the oxygen pressure and the hydrogen pressure are 1 atmosphere. The electrolyzer operates at 298 K.

A 1.333 V constant-voltage power supply is connected to the electrolyzer, in series with the 0.01 Ω resistor. The current is monitored.

1. What is the initial current?
2. As time goes on, the pressures in the gas containers rise. Assuming these containers do not rupture, what is the maximum hydrogen pressure that the system will reach?

10.10   A spherical balloon is filled with hydrogen until the internal pressure equals that of the surrounding air. The bag is made of a material that can expand or contract but contributes negligibly to the pressure of the gas. The temperature of the hydrogen is equal to that of the air. The balloon, when empty, weighs 32 kg. When the air pressure is exactly one atmosphere and its temperature is 0 C, the balloon has a diameter of 10 m.

1. What is its net lifting force? How does the lift force depend on the external pressure and temperature?
2. To fill the balloon, a water electrolyzing plant is employed. The plant consists of 100 cells, each operated at 2000 A and 1.92 V. How long must this plant operate to produce the required amount of hydrogen?
3. How much feed water is used per second (in liters per second)?
4. What is the rate of heat production by the plant?

10.11   A battery of 12 water electrolyzers connected in series (so that the same current flows through each unit) operates in a room maintained at 298 K. The product gases are withdrawn under a pressure of 1 atmos. A voltage of 17.784 V is applied and the resultant current is 1200 A.

1. Calculate the production rate of hydrogen in kg/day.
2. Calculate the water consumption in liters/day.
3. Calculate the operating temperature of the electrolyzers.

10.12   How much power does an ideal electrolyzer operating at 298.2 K require to produce 1 kg of hydrogen per hour at a pressure of 400 atmospheres? The oxygen is produced at 370 atmospheres.

10.13   You have been hired to run a hydrogen production plant. The fixed cost (amortization of the equipment and the buildings, salaries, taxes, etc.) amounts to $c_F = \$2000$/day regardless of how much hydrogen is produced.

The electrolyzer consists of N cells connected in series, each having the characteristic,

$$V = V_0 + R_{int}I.$$

The utility will provide electric energy at a price, c (in \$/kWh), that can vary from day to day, depending on availability.

Each day you must adjust the hydrogen production rate, $H_{prod.rate}$ (kg/day) so as to minimize the cost of the gas. To this end, you must develop a formula that tells you the optimal production rate as a function of c.

Calculate the hydrogen production rate that leads to the cheapest gas for the case in which there are 250 series connected cells each one having the characteristic

$$V = 1.420 + 20 \times 10^{-6}I.$$

Do this for the an electric energy cost of 2 cents per kWh.

10.14   Consider the electrolyzer of Problem 10.13, operating at 20,000 A.
Calculate:

1. The total voltage that must be applied.
2. The hydrogen production rate in kg (H_2)/day.
3. The rate of water consumption in m^3/day.
4. The heat power rejected.

10.15   A hydrogen production system using direct dissociation of water is
to operate at 1500 K. To have adequate hydrogen flux through a palladium
filter used to separate the gas from oxygen and water vapor, it is necessary to have a hydrogen pressure differential of 5 atmospheres across the
membrane. Assume that the pure hydrogen side is at 1 atmosphere. What
pressure must be used on the water vapor side? Repeat for $T = 3000$ K.

10.16   An electrolyzer is made of a MEA that has a specific resistance,
$\Re = 65$ $\mu\Omega$ m^2. Each cell is to produce 100 grams of hydrogen per hour.
The current efficiency is 100%.

1. What is the current through each cell?
2. The V-I characteristic is

$$V = V_{rev} + V_{offset} + RI,$$

where V_{offset} is a constant offset voltage of 0.1 V.
    The price of electricity is \$0.05/kWh. The price of the electrolyzer is
proportional to the active area of the cell and is \$10,000 per square meter
of active area. The cost of money is 12% per year. The only costs to be
considered are those of investment and of the electric energy.
    Temperature of operation is 298 K. The plant operates continuously
throughout the year. In assembling the electrolyzer, you can choose the
active area of the cell. What current density results in the most economical
hydrogen production?

10.17   It is difficult to design containers that can operate at the extreme
high temperature required to thermally dissociate water. Let us assume,
optimistically, that temperatures of 2800 K can be handled. The canister
contains 100 g of liquid water when cold. It is, then, heated to 2800 K and
part of the water will dissociate into hydrogen.

1. How many grams of hydrogen are formed?
2. The amount of free hydrogen is going to be small. What would happen
   if instead of 100 g of water, you had used 10 kg?

10.18   Ammonia is perhaps the most important of fertilizers. It provides
nitrogen essential to the growth of plants. In 2000, world-wide production
of ammonia exceeded 120 million tons per year.

All ammonia is produced by the Harber-Bosch process:
$$3H_2 + N_2 \leftrightarrow 2NH_3. \tag{75}$$
Enthalpy of formation: $\Delta \bar{h}_{f^\circ_{NH_3}} = -46.19$ MJ/kmole.

1. Is the equilibrium toward the ammonia side favored by raising or by lowering the pressure in the reactor?
2. Is the equilibrium toward the ammonia side favored by raising or by lowering the temperature of the gases? Clearly, it is not possible to take the favorable conditions of pressure and temperature to an extreme. Give one good reason for limiting pressure and temperature extremes.

10.19   Nitric oxide, NO, is a gas of great importance in the biology of mammals where it plays a central role in the operation of cells. It is also a serious pollutant because it readily combines with oxygen to form the toxic gas, NO_2. The nitrogen in the air can react with the oxygen according to
$$\frac{1}{2}N_2 + \frac{1}{2}(O)_2 \leftrightarrow NO. \tag{76}$$
This causes the formation of the undesired NO which, reacting with atmospheric oxygen and water vapor, is converted to nitric acid and causes acid rain to fall. NO is also a source of photochemical smog and a destroyer of the ozone layer.

Here are some pertinent thermodynamic data:

Enthalpies of formation:

	298.15 K	6000 K	
NO	90.37	298.87	MJ/kmole

Entropies:

	298.15 K	6000 K	
N_2	95.7	146.3	kJ K^{-1}kmole^{-1}
O_2	205.0	313.3	kJ K^{-1}kmole^{-1}
NO	210.6	369.4	kJ K^{-1}kmole^{-1}

Equilibrium constant, $K_p = 4.522 \exp - \dfrac{90.58 \times 10^6}{RT}$.

1. Does the reaction that generates NO from N_2 and O_2 proceed spontaneously at room temperature? If you say it does, then you have to explain how come there is so little NO in the air. If you say it does not, then how come there is a problem with NO pollution?
2. What happens when the gases are at 6000 K? If the reaction converts nitrogen and oxygen into nitric oxide, what fraction of the gas mixture, at equilibrium, consists of NO?
3. You will have found from Item 2 that at 6000 K (easily reached during a lightning stroke) substantial amounts of NO are produced. But the air quickly cools down and as we saw from Item 1, at low temperature the equilibrium is toward complete dissociation of NO. So, how come NO lingers around constituting a serious pollutant?

Chapter 11
Hydrogen Storage

As one approaches the city, one sees a landscape unmarred by unsightly towers and dangling transmission lines. Suburbs and center are devoid of wire-carrying poles, and the electric system is totally immune to storm-felled trees and branches. The only carbon dioxide produced has been exhaled by men, animals, and, at night, by plants. Cooking fires and heating systems release only water vapor, as do cars and buses. No ozone, no carbon monoxide, no nitrogen oxides can trace their birth to devices created by humans.

Each factory, each office, each house derives its energy from ultra-pure hydrogen either generated in situ or brought in by reliable underground pipelines. During the day, gleaming roofs collect sunlight and generate more electricity than can be used. The excess is sent to electrolyzers and stored locally as high-pressure hydrogen to be used at night or as fuel for cooking and heating. High-precision, inexpensive oscillators keep the fuel cell generating electricity at the exact prescribed frequency, its phase updated periodically by low frequency radio signals.

When needed, additional hydrogen, from nonpolluting sources, is imported through existing pipes. The energy is totally ecologically friendly. Utopic? Yes, but entirely realizable with current technology, yet impossible with current economics!

To make this picture a reality, great advances must be made in bringing down the cost of numerous processes, not the least of which is the technique for storing hydrogen, the topic of this chapter.

The qualities of hydrogen have been extolled by enthusiasts. Indeed, once it has been produced (a process that can occur without pollution), you have the most ecologically friendly fuel possible. It suffers, however, from a major drawback—its extremely low density. Per cubic meter, hydrogen has but one-third of the combustion energy of methane. Nevertheless, a given pipeline has about the same power carrying capacity with the two gases: the lower specific energy of hydrogen is almost exactly compensated by its lower viscosity. Thus, the bulk distribution of hydrogen should not present insurmountable problems. A more serious problem arises when hydrogen has to be supplied to a moving vehicle. One must either generate it in the vehicle itself or find a convenient way to store it aboard. In this chapter, we will examine the latter alternative.

Hydrogen can be stored as an element or it can be extracted, as needed, from some hydrogen-rich substance using an onboard extraction process:

1. processes that alter the state or the phase of hydrogen (**hydrogen only systems**):

   1.1 compression of the gas (see Subsection 11.1), or a combination of compression and refrigeration;

   1.2 liquefaction of the element (see Subsection 11.2). Owing to its low critical temperature, hydrogen must be cooled to some 20 K to remain liquid in nonpressurized vessels;

2. processes that associate hydrogen to other substances:

   2.1 adsorption of the gas on some appropriate substrate such as activated carbon;

   2.2 chemical combination of hydrogen so as to create a hydrogen-rich compound. Such compounds can be:

   2.2.1   compounds in which H_2 is tightly bound requiring a relatively complex chemical process for the recovery of the gas. Included there are, for instance, substances like methanol (discussed in Chapter 10), ethanol, ammonia, and water itself that can be thought of as "carriers" of hydrogen;

   2.2.2   compounds that can be reversibly transformed into another substance with a higher (or lower) hydrogen content;

   2.2.3   metal hydrides that are metal-hydrogen compounds that can release and absorb hydrogen reversibly by a simple change of temperature.

A number of characteristics have to be considered when evaluating hydrogen storage systems. They include

1. **Gravimetric concentration, GC.** This is the ratio of the mass of the stored hydrogen to the overall mass of the (loaded) storage and retrieval system. The dimensions are kg per kg—that is, it is a dimensionless parameter.
2. **Volumetric concentration, VC.** This is the ratio of the mass of the stored hydrogen to the total volume of the storage and retrieval systems. The dimensions are kg per m^3 $[ML^{-3}]$.
3. **Turnaround efficiency.** This can be the ratio of the retrieved hydrogen to the amount of input hydrogen, or the ratio of the energy retrieved to the input energy.
4. **Dormancy.** The ability of the system to retain its hydrogen over a long period of time.

In addition, there are considerations of cost (capital, maintenance, operating, and replacement costs), safety, ease of utilization, and so on.

11.1 Compressed Gas

For compressed gas containers, the main quantity of interest is the gravimetric concentration—that is, the ratio of the mass of the maximum amount of gas that can be stored to the mass, M_{cont} of the container, where the maximum amount of gas corresponds to gas at just under the burst pressure, p_{burst}, of the vessel. This ratio is proportional to the **performance factor**, PF, of the container:

$$PF \equiv \frac{p_{burst}V}{M_{cont}}. \tag{1}$$

In SI, the performance factor has the dimensions of joules per kilograms.

For a given material and a given technology for building a compressed gas container, the mass of the container is proportional to the pressure so that the ratio of stored gas mass to the mass of the container is independent of the storage pressure. Hence, the only way to improve the PF is to use better materials and better technology in the construction.

Small quantities of hydrogen, as used in chemical laboratories, can be conveniently stored in simple steel pressure cylinders, usually at 150 atmospheres.

For fuel cell vehicles (FCV), compressed hydrogen may be a practical way to carry the necessary fuel. It is certainly the simplest storage system and it requires no special equipment to retrieve the gas. What is needed are containers with a good PF.

Aluminum canisters of modern design, reinforced by a wrapping of carbon fibers, look promising. They can store the gas at 500 atmospheres and are relatively light—a 0.15 m^3 (150 liter) canister can hold 6 kg of H_2 (860 MJ) and masses less than 90 kg (see Williams, 1994). This 6.7% gravimetric concentration compares favorably with hydride storage systems discussed later in this chapter. The failure mode of these canisters is relatively benign—they do not explode in shrapnel but fail by delamination of the wrapping. They are designed to leak before bursting.

The 150-liter canister mentioned could be a cylinder 1.5 m long with 0.36 m diameter and would not be difficult to fit into a vehicle. Since at the moment, the recommended safety factor is 3:1, its burst pressure should be 1500 atmospheres and its PF, 250,000 J/kg.

860 MJ of hydrogen correspond (in energy) to some 20 liters of gasoline, a modest amount of fuel, considering that a normal passenger car carries typically a 50-liter gas tank. However, a FCV has more than twice the efficiency of an ICV (internal combustion vehicle).

For very large scale storage, it may be possible to keep hydrogen in underground formations, such as porous rocks, old mines, caves, aquifers, and exhausted natural gas deposits.

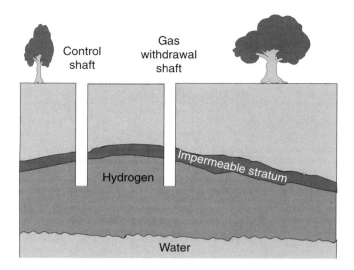

Figure 11.1   Hydrogen storage in an aquifer.

At present there is little experience with underground storage of hydrogen. However, results of the helium storage in Amarillo, Texas, suggest that there will be little difficulty with this technology.

Figure 11.1 depicts an arrangement for keeping large amounts of gas in an aquifer, provided an impermeable layer of rocks forms an adequate roof over the structure.

In Amarillo, 8.5×10^8 m^3 of helium are stored without problems. It should be noted that helium has leakage characteristics similar to those of hydrogen. At STP, 8.5×10^8 m^3 of hydrogen correspond to 10,000 TJ of stored energy.

To gain an idea of how much this storage capacity is, one can compare it with that of one of the world's largest pumped storage facilities[†] in Ludington, Mich. This facility has a capacity of 54 TJ, or nearly 200 times less than that of a hydrogen filled Amarillo reservoir.

Another storage arrangement for hydrogen would be the very pipelines used for transporting the gas. A typical trunk pipeline for natural gas can be over 1000 km long. It may have a 1.2 m diameter and operate at 6 MPa (60 atmospheres). The hydrogen stored in such a pipeline would correspond to an energy of 1000 TJ, nearly 20 times the capacity of the Ludington reservoir.

[†] In pumped storage plants, normal hydroelectric machinery can be reversed so as to pump water back into the reservoir during times when there is surplus electric energy in the system.

11.2 Cryogenic Hydrogen

Although hydrogen was first liquefied in 1898, it was only recently, through the efforts of NASA, that the technology for production and storage of large quantities of the liquid was developed.

The largest storage unit in existence is one at Cape Canaveral, with a capacity of 3375 m^3. Since the density of liquid hydrogen is 71 kg m^{-3}, the facility can accumulate 240,000 kg of liquid hydrogen, or 34 TJ, just a little less than the capacity of the Ludington reservoir we have been using for comparison.

There are two different species of hydrogen molecules: para- and ortho-hydrogen. In the first, the spin in the two atoms that constitute the molecule are in opposite directions, while in the second, the spins are in the same direction.

In the liquid state, para-hydrogen (p-H$_2$) has lower enthalpy than the ortho form (o-H$_2$). At the boiling point of H$_2$ (20.4 K at 0.1 MPa), the difference is 1.406 MJ kmole $^{-1}$.

In hydrogen, gaseous or liquid, the reaction

$$\text{p-H}_2 \leftrightarrow \text{o-H}_2 \qquad (2)$$

goes on continuously and an equilibrium concentration of each species is established. The equilibrium at STP corresponds to 25% p-H$_2$ and 75% o-H$_2$, whereas at 20.4 K, the equilibrium shifts to 99.79% p-H$_2$. Owing to the slow kinetics of the reaction, freshly cooled hydrogen tends to have an excess of the ortho variety and its transformation into the para variety results in the release of heat causing the liquid to boil even though no external heat is supplied.

Freshly condensed hydrogen, even if kept in a perfectly adiabatic container, will lose 1% of its mass during the first hour and 50% during the first week. To minimize such losses, o-H$_2$ is catalytically converted to p-H$_2$ during the liquefaction process. Levels of 95% p-H$_2$ are desirable.

Liquid hydrogen has been considered as a fuel for aircraft. Lockheed investigated the performance of a supersonic airplane designed to carry 234 passengers 7800 km at Mach 2.7. A kerosene-powered plane with such a capability would have a gross weight of 232 tons of which 72 tons would be fuel. A hydrogen plane with equivalent performance would have a gross weight of only 169 tons of which less than 22 tons would be fuel.

Hydrogen driven commercial airplanes will probably not be seen in the near future. Present day design efforts for transport planes in the Mach 3 range are based on jet fuel (kerosene) engines. However, the proposed space plane will probably need hydrogen as a fuel. It will be a hypersonic transport (perhaps Mach 8) capable of taking off from a conventional runway and achieving orbital flight.[†]

One of the problems of high speeds while inside the atmosphere is the high temperatures generated. The **stagnation temperature** of a body moving through a gas (the temperature reached by the gas at the point in which its flow speed relative to the body is zero) is given by

$$\frac{T}{T_{amb}} = 1 + \frac{\gamma - 1}{2} M^2 \tag{3}$$

See insert.

For air, $\gamma = 1.4$ and our equation becomes

$$\frac{T}{T_{amb}} = 1 + 0.2M^2 \tag{4}$$

For M = 2.5 and $\gamma = 1.4$, $T/T_{amb} = 3.25$ and for M = 25, it is 226. This means that, at the latter velocity, if the ambient temperature is 300 K, the stagnation temperature is 67,800 K.

Clearly, no material exists that can operate at such temperatures. This means that the heat developed at the leading edges of the fuselage, wings, and control surfaces must be efficiently removed. Part of this can be achieved by radiation and conduction and in part by refrigeration. The liquid hydrogen fuel can be used to cool critical regions of the plane prior to being conveyed to the engine in gaseous form.

Hydrogen to be liquefied must be of high purity. Most other gases will freeze during the process and will tend to clog the pipes. If oxygen ice is formed, explosions may result. Specifications generally call for less than 10 ppm of O_2.

Practical liquefaction machines require about 40 MJ to condense one kg of H_2. This energy cannot be recovered; the resulting storage turn-around efficiency is $143/(143 + 40) = 0.78$ or 78%.

The cost of a cryogenic plant does not scale linearly with the rate of production; it grows with $\dot{M}^{0.7}$, where $\dot{M}$ is the rate of production.

Stagnation temperature

The sum of the enthalpy and the kinetic energy of a flowing gas is constant:

$$c_p T + \frac{1}{2} u^2 = c_p T_\infty + \frac{1}{2} u_\infty^2. \tag{5}$$

(continued)

[†] A supersonic combustion ramjet (scramjet) engine seems to be essential for the success of hypersonic transport planes. On March 27, 2004, an experimental scramjet engine was tested in free flight. It operated for 10 seconds and accelerated to Mach 7 (8000 km/hr).

(continued)

The preceding is, of course, per unit mass. Let T_∞ be the undisturbed—that is, ambient temperature and $u_\infty \equiv v$ be the undisturbed wind velocity—that is, the velocity of the object. At the **stagnation point**, the velocity is, by definition, zero. Hence,

$$c_p T = c_p T_{amb} + \frac{1}{2}v^2, \tag{6}$$

$$\frac{T}{T_{amb}} = 1 + \frac{1}{2}\frac{v^2}{c_p T_{amb}}. \tag{7}$$

The speed of sound is $c = \sqrt{\gamma R T_{amb}}$, therefore

$$\frac{T}{T_{amb}} = 1 + \frac{\gamma R}{2c_p}\frac{v^2}{c^2} = 1 + \frac{1}{2}\gamma R M^2 \tag{8}$$

where $M \equiv v/c$ is the **Mach number**.
But,

$$\gamma = \frac{c_p}{c_v} \tag{9}$$

and

$$c_p = c_v + R, \tag{10}$$

thus

$$c_p = \frac{\gamma R}{\gamma - 1} \tag{11}$$

$$\frac{T}{T_{amb}} = 1 + \frac{\gamma - 1}{2}M^2 \tag{12}$$

11.3   Storage of Hydrogen by Adsorption

Both hydrogen molecules and methane can be readily adsorbed on carbon. The gases are held in place by weak Van der Waals forces so that the energy necessary to retrieve the fuel is small.

Carbon systems can be combined with other techniques: the gas can be pressurized and the temperature can be lowered. Typically, an adsorption system using activated carbon requires 20 to 40 atmosphere pressure and liquid nitrogen temperature to hold the hydrogen. These requirements severely limit the practical application of such systems. They achieve 5% to 6% gravimetric concentration (GC), just about the same as that of good metal hydrides systems (discussed further on).

Carbon nanotubes promise to uptake hydrogen much more effectively. Single wall carbon nanotubes have been reported as yielding gravimetric concentrations of up to 10% when operated at 120 K and 0.4 atmospheres. N. M. Rodriguez (Northeastern University) claims 40% GC in graphite nanotubes. Chambers et al. report 67% GC at room temperature and 120 atmospheres in graphite nanofibers with herringbone structure.

All these systems require low temperatures or high pressures. Chen et al. describe a system that operates at 1 atmosphere and at some 200 C to 400 C. It uses lithium doped carbon nanotubes and achieves a GC of 20%.

It should be noted at this point that if each carbon atom were to attach 1 hydrogen atom, then the GC would be only 8%.

11.4 Storage of Hydrogen in Chemical Compounds

11.4.1 Generalities

The main difficulty encountered in the storage of hydrogen is, as pointed out, the low density of the gas. It is possible to substantially increase the packing density by associating hydrogen with other substances. The storage and retrieval processes consist then of the synthesis of a hydrogen-rich compound followed, when the gas is needed, by its dissociation.

The requirements of a practical hydrogen storing compound include:

1. High storage capacity
   The density of liquid hydrogen is 71 kg m^{-3}. Many hydrogen-rich compounds have packing densities that exceed this value. As an example, consider three common hydrides listed in Table 11.1.

Table 11.1
Hydrogen Packing Densities in Different Compounds

Technical [†] Name	Common Name	kg(H_2)m^{-3}	kg(H_2)kg^{-1}
Oxygen di-hydride	water	111	0.111
Nitrogen tri-hydride	ammonia	113	0.147
Di-nitrogen tetra-hydride	hydrazine	126	0.125

[†] With an apology to Chemists, we have characterized the three substances as "hydrides," to emphasize their hydrogen carrying properties.

Table 11.2
Enthalpies of Formation of Some Hydrogen-Rich Compounds

COMPOUND	ENERGY OF FORMATION (per kg of H_2) (MJ)
Water(ℓ)	143
Ammonia(g)	15.4
Hydrazine	-12

To achieve high **volumetric storage capacity**, the hydride must have a high hydrogen-packing density. To achieve a high **gravimetric storage capacity**, the hydride must be of relatively low density.

2. Low reaction energy

Hydrogen as a fuel is used, almost always, by combining it with oxygen and producing water. This releases 143 MJ per kilogram of liquid water. Clearly, the energy of formation of the hydride in which hydrogen is stored must be substantially lower than this value for the storage system to be useful. For instance, 15.4 MJ are necessary to dissociate enough ammonia to produce 1 kg of hydrogen—a theoretical turnaround efficiency of $143/(143 + 15.4) = 90\%$. This is acceptable. On the other hand, water cannot be used because it has a theoretical turnaround efficiency of 0%.

The enthalpies of formation of the three compounds in the preceding table are shown in Table 11.2.

3. Reversibility

The reaction should be easily reversible, i.e., it should be easy to shift the equilibrium to either the H_2 or the hydride side.

4. Kinetics

The hydrogen fixing and releasing reactions must occur rapidly, at relatively low temperatures and without requiring expensive catalysts.

5. Separability

It should be easy to separate the dissociation products. Ideally, the products should be gaseous hydrogen plus solid residues.

6. Low corrosiveness

11.4.2 Hydrogen Carriers

Clearly, one way to store and transport hydrogen is to synthesize a hydrogen-rich substance and then, as needed, generate hydrogen by a chemical "reforming" process like some of those described in Chapter 10. Of great interest for fuel cell vehicles (FCV) is the use of methanol, a liquid fuel, which can be reformed with relative ease. Considerable effort is being devoted to the design of simple onboard reformers, but there are difficulties:

1. Efficiency
   Assume natural gas as the feedstock. If hydrogen is produced from this gas in a practical device, $\approx 90\%$ of the heat of combustion of the initial raw material appears in the H_2. If natural gas is converted to methanol, only 71% of the heat of combustion of the raw material is available from the alcohol produced. While hydrogen can be used directly by a fuel cell, methanol has to be reformed onboard and only 77% of the heat of combustion of the alcohol becomes available as hydrogen for the cell. Thus, the relative efficiencies are 90% for hydrogen and only 77% of 71% or an overall 55% for methanol.

2. CO_2 emission
   Again, assume that both methanol and hydrogen are derived from natural gas. Fueling an FCV with methanol results in an overall emission of CO_2 1.5 times that resulting from using hydrogen.

3. Contaminants
   Fuel cells for automotive use will, most likely, be low temperature SPFCs that are sensitive to the presence of impurities, such as CO, that degrade the catalyst. Such impurities will be present in the hydrogen whether it is produced directly from natural gas or through an intermediate methanol step. The latter requires reforming aboard the vehicle where, plausibly, good purification of the hydrogen will be more difficult and more expensive than at a central hydrogen generating plant.

3. Environmental danger
   If the use of methanol ever becomes as widespread as that of gasoline today, there will, unavoidably, be some spills as happens occasionally with petroleum products. The consequences of methanol spills may, however, be much more serious than those of oil or gasoline. The last two fuels do not mix with water and float on its surface. Methanol mixes with water in any proportion and a major spill may contaminate an aquifer. This will render the water of a given region undrinkable because methanol is quite poisonous, blinding or killing those who

ingest a sufficient quantity of it.

11.4.3 Water Plus a Reducing Substance

In Chapter 10, it was mentioned that balloonists of the last century produced hydrogen by passing steam through a bed of iron filings. The iron oxidizes into rust and the water is reduced to hydrogen in a reversible reaction:

$$3Fe + 4H_2O \leftrightarrow Fe_3O_4 + 4H_2. \tag{13}$$

Driving this reaction toward the left constitutes a way of storing hydrogen. There are, however, substantial difficulties in this scheme.

Unfortunately the reaction cannot be driven to completion either in the forward, or hydrogen producing, direction or in the reverse direction.

The temperatures that must be used are quite high (above 1000 C).

Although iron oxidation is exothermic, the heat released is far less than that needed to boil the required water and superheat the resulting steam.

11.4.4 Metal Hydrides[†]

A majority of the requirements listed in Subsection 11.4.1 can be met by a class of substances loosely called **metal hydrides**, or simply hydrides.

A number of elements form unstable hydrides (hydrides that can easily be reversed). Magnesium, iron, titanium, zirconium, yttrium, lanthanum, and palladium are examples. Hydrides of elements are called **binary**. **Ternary hydrides**—hydrides formed by a combination of hydrogen with a binary compound—are more promising. A typical example are the hydrides of TiFe. The addition of a third element, leading to **quaternary hydrides**, increases even further the degree of freedom in choosing the characteristics of the system. As can be seen, a large number of combinations are possible. Research in this area has only scratched the surface.

The exchange of hydrogen (or any other gas) with a solid is called **sorption**. If hydrogen is being fixed, the reaction is called **absorption**; if hydrogen is being released, **desorption**.

Hydride characteristics are best inspected by examining the pressure versus hydrogen concentration isotherms for the material. Such data are measured by apparatus similar to the one shown schematically in Figure 11.2.

[†] It is important to distinguish the noun "hydrate" from the noun "hydride." The former describes a combination of an element or a radical with water, while the latter describes the combination with hydrogen. "To hydrate" is the action of creating a hydrate. The *Oxford English Dictionary* does not list the verb "to hydride," which we will use in this chapter to indicate the action of creating a hydride.

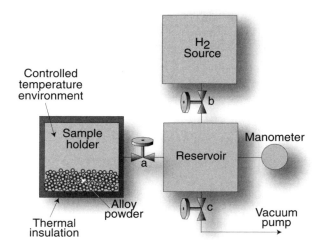

Figure 11.2   Apparatus for the determination of pressure
versus concentration isotherms.

A certain amount of **activated** (see further on) granules composed, for
example, of elements A and B forming an alloy, AB, is placed in the sample
holder. The exact number of kilomoles of the alloy, μ_{AB}, has previously
been determined. Valves $\underline{a}$ and $\underline{c}$ are open and valve $\underline{b}$ is closed. The
sample is degassed by heating it to a high temperature and extracting the
released vapors by means of a vacuum pump.

Valves $\underline{a}$ and $\underline{c}$ are then closed and $\underline{b}$ is opened filling the reservoir with
a known volume of hydrogen at known pressure and temperature—that is,
with a known amount of the gas. With $\underline{b}$ and $\underline{c}$ closed, $\underline{a}$ is opened and
hydrogen is absorbed. This causes the temperature of the powder to rise
because the absorption is exothermic. The system is then returned to the
initially selected temperature and the equilibrium pressure is observed.

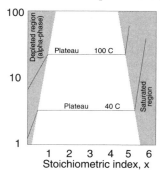

Figure 11.3   Idealized pressure-stoichiometry plot for hydrides.

11.12

In this manner, the amount, μ_H, of hydrogen taken up by the granules can be calculated. Either the **hydrogen/metal ratio**, H/M, or the stoichiometric index, x, of the formula ABH_x, is determined. For instance, $ABH_{0.4}$ has a H/M ratio of $0.4/(1+1) = 0.2$. Many pressure/composition diagrams in this chapter have ordinates scaled in both x and H/M.

The procedure is repeated with increasing amounts of hydrogen and H/M (or x) is tabulated versus the equilibrium pressure, p. In an extremely idealized way, the resulting plot may look like the one shown in Figure 11.3.

The solid material in the system can be in one of several states:

1. It may be simply the initial metallic alloy (called the **alpha-phase**),
2. it may be a single hydride of the alloy (called the **beta-phase**), or
3. it can, in some cases, be a double hydride of the alloy (called the **gamma-phase**).

At low hydrogen concentrations, the pressure depends strongly on x: hydrogen either occupies the interstices between the granules or is dissolved in the alloy without altering its chemical composition. The corresponding region on the pressure-stoichiometry diagram is called the **depleted region**.

As the concentration increases, a **plateau** region is reached where pressure is nearly independent of concentration. This is a region of equilibrium between the alpha- and the beta-phase. The higher the hydrogen concentration the larger the fraction of the alloy in the beta-phase and the smaller in the alpha-phase. In this region the alloy is hydrided at discrete locations where its composition becomes ABH. This occurs because it is energetically more favorable to react chemically than to dissolve. An equilibrium between hydrided sites, ABH, and nonhydrided sites, AB, is established. The more hydrogen is introduced, the larger the ratio of the former to the latter. The exact proportion is indicated by the value of x in ABH_x.

The plateau persists until all the alloy is hydrided—that is, until all the material has the composition ABH—it is all in the beta-phase. If the amount of hydrogen is increased further, the excess will dissolve in the beta-phase and the pressure again rises rapidly. This is the **saturated region**. In some cases, a dihydride (gamma-phase) is formed and a second plateau may appear.

Real pressure-stoichiometry curves depart from the ideal shape described above. The transition from the alpha- to the beta-phase is smooth, not abrupt. The plateau may exhibit a small slope and a sorption hysteresis may become evident: the plateau pressure can be noticeably higher during absorption than during desorption. In some cases, a second plateau may be reached where a dihydride, ABH_2, is formed. This second plateau is called the **beta prime-phase**. See Figure 11.4.

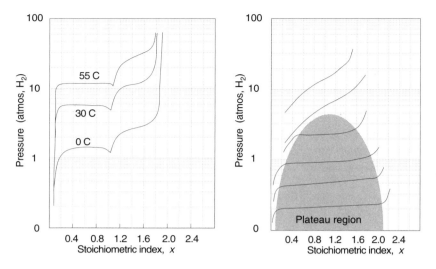

Figure 11.4   Pressure vs stoichiometric index, x, for TiFe.

Figure 11.5   The plateau region becomes progressively narrower as the temperature increases.

Different plateau pressures correspond to different temperatures. The width of the plateau becomes narrower as the temperature increases. Above a given temperature, no plateau is formed as indicated in Figure 11.5.

Figure 11.6 shows how the plateau pressure can be tailored to a given specification by a relatively small change in alloy composition. Three different alloys, Ni_5La, $Ni_5La_{0.8}Nd_{02}$, and $Ni_5La_{0.8}Er_{02}$, display, at 40 C, plateau pressures of about 3, 5, and 10 atmospheres, respectively.

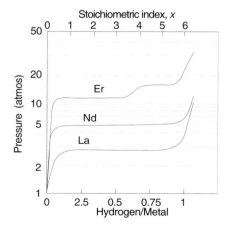

Figure 11.6   The plateau pressure can be tailored by relatively small composition changes of the alloy.

11.14

When a metallic compound is first prepared, it will, usually, not react well with hydrogen at room temperature, probably owing to an oxide layer on the surface.

Many alloys require activation which consists of heating granules of the material to some 300 to 500 C in high vacuum to outgas them, and, afterward, expose them to high purity hydrogen.

Not surprisingly, the lattice constant of the hydride is not the same as that of the parent material—absorption of hydrogen is accompanied by an increase in volume.

In the $LaNi_5$ case, for instance, there is a 25% expansion. Since hydrides tend to be brittle, the expansion causes the material to break into a fine powder. The breaking up of particles helps the activation because it exposes new clean surfaces. TiFe needs a thorough activation, while $LaNi_5$ needs only a very simple one.

The tendency to expand during absorption and contract during desorption can lead to mechanical difficulties. During desorption, the contracting particles constitute a more compact powder that can accommodate itself into new positions in the container and, upon subsequent expansion, can provoke serious strains on its walls.

The theoretical volumetric concentration, VC, of hydrogen in hydrides can exceed that of liquid hydrogen (see Table 11.3). These data, however, overestimate the concentration and, consequently, overestimate the corresponding volumetric energy concentration because the bulk density of the hydride powder is much less than that of the hydride itself owing to the imperfect packing of the individual granules.

For instance, the realizable hydrogen concentration in $LaNi_5H_5$ is about 45 kg of hydrogen per cubic meter, about half of the number that appears in the table. Nevertheless, in a per volume basis, hydrides can pack roughly the same energy as liquid hydrogen, whereas in a per mass basis, hydrides pack only a few percent of the liquid hydrogen energy. For applications, such as airplanes, in which weight considerations are of primary importance, hydrides do not loom as promising energy storage methods. For locomotives, barges, trucks, and buses, they may be useful. Hydrides may even become attractive for automobiles.

Compare Mg_2NiH_4 with gasoline. 100 kg of hydride store, at best, $3.6 \times 143 = 515$ MJ of fuel, whereas 100 kg of gasoline represent 4700 MJ, almost a 10-fold advantage. However, the usual efficiency of a gasoline engine is around 20% while that of a hydrogen fuel cell is around 60%; this reduces the gasoline advantage by a factor of 3.

A hydride fuel cell car (of current technology) has a weight advantage over an electric car using advanced batteries.

Table 11.3
Hydrogen Concentration in Various Hydrides

	MASS OF H_2 in HYDRIDE %	MASS OF H_2 in HYDRIDE (kg/m^3)	ENERGY DENSITY (MJ/kg)	ENERGY DENSITY (GJ/m^3)
$H_2(\ell)$	100	71	143	10.2
$H_2(g)$, STP	100	0.089	143	0.013
LaH_3	2.1	108	3.0	15.4
MgH_2	7.6	101	10.0	14.4
TiH_2	4.0	153	5.7	21.9
VH_2	3.8	95	3.0	13.6
ZrH_2	2.1	122	3.0	17.4
$LaNi_5H_5$	8.7	89	2.0	12.7
Mg_2NiH_4	3.6	81	4.5	11.6
$TiFeH_{1.95}$	1.85	101	2.6	14.4

One additional practical consideration is the rapidity with which a hydride storage system can be charged and discharged. This depends, of course, on the kinetics of the reaction. Magnesium, owing to its small density would be attractive were it not for its inability to absorb hydrogen under normal conditions. The kinetics of the reaction can be improved by the addition of nickel as a catalyst. A 5% nickel doping already helps noticeably, but a much larger amount of nickel is usually used, such as in the Mg_2Ni alloy.

If fast recharge rates are required, the total amount of absorbed hydrogen may be smaller than when a slow recharge is used. FeTi can be (relatively quickly) charged to $FeTiH_{1.6}$, while to achieve $FeTiH_{1.95}$, an overnight charge may be needed.

The capacity of the storage system is also determined by the maximum amount of hydrogen that can be dissolved in the alloy (alpha-phase). The 298 K plateau for FeTi ends when $FeTiH_{0.4}$ is reached. Any attempt to extract more hydrogen is accompanied by a great reduction in hydrogen pressure. Thus, for quick charge-discharge cycles, this particular alloy can only be operated between $FeTiH_{0.4}$ and $FeTiH_{1.6}$ realizing a turnaround capacity only 60% of the total theoretical capacity of the system.

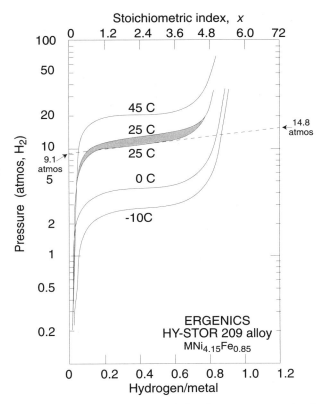

Figure 11.7   Measurement of the slope parameter in the pressure-composition isotherm.

11.4.4.1 Characteristics of Hydride Materials[†]

Some of the distinguishing characteristics of materials used in hydride systems are

– *Plateau slope*

Simple thermodynamic considerations, as discussed in the next Section, predict horizontal plateaus. In practice the plateau pressure rises somewhat with increasing hydrogen content.

[†] All alloys using the HY-STOR trademark are produced by Ergenics, Inc, and much of the data used in this Text are from a paper by Houston and Sandrock. In the formulas, "M" stands for Mischmetall (German), a mixture of rare earths generally extracted from monazite sands. Their effect on the plateau pressure depends strongly on their cerium-to-lanthanum ratio.

The slope of the plateau pressure can be represented by the **slope parameter**, $d\ln(p_d)/d(H/M)$, where p_d is the plateau pressure of the desorption isotherm. In Figure 11.7, the dashed line through the 25 C desorption isotherm intercepts the $H/M = 0$ axis at 9.1 atmospheres and the $H/M = 1.2$ axis, at 14.8 atmospheres.

Hence

$$\frac{d\ln p_d}{d(H/M)} = \frac{\ln 14.8 - \ln 9.1}{1.2} = 0.41. \tag{14}$$

This is a reasonably large slope parameter. TiFe, for instance, tends to have a slope parameter of 0.00, whereas some (but not all) calcium based alloys have parameters larger than 3. When an alloy solidifies (during the initial manufacture), the component elements sometimes tend to segregate—that is, to settle out. This seems to be the main reason for the appearance of a plateau slope, because, from thermodynamic grounds, a perfectly uniform alloy should exhibit no slope at all. Annealing of the material *prior* to crushing, can reduce the plateau slope. Slope parameters and other characteristics are listed in Tables 11.4, 11.5, and 11.6.

– *Sorption hysteresis*

As mentioned previously, the plateau pressure during absorption is usually somewhat larger than that during desorption. In other words, there is a **sorption hysteresis** when the alloy is cycled. See, for instance, Figures 11.7, 11.8, 11.10, and 11.11.

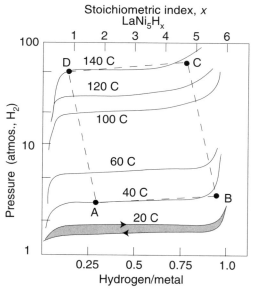

Figure 11.8   Hysteresis in the sorption isotherms of the $LaNi_5$.

Hysteresis is an irreversible process caused by the heat generated by the plastic deformation of the crystal as the lattice expands or contracts during the sorption cycle.

The degree of hysteresis is indicated by the ratio of the absorption plateau pressure, p_a, to the desorption pressure, p_d, measured at a H/M value of 0.5 and, frequently, at 25 C.

It is commonly assumed that the degree of hysteresis is fairly temperature independent.

– *Usable capacity*

The usable capacity is defined, somewhat optimistically, as the change in the H/M ratio in a hydride when the pressure is lowered from 10 times the plateau pressure to 0.1 times this plateau pressure. This is a 100:1 pressure range. A more realistic definition would be based on a much smaller pressure range.

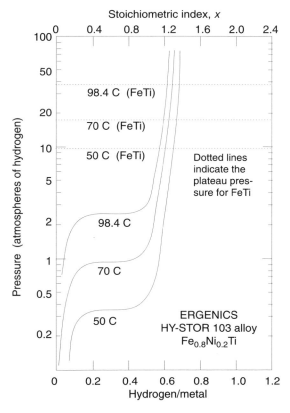

Figure 11.9   Addition of small amounts of Ni causes a substantial change in the characteristics of the FeTi alloy of Figure 11.4.

11.19

In Figure 11.9 (Alloy $Fe_{0.8}Ni_{0.2}Ti$), the plateau pressure of the 70 C isotherm is about 0.9 atmosphere. At 10 times this pressure, the H/M ratio is 0.65, whereas, at 0.1 times the plateau pressure, $H/M = 0.02$. This leads to $\Delta(H/M) = 0.63$. In other words, the hydride releases 0.63 kilomoles of H (0.63 kg) for each kilomole of metal.

The alloy consists of 0.80 kilomoles of iron (massing $0.8 \times 55.8 = 44.6$ kg), 0.2 kilomoles of nickel (massing $0.2 \times 58.7 = 11.7$ kg), and 1 kilomole of titanium (massing 47.9 kg). This adds up to 2 kilomoles of metal massing $44.6 + 11.7 + 47.9 = 104.2$ kg, or 52.1 kg/kmole. Thus, the released hydrogen represents $0.62/52.1 = 0.0012$ of the mass of the hydride or 1.2%.

– *Heat capacity*

Hydride systems are activated by temperature changes. It is important to know the heat capacity of the different alloys in order to properly design the systems. Values of heat capacities of a number of alloys are listed in Table 11.6.

– *Plateau pressure dependence on temperature*

The temperature dependence of the plateau pressure is a function of the thermodynamic properties of the material and is examined in more detail in the next section. By altering the composition of the alloy, it is possible to taylor the characteristics of the material to suit many different applications.

From Figure 11.4, it can be seen that the popular FeTi system has a plateau pressure above 10 atmospheres when the temperature is 55 C.

The substitution of Ni for some of the Fe dramatically reduces the plateau pressure. See Figure 11.9. 20% Ni is sufficient to lower the pressure (at 50 C) to 0.3 atmospheres. In addition, the nickel greatly reduces the hysteresis (although this is not shown in the figures) and facilitates the activation of the material. The $Ca_{0.2}M_{0.8}Ni_5$ system (see Figure 11.10) exhibits a plateau pressure of some 30 atmospheres at 25 C, whereas the calcium-free $MNi_{4.15}Fe_{0.85}$ system has a plateau pressure of only 10 atmospheres at the same temperature (Figure 11.7). Lower than atmospheric plateau pressure at room temperature can be achieved, for instance, by adding aluminum to the $LaNi_5$ system (Figure 11.11).

The reaction of hydrogen with metallic compounds involves, of course, enthalpy changes: absorptions are exothermic and desorptions are endothermic. The enthalpy changes can be determined by calorimetric techniques. More often, they are determined from the pressure versus concentration isotherms.

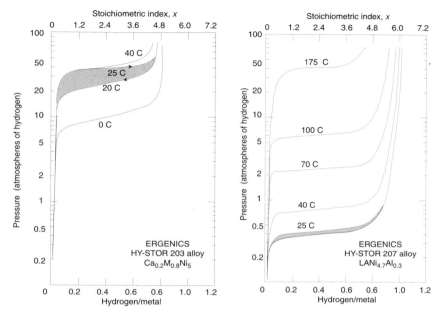

Figure 11.10   Characteristics of the $Ca_{0.2}M_{0.8}Ni_5$ system.

Figure 11.11   Characteristics of the $LaNi_{4.7}Al_{0.3}$ system.

11.4.4.2   Thermodynamics of Hydride Systems

By scaling the data of Figure 11.4, (using more isotherms than shown), one can obtain an empirical relationship between the plateau pressure, p (expressed in atmospheres), and the temperature, T:

$$\ln p = 12.7 - 3360\frac{1}{T}. \tag{15}$$

This relationship can be written in the form of Boltzmann's equation:

$$p = p_0 \exp\left(-\frac{W_a}{RT}\right) \tag{16}$$

or

$$p = 328 \times 10^3 \exp\left(-\frac{28 \times 10^6}{RT}\right). \tag{17}$$

$R = 8314 \ JK^{-1}kmoles^{-1}$ is the gas constant. The enthalpy change of hydriding, ΔH, in this case is equal to $3360 \times 8314 = 28$ MJ $kmole^{-1}$, a value close to the enthalpy change of the sorption reaction determined calorimetrically at STP.

Another way of interpreting the empirical relation of Equation 15 is to observe that the change in free energy in a gas compressed isothermally is

$$\Delta G = RT \ln r \quad \text{J/kmole,} \tag{18}$$

where $r = p/p_0$ is the compression ratio. If the pressures are expressed in atmospheres and the reference pressure, p_0, is one atmosphere, then

$$\Delta G = RT \ln p. \tag{19}$$

But

$$\Delta G = \Delta H - T\Delta S = RT \ln p. \tag{20}$$

$$\ln p = -\frac{\Delta S}{R} + \frac{\Delta H}{RT}. \tag{21}$$

Comparing Equations 15 and 21,

$$\Delta H = -3360 \times 8314 = -28 \quad \text{MJ kmole}^{-1}, \tag{22}$$

$$\Delta S = -12.7 \times 8314 = -106 \quad \text{kJ K}^{-1}\text{kmole}^{-1}. \tag{23}$$

The negative sign corresponds to the case of absorption: the enthalpy change is negative because absorptions are exothermic. The entropy change is negative because hydrogen is in a more ordered state in the hydride than in the gas. If the order in the hydride were perfect, the entropy change would be -130 kJ K^{-1} kmole^{-1}, the negative of the entropy of the gas at STP. It can be seen that hydrogen is nearly perfectly ordered in the hydride. Indeed, all metallic hydrides have roughly the same ΔS for the reaction, as can be seen from Table 11.4.

Equation 21 as well as the data in Table 11.4 would only be valid if ΔH and ΔS were temperature independent, which they are not, and if the plateau slope were zero (plateau pressure independent of stoichiometry) and in absence of hysteresis (i.e., if the plateau pressure for absorption were equal to that for desorption). Such conditions do not occur in practice and, consequently, the data in the table are to be used only for a first estimation of the hydride performance.

Deviation from the ideal behavior described in the preceding paragraph are illustrated in Table 11.5.

Table 11.4
Thermodynamic Data
Selected Hydrides

Hydride	HY-STOR alloy*	ΔH_f MJ kmole^{-1} of H_2	ΔS_f kJ K^{-1}kmole^{-1} of H_2
MNi_5	204	-20.9	-96.8
$Ca_{0.2}M_{0.8}Ni_5$	203	-24.3	-108.7
$MNi_{4.15}Fe_{0.85}$	209	-25.1	-104.8
$Ca_{0.7}M_{0.3}Ni_5$	202	-26.8	-100.4
FeTi	101	-28.0	-106.1
$MNi_{4.5}Al_{0.5}$	208	-28.0	-104.8
$Fe_{0.9}Mn_{0.1}Ti$	102	-29.3	-107.0
$LaNi_5$	205	-31.0	-107.7
$CaNi_5$	201	-31.8	-101.2
$LaNi_{4.7}Al_{0.3}$	207	-33.9	-106.8
$Fe_{0.8}Ni_{0.2}Ti$	103	-41.0	-118.8
$ZrCr_2$		-46.0	-98.3
Mg_2Ni	301	-64.4	-122.3
Mg_2Cu	302	-72.8	-142.3
Mg		-77.4	-138.3

* HY-STOR is a trademark of Energics, Inc.

Table 11.5
Deviations from the Ideal
Selected Hydrides

Hydride	HY-STOR alloy*	Plateau Slope[a] $\frac{d \ln p_d}{d(H/M)}$	Hysteresis Parameter P_a/P_d
FeTi	101	0.00 [c]	1.89
$Fe_{0.9}Mn_{0.1}Ti$	102	0.65 [d]	1.85
$Fe_{0.8}Ni_{0.2}Ti$	103	0.36 [e]	1.05
$CaNi_5$	201	0.19	1.17
$Ca_{0.7}M_{0.3}Ni_5$	202	3.27	1.11
$Ca_{0.2}M_{0.8}Ni_5$	203	0.98	1.48
MNi_5	204	0.54	5.2
$LaNi_5$	205	0.09	1.21
$LaNi_{4.7}Al_{0.3}$	207	0.48	1.05
$MNi_{4.5}Al_{0.5}$	208	0.36	1.12
$MNi_{4.15}Fe_{0.85}$	209	0.43	1.18
Mg_2Ni	301	0.02 [f]	—
Mg_2Cu	302	0.17 [g]	—

* HY-STOR is a trademark of Energics, Inc.
a – 25 C plateau   c – 30 C.
d – 40 C.   e – 70 C.
f – 25 C.   g – 325 C.

Table 11.6
Hydrogen Capacity and Heat Capacity
Selected Hydrides

Hydride	HY-STOR alloy*	Hydrogen Capacity Δ(H/M)	wt.%	Heat Capacity J/(kg K)
FeTi	101	0.90 [c]	1.75	540
$Fe_{0.9}Mn_{0.1}Ti$	102	0.92 [d]	1.79	540
$Fe_{0.8}Ni_{0.2}Ti$	103	0.62 [e]	1.21	500
$CaNi_5$	201	0.71	1.39	540
$Ca_{0.7}M_{0.3}Ni_5$	202	0.96	1.60	500
$Ca_{0.2}M_{0.8}Ni_5$	203	0.74	1.08	440
MNi_5	204	1.01	1.41	420
$LaNi_5$	205	1.02	1.43	420
$LaNi_{4.7}Al_{0.3}$	207	0.95	1.36	420
$MNi_{4.5}Al_{0.5}$	208	0.83	1.20	420
$MNi_{4.15}Fe_{0.85}$	209	0.82	1.15	—
Mg_2Ni	301	1.31 [f]	3.84	750
Mg_2Cu	302	0.75 [g]	2.04	750

* HY-STOR is a trademark of Energics, Inc.

c – 30 C.

d – 40 C.   e – 70 C.

f – 25 C.   g – 325 C.

At 1 atmosphere, $p/p_0 = 1$, and $\ln p/p_0 = 0$. Consequently, from Equation 21

$$T = \Delta H / \Delta S. \tag{24}$$

But all hydrides have roughly the same ΔS of about 100 kJ K^{-1} kmole^{-1}, so that

$$T \approx 10^{-5} \Delta H \tag{25}$$

or, expressing the enthalpy change in MJ,

$$T \approx 10 \Delta H. \tag{26}$$

This formula yields the temperature at which a given hydride, in its plateau region, is in equilibrium with 1 atmosphere of hydrogen. It is an indication of the stability of the material. Yttrium and cerium hydrides are the most stable of all. They have to be heated to some 1400 K for their plateau pressure to reach 1 atmosphere.

Plateau pressures can be tailored by suitably doping certain alloys. $LaNi_4Al$ has a plateau pressure of 0.002 atmospheres at 298 K while at the same temperature, $GdNi_5$ has a plateau pressure of 150 atmospheres. It is also possible to "design" materials with fairly flat plateaus (useful in compressors and heat pumps). Mischmetall is inexpensive but results in markedly sloping plateaus.

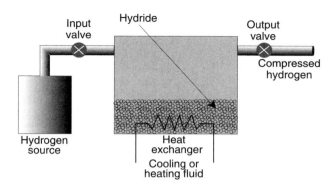

Figure 11.12   Schematic diagram of a hydrogen compressor.

11.5 Hydride Hydrogen Compressors

Metallic hydride systems lend themselves to the construction of hydrogen compressors that have no moving parts (other than some valves). Since such compressors also store a substantial amount of gas, they may be of particular interest in processes in which it is necessary to both compress and store the gas. This is, for instance, the case of electrolytic ammonia plants operating with intermittent electricity supplies or of "hydrogen gas stations."

A hydrogen compressor may consist of a hydride-containing vessel equipped with a heat exchanger through which either a heating or a cooling fluid circulates. See Figure 11.12. Two external valves permit the control of inflow and outflow of the gas. Heating up the saturated hydride causes the pressure to rise. Moderate changes in temperature result in substantial increases in pressure owing to the exponential relationship between these variables.

When the compressor delivers hydrogen, it needs additional heat to supply the desorption energy and maintain the pressure constant. Having exhausted the hydride, the system pumps cooling fluid into the heat exchanger so as to permit the recharging at a low pressure.

To understand the operation of the compressor, refer back to Figure 11.8. Consider a system at low temperature (40 C) and saturated with hydrogen ($x \approx 5$ in $LaNi_5H_x$). This corresponds to a hydrogen pressure of 0.3 MPa (3 atmospheres); refer to Point B in the figure. The cycle starts by heating the hydride to 140 C (Point C). An amount of heat, $\Delta H_{B \to C}$ must be added. The free hydrogen pressure increases to 5 MPa (50 atmospheres) and, consequently, the amount of the gas filling the space not occupied by the hydride increases. Some hydrogen desorption occurs: heating the system causes a (small) reduction in the value of x.

As hydrogen is removed, the heat of desorption of the gas withdrawn from the vessel would cause the system to cool down. An external source supplies compensating heat maintaining the temperature at a constant 140 C. The pressure will remain constant until the low end of the plateau, Point D, is reached. The total heat supplied for desorption is $\Delta H_{C \to D}$.

Upon reaching Point D, the control unit circulates cooling fluid through the exchanger. The temperature falls back to 40 C, the pressure falls to 0.3 MPa and the system reaches Point A with an x somewhat larger than at Point D because of the reabsorption of some hydrogen. To cool the system down, an amount of heat, $\Delta H_{D \to A}$, must be withdrawn.

Hydrogen is now introduced into the system at a pressure a little larger than 0.3 MPa and is absorbed at essentially constant pressure until the cycle is completed reaching Point B. The cooling fluid must remove $\Delta H_{A \to B}$ units of heat of absorption during this recharging phase.

To roughly estimate the upper limit of the efficiency of the compressor, one notices that the useful work done is the isothermal[†] compression of hydrogen from p_A to p_C:

$$W = RT_A \ \ln \frac{p_C}{p_A} \qquad \text{(per kilomole)}, \qquad (27)$$

while the energy input is mainly the heat of desorption, $\Delta H_{C \to D}$. Thus,

$$\eta = \frac{RT_A \ \ln(p_C/p_A)}{\Delta H_{C \to D}} \qquad (28)$$

Example

A LaNi$_5$ system compresses hydrogen from 0.3 to 5 MPa with a temperature change from 40 to 140 C. The work done is

$$W = 8314 \times (273 + 40) \ \ln \frac{5}{0.3} = 7.3 \qquad \text{MJ kmole}^{-1}. \qquad (29)$$

The enthalpy change for the desorption reaction is 30.2 MJ/kilomole. The resulting theoretical efficiency is

$$\eta = 7.3/30.2 = 0.24 \quad \text{or} \quad 24\%. \qquad (30)$$

(continued)

[†] Although the gas leaves the compressor at a temperature higher than that of the input, this temperature elevation is generally of no interest to the user.

(continued)

Table 11.7
Stoichiometric Indices
throughout the
Compressor Cycle

Point	Stoichiom. Index
A	1.75
B	5.64
C	4.69
D	0.80

The practical efficiency is considerably lower than the one estimated above. The factors that contribute to a reduced performance include:

1. Ignoring the heat, $\Delta H_{B \to C}$, necessary to raise the temperature of the system from 40 to 140 C. The components of $\Delta H_{B \to C}$ are

   1.1. The heat of desorption of some hydrogen. It is proportional to the heat of formation of the hydride and the fraction of the total hydrogen that was desorbed. This latter amount depends, of course, on the pressure step and on the relative magnitude of the "dead volume" inside the compressor vessel, and this, in turn, depends on the grain size and the packing of the alloy. Typically the "dead volume" amounts to some 60% of the volume of the vessel. In Figure 11.8 the stoichiometric index varies as shown in Table 11.7:

   Notice that the difference between the values of the stoichiometric index, x, at Points C and D is the same as that between Points B and A because the amount of hydrogen desorbed must equal that absorbed.

   The number of kilomoles of hydrogen pumped per cycle per kilomole of alloy is $\frac{1}{2}(x_C - x_D)$, the factor $\frac{1}{2}$ corresponding to the diatomic nature of the hydrogen molecule. This amounts to 1.95 kilomoles.

   The number of kilomoles of hydrogen desorbed when the temperature is raised is $\frac{1}{2}(x_B - x_C)$, or 0.48 kilomoles. Thus, one must add $(0.48/1.95) \times 30.2 = 7.4$ MJ of heat to pump 1 kilomole of hydrogen. This brings the efficiency down to $7.3/(30.2 + 7.4)$ or 19.4%.

(continued)

(continued)

   1.2. The heat capacity of the hydride whose temperature must also be raised from 40 to 140 C. The heat capacity of many metals and alloys is roughly 30 kJ/K per kilomole. That of the hydride is somewhat higher. For each kilomole of hydrogen pumped, about 0.5 kilomoles of alloy are need. The heat required to elevate its temperature is then $0.5 \times 30 \times (140 - 40) = 1.5$ MJ per kilomole of H_2. The efficiency is now $7.3/(30 + 7.4 + 1.5)$ or 18.8%.

   1.3. The heat capacity of the vessel itself.

2. Ignoring the heat losses of the system to the environment. The equipment must be well insulated to reduce such losses. Increasing the size of the vessel—that is, the system capacity, increases the volume-to-surface ratio and reduces the relative heat losses. Larger compressors operate with proportionally smaller heat losses.

3. Failing to take into account the need to have the temperature of the heating fluid higher than that of the hydride. There must be a temperature difference across the walls of the heat exchanger. Thus, the temperature swing of the hydride is smaller than that of the heat exchanger fluid.

4. Disregarding hysteresis. As pointed out previously, the absorption plateau pressure tends to exceed by an amount, Δp, the desorption pressure. The two 20 C isotherms in Figure 11.8 illustrate this phenomenon. The area defined by the absorption-desorption contour represents wasted energy. Since Δp is nearly independent of pressure, the hysteresis is, proportionally, more important at lower temperatures where the pressures are smaller.

Laboratory built $LaNi_5$ compressors exhibit efficiencies of some 6% showing that the contribution of the second order loss mechanisms is significant. Compared with mechanical hydrogen compressors (typical efficiency of 60%), hydride compressors operate with modest efficiencies. They use, however, low grade heat for driving energy, heat frequently available from some cooling process in another part of the plant.

$LaNi_5$ has been suggested as the working alloy owing to its large absorption capacity, its fast reaction rate, the levelness of its plateaus, and the favorable pressure ranges.

Three units of the type described above can be combined so that while one is delivering the gas, the others are at different phases of the recharging process. This insures a steady delivery of hydrogen. Typically, an individual cycle will last 10 or more minutes.

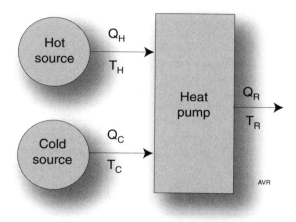

Figure 11.13   A thermally driven heat pump.

11.6 Hydride Heat Pumps

A hydride absorbs heat when it releases hydrogen. This suggests its use as a heat pump.

Consider the thermally driven heat pump shown schematically in Figure 11.13. We want to pump an amount of heat, Q_C, from a cold source at a temperature, T_C, into an environment at a higher temperature, T_R.

Since heat will not flow spontaneously from a colder reservoir to a hotter one, the pumping requires the expenditure of a certain amount of additional (heat) energy, Q_H, from a hot source. The total amount of heat dumped into the warm environment is

$$Q_R = Q_C + Q_H. \tag{31}$$

The cold source may be the outside of a house in winter, while the hot source may be from collected sunlight, burning fuel, or an electric heater. Q_R may be used to warm the house.

In the reversible case

$$\frac{Q_R}{T_R} = \frac{Q_C}{T_C} + \frac{Q_H}{T_H}, \tag{32}$$

from which the Carnot efficiency of the process can be derived:

$$\eta_{CARNOT} = \frac{Q_R}{Q_H} = \frac{1 - T_C/T_H}{1 - T_C/T_R}. \tag{33}$$

With an outside temperature of 0 C (273 K), a house temperature of 25 C (298 K), and a hot source at 100 C (373 K) the Carnot efficiency of the heat pump is

$$\eta_{CARNOT} = \frac{1 - 273/373}{1 - 273/298} = 3.2. \tag{34}$$

To avoid talking about efficiencies larger than unity, the value above is called the **coefficient of performance** (COP). A COP of 3.2 means that for each joule of input to the heat pump from the hot source, 3.2 joules of heat are delivered to the house.

A heat pump can be built by using two hydrides with matching properties. Consider two containers, one of which (Container B) is inside the house and at a temperature T_R—the room temperature. The type of alloy is such that at the system pressure of the moment, p_0, it is fully saturated with hydrogen. Outside the house, there is another container (Container A) interconnected with B so that both experience the same pressure, p_0. The material in A is chosen so that, although the temperature is lower (it is the outside temperature, T_C), the alloy is completely depleted.

As an example, let

Room temperature, T_R	300 K,
Outside temperature, T_C	270 K,
Hot source temperature, T_H	400 K,
Entropy change (absorption) (Alloy A), ΔS	−110 kJ K^{-1} per kilomole of H_2,
Entropy change (absorption) (Alloy B), ΔS	−110 kJ K^{-1} per kilomole of H_2,
Enthalpy change (absorption) (Alloy A), ΔH	−26 MJ per kilomole of H_2,
Enthalpy change (absorption) (Alloy B), ΔH	−30 MJ per kilomole of H_2.

By applying Equation 21, one calculates for this initial step the plateau pressures for Hydride A at 270 K as 5.2 atmospheres and for Hydride B at 300 K as 3.3 atmospheres.

Assume that enough hydrogen exists in the system to make its pressure $p_0 = 4$ atmospheres. This pressure is higher than the plateau pressure of hydride B. Thus, the hydride is saturated. However, 4 atmospheres is less than the plateau pressure of Hydride A which consequently must be depleted (it is in its alpha-phase).

The first step in the operation consists of moving Container A inside (its temperature rises to T_R and it continues depleted).

An amount of heat, ΔH_B, is then delivered by a heat source to Container B raising its temperature to T_H causing the desorption of the hydrogen. The pressure is high enough to force the absorption of hydrogen by the alloy in Container A which sheds into the room ΔH_A units of heat for each kilomole of gas absorbed.

In our example, when the temperature of Alloy B is raised to 400 K, its plateau pressure is raised to about 67 atmospheres, while that of Alloy

A (now at 300 K) is 16 atmospheres. Of course, the hydrogen pressure in the system does not reach the 67 atmospheres; as Alloy B is heated (taking in 30 MJ per kilomole from the hot source), it releases some gas and the system pressure starts going up. When it reaches a value slightly above 16 atmospheres, the gas is absorbed by Alloy A and the pressure remains constant unless the alloy saturates. This absorption causes the shedding of 26 MJ of heat energy into the room for each kilomole absorbed.

The second step consists of removing the heat source from Container B and moving Container A outside again. The temperature in B falls to T_R and that in A, to T_C. But these are exactly the initial temperatures that led to B being saturated and A depleted. The saturation of B causes this container to shed ΔH_B units of heat into the room while the depletion of A causes it to absorb ΔH_A units of heat from the cold outside.

In this step, Alloy B has its plateau pressure lowered back to 3.3 atmosphere while the system pressure is (initially) still 16 atmospheres. Hydrogen is absorbed and the absorption heat of 30 MJ/kilomole is returned to the room.

The cycle has been completed. The room received the heat ΔH_A from Container A during Step 1 and ΔH_B from Container B in Step 2. This latter amount of heat came from the hot source but ΔH_A came from the cold outside.

The coefficient of performance is

$$\mathrm{COP} = \frac{\Delta H_B + \Delta H_A}{\Delta H_B} = 1 + \frac{\Delta H_A}{\Delta H_B}. \tag{35}$$

The room received 26 MJ during Step 1 from Alloy A and 30 MJ during Step 2 from Alloy B, a total of 56 MJ all at a cost of 30 MJ from the hot source. The coefficient of performance is $56/30 = 1.86$.

Clearly, there is no necessity of moving containers physically from the outside to the inside of the house. Equivalent effect can be achieved by circulating inside and outside air at the appropriate moments during the cycle.

It is obvious that a similar system can be used as a refrigerator.

References

Although there are numerous works published on hydrides and their applications, there is one, fairly ancient paper, that is widely regarded as the "Bible" of this topic. It is van Mal's Ph.D. dissertation (notice the latinized form of the Dutch name, a charming medieval practice still observed in Europe[†]):

Van Mal, Harmannus Hinderikus, Stability of ternary hydrides and some applications, *Ph.D. dissertation*, Technische Hogeschool Delft (Netherlands), May **1976**.

Williams, Robert H., The clean machine, *Technology Review*, April, **1994.**

Amankwah, K.A.G. et al. Hydrogen storage on superactivated carbon at refrigeration temperatures, *International Journal of Hydrogen Energy*, Vol. 14 No. 7 pages 437–447, **1989**.

Chambers, A, C. Park, R.T.K. Baker, *J. Phys. Chem. B* **102**, 4253 (**1998**).

Chen, P., X. Wu, J. Lin, K.L. Tan, *Science* **285** 91, July 2 **1999**.

Houston, E. L., and G. D. Sandrock, Engineering properties of metal hydrides, *J. Less-Common Metals*, **74** p 435–443, **1980**.

[†] Remember that the famous Mercator projection so frequently used in map making is due to Gerhard Kremer, the Flemish mathematician. Compare Kremer (Dutch), Krämer (German = shopkeeper), and Mercator (Latin).

PROBLEMS

11.1   An alloy, Mg_2Ni, interacts with hydrogen so that the following reversible reaction takes place:

$$Mg_2Ni + 2H_2 \leftrightarrow Mg_2NiH_4.$$

The thermodynamic data for the absorption reaction are (per kilomole of H_2):
$\Delta H f^0 = -64.4$ MJ,
$\Delta S f^0 = -122.3$ kJ/K.
The relevant atomic masses (in daltons) are

H	1
Mg	24.3
Ni	58.7

In the questions below, all pressures are in the plateau region.

1. What is the dissociation pressure at 300 K?
2. At what temperature is the equilibrium pressure 1 MPa?
3. What are the proportions, by mass, of the elements in Mg_2Ni?
4. A vessel with an internal volume of 1000 cm^3 contains 1.56 kg of Mg_2NiH_4. The density of this material is 2600 kg/m^3. The saturated hydride is placed in the evacuated vessel. The temperature is raised to 400 C, but no hydrogen is allowed to escape. How many kg of hydrogen are desorbed? How many remain in the hydride?
5. How much heat must be added to desorb the rest of the hydrogen? Assume that desorption stops when the empirical formula of the material reaches the $Mg_2NiH_{0.4}$ composition. The hydrogen desorbed is removed from the vessel so that the pressure remains constant.
6. If a hydrogen compressor were built using the vessel above, and if the hydrogen were fed into the system at 10^5 Pa and 85 C, what would the theoretical efficiency of the compressor be? The output gas is at 85 C and 5 MPa. Neglect all losses.

11.2   Figure 11.8 (Text) shows the pressure-composition isotherms for the reaction of hydrogen with $LaNi_5$. Assume that, in the plateau region, the isotherms are horizontal—that is, that the pressure does not depend on the composition. Assume further that the pressure at 40 C is 0.3 MPa and at 140 C is 5 MPa.

One kg of $LaNi_5H_5$ is placed inside a pressure vessel and heated to 140 C. Hydrogen is slowly withdrawn until the hydride is left with the empirical formula $LaNi_5H_2$.

Molecular masses: La, 138.9; Ni, 58.7.

The ΔH for absorption for this alloy is -32.6 MJ per kilomole of H_2, and the ΔS is -107.7 kilojoules per kelvin per kilomole.

1. To maintain the temperature at 140 C during the desorption, must heat be supplied to or withdrawn from the hydride?
2. How many joules of heat must be exchanged with the hydride to perform the desorption?
3. What is the change in entropy of the system during desorption? Assume that ΔS is temperature independent. Does the entropy increase or decrease during the desorption?
4. How many kilograms of hydrogen were withdrawn?

11.3 Hydrogen is to be transported. Two solutions must be considered:

1. Liquefy it.
2. Convert it to ammonia. Later, when hydrogen is to be used, the ammonia can be catalytically cracked with 85% recovery of hydrogen. 46.2 MJ per kg of ammonia are required to dissociate the gas.

Purely from the energy point of view, which is the more economical solution?

11.4 Calorimeter measurements show that when a compound, AB, reacts with H_2 forming a hydride ABH, 18.7 MJ of heat are released for each kilomole of H_2 absorbed. The manufacturer of this hydride consults you about the advisability of shipping it (saturated with hydrogen) inside a normal gas pressure cylinder (which is rated at 200 atmospheres). During shipment the cylinder may be exposed to the sun. Without having any additional data on the hydride, what would you advise the manufacturer? Could you tell him the maximum temperature that the hydride can safely reach?

11.5
1. Estimate the enthalpy change for the hydrogen absorption reaction of an alloy that has a plateau pressure of 1 atmosphere when $T = 0$ C.
2. Estimate the enthalpy change for the hydrogen absorption reaction of an alloy that has a plateau pressure of 1 atmosphere when $T = 30$ C. Assume that Alloy A is in an environment that causes its temperature to be 0 C (perhaps the outside of a house) and that Alloy B is at 28 C (say, inside a room). The alloy containers are interconnected by means of a pipe so that the hydrogen pressure is the same in both. The amount of hydrogen in the system is such that, in any phase of the cycle, when one alloy is saturated, the other is depleted.
3. What is the system pressure?

11.34

4. Which alloy is depleted and which contains most of the hydrogen?
5. What is the plateau pressure? What is the actual hydrogen pressure? Circulate hot water (from a hot source) through the heat exchanger of Alloy B and heat it up to 100 C (373 K). Maintain this temperature.
6. What is the pressure of the system and what happens to the two alloys? Return the temperature of Alloy B to 28 C, and that of Alloy A to 0 C (by putting it in contact with the exterior).
7. What happens?
8. Tabulate all the sorption heat inputs and outputs of the system and the corresponding temperatures. Ignore all other heat inputs and outputs. For example, ignore the heat necessary to change the temperature of the system.
9. How much heat is delivered (per kmole of hydrogen) to the environment at 28 C in a complete cycle as described above?
10. Of this amount of heat, how much must come from the hot water source (and must be paid for)?
11. What is the coefficient of performance of this heat pump—that is, what is the ratio of the heat delivered to the environment at 28 C to the heat required from the hot water?

11.6   A recently developed binary compound is to be used as a hydrogen storage medium because it forms a reversible (ternary) monohydride.
    The data of the system include:

Molecular mass of the compound (no hydrogen): 88 daltons.
Density of the compound (hydrided or depleted): 8900 kg m^{-3}.
Enthalpy of hydriding: -26.9 MJ per kilomole of H_2.
Change of entropy owing to absorption of hydrogen: -100 kJ K^{-1} per kilomole of H_2.
Type of isotherm: single plateau.
Heat capacity of the compound: 200 J K^{-1} kg^{-1}.
Heat capacity of the container and of the hydrogen gas: negligible.
Heat capacity of water: 4180 J K^{-1} kg^{-1}.
Density of hydrogen: 0.089 kg m^{-3}.

2.5 kg of the compound (activated and in a fine powder at a temperature of 0 C) are placed inside a container measuring internally 10 by 10 by 10 cm.
    We want to adjust the system so that the alloy is saturated and the gas pressure is exactly the plateau pressure. To find this point we will have to observe the behavior of the pressure in the alloy canister as hydrogen is added. The pressure initially rises, then as the plateau is reached, it stabilizes until saturation is achieved. We will first overshoot this point, then, having removed the hydrogen source, we will purge some of the gas,

observing the pressure and stopping the purge just when the pressure falls to the previously observed plateau pressure. During this whole operation the temperature is carefully maintained at 0 C.

A 5-atmosphere hydrogen source is used to fill the container. The gas is applied for a time long enough to allow equilibrium to be established. Any heat absorbed or released by this operation is removed or added to the system so that at the end of the operation, the pressure is still 5 atmospheres and the temperature is still 0 C. Next the hydrogen source is removed. The pressure inside the container remains at 5 atmospheres. A valve is cracked open and hydrogen leaks out while the pressure is monitored. As soon as the pressure stabilizes, the valve is closed.

1. How many grams of hydrogen were lost in the bleeding above?
2. How much hydrogen remains in storage? Express the loss as a percentage of the stored gas.
3. What is the pressure of the stored gas?

The container is now immersed in a tank of water at 40 C. This tank contains 0.3 liters of water and is entirely adiabatic. Its walls have negligible heat storage capacity.

4. What is the temperature of the system (water tank plus alloy container) after equilibrium is reached?

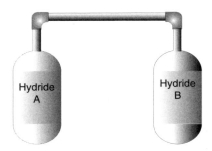

11.7   An inventor proposes the following device to cool drinks at a picnic. It consists of two sturdy containers (something like small portable oxygen bottles) one of which (Container A) can be placed inside a styrofoam box in which there are 12 beer cans. The other (Container B) is outside the box and can be placed over a fire. A pipe interconnects the two containers. See the figure. The containers are filled with different alloys capable of absorbing hydrogen. Alloy A is TiFe. Alloy B has to be described.

When both containers are at the same temperature, 298 K, Alloy A is depleted and B is saturated.

11.36

1. For this to happen, what are the required characteristics of Alloy B? Establish a relationship between the thermodynamic parameters of the two alloys.

2. Which of the alloys from the table below can be used as Alloy B in the device under consideration?

Hydride	ΔH MJ/kmole	ΔS kJ/K per kmole
AB	−21.0	−96.5
CD	−26.1	−99.4
EF	−27.9	−106.8
GH	−32.1	−101.8
IJ	−32.6	−110.5
KL	−33.4	−98.3

To operate the system, Container A is placed in a bucket of water and kept at 25 C. This requires refreshing the water occasionally. Container B is placed over the picnic fire so that hydrogen is transferred to Container A whose hydride becomes saturated.

Next, Container A is placed inside the styrofoam box, in contact with the 355-ml beer cans which, for good thermal contact, are immersed in 4.5 liters of water.

Container B is now cooled to 298 K, returning the system to its original state.

The heat required to desorb the hydrogen from A will cool the 12 cans of beer from 25 C to 10 C.

Assume that beer behaves as water, at least as far its thermal capacity is concerned. The styrofoam box is, essentially, adiabatic. Assume that during the cycle the composition of the hydride in A varies from $TiFeH_{0.95}$ to $TiFeH_{0.4}$

The atomic mass of Ti is 47.9 daltons and that of Fe is 55.8 daltons.

3. Estimate the minimum mass of TiFe required.

11.8   Two canisters are interconnected by a pipe. Canister "A" contains TiFe and is at 300 K, while Canister "B" contains $CaNi_5$ and is at 350 K. The system is filled with hydrogen at a pressure of 4 atmospheres.

In which canister is the bulk of the hydrogen? No guesses, please! Use the thermodynamic data of Table 11.4 of the Text.

11.9   *WARNING: This problem contains units, such as grams, centimeters, and so on, that are not of the SI.*

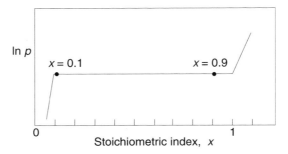

Consider a vessel containing a hydrogen-absorbing alloy, AB.

Vessel:

Volume: 200 cm^3,

Thermal insulation: adiabatic,

Heat capacity: negligible.

Alloy:

Formula mass: 120 daltons,

Amount: 200 g,

Heat capacity: 1700 J/K per kg of alloy (same for hydrided alloy).

Hydride:

Heat of formation of ABH (absorption): -30.0 MJ per kmole of H$_2$,

Entropy change owing to absorption: -110 kJ/K per kmole of H$_2$,

Density (of ABH$_{0.9}$): 1600 kg/m^3.

System:

Initial temperature: 300 K,

Initial hydrogen pressure: equilibrium.

Hydrogen is forced into the vessel until the average alloy composition is ABH$_{0.9}$ What is the minimum pressure required to force all the needed hydrogen into the vessel? How much hydrogen is forced in?

11.10   A perfectly adiabatic (heat insulated) vessel has an internal volume of 100 cm^3 and contains 240 g of an alloy powder, AB, that forms a monohydride, ABH. The thermodynamic data for absorption are:

$\Delta H = -28$ MJ per kilomole of H$_2$,

$\Delta S = -100$ kJ/K per kilomole of H$_2$,

Heat capacity, $c_v = 400$ J/K per kg of alloy.

Additional data include

Formula mass of the alloy $= 150$ daltons,

Density of the alloy $= 8000$ kg/m^3,

Bulk density of the alloy powder $= 4000$ kg/m^3.

The vessel has been charged with hydrogen so that the pressure is 10 atmospheres. The system is at 30 C.

11.38

1. Hydrogen is withdrawn. How many milligrams of the gas can be removed without causing a change in the temperature of the hydride? Remember that the container is adiabatic. The heat owing to the work that the withdrawn hydrogen may exert is exchanged with the gas outside the hydride container and does not influence the temperature of the latter.
2. If more hydrogen is released, it will cause the cooling of the hydride. Assume that the vessel has no heat capacity. How much hydrogen is released if the pressure falls to 1 atmosphere?
3. What is the value of x in the empirical formula ABH_x after the above desorption?

11.11   The Pons and Fleishman cold fusion experiment employs an electrolytic cell consisting of a palladium negative electrode and a platinum positive electrode. The electrolyte is a concentrated solution of LiOH in D_2O. The palladium electrode is a cylindrical rod 10 cm long and 1.2 cm in diameter. Just prior to the experiment, the rod is completely degassed by heating it up in a vacuum.

When a 0.5 A current is forced through the cell, nothing unusual happens for a long time. To be sure, normal electrolysis occurs with D_2 evolving at the palladium and 0_2 at the oxygen electrodes. The D_2O used up is continually replenished.

In some rare instances, it is claimed, after the electrolysis has proceeded for a long time, heat suddenly begins to be produced in substantial amounts—73 W, in this case. This heat production rate is sustained for 120 hours after which the cell is disconnected.

If you look up the enthalpies of formation of all palladium compounds, you will find that the largest value is associated with the formation of palladium hydroxide: 706 MJ/kmole.

You will also find that the atomic mass of palladium is 106 daltons and that the density of the metal is 12 g cm^{-3}.

Can you prove that the energy generated is not chemical in nature?

To explain the delay, assume that D_2-D_2 fusion occurs at a rapid rate only if the deuterium is packed with sufficient density and that this will happen only when the palladium is completely saturated with deuterium and the formation of palladium di-deuteride begins. How long would you expect the cell to operate before it heats up?

11.12   A canister contains a mixture of two alloys (Alloy 1 and Alloy 2). A hydrogen source equipped with a valve is connected to this canister. A measured amount of the gas can be delivered to it.

Describe the behavior of p_{system} (the hydrogen pressure, in pascals, read by a manometer connected to the canister) versus the amount, μ_{H_2} of

H_2 (in moles) introduced into the system. Sketch a rough p_{system} vs μ_{H_2} graph.

Estimate all the break points in the above sketch—that is, all the values of μ_{H_2} at which the curve changes abruptly its character.

Do the above for $T = 400$ K, following the detailed instructions in Items 1 through 5.

Here are some data:

Internal volume of the canister: 1 liter.

The idealized p vs x characteristics of the alloys are as follows:

Region 1 – The equilibrium pressure is proportional to the stoichiometric coefficient, x, for $0 < p < p_{plateau}$. When $x = x_{crit_\ell}$, p reaches $p_{plateau}$.

Region 2 – The plateau pressure is perfectly constant until x reaches a critical value, x_{crit_u}.

Region 3 – For $x > x_{crit_u}$, the pressure rises linearly with x with the same slope as that of Region 1.

Alloy 1: 0.8 kg of alloy AB having a density of 2750 kg/m^3, $\Delta S = -110$ kJ K^{-1}kmole^{-1}, $\Delta H = -35$ MJ kmole^{-1}.

At 400 K, $x_{crit_{\ell 1}}$ (the minimum value of x in the plateau region) is 0.3, and $x_{crit_{u1}}$ (the maximum value of x in the plateau region) is 3.55.

Alloy 2: 1.0 kg of alloy CD having a density of 2750 kg/m^3, $\Delta S = -90$ kJ K^{-1}kmole^{-1}, $\Delta H = -28$ MJ kmole^{-1}.

At 400 K, $x_{crit_{\ell 2}}$ (the minimum value of x in the plateau region) is 0.3, and $x_{crit_{u2}}$ (the maximum value of x in the plateau region) is 4.85.

The formula masses are

A – 48 daltons,

B – 59 daltons,

C – 139 daltons,

D – 300 daltons.

The density of the above alloys is (unrealistically) the same whether hydrided or not.

Define "gas-space" as the space inside the canister not occupied by the alloys.

1. Calculate the volume, $V_{gas-space}$ of the gas space.
2. What is the number of kilomoles of each alloy contained in the canister?
3. Tabulate the plateau pressures of the two alloys for 300 and 400 K.
4. Describe, in words, the manner in which the system pressure, p_{system}, varies as hydrogen is gradually introduced in the system. Use μ_{H_2} as the measure of the number of moles of H_2 introduced. Sketch a p_{system} vs μ_{H_2} graph.
5. Calculate the values of μ_{H_2} that mark the break-points (points of abrupt change in the p_{system} behavior—that is, points where the pressure changes from growing to steady, or vice-versa).

11.13   A container able to withstand high pressures has an internal capacity of 0.1 m^3. It contains 490 kg of an alloy, AB, used to store hydrogen.

The properties of this alloy are:

Atomic mass of A	60	daltons
Atomic mass of B	70	daltons
Density of AB	8900	kg/m^3
Heat capacity	1	kJ K^{-1}kg^{-1}
ΔH	-25	MJ /kmole of H$_2$
ΔS	-105	kJ K^{-1} /kmole of H$_2$
Depletion end of plateau		x = 0.01
Saturation end of plateau		x = 1

x is the stoichiometric coefficient of hydrogen in ABH$_x$. The values correspond to 300 K. The plateau pressure is essentially independent of x in the above interval. Above $x = 1$, the equilibrium pressure rises very rapidly with x so that x does not appreciably depend on the hydrogen pressure.

1. What volume inside the container can be occupied by gas?

2. How many kilomoles of alloy are inside the container?

3. What is the plateau pressure of the hydrogen in the alloy at 300 k?

4. Introduce 10 g of H$_2$. Give an upper bound for the pressure of the gas in the container.

5. Now introduce additional hydrogen so that the total amount introduced (Steps 4 and 5) is 100 g. The temperature is kept at 300 K. What is the pressure of the gas?

6. What is the mean stoichiometric index, x, of the hydride, ABH$_x$?

7. Finally, introduce sufficient hydrogen so that the total amounts to 4 kg. The temperature remains at 300 K. What is the pressure of the gas?

8. Assume that the container has negligible heat capacity. Withdraw adiabatically (i.e., without adding any heat to the system) 3 kg of hydrogen. What will be the pressure of the gas inside the container after the 1 kg of hydrogen has been removed?

11.14   Two hydrogen storing alloys have the following properties:

Hydride	ΔH MJ kmole^{-1}	ΔS kJ K^{-1} kmole^{-1}
A	−28	−100
B	−20	?

What must the value of ΔS_B be to make the plateau pressure of the two hydrides be the same when $T = 400$ K?

ΔS_B is, of course, the ΔS of Alloy B.

How do you rate your chances of finding an alloy with the properties of Alloy B? Explain.

11.15 TiFe is sold by Energics, Inc under the label HY-STOR 101. Pertinent data are found in Table 11.4 through 11.6 in the Text. Treat this alloy as "ideal" (no hysteresis).

The atomic mass of iron is 55.8 daltons and that of titanium is 47.9 daltons. The saturated alloy (TiFeH$_{0.95}$) is at 350 K and is in a perfectly adiabatic container with negligible heat capacity.

A valve is opened and hydrogen is allowed to leak out until the pressure reaches 2 atmospheres. What is the composition of the hydride at the end of the experiment, i.e., what is the value of x in TiFeH$_x$?

To simplify the problem, assume that there is no "gas space" in the container (patently impossible). Also, neglect the heat capacity of the hydrogen gas and any Joule-Thomson heating owing to the escaping gas.

11.16 A canister contains an alloy, AB, that forms a monohydride, ABH. The canister is perfectly heat-insulated—that is, adiabatic and contains 0.01 kmoles of the alloy and a free space of 600 ml which is, initially, totally empty (a vacuum).

The molecular mass of the alloy is 100 daltons and its thermodynamic characteristics for absorption are $\Delta H = -25$ MJ and $\Delta S = -100$ kJ/K all per kilomole of H_2. For simplicity, make the unrealistic assumption that the plateau extends from $x = 1$ all the way to $x = 0$. x is the stoichiometric coefficient in ABH$_x$. Assume also that the plateau is perfectly horizontal and that there is no hysteresis. The heat capacity of the alloy is 500 J kg^{-1}K^{-1}. Again, for simplicity, assume that neither the canister itself nor the hydrogen gas has significant heat capacity. Finally, assume that the volume of the alloy is independent of x.

Introduce 0.002763 kmoles of H_2 into the canister. Both canister and hydrogen are at 300 K. What will the gas pressure be inside the canister?

11.17 HELIOS is an electric airplane developed by AeroVironment to serve as a radio relay platform. It is supposed to climb to fairly high altitudes (some 30 km) and orbit for a prolonged time (months) over a given population center fulfilling the role usually performed by satellites.

Although its geographic coverage is much smaller, HELIOS promises to be substantially more economical.

The plane is propelled by 14 electric motors of 1.5 kW each. Power is derived from photovoltaic cells that cover much of the wing surface. In order to stay aloft for many days, the plane must store energy obtained during the day to provide power for nighttime operation. The solution to this problem is to use a water electrolyzer that converts the excess energy, provided during daytime hours by the photovoltaics, into hydrogen and oxygen. The gases are then stored and, during darkness feed a fuel cell that provides the power required by the airplane.

Although the specifications of the HELIOS are not known, let us take a stab into providing the outline of a possible energy storage system. To that end, we will have to make a number of assumptions that may depart substantially from the real solution being created by AeroVironment.

1. Calculate the amount of hydrogen and of oxygen that must be stored. Assume that during take off and climbing to cruise altitude, the full 1.5 kW per motor is required, but, for orbiting at altitude only half the above power is required. Assume also that the power needed for the operation of the plane (other than propulsion, but including the energy for the radio equipment) is 3 kW. Assume also that the longest period of darkness last 12 hours. The fuel cell has an efficiency of 80%.

2. The amount of fuel calculated in Item 1 must be stored. Assuming STP conditions, what is the volume required?

3. Clearly, the volumes calculated in Item 2 are too large to fit into HELIOS. The fuel cell busses being operated experimentally in Chicago have hydrogen tanks that operate at 500 atmospheres and that allow a gravimetric concentration of 6.7% when storing hydrogen. If such tanks were adopted for the HELIOS, what would be the mass of the fully charged fuel storage system?

4. Assume that the efficiency of a mechanical hydrogen compressor is 60% and that of an oxygen compressor is 80%. How much energy do these compressors require to compress the gases isothermally to their 500 atmosphere operating temperature. Assume for simplicity that the electrolyzer that produces these gases is pressurized to 5 atmospheres.

5. While orbiting, during daylight, what is the total energy that the photovoltaic collectors have to deliver? The electrolyzer is 80% efficient.

11.18   Two 100-liter canisters are interconnected by a pipe (with negligible internal volume). Canister "A" contains 37.2 kg of FeTi and Canister "B," 37.8 kg of $Fe_{0.8}Ni_{0.2}Ti$.

Although the gas can freely move from one canister to the other, there is negligible heat transfer between them. Thus the gas can be at different temperatures in the two canisters. The gas always assumes the temperature of the alloy it is in contact with.

The pertinent data are summarized in the two boxes below:

Element	Atomic Mass (daltons)	Density (kg/m^{-3})
Ti	47.90	4540
Fe	55.85	7870
Ni	58.71	8900

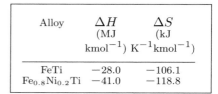

Alloy	ΔH (MJ kmol^{-1})	ΔS (kJ K^{-1}kmol^{-1})
FeTi	-28.0	-106.1
Fe$_{0.8}$Ni$_{0.2}$Ti	-41.0	-118.8

To simplify the solution, assume that the $\ln p$ vs x characteristics of the alloys consist of a perfectly horizontal plateau followed by a vertical line in the saturated region. In other words, the characteristics look as sketched in the figure. The alloys are depleted when $x = 0$ and are saturated when $x = 1$. Also, neglect all the hydrogen dissolved in the saturated alloy.

Initially, Canister A is at 300 K and Canister B, at 400 K. They have been carefully evacuated (the gas pressure is zero).

Assume that the density of the alloys is the average of that of the component elements.

Enough hydrogen is introduced into the system so that one of the alloys becomes saturated. This will cause the temperature of the alloys to change.

To simplify things, assume that the alloys and the canisters themselves have negligible heat capacity.

1. Does the temperature increase or decrease?
2. The temperature is now adjusted to the values of 300 K and 400 K, as before. How many kg of hydrogen had to be introduced into the system to make the gas pressure, at this stage, 10% higher than the plateau pressure of the saturated alloy while leaving the other alloy depleted? Please be accurate to the gram.
3. Now raise the temperature of Alloy A to 400 K. Describe what happens:

3.1 what is the final gas pressure,

3.2 what is the stoichiometric value of H in each alloy,

3.3 how many joules of heat had to be added,

3.4 how many joules of heat had to be removed.

11.19   A canister with 1 m^3 capacity contains 3000 kg LaNi$_5$. Although initially the system was evacuated, an amount, μ_1, of hydrogen is introduced so that the pressure (when the canister and the alloy are at 298 K) is exactly 2 atmospheres. Assume that once the alloy is saturated, it cannot dissolve any more hydrogen, i.e., assume that the p-x characteristic just after the beta-phase is vertical.

The density of lanthanum (atomic mass 138.90 daltons) is 6145 kg/m^3 and that of nickel (atomic mass 58.71 daltons) is 8902 kg/m^3. Assume that the density of the alloy is equal to that of a mixture of 1 kilomole of lanthanum with 5 kilomoles of nickel.

1. What is the value of μ_1?
2. 100 MJ of heat are introduced into the canister whose walls are adiabatic and have no heat capacity. Ignore also the heat capacity of the free hydrogen gas.

Calculate the pressure of the free hydrogen gas.

11.20   Please use $R = 8314$.

Here is the setup:

A hydrogen source is connected to a canister containing an alloy, A, that can be fully hydrided to AH.

The hydrogen source is a high pressure container with an internal capacity of $V_s = 1$ liter. It is charged with enough hydrogen to have the gas at $p_0 = 500$ atmospheres when the temperature is $T_0 = 300$ K. The container is in thermal contact with the hydride canister. In steady state, the container, the hydrogen, and the canister are all at the same temperature. The heat capacity of the container is 300 J/K.

The whole system— container and canister—is completely adiabatic: no heat is exchanged with the environment.

There is a pipe connecting the hydrogen source to the hydride canister. A valve controls the hydrogen flow. According to the Joule-Thomson law, the gas escaping from the source and flowing into the canister will warm up. However, in this problem, assume that there is no Joule-Thomson effect.

Initially, while the hydrogen delivery valve is still shut off, there is no gas pressure inside the canister and canister and contents are at 300 K.

Canister:

Volume, $V = 1$ liter.

Heat capacity, $c_{can} = 700$ J/K.

The alloy has the following characteristics:

Amount of hydride, $m = 5.4$ kg.
Density of hydride, $\delta = 9000$ kg/m^3.
Molecular mass of alloy, A, 100 daltons.
Heat of absorption: -28 MJ/kmoles.
Entropy change of absorption: -110 kJ K^{-1}kmole^{-1}.
Heat capacity, $c_{hyd} = 500$ J kg^{-1} K^{-1}.
The $\ln p$ versus x characteristics of the alloy are a perfectly horizontal plateau bound by vertical lines corresponding to the depleted and the saturated regions.

The valve is opened and hydrogen flows into the canister. If you wait long enough for the transients to settle down, what is the pressure of the gas in the hydrogen source? Is the hydride in the plateau region?

11.46

Part III

Energy from the Sun

Chapter 12
Solar Radiation

12.1 The Nature of the Solar Radiation

The sun radiates in all regions of the spectrum, from radio waves to gamma rays. Our eyes are sensitive to less than one octave of this, from 400 to 750 THz (750 to 400 nm), a region known, for obvious reasons, as **visible**. Although narrow, it contains about 45% of all radiated energy. At the distance of one astronomical unit, the power density of the solar radiation is about 1360 W m^{-2}, a value called **solar constant**, which is not really constant—it varies a little throughout the year, being largest in January when the Earth is nearest the sun.

The expression **power density** is used to indicate the number of watts per square meter. This is also known as **energy flux**. We will use the expression **spectral power density** to indicate the power density per unit frequency interval or per unit wavelength interval.

Roughly, the distribution of energy over different spectral regions is

Infrared and below($f < 400$ THz, $\lambda > 750$ nm)	46.3%
Visible (400 THz $< f <$ 750 THz, 400 nm $< \lambda <$ 750 nm)	44.6%
Ultraviolet and above (f > 750 THz, $\lambda < 400$ nm)	9.1%

A much more detailed description of the solar radiation is given in Table 12.1 which shows the fraction, G, of the solar constant associated with frequencies larger than a given value, f. These data, plotted in Figure 10.13 (see the chapter on Production of Hydrogen), correspond to the spectral power density distribution shown in Figure 12.1. For comparison, the spectral power density distribution of black body radiation (6000 K) is plotted for constant wavelength intervals and constant frequency intervals. Notice that these two distributions, although describing the very same radiation, peak at different points of the spectrum. To understand the reason for this apparent paradox, do Problem 12.5.

All of the above refers to radiation outside Earth's atmosphere. The power density of solar radiation on the ground is smaller than that in space owing to atmospheric absorption. Radiation of frequencies above 1000 THz ($\lambda < 300$ nm) is absorbed by the upper atmosphere causing photochemical reactions, producing photoionization, and generally heating up the air. However, this part of the spectrum contains only 1.3% of the solar constant. The ozone layer near 25 km altitude absorbs much of it. Ozone is amazingly opaque to ultraviolet. If the atmosphere were a layer of gas of uniform sea level density, it would be 8 km thick. The ozone layer would then measure 2 mm.

Table 12.1
Cumulative Values of Solar Power Density
Fraction of Total Power.
Data from F. S. Johnson.

f (THz)	G	f (THz)	G	f (THz)	G	f (THz)	G
43	0.9986	176	0.9083	536	0.3180	779	0.0778
50	0.9974	188	0.8940	541	0.3120	789	0.0735
60	0.9951	200	0.8760	545	0.3050	800	0.0690
61	0.9948	214	0.8550	550	0.2980	811	0.0642
63	0.9945	231	0.8290	556	0.2900	822	0.0595
64	0.9941	250	0.7960	561	0.2830	833	0.0553
65	0.9938	273	0.7570	566	0.2760	845	0.0510
67	0.9933	300	0.7090	571	0.2690	857	0.0469
68	0.9929	316	0.6810	577	0.2630	870	0.0427
70	0.9923	333	0.6510	583	0.2560	882	0.0386
71	0.9918	353	0.6170	588	0.2490	896	0.0346
73	0.9913	375	0.5790	594	0.2420	909	0.0308
75	0.9905	400	0.5370	600	0.2350	923	0.0266
77	0.9899	405	0.5270	606	0.2280	938	0.0232
79	0.9891	411	0.5180	612	0.2200	952	0.0233
81	0.9883	417	0.5080	619	0.2130	968	0.0166
83	0.9874	423	0.4980	625	0.2060	984	0.0150
86	0.9863	429	0.4880	632	0.1980	1000	0.0130
88	0.9852	435	0.4780	638	0.1900	1017	0.0106
91	0.9839	441	0.4670	645	0.1820	1034	0.0085
94	0.9824	448	0.4560	652	0.1750	1053	0.0070
97	0.9808	455	0.4450	659	0.1670	1071	0.0059
100	0.9790	462	0.4330	667	0.1590	1091	0.0051
103	0.9772	469	0.4210	674	0.1510	1111	0.0042
107	0.9747	476	0.4090	682	0.1440	1132	0.0035
111	0.9721	484	0.3970	690	0.1370	1154	0.0029
115	0.9690	492	0.3840	698	0.1300	1176	0.0025
120	0.9657	500	0.3720	706	0.1240	1200	0.0021
125	0.9618	504	0.3650	714	0.1170	1224	0.0018
130	0.9571	508	0.3590	723	0.1100	1250	0.0016
136	0.9520	513	0.3520	732	0.1030	1277	0.0014
143	0.9458	517	0.3450	741	0.0970	1304	0.0011
150	0.9387	522	0.3390	750	0.0908	1333	0.0008
158	0.9302	526	0.3320	759	0.0860	1364	0.0006
167	0.9203	531	0.3250	769	0.0819		

12.2

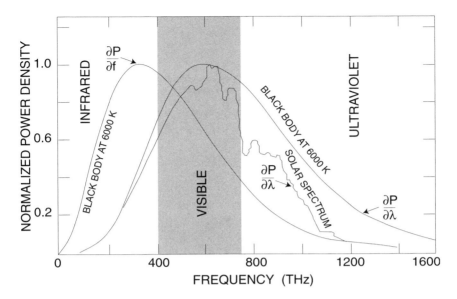

Figure 12.1   The solar power density spectrum compared with
that of a black body.

Although solar radiation is generated by several different mechanisms, the bulk of it is of the black body type. The energy per unit volume per unit frequency interval inside a hollow isothermal black body is given by Plank's law:

$$\frac{dW}{df} = \frac{8\pi h}{c^3} \frac{f^3}{\exp(hf/kT) - 1} \qquad \text{J m}^{-3}\text{Hz}^{-1}. \qquad (1)$$

In the preceding expression, W is the energy concentration. The energy flux is equal to the energy concentration times the speed of light (just as a particle flux is equal to the particle concentration times the speed of the particle). Energy flux is, as we stated, the same as power density, P:

$$\frac{dP}{df} \propto \frac{f^3}{\exp(hf/kT) - 1} \qquad \text{W m}^{-2}\text{Hz}^{-1}. \qquad (2)$$

In terms of wavelength,

$$\frac{dP}{d\lambda} \propto \frac{\lambda^{-5}}{\exp(hc/kT\lambda) - 1} \qquad \text{W m}^{-2} \text{ per m of wavelength interval.} \qquad (3)$$

When one tries to match a black body spectrum to that of the sun, one has the choice of picking the black body temperature that best fits the *shape* of the solar spectrum (6000 K) or the temperature that, at one astronomical unit, would produce a power density of 1360 W m^{-2} (5800 K).

12.2 Insolation

12.2.1 Generalities

Insolation[†] is the power density of the solar radiation. In Section 12.1, we saw that the insolation on a surface that faces the sun and is just outside Earth's atmosphere is called the solar constant. It has a value of 1360 W m^{-2}.

It is convenient to define a **surface solar constant**—that is, a value of insolation on a surface that, at sea level, faces the vertical sun on a clear day. This "constant" has the convenient value of about 1000 W m^{-2} or "one sun." At other than vertical, owing to the larger air mass through which the rays have to pass, the insolation is correspondingly smaller.

American meteorologists depart from the SI and define a new—and unnecessary—unit called a langley. It is one gram calorie per cm^2 per day. To convert langleys to W m^{-2}, multiply the former by 0.4843.

The insolation depends on:

1. the orientation of the surface relative to the sun, and
2. the transparency of the atmosphere.

> Caveat
>
> In the discussion below, we will make a number of simplifications that introduce substantial errors in the results but still describe in general terms the way insolation varies throughout the year and during the day. Some of these errors can, as indicated further on, be easily corrected and those that remain are of little consequence for the planners of solar energy collection systems.
>
> The major sources of errors are:
>
> 1. We assume that the length of the period between successive sunrises is constant throughout the year. That is not so—see "Equation of Time" in Appendix B to this chapter.
> 2. The time used in our formulas is the "Mean Local Time" and differs from the civil time which refers to the time measured at the center of each time zone.
>
> (continued)

[†] One should not confuse *insolation*, from the Latin *sol* = *sun*, with the word of essentially the same pronunciation, *insulation*, from the Latin *insula* = *island*.

(continued)

Corrections for effects a and b can easily be made by introducing the "Time Offset." (See Appendix A.)

3. Our formulas consider only the geometry of the situation. The presence of the atmosphere causes diffraction of the light so that the sun is visible even when it is somewhat below the geometric horizon. This causes the apparent sunrise to be earlier than the geometric one and the apparent sunset to be later. This effect can be somewhat corrected by using a solar zenith angle at sunrise and sunset of 90.833° instead of the geometric 90°.

However, it must be remembered that the refraction correction is latitude dependent and becomes much larger near the polar regions. Much more detailed information on the position of the sun can be found at www.srrb.noaa.gov/highlights/sunrise .

4. The insolation data assume perfectly transparent atmosphere. Meteorological conditions do, of course, alter in a major way the amount of useful sun light.

In this part of this book, the position of the sun is characterized by the **zenith angle**, χ (the angle between the local vertical and the line from the observer to the sun), and by the azimuth, ξ, measured clockwise from the north. This is a **topocentric** system—the observer is at the origin of the coordinates. In Appendix B, we will use two different points of view: a **geocentric** system (origin at the center of Earth) and a **heliocentric** system (origin at the center of the sun). In our topocentric system, both χ and ξ are functions of:

1. the local time of day, t,[†]
2. the day of the year, d, and
3. the latitude of the observer, λ.

Observe that the time, t, in our formulas is not the time shown in your watch. These times differ by the **time offset** which has two components, one related to the difference in longitude of the place of interest from that of the center of the time zone, and one owing to the **equation of time, EOT** (see appendix). To get a better feeling about these times, do Problem 14.24. t in our formulas is exactly 12:00 when the *mean* sun crosses the local meridian—that is, when the solar zenith angle is at a minimum. The real

---

[†] For comments on time measurement, please refer to Appendix A to this chapter.

sun is in general either ahead or behind the mean sun by an amount called the EOT.

The time of day is represented by the **hour angle**, α, a usage borrowed from astronomers who in the past worked mostly at night and thus preferred to count a new day from noon rather than from midnight. They define the hour angle as

$$\alpha \equiv \frac{360}{24}(t - 12) \qquad \text{degrees } (t \text{ in hours, 24-h clock)} \qquad (4)$$

$$\alpha \equiv \frac{2\pi}{86400}(t - 43200) \qquad \text{radians } (t \text{ in seconds, 24-h clock)} \qquad (4a)$$

The day of the year or "season" is represented by the solar declination, δ—by the latitude of the sun.

The solar declination can be found, for any day of a given year, in the **Nautical Almanac** or by consulting the NOAA or the Naval Observatory website. It also can be estimated with sufficient precision for our purposes by the expression:

$$\delta = 23.44 \sin\left[360\left(\frac{d - 80}{365.25}\right)\right] \qquad \text{degrees,} \qquad (5)$$

where d is the day number.

The solar zenith angle and the solar azimuth are given by

$$\cos \chi = \sin \delta \sin \lambda + \cos \delta \cos \lambda \cos \alpha \qquad (6)$$

and

$$\tan \xi = \frac{\sin \alpha}{\sin \lambda \cos \alpha - \cos \lambda \tan \delta}, \qquad (7)$$

where λ is the latitude of the observer.

To find the value of ξ, we have to take $\arctan(\tan \xi)$. Notice, however, that $\arctan(\tan \xi)$ is not necessarily equal to ξ. Consider, for instance, the angle 240° whose tangent is 1.732. A calculator or a computer will tell you that $\arctan 1.732 = 60°$ because such devices yield the **principal value** of arctanξ which, by definition, lies in the range from $-90°$ to $90°$. The following rule must be observed when obtaining ξ from Equation 7:

Sign(α)	Sign(tanξ)	ξ
+	+	180° + arctan(tanξ)
+	−	360°+ arctan(tanξ)
−	+	arctan(tanξ)
−	−	180°+ arctan(tanξ)

An alternative formula for determining the solar azimuth is

$$\cos(180° - \xi) = -\frac{\sin \lambda \cos \chi - \sin \delta}{\cos \lambda \sin \chi}.$$

At both sunrise and sunset, $\chi = 90°^\dagger$; thus, $\cos \chi = 0$. From Equation 6,

$$\cos \alpha_R = \cos \alpha_S = -\tan \delta \tan \lambda \qquad (8)$$

where $\alpha_{R,S}$ is the hour angle at either sunrise or sunset. The hour angle α_R, at sunrise, is negative and α_S, at sunset, is positive.

$$\alpha_R = -\alpha_S. \qquad (9)$$

12.2.2 Insolation on a Sun-Tracking Surface

If a flat surface continuously faces the sun, the daily average insolation is

$$<P> = \frac{1}{T} \int_{t_R}^{t_S} P_S dt \quad \text{W m}^{-2}, \qquad (10)$$

where t_R and t_S are, respectively, the times of sunrise and sunset, T is the length of the day (24 hours), and P_S is the solar power density, which, of course, depends on the time of day and on meteorological conditions. Assuming (unrealistically) that P_S is a constant from sunrise to sunset, the average insolation, $<P>$, in terms of the hour angle, is

$$<P> = \frac{1}{\pi} \alpha_S P_S \quad \text{W m}^{-2}. \qquad (11)$$

At the equinoxes, $\delta = 0$, and, consequently, $\alpha_S = \pi/2$, and

$$<P> = \frac{1}{2} P_S \approx 500 \quad \text{W m}^{-2} \approx 43.2 \quad \text{MJ m}^{-2}\text{day}^{-1}. \qquad (12)$$

12.2.3 Insolation on a Stationary Surface

Instantaneous insolation on surfaces with elevation, ϵ, azimuth, ζ, is

$$P = P_S[\cos \epsilon \cos \chi + \sin \epsilon \sin \chi \cos(\xi - \zeta)]. \qquad (13)$$

Care must be exercised in using Equation 13. The elevation angle, ϵ, is always taken as positive. See Figure 12.2. It is important to check if the

---

† This will yield the geometric sunrise. As explained before, to correct for atmospheric refraction, χ at sunrise and sunset is taken as $90.833°$.

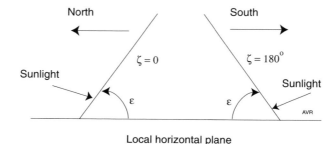

Figure 12.2   The two surfaces above have the same elevation but different azimuths.

sun is shining on the front or the back of the surface. The latter condition would result in a negative sign in the second term inside the brackets. Negative values are an indication that the surface is in its own shadow and that the insolation is zero.

The daily average insolation is

$$<P> = \frac{1}{T} \int_{t_R}^{t_S} P dt = \frac{1}{2\pi} \int_{\alpha_R}^{\alpha_S} P d\alpha. \tag{14}$$

For the general case, the preceding integral must be evaluated numerically. Figures 12.3 through 12.5 show some of the results. Figure 12.3 shows the insolation on south-facing surfaces located at a latitude of 40° north and with various elevation angles, all as a function of solar declination. A horizontal surface ($\epsilon = 0$) receives a lot of sunlight during the summer ($\delta = +23°$). At the height of summer, it receives more light than at the equator where the average (normalized) insolation is, by definition, 1, independently of the season. As the seasons move on, the insolation diminishes and, in winter, it is less than 40% of that in summer.

A vertical surface facing the equator receives more insolation in winter than in summer. There is an optimum elevation that yields maximum year around insolation and, incidentally, minimum seasonal fluctuation. For the latitude of Figure 12.3 (40°), the optimum elevation is 42°—that is, 2° more than the latitude.

The difference between the optimum elevation and the latitude is plotted, as a function of latitude, in Figure 12.4.

The latitude has small influence on the annual insolation for a surface at optimum elevation angle, as can be seen from Figure 12.5. At 67°, the highest latitude at which the sun comes up every day, the yearly insolation is still well above 80% of that at the equator. This assumes an atmosphere whose transparency does not depend on season and on elevation angles. Actually, the farther north, the larger the atmospheric absorption owing to the smaller average solar elevation angle.

12.8

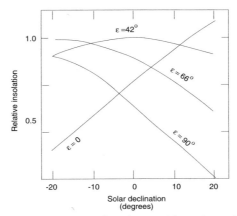

Figure 12.3   Relative insolation of surfaces with various elevations at a latitude of 40°. Each curve has been normalized by comparing the insolation with that on a horizontal surface at the equator.

On the other hand, in many equatorial regions, such as the Amazon valley, the cloud cover is so frequent that the average insolation is only some 60% of that under ideal meteorological conditions.

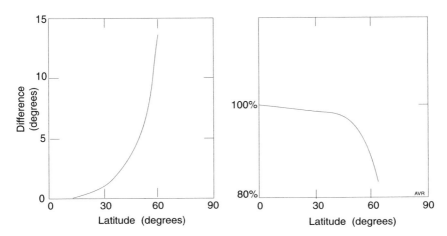

Figure 12.4   Difference between the optimum elevation angle of a solar collector and its latitude.

Figure 12.5   Effect of latitude on the annual insolation of a solar collector at optimum elevation.

Table 12.2
Estimated Insolation
South-Facing Surface
Elevation Equal to Latitude

12.9

City	Average Insolation W/m^2
Bangor, ME	172
Boston, MA	177
Buffalo, NY	161
Concord, NH	171
Hartford, CT	149
Honolulu, HI	230
Los Angeles, CA	248
Newark, NJ	186
New York, NY	172
Philadelphia, PA	185
Phoenix, AZ	285
San Francisco, CA	246
Tucson, AZ	286

It can be seen that the yearly average insolation depends on the orientation of the surface, the latitude, and the prevailing meteorological conditions.

Table 12.2 shows the yearly average insolation at different cities of the USA for south-facing surfaces with an elevation equal to the local latitude. The table was adapted from one published in IEEE SPECTRUM, October 1996, page 53 whose source was "Optimal BIPV Applications," Kiss and Co., Architects, November 1995.

Clearly, these estimated insolation values vary from year to year owing to the variability of the cloud cover.

12.2.4 Horizontal Surfaces

For horizontal surfaces ($\epsilon = 0$), Equation 13 reduces to

$$P = P_S \cos \chi. \tag{15}$$

Consequently,

$$
\begin{aligned}
<P> &= \frac{1}{2\pi} \int_{\alpha_R}^{\alpha_S} P_S \cos \chi \, d\alpha \\
&= \frac{P_S}{2\pi} [\sin \delta \sin \lambda (\alpha_S - \alpha_R) + \cos \delta \cos \lambda (\sin \alpha_S - \sin \alpha_R)] \\
&= \frac{P_S}{2\pi} \cos \delta \cos \lambda (2 \sin \alpha_S + 2\alpha_S \tan \delta \tan \lambda) \\
&= \frac{P_S}{\pi} \cos \delta \cos \lambda (\sin \alpha_S - \alpha_S \cos \alpha_S).
\end{aligned}
\tag{16}
$$

At the equinoxes, $\delta = 0$, $\alpha_S = \pi/2$, therefore

$$<P> = \frac{P_S}{\pi} \cos \lambda. \tag{17}$$

At the equator, regardless of δ, $\alpha_S = \pi/2$, therefore

$$<P> = \frac{P_S}{\pi} \cos \delta. \tag{18}$$

12.3   Solar Collectors

Methods for collecting solar energy for the production of either heat or electricity include:

1. appropriate architecture,
2. flat collectors,
3. evacuated tubes,
4. concentrators, and
5. solar ponds.

12.3.1 Solar Architecture

Proper architecture is an important energy saving factor. Among others, the following points must be observed:

12.3.1.1 Exposure Control

Building orientation must conform to local insolation conditions. To provide ambient heating, extensive use can be made of equatorward facing windows protected by overhangs to shield the sun in the summer. Reduction or elimination of poleward facing windows will diminish heat losses. Shrubs and trees can be useful. Deciduous trees can provide shade in the summer while allowing insolation in the winter.

12.3.1.2 Heat Storage

Structures exposed to the sun can store heat. This may be useful even in summer—the heat stored may be used to pump cool air by setting up convection currents.

Roof ponds can contribute to both heating and cooling. Any part of the building (walls, floor, roof, and ceiling) can be used for heat storage.

Figure 12.6 shows an elaborate wall (proposed by Concept Construction, Ltd., Saskatoon, Saskatchewan) that acts as a heat collector. It consists of a 25-cm thick concrete wall in front of which a glass pane has been installed leaving a 5-cm air space.

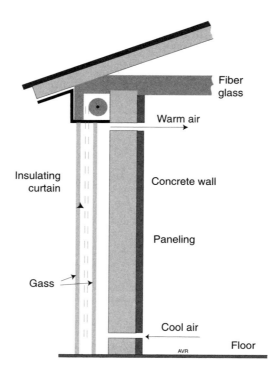

Figure 12.6   Heat storing wall. (Concept Construction, Ltd.)

During the summer, the warmed air is vented outside and the resulting circulation causes cooler air from the poleward face of the house to be taken in.

Instead of a concrete wall, the heat-storing structure can be a large stack of "soda" cans full of water (or, for that matter, full of soda). This takes advantage of the high heat capacity of water. The glazing is placed in contact with the cans so as to force the air to seep through the stack, effectively exchanging heat with it. (You may as well paint the cans black.)

12.3.1.3 Circulation

Heat transfer can be controlled by natural circulation set up in a building by convection currents adjusted by vents. Circulation is important from the health point of view—attempts to save energy by sealing houses may cause an increase in the concentration of radon that emanates from the ground in some places but that is normally dissipated by leakages. Other undesirable gases accumulate, one being water vapor leading to excessive moisture. In addition, noxious chemicals must be vented. This is particularly true of formaldehyde used in varnishes and carpets.

To reduce heat losses associated with air renewal, air-to-air heat exchangers or **recuperators** are used. In this manner, the outflowing air preheats (or pre-cools) the incoming fresh air. About 70% of the heat can be recovered.

12.3.1.4 Insulation

The heat power, P, conducted by a given material is proportional to the heat conductivity, λ, the area, A, and the temperature gradient, dT/dx:

$$P = \lambda A dT/dx. \tag{19}$$

In the SI, the unit of heat conductivity is $WK^{-1}m^{-1}$. Many different nonmetric units are used in the USA. Commonly, insulating materials have their conductivity expressed in BTU hr^{-1} ft^{-2} $(F/inch)^{-1}$, a unit that is a good example of how to complicate simple things. To convert from this unit to the SI, multiply it by 0.144.

Under constant temperature gradient, $dT/dx = \Delta T/\Delta x$ where ΔT is the temperature difference across a material Δx units thick.

One has

$$P = \frac{\lambda A \Delta T}{\Delta x} = \frac{A \Delta T}{R}, \tag{20}$$

where

$$R \equiv \Delta x/\lambda. \tag{21}$$

R has units of m^2K W^{-1} or, in the USA, hr ft^2F/BTU. Again, to convert from the American to the SI, multiply the former by 0.178.

Insulating materials are rated by their R-values. Fiberglass insulation, 8 cm thick, for instance, is rated R-11 (in the American system).

Consider a house with inside temperature of 20 C and an attic temperature of 0 C. The ceiling has an area of 100 m^2 and is insulated with R-11 material. How much heat is lost through the attic?

In the SI, R-11 corresponds to $11 \times 0.178 = 2$ m^2 K W^{-1}. Thus, the heat loss under the assumed 20 K temperature difference is

$$P = \frac{100 \times 20}{2} = 1000 \quad W. \tag{22}$$

Thus, if the only heat losses were through the ceiling, it would take little energy to keep a house reasonably warm. There are, of course, large losses through walls and, especially, through windows. A fireplace is a particularly lossy device. If the chimney is left open, warm air from the house is rapidly syphoned out. If a metallic damper is used to stop the convection current, then substantial heat is conducted through it.

12.3.2 Flat Collectors

Flat collectors work with both direct and diffused light. They provide low temperature heat (less than 70 C) useful for ambient heating, domestic hot water systems, and swimming pools. This type of collector is affected by weather and its efficiency decreases if large temperature rises are demanded.

For swimming pools in the summer, when only a small temperature increase is needed, flat collectors can be over 90% efficient. It is necessary to operate them so that large volumes of water are only slightly heated rather than heating small amounts of water to a high temperature and then mixing them into the pool.

Simple collectors are black plastic hoses exposed to the sun. More elaborate collectors use both front and back insulation to reduce heat losses.

Collectors may heat water directly or may use an intermediate heat transfer fluid.

Figure 12.7 shows a cross-section through a typical flat collector. Light and inexpensive aluminum is used extensively; however, it tends to be corroded by water. Copper is best suited for pipes. If an intermediate heat exchange fluid is employed, aluminum extrusions that include the channels for the liquid are preferred.

Some panels use a thin copper sleeve inserted into the aluminum tubing.

Panels can be black-anodized or painted. There is some question about the lifetime of paints exposed to solar ultraviolet radiation.

The front insulation can be glass or plastic. The former is fragile but the latter does not withstand ultraviolet well. To avoid heat losses through the back of the panel, insulation such as fiberglass mats or polyurethane foam is used. The latter imparts good rigidity to the panel allowing a reduction in the mass of the material employed.

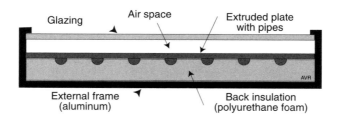

Figure 12.7   Cross-section through a typical flat collector.

12.3.3 Evacuated Tubes

This type of collector consists of two concentric cylinders, the outer one of glass and the inner, a pipe through which the liquid flows. They bear a superficial resemblance to fluorescent lamps. A vacuum is established between the two cylinders, reducing the convection heat losses.

Evacuated tubes are nondirectional and can heat liquids to some 80 C. They are usually employed in arrays with spacing equal to the diameter of the outer tube. It is customary to place a reflecting surface behind the array.

12.3.4 Concentrators

Concentrators can be of the non-imaging or of the focusing type. Either can be line concentrators (2-D) or point concentrators (3-D).

A solar collector consists of a concentrator and a receiver. The concentrator may be of the refracting (lens) or of the reflecting (mirror) type. The receiver may be thermal or photovoltaic.

Two important parameters describe the collector performance:

1. the concentration, C, and
2. the acceptance angle, θ.

The concentration can be defined as either the ratio of the aperture area to the receiver area or as the ratio of the power density at the receiver to that at the aperture. These definitions are not equivalent; the latter is preferable.

The acceptance angle is the angle through which the system can be misaimed without (greatly) affecting the power at the receiver (see Figure 12.10).

There is a theoretical relationship between the concentration and the acceptance angle for the ideal case:

$$C_{ideal} = (\sin\theta)^{-1} \qquad \text{for a 2-D concentrator,} \qquad (23)$$
$$C_{ideal} = (\sin\theta)^{-2} \qquad \text{for a 3-D concentrator.} \qquad (24)$$

It is instructive to calculate the maximum temperature that a receiver can attain as a function of concentration. An ideal receiver will work in a vacuum (no convection losses) and will be perfectly insulated (no conduction losses). Nevertheless, radiation losses are unavoidable. They will amount to

$$P_r = \sigma\epsilon T^4 \qquad \text{Wm}^{-2}, \qquad (25)$$

where ϵ is the emissivity (taken as unity)[†] and σ is the Stefan-Boltzmann

---

[†] It is, of course, desirable to have a low emissivity at the temperature at which the receiver operates and a high absorptivity at the region of the spectrum dominated by the incident sunlight.

constant $(5.67 \times 10^{-8}$ W m^{-2}K$^{-4})$. See Chapter 6.

The power density at the receiver (assuming no atmospheric losses) is

$$P_{in} = 1360 \ C \qquad \text{Wm}^{-2}. \qquad (26)$$

In equilibrium, $P_r = P_{in}$,

$$\sigma T^4 = 1360 \ C \qquad (27)$$

or

$$T = (2.4 \times 10^{10} C)^{1/4} = 394 \ C^{1/4}. \qquad (28)$$

With unity concentration (flat plate collector), $T = 394$ K or 120 C.

When the concentration is raised to 1000, the maximum temperature theoretically attainable is 2200 K. Were it possible to construct a collector with a concentration of 1 million, the formula would predict a receiver temperature of 12,400 K. This would violate the second law of thermodynamics because heat would be flowing unaided from the cooler sun (6000 K) to the hotter receiver. Clearly, the upper bound of the receiver temperature must be 6000 K.

One can arrive at the same conclusion by considering the expression for C_{ideal}. The solar angular radius, as seen from Earth, is 0.25°. Thus, the minimum accept, and this leads to a $C_{max} = 52,000$ for a 3-D collector and 230 for a 2-D collector as calculated from Equations 23 and 24. A concentration of 52,000 corresponds to a T_{max} of 5900 K (Equation 28), just about right.

Numerous reasons cause the concentration (in terms of power densities) to be less than ideal:

1. reflector shape and alinement errors,
2. less than perfect reflector surface reflectivity,
3. tracking errors,
4. atmospheric scatter, and
5. atmospheric absorption.

12.3.4.1 Holographic Plates

Just as a flat plastic sheet with appropriate grooves (a **Fresnel** lens) can concentrate light, a holographic plate can be fashioned to do the same. The advantage of the holographic approach is that the plate can simultaneously diffract and disperse, creating a rainbow of concentrated light with each region of the spectrum directed to a collector optimized for that particular color range. This is a significant advantage when using photovoltaic cells. This technology, which looks promising, is discussed in Chapter 14.

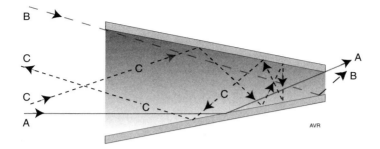

Figure 12.8   Ray paths in a conical concentrator.

12.3.4.2 Nonimaging Concentrators

The simplest nonimaging concentrator is a cone. In a properly designed cone, all rays parallel to its axis will be reflected into the exit aperture. See ray A in Figure 12.8. The acceptance angle, however, is small. Ray B makes it through the exit, but ray C, although parallel to B, does not—it bounces around a number of times and finally returns to the entrance.

The performance of nonimaging concentrators is improved by the use of a **compound parabolic concentrator**, or CPC.

This device consists of parabolic surfaces as shown in Figure 12.9 (from Welford and Winston). Notice that the section is not that of a truncated parabola but rather that of two independent parabolas, mirror images of one another. The CPCs in Figure 12.9 have identical exit apertures. The largest has the biggest collecting area and consequently the largest concentration; the acceptance angle, however, is small compared with that of the CPCs with smaller concentration.

The area of the reflecting surfaces of a CPC is much larger than that of a focusing paraboloid of the same concentration. This makes CPCs heavy, expensive, and difficult to mount.

Ideally, a plot of the normalized light power density that leaves the exit aperture as a function of the **aiming angle**, Θ, should be a rectangle: unity for $\Theta < \Theta_i$ and zero for $\Theta > \Theta_i$. Here, Θ_i is the acceptance angle. This ideal characteristic is poorly approached by a conical concentrator, but reasonably well achieved by a CPC. See Figure 12.10.

2-D nonimaging collectors with concentrations up to 2 or 3 need not track the sun. Devices with larger concentrations require a once a month re-aiming to compensate for variations in solar declination.

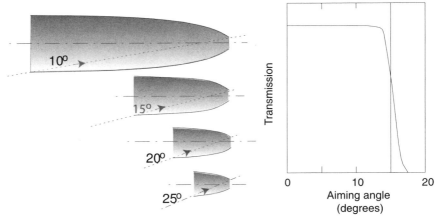

Figure 12.9   For the same exit aperture, the larger the entrance aperture, the smaller the acceptance angle.

Figure 12.10   A CPC approaches the ideal transmission versus aiming angle characteristics.

12.4   Some Solar Plant Configurations

12.4.1   High Temperature Solar Heat Engine

A straightforward method of generating electricity from solar energy is to use concentrators to produce high temperatures that can drive either a Stirling or a Rankine (steam) engine.

Southern California Edison Company had a 10 MW facility in Barstow, in the Mojave Desert. It cost $140 million—that is, $14,000/kW. Since fossil-fueled plants may cost around $1000/kW, it can be seen that this particular solar thermal installation can be justified only as a development tool. Since the "fuel" is free, it is important to determine the plant life, and the operation and maintenance cost so as to be able to compare the cost of electricity over the long run with that from traditional sources.

The plant operated in the manner indicated schematically in Figure 12.11. The collecting area was roughly elliptical in shape and covered some 300,000 m^2 (30 hectares). Insolation, averaged over 24 hours, is probably around 400 W/m^2 for sun-tracking collectors, leading to an efficiency of the order of 8%.

The collector consisted of 1818 sun-tracking flat mirrors forming a gigantic heliostat capable of focusing the sun's energy on a boiler.

The boiler was a cylinder 7 m in diameter and 14 m in height. It was operated at 788 K (516 C) and 10.7 MPa (105 atmospheres).

A thermal storage system was incorporated having a 100 GJ (electric) capacity permitting the plant to deliver some 7 MW for 4 hours during nighttime.

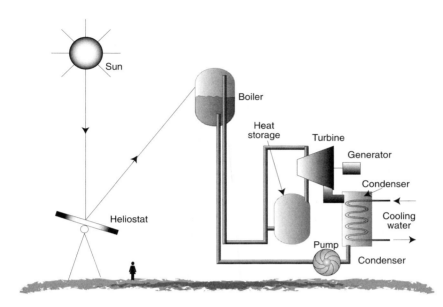

Figure 12.11   Solar One power plant of the Southern California Edison Co., in the Mojave Desert.

Solar One was decommissioned and an updated installation, Solar Two, took its place starting operation in July 1996. The new plant cost $48.5 million but inherited considerable assets from the previous effort (among other, the many heliostats) so that its true investment cost is hard to estimate.

Solar Two has the same 10 MW rating of Solar One. The main difference is the use of an intermediate working fluid—a $NaNO_3/KNO_3$ mix containing 40% of the potassium salt—that is heated in the solar tower and transfers its energy to the turbine operating steam by means of a heat exchanger. The salt leaves the tower at 565 C and, after delivering part of its heat energy to the steam, returns to the tower at 288 C.[†]

The salt mixture is somewhat corrosive requiring low grade stainless steel in pipes and containers. At the operating temperature the mixture is quite stable and has a low vapor pressure.

[†] Both sodium nitrate and potassium nitrate go by the common name of **niter** or **saltpeter**. Sodium nitrate (a common fertilizer and oxidant) melts at 306.8 C and potassium nitrate (used in the manufacture of black powder) melts at 334 C. Usually, alloys and mixtures melt at a lower temperature than their constituents. **Eutectic** mixture is the one with the lowest melting point. Solar Two uses less potassium than the eutectic, because the potassium salt is more expensive than the sodium one. The mixture used in the plant melts at 288 C.

The mirrors are made of a sandwich of two glass panes with a silver layer between them. This protects the reflecting layer from corrosion. When not in use, the mirrors are placed in a horizontal position to protect them from the destructive action of wind storms. This also reduces abrasion from wind carried sand.

12.4.2 Solar Chimney

A circular tent with 121 m radius made mostly of plastic material (and, partially, of glass), was built in Manzanares, Spain. The height of the tent is 2 m at the circumference and 8 m at the center where a 194-m tall chimney has been erected.

The air under the tent is heated by the greenhouse effect and rises through the 10-m diameter chimney driving a 50 kW turbine.

The installation, owned by the Bundesministerium für Forschung und Technologie, Bonn, Germany, was built by Schlaich Bergermann und Partner. It operated from 1989 through 1996 and served to demonstrate the principle.

In 2002, the Australian Minister for Industry, Tourism and Resources gave its support to the firm EnviroMission to work on a much larger plant of the Manzanares type. The plant would generate at peak 200 MW of electricity and would cost $800 million. This corresponds to $4000/kW, a reasonable amount for a development project (typically, electric generating plants cost about $1000/kW). The tower is supposed to generate 650 GW-hours per year (2.3×10^{15} J/year). This represents an optimistic 36% utilization factor.

The proposed plant would have a collecting tent of 7 km diameter— that is, a collecting area of some 38×10^6 m^2. At a peak insolation of 1000 W/m^2, the efficiency would be about 0.5%.

Since the efficiency of the system depends on the height of the chimney, the design requires a 1000-m high structure. Such a tall structure will certainly be a challenge to civil engineers because (among other factors) it would be subject to enormous wind stresses. The tallest existing tower (excluding radio towers) is the 553-m Canada's National Tower in Toronto.

12.4.3 Solar Ponds

The OTEC principle discussed in Chapter 4 can be used to generate electricity from water heated by the sun. Surprisingly high temperatures can be obtained in some ponds.

In shallow ponds with a dark colored bottom, the deep layers of water are warmed up and rise to the surface owing to their lower density. This causes mixing that tends to destroy any temperature gradient. In such ponds the temperature rise is modest because most heat is lost through the surface by the evaporation of water.

The solution is, of course, to cover the pond with an impermeable light-transmitting heat-insulating layer as it is done in swimming pools with plastic covers.

It is interesting to observe that the insulating layer can be the water itself. If a vertical salinity gradient is created in the pond, so that the deeper layers contain more salt and become correspondingly denser, it is possible to impede convection and achieve bottom temperatures as high as 80 C.

Working against a 20 C cold sink, the Carnot efficiency of an OTEC would be 20% and practical efficiencies of 10% do not seem impossible.

Difficulties involve:
1. mixing owing to wind action and other factors,
2. development of turbidity owing to collected dirt and the growth of microorganisms.

It may be possible to overcome such difficulties by using **gel ponds** in which the water is covered by a polymer gel sufficiently viscous to impede convection. Such gel must be

1. highly transparent to sunlight,
2. stable under ultraviolet radiation,
3. stable at the operating temperature,
4. insoluble in water,
5. nonbiodegradable,
6. nontoxic,
7, less dense than the saline solution, and
8. inexpensive.

E.S. Wilkins and his colleagues at the University of New Mexico (Albuquerque) claim to have developed such a gel.

To keep the surface clean, a thin layer of water runs on top of the gel sweeping away dirt and debris.

12.21

Appendix A (The Measurement of Time)

The Duration of an Hour[†]

How long is an hour?

In Roman times, the hour was defined as 1/12 of the time period between sunrise and sunset. Since this interval varies with seasons, the "hour" was longer in the summer than in the winter.

At the latitude of Rome, about 42° N, one hour would last anywhere between 44.7 modern minutes (in the end of December) and 75.3 (in mid-June). This variability was a major problem for clockmakers who had to invent complicated mechanisms to gradually change the clock speed according to the season of year. See *On Architecture* by Vitruvius (Marcus Vitruvius Pollio, a book still being sold).

Much of the variability is eliminated by defining the hour as 1/24 of the interval between two consecutive noons, i.e., two consecutive solar crossings of the local meridian. Unfortunately, this also leads to an hour whose length varies throughout the year, albeit much less than the Roman one. See detailed explanation in Appendix B to this chapter. The obvious solution is to define a **mean solar hour** as the average value of the **solar hour** taken over one year interval. But again, owing to the very slow changes in astronomical constants (eccentricity, semi major axis, argument of perihelion, etc.), this definition of an hour will not be constant over long periods of time. The final solution is to define arbitrarily an **ephemeris hour** referred to an atomic clock. At present, the ephemeris hour is very close to the mean solar hour.

Astronomy and chronology being ancient sciences have inherited ancient notions and ancient terminology. The division of the hour into minutes and seconds is one example.

Using the Babylonian sexagesimal system, the hour was divided into "minute" parts (*pars minuta prima* or first minute part or simply "minutes") and then again into *pars minuta secunda* or simply, "seconds." The latter is the official unit of time for scientific purposes, and has a value of 1/86,400 of a mean solar day.

The times obtained from the approximate formulas in this book are the mean solar times and may differ by as much as ±15 minutes from the true solar times.

Time Zones

The local mean solar time is not a convenient measure of time for every day use because it depends on the longitude of the observer. It varies by

---

[†] Fore more information on the history of the hour, read Dohrn-van Rossum. See Further Reading at the end of this chapter.

1 hour for every 15° of longitude. This means, of course, that the time in San Francisco is not the same as in Sacramento. The use of time zones, 1 hour or 15° wide, circumvents this difficulty. In each zone, the time is the same regardless of the position of the observer. At the zone boundaries, the time changes abruptly by 1 hour. The center meridian of any time zone is a multiple of 15°; the first zone is squarely centered on the zeroth meridian, that of Greenwich, and the time there is called **Greenwich mean time, GMT** (or, to astronomers, **universal time, UT**). The zone time is called **standard time** (such as, for instance, PST, for Pacific Standard Time, the −8 time zone centered at 120° W).

Time Offset

The true solar time, t_{true}, at any given longitude, L, can be found from

$$t_{true} = t_{local\ mean} + t_{offset},\tag{29}$$

where t_{true} and t_{local} are expressed in hours and minutes but t_{offset} is in minutes only,

$$t_{offset} = EOT - 4L + 60t_{zone}\qquad\text{minutes},\tag{30}$$

where EOT is the equation of time (in minutes) discussed in Appendix B, L is the longitude in degrees (east > 0, west < 0), and t_{zone} is the number of hours of the local time zone referred to the UT (east > 0, west < 0).

A simple example:
What is the true solar time on February 20, at Palo Alto, CA (125° W) when the local mean time is 12:00 (noon)?

The EOT for February 20 (scaled from Figure 12.11) is +14 minutes. The time zone of Palo Alto is Pacific Standard Time, i.e., it is −8 h. We have

$$t_{offset} = 14 - 4 \times (-125) + 60(-8) = 34\qquad\text{minutes}\tag{31}$$

$$t_{true} = 12^{\text{h}}{:}00^{\text{m}} + 34^{\text{m}} = 12^{\text{h}}{:}34^{\text{m}}.\tag{32}$$

The Calendar

There are a few recurring astronomical features that serve quite obviously as measurements of time. There is the daily rise of the sun that leads to the definition of a **day** and its divisions (hours, minutes, and seconds). There are also the phases of the moon which are repeated (approximately) every 28 days leading to the notion of **month** and its subdivision, the **week**.

And then there is the time it takes the Earth to complete one orbit around the sun which leads to the definition of **year** and to the recurrence of the seasons.

Unfortunately the number of days in a year or in a month is not an exact integer and this complicates the reckoning of the date. If one month were exactly 4 weeks (28 days), and if the year were exactly 12 months (336 days) there would be no difficulty. However, the year is closer to thirteen 28-day months. This leads to 364 days per year. The extra day could be declared a universal holiday. The trouble with this scheme is that it does not lend itself to an easy divisions in quarters. Thus, 12 months per year is the choice.

The first month of the Roman calendar used to be *Martius*, our present March, the fifth month of the year was *Quintilis*, the sixth, *Sextilis*, the seventh, *September*, and so on. From 153 B.C. on, *Ianuarius* was promoted to first place leading to September being the 9th month, not the 7th as before.

The exact date of the equinox could be easily measured (as was probably done at Stonehenge and at other much older observatories), and thus, one could easily establish that the Vernal Equinox should always occur at the same date (March, 21, for instance). This meant that occasionally one additional day would have to be added to the year. In this respect, the Romans were a bit sloppy and left the slippage accumulate until it became painfully obvious that the seasons were out of phase. If a given crop had to be planted at, say, the first day of spring, you could not assign a fixed date for this seeding day.

By the time the error became quite noticeable, the Pontifex maximus[†] would declare a *mensis intercalaris*, an intercalary month named *Mercedonius* and stick it somewhere toward the end of February, which was then at the end of the year. With time, the position of pontifex became a sinecure and the calendar adjustment became quite erratic and subject to political corruption. To correct things, in 46 B.C,, Julius Caesar declared a year 445 days long (three intercalary months were added). He also decreed a new calendar establishing that each year would be 365 days long, and, to account for the extra 1/4 day, every four years an additional day would be added. He commemorated this achievement by naming Quintilis after himself—thus introducing the name July. Not to be outdone, his nephew Octavian (Augustus) insisted in changing Sextilis to August. Of course, both July and August are 31-day months. What else?

The corrections worked for a while but not perfectly (because the year is not *exactly* $365\frac{1}{4}$ days). In March 1582, Gregory XIII introduced a fix— the Gregorian calendar used today. A number of European nations immediately adopted the new calendar; others resisted. England, always bound by

---

[†] Chief bridge builder, possible in charge of bridge maintenance in old Rome.

tradition, only adopted the Gregorian on September 2, 1752. The next day became September 14, 1752. So the answer to the trivia question "What important event occurred in the United States on September 10, 1752?" is "Nothing!" By the way, Russia only converted to the Gregorian calendar on February 1, 1918, and this is why the "October Revolution" actually occurred in November.

The Julian Day Number

In many astronomical calculations, it proves extremely convenient to have a calendar much simpler than any of those in common use. The simplest way to identify a given day is to use a continuous count, starting at some arbitrary origin in the past, totally ignoring the idea of year, month, and week. The astronomical Julian day number is such a system. We will define it as the day count starting with the number 2,400,000 corresponding to November 16, 1858. Thus, for instance, the next day, November 17, 1858, has a Julian day number of 2,400,001. The Julian day number starts at noon.

To determine the Julian day number corresponding to a given Gregorian date, it is sufficient to count the number of days after (or before) November 16, 1858 and add this to 2,400,000. Easier said than done. It is a pain to count the number of days between two dates. We suggest the following algorithm taken from *http://webexhibits. org/calendars/calendar-christian.html*:

$$JD = d + \mathrm{INT}((153m + 2)/5) + 365y + \mathrm{INT}(y/4)$$
$$- \mathrm{INT}(y/100) + \mathrm{INT}(y/400) - 32045.$$

In the above, y stands for the year (expressed in four figures), m is the month, and d is the day.

Notice that there is no year zero in the Julian or Gregorian calendars. The day that precedes January 1, 1 A.D. is December 31, 1 B.C. If you want a Julian day number of a B.C. date, to use the formula above, you must convert to negative year numbers, as, for instance, 10 B.C. must be entered as -9.

If you are dealing with ancient dates, you must realize that they are given, most frequently, in the Julian calendar, not the Gregorian, even if they are earlier than the year (45 BC) when the Julian calendar was established. This use of a calendar to express dates before it was established is an "anticipation" or "prolepsis" and is called a **proleptic** calendar. Do not confuse the Julian calendar with the Julian day number (different Juliuses, almost certainly).

For more information on Julian day numbers and on algorithms to convert Julian day numbers to Gregorian or to Julian dates, please refer to the URL given above.

Appendix B (Orbital Mechanics)
Sidereal versus Solar

The most obvious measure of time is the interval between two consecutive **culminations** of the sun, called a **solar day**. The sun culminates (reaches the highest elevation during a day—the smallest zenith angle) when it crosses the local meridian and is, consequently, exactly true south of an observer in the northern hemisphere. As will be explained later, there is an easily determined moment in the year when equinoxes occur. The time interval between two consecutive vernal equinoxes—that is, the length of the **tropical year** has been measured with—what else?— astronomical precision. It is found that during one tropical year, the sun culminates 365.24219878 times (call it 365.2422 times),[†]—in other words, there are, in this one-year interval, 365.2422 solar days. We define the **mean solar hour** as 1/24 of a mean solar day (a day that lasts 1/365.2422 years). Unfortunately, the length of a solar day as measured by any reasonably accurate clock changes throughout the year. The change is not trivial—the actual solar day (the time between successive culminations) is 23 seconds shorter than the mean solar day on September 17, and it is 28 seconds longer on December 22. These differences accumulate—in mid-February the sun culminates some 14 minutes after the mean solar noon and in mid-November, some 15 minutes before mean solar noon. We will explain later what causes this variability, which renders the interval between consecutive culminations an imprecise standard for measuring time.

If we were to measure the time interval between successive culminations of a given star, we would find that this time interval is quite constant—it does not vary throughout the year. We would also find that a given star will culminate 366.2422 times in a tropical year. The number of "star," or **sidereal**, days in a year is precisely one more day than the number of solar days.

The discrepancy between the sidereal and the solar time is the result of the Earth orbiting *around* the sun. Refer to Figures 12.12 and 12.13.

In a planetary system in which a planet does not spin, its motion around the sun causes an observer to see the sun move *eastward* throughout the year resulting in one apparent day per year. If, however, the planet spins (in the same direction as its orbital motion) at a rate of exactly 360 degrees per year, then the sun does not seem to move at all—the planetary spin exactly cancels the orbitally created day.

[†] The length of a tropical year is decreasing very slightly, at a rate of $169 \times 10^{-10}\%$ per century. This means that there is a very small secular reduction in the orbital energy of Earth.

12.26

Table 12.3

Time Definitions

Year (tropical)	Interval between two consecutive vernal equinoxes
Mean solar day	1/365.24219878 years
Mean solar hour	1/24 mean solar days
Minute	1/60 mean solar hours
Second	1/60 minutes
Sidereal day	1/366.24219878 years
Mean solar hours/year	8,765.81
Minutes per year	525,948.8
Seconds per year	31.5569×10^6
Length of sidereal day	23:56:04.09

As explained, the completion of a full orbit around the sun will be perceived from the surface of a nonspinning planet as one complete day— the sun will be seen as moving in a complete circle around the planet. If we count 24 hours/day, then each 15° of orbital motion causes an apparent hour to elapse, and 1° of orbital motion corresponds to 4 minutes of time.

Thus, in the case of the planet Earth which spins 366.2422 times a year, the number of solar days is only 365.2422 per year because the orbital motion cancels one day per year.

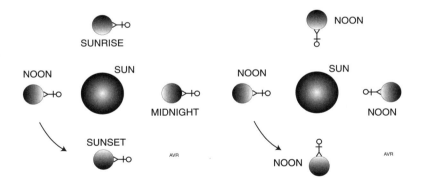

Figure 12.12   In a planetary system in which the planet does not spin, the orbital motion introduces one apparent day per year.

Figure 12.13   If the planet spins exactly once per year, the sun appears not to move.

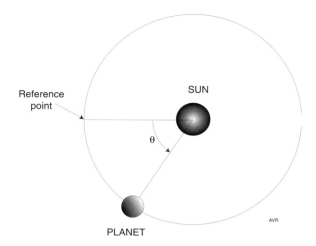

Figure 12.13   The angular position of a planet in its orbit in the ecliptic plane is called the true anomaly, θ.

Orbital Equation

The angle *reference point-sun-planet* is called the **true anomaly**. See Figure 12.14. Notice the quaint medieval terminology frequently used in astronomy.

In a circular orbit, there is no obvious choice for a reference point. However, since most orbits are elliptical (or, maybe, hyperbolic) the periapsis (nearest point to the attracting body)[†] is a natural choice. The apoapsis is not nearly as convenient because, in the case of long period comets, it cannot be observed.

Thus, the origin for the measurement of the true anomaly is the periapsis. The anomaly increases in the direction of motion of the planet.

Consider a planetary system consisting of a sun of mass, M, around which orbits a planet of mass, m. Assume that the orbital velocity of the planet is precisely that which causes the centrifugal force, $mr\omega^2$ (where ω is the angular velocity of the planet in radians/second) to equal the attracting force, GmM/r^2, where G is the gravitational constant (6.6729×10^{-11} $\mathrm{m}^2\mathrm{s}^{-2}\mathrm{kg}^{-1}$). In this particular case, these two forces, being exactly of the same magnitude and acting in opposite directions, will perfectly cancel one another—the distance, r, from sun to planet will not change and the orbit is a circle with the sun at the center:

$$mr\omega^2 = G\frac{Mm}{r^2}, \tag{33}$$

[†] Here we have a superabundance of terms with nearly the same meaning: **periapsis, perihelion, perigee, periastron, pericenter,** and **perifocus** and, **apoapsis, aphelion, apogee, apastron, apocenter,** and **apofocus.**

from which

$$r = \frac{GM}{\omega^2}^{1/3}.$$
(34)

For the case of the sun whose mass is $M = 1.991 \times 10^{30}$ kg and the Earth whose angular velocity is $2\pi/365.24$ radians per day or 199.1×10^{-9} radians per second, the value of r comes out at 149.6×10^9 m. This is almost precisely correct, even though we know that the orbit of Earth is not circular.

In fact, it would be improbable that a planet had exactly the correct angular velocity to be in a circular orbit. More likely, its velocity may be, say, somewhat too small, so that the planet would fall toward the sun, thereby accelerating and reaching a velocity too big for a circular orbit and, thus, would fall away from the sun decelerating, and so on—it would be in an elliptical orbit.

The orbital equation can be easily derived, albeit with a little more math than in the circular case.

We note that the equations of motion (in polar coordinates) are

$$m\left[-\frac{d^2r}{dt^2} + r\left(\frac{d\theta}{dt}\right)^2\right] = G\frac{mM}{r^2}$$
(35)

and

$$m\left(2\frac{dr}{dt}\frac{d\theta}{dt} + r\frac{d^2\theta}{dt^2}\right) = 0$$
(36)

for, respectively, the radial and the tangential components of force.

Equation 36 can be multiplied by r/m yielding

$$2r\frac{dr}{dt}\frac{d\theta}{dt} + r^2\frac{d^2\theta}{dt^2} = 0$$
(37)

Note that Equation 35 becomes Equation 33 for the circular case (in which d^2r/dt^2 is zero). In this case Equation 36 reduces to $d^2\theta/dt^2 = 0$ (because dr/dt is also zero) showing, as is obvious, that the angular velocity is constant.

Note also that

$$\frac{d}{dt}\left(r^2\frac{d\theta}{dt}\right) = 2r\frac{dr}{dt}\frac{d\theta}{dt} + r^2\frac{d^2\theta}{dt^2} = 0.$$
(38)

Integrating Equation 38,

$$r^2\frac{d\theta}{dt} = \text{constant.}$$
(39)

It is possible to show that Equations 35 and 36 represent, when t is eliminated between them, an equation of a conical section of eccentricity, ϵ (which depends on the total energy of the "planet"). When $\epsilon < 1$, the path or orbit is an ellipse described by

$$r = \frac{a(1 - \epsilon^2)}{1 + \epsilon \cos \theta}. \tag{40}$$

In the preceding, a is the semi major axis and θ is the true anomaly. Given the position, θ, of the planet in its orbit, it is easy to calculate the radius vector, provided the major axis and the eccentricity are known.

Of greater interest is the determination of the position, θ, as a function of time. From Equations 39 and 40,

$$\frac{d\theta}{dt} \propto \frac{(1 + \epsilon \cos \theta)}{a^2(1 - \epsilon^2)^2} = A(1 + \epsilon \cos \theta)^2. \tag{41}$$

If the orbit of Earth were circular, i.e., if $\epsilon = 0$, then the angular velocity, $d\theta/dt$, would be constant;

$$\frac{d\theta}{dt} = \frac{360°}{365.2422 \text{ days}} = 0.98564733 \quad \text{degrees/day}. \tag{42}$$

This is the mean angular velocity of Earth. Thus,

$$A = 0.98564733 \quad \text{degrees/day}. \tag{43}$$

and, because the eccentricity of Earth is, at present, approximately 0.01670,

$$\frac{d\theta}{dt} = 0.98564733(1 + 0.0167 \cos \theta)^2. \tag{44}$$

Consequently, when $\theta = 0$—that is, at the moment of perihelion passage, the angular velocity of Earth is 1.01884 degrees/day, while at aphelion the angular velocity is down to 0.95300 degrees/day.

A description of the motion of Earth—that is, a table of θ as a function of time, can be obtained by evaluating

$$\theta_i = \theta_{i-1} + A(1 + \epsilon \cos \theta_{i-1})^2 \Delta t. \tag{45}$$

θ_0 is made equal to zero, corresponding to the perihelion.

Good accuracy is achieved even when the time increment, Δt, is as large as 1 day.

It turns out that the perihelion is not a very useful point of reference. It is more convenient to use a more obvious direction, the vernal equinox, as the initial point. For this, an **ecliptic longitude** is defined as the angle, measured along the ecliptic plane, *eastward* from the vernal equinox. The tabulation resulting from Equation 45 can be used provided the **longitude of the perihelion** also called the **argument of perihelion**, is known. This quantity varies slowly but is, at present, close to $-77°$.

The heliocentric polar coordinate system used to derive the orbital motion of Earth is not very convenient for describing the position of a celestial body as seen from Earth. For this, the latitude/longitude system used in geography can easily be extended to astronomy. The various positions are described by a pair of angles: the **right ascension** equivalent to longitude and the **declination** equivalent to latitude. We are back to the old geocentric point of view although we recognize, as Aristarchus of Samos did back in around 200 B.C., that we are not the center of the universe. For such a system, the reference is the spin axis of the planet, a direction perpendicular to the **equatorial plane**, which passes through the center of Earth. Only by extreme coincidence would the equatorial plane of a planet coincide with the **ecliptic** (the plane that contains the orbit of the planet). Usually, these two planes form an angle called **obliquity** or **tilt angle**, τ. It is possible to define a **celestial equator** as a plane parallel to the terrestrial equator but containing the center of the sun, not that of the Earth.

Celestial equator and ecliptic intersect in a line called the **equinoctial line**. When Earth crosses this line coming from south to north (**the ascending node**), the **vernal equinox** occurs. At the descending node, when Earth crosses the line coming from the north, the **autumnal equinox** occurs.

The vernal equinox is used as a convenient origin for the measurement of both the ecliptic longitude and the right ascension. Remember that the former is an angle lying in the ecliptic plane while the latter lies in the equatorial plane.

The time interval between two consecutive vernal equinoxes is called the tropical year, referred to at the beginning of this appendix. In our derivation, we used the perihelion as the origin for measuring the true anomaly. The time interval between two consecutive perihelion passages is called the **anomalistic year**, which, surprisingly, is slightly longer than the tropical year. How can that be? The reason for this discrepancy is that the line of apsides slowly changes its orientation completing 360° in (roughly) 21,000 years. The corresponding annual change in the longitude of the perihelion is $360/21{,}000 = 0.017$ degrees per year.

Since the orbital angular velocity of Earth is roughly 1 degree per day, this means that the anomalistic year will be about 0.017 days (25 minutes) longer than the tropical one which is 365.242 days long. Hence, the anomalistic year should be about 365.259 days long (more precisely, 365.25964134 days).

12.31

Relationship Between Ecliptic and Equatorial Coordinates

Celestial longitude is measured along the ecliptic while right ascension —which is also a measure of longitude—is measured along the equatorial plane. Clearly, a simple relationship must exist between these coordinates.

Consider a right handed orthogonal coordinate system with the center of Earth at the origin, in which the x-y plane coincides with the equatorial plane and the y-axis is aligned with the equinoctial line. The z-axis points north.

Let $\vec{s}$ be a unit vector, starting from the origin and pointing toward the sun,

$$\vec{s} = \vec{i}s_x + \vec{j}s_y + \vec{k}s_z. \tag{46}$$

This vector must lie in the ecliptic plane—that is, it must be perpendicular to the spin axis whose unit vector is

$$\vec{u} = -\vec{i}\sin\tau + \vec{k}\cos\tau. \tag{47}$$

If $\vec{s}$ is perpendicular to $\vec{u}$, then their dot product must be zero:

$$\vec{s}\cdot\vec{u} = 0 = -s_x\sin\tau + s_z\cos\tau, \tag{48}$$

$$s_z = s_x\tan\tau \tag{49}$$

The reference for measuring longitude and right ascension is the direction of the vernal equinox, which, in our co-ordinate system has a unit vector $-\vec{j}$. Hence, the longitude, Λ, is given by

$$\cos\Lambda = -s_y. \tag{50}$$

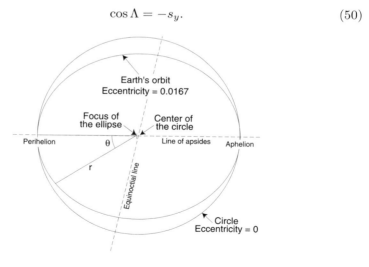

Figure 12.15   The orbit of Earth is slightly elliptical. The view is normal to the ecliptic.

Since $\vec{s}$ is a unit vector,

$$s_x^2 + s_y^2 + s_z^2 = s_x^2 + \cos^2 \Lambda + s_x^2 \tan^2 \tau = 1, \tag{51}$$

$$s_x^2(1 + \tan^2 \tau) = 1 - \cos^2 \Lambda, \tag{52}$$

$$s_x = \sin \Lambda \cos \tau. \tag{53}$$

But on the equatorial plane, the right ascension, $\Re$, is given by

$$\tan \Re = \frac{s_x}{-s_y} = \frac{\sin \Lambda \cos \tau}{\cos \Lambda} = \cos \tau \tan \Lambda. \tag{54}$$

$$\Re = \arctan(\cos \tau \tan \Lambda). \tag{55}$$

As usual, when one takes reverse trigonometric functions, the answer is ambiguous—the calculator or the computer gives only the principal value and a decision must be made of what the actual value is. In the present case, the following Boolean statement must be used:

If $\Lambda >= 90°$ AND $\Lambda < 270°$ THEN $\Re = \Re + 180°$

ELSE IF $\Lambda >= 270°$ THEN $\Re = \Re + 360°.$ $\qquad$ (56)

This relates the longitude of the sun to its right ascension.
The declination of the sun is

$$\sin \delta = s_z = s_x \tan \tau = \sin \Lambda \cos \tau \tan \tau = \sin \Lambda \sin \tau. \tag{57}$$

The Equation of Time

We have now to explain why the time between successive solar culminations varies throughout the year:

Assume you have on hand two instruments:

1. An accurate clock that measures the uniform flow of time calibrated in mean solar time. From one vernal equinox to the next, it registers the passage of $365.2422 \times 24 \times 60 \times 60 = 31.55692 \times 10^6$ seconds.
2. A sundial that can measure the solar time with a resolution of at least one minute.[†]

Set your clock to start at exactly noon (as seen in the sundial) on any arbitrary date. It will be noted that, although by the same date of the next year, the clock and the sundial are again synchronized, throughout

---

[†] Owing to the finite angular diameter of the sun $(0.5°)$, it is difficult to read a sundial to greater precision than about 1 minute of time.

the year, the sundial seems sometimes to be slow and, at other periods of the year, to be fast. The difference between the sundial time and the clock time—between the solar time and the mean solar time—may reach values of up to 15 minutes fast and 15 minutes slow. This difference is called the **Equation of Time, EOT**.

Figure 12.19 shows how the EOT varies along the year. For planning solar collectors, it is sufficient to read the value of the EOT off the figure. For such an application, the empirical formula below (Equation 59) is an overkill.[†] It yields the EOT in minutes when the day of the year, d, is expressed as an angle, d_{deg}:

$$d_{deg} \equiv \frac{360}{365}d \quad \text{degrees.} \tag{58}$$

Greater precision may be useless because the EOT varies somewhat from year to year with a 4 year period, owing to the leap years.

$$EOT = -0.017188 - 0.42811\cos(d_{deg}) + 7.35141\sin(d_{deg})$$
$$+ 3.34946\cos(2d_{deg}) + 9.36177\sin(2d_{deg}) \quad \text{minutes} \tag{59}$$

We recall that there are two measures of solar longitude, both increasing eastward from the direction of the vernal equinox: one measure is along the ecliptic and is called the ecliptic longitude, Λ, the other is measured along the equatorial plane and is called the right ascension, $\Re$. These two longitudes are exactly the same at the equinoxes and at the solstices, but are different anywhere in between. See Table 12.4.

Table 12.4
Difference Between
Longitude
and Right Ascension

Season	$\Lambda - \Re$
Spring	> 0
Summer	< 0
Autumn	> 0
Winter	< 0

[†] http://www.srrb.noaa.gov/highlights/sunrise/program.txt. is the source of the NOAA formula mentioned.

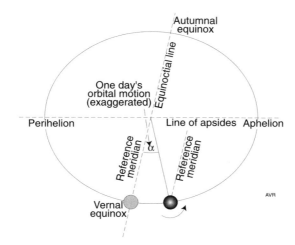

Figure 12.16   As the Earth moves in its orbit, the sun appears to move eastward.

The discrepancy between Λ and $\Re$ increases with the obliquity of the orbit. If the obliquity of Earth's orbit were zero, then $\Lambda = \Re$ under all circumstance.

Assume that you are on the surface of Earth at an arbitrary latitude but on a meridian that happens to be the one at which the sun culminates at the exact moment of the vernal equinox. At this moment $\Lambda = \Re = 0$.

$23^{\mathrm{h}}: 56^{\mathrm{m}}: 04^{\mathrm{s}}.09$ or 23.93447 h later, Earth has completed a full rotation and your meridian faces the same direction it did initially.

However, the sun is not exactly at culmination. The reason is that the orbital motion of Earth has caused an apparent eastward motion of the sun. The Earth has to spin another α degrees for the reference meridian to be facing the sun again. See Figure 12.16.

If the orbit were circular and the obliquity zero, then the uniform eastward motion of the sun would be $360°/365.2422 = 0.985647$ degrees/mean solar day. In more technical terms, the rate of change of the anomaly would be constant and the **mean anomaly** would be

$$< \theta > = 0.985647t, \tag{60}$$

where t is the time (in mean solar days) since perihelion passage. The ecliptic longitude changes at the same rate as the anomaly. Since the spin rate of Earth is $360°$ in 23.93447 hours or $15.04107°$/hour, or $0.2506845°$ per minute, it takes $0.985647/0.2506845 = 3.931823$ minutes (0.06553 hours) to rotate the $0.985647°$ needed to bring the reference meridian once again under the sun—that is, to reach the next culmination or next noon. Not surprisingly, $23.93447 + 0.06553 = 24.00000$ hours. This means that in this simple case, consecutive noons are evenly spaced throughout the year and occur exactly 24 hours apart, a result of the definition of the solar mean hour.

In reality noons do not recur at uniform intervals—that is, the sun dial time does not track exactly the clock time. Twenty-four hours between noons is only the value averaged over one year. The difference between the clock time and the sun dial time is called, as stated before, the equation of time, EOT. Two factors cause this irregularity: the eccentricity of the orbit (leading to $EOT_{eccentricity}$) and the obliquity of the orbit (leading to $EOT_{obliquity}$).

Orbital Eccentricity

If the orbit is not circular, the rate of solar eastward drift, (the rate of change of the anomaly) is not constant. It changes rapidly near perihelion and more slowly near aphelion. Hence, the true anomaly differs from the mean anomaly so that, after each 24 mean solar hours period, the Earth has to spin an additional $\theta - <\theta> \equiv C$ degrees. Here the difference, C, between the true anomaly and the mean anomaly is called the **equation of center**, another example of medieval terminology.

Since the spin rate of Earth is very nearly 1 degree in 4 minutes of time, the time offset between the true noon and the mean noon owing to the eccentricity of the orbit is

$$EOT_{eccentricity} = 4C \quad \text{minutes of time.} \qquad (61)$$

In the preceding, the angle, C, is in degrees.

The eccentricity component of the equation of time varies throughout the year in a sinusoidal fashion with zeros at the perihelion and aphelion and with extrema of 8 minutes of time midway between these dates. Figure 12.17 shows these variations.

The value of C for any time of the year can be found by calculating θ using Equation 44 and subtracting the mean anomaly obtained from Equation 60. For many applications, it may prove more practical to calculate C using the empirical equations below:[†]

$$<\theta> = 357.52911 + 35999.05029T - 0.0001537T^2 \qquad (62)$$

$$
\begin{aligned}
C = {} & (1.914602 - 0.004817T - 0.000014T^2) \sin <\theta> \\
& + (0.019993 - 0.000101T) \sin(2 <\theta>) \\
& + 0.000289 \sin(3 <\theta>) \quad \text{degrees.} \qquad (63)
\end{aligned}
$$

Here, T, is the number of Julian centuries since January 1, 2000. See farther on for explanation of Julian dates.

[†] The formulas are from a NOAA program:
http://www.srrb.noaa.gov/highlights/sunrise/program.txt.

Orbital Obliquity

If the orbit is circular but the obliquity is not zero, then although the rate of solar ecliptic longitude increase (or the rate of the anomaly increase) is constant, the rate of change of the right ascension is not. The moment of solar culmination is related to the right ascension.

On the day after the vernal equinox, as seen by an observer on the reference meridian on Earth, the sun is at a right ascension of

$$\Re = \arctan(\cos\tau\tan\Lambda) = \arctan(\cos(23.44°)\tan(0.985647°))$$

$$= \arctan(0.91747 \times 0.017204) = \arctan\big(0.015785\big), \tag{64}$$

$$\Re = 0.904322° \tag{65}$$

For the sun to culminate, the Earth has to spin an additional $0.904322°$ rather than $0.985647°$ as in the case of zero obliquity. Thus, noon will occur somewhat earlier than in the zero obliquity situation. In fact it will occur $4 \times (0.985647 - 0.904322) = 0.325$ minutes earlier.

Generalizing,

$$EOT_{obliquity} = 4(\Lambda - \Re). \tag{66}$$

The obliquity component varies, as does the eccentricity one, in a sinusoidal fashion but completes two cycles rather than one in the space of one year. The zeros occur at the equinoxes and solstices instead of at the perihelion and aphelion as they do in the eccentricity case. The amplitude of $EOT_{obliquity}$ is 10 minutes. This behavior is depicted in Figure 12.18, while the behavior of the full EOT (the sum of the eccentricity and the obliquity components) is depicted in Figure 12.19.

It is important not to confuse perihelion (when the Earth is closest to the sun) or aphelion (when it is farthest) with the solstices which occur when the solar declination is an extremum—that is, when $\delta \pm 23.44°$. As it happens, the dates of the solstices are near those of perihelion and aphelion, but this is a mere coincidence. There are roughly 12 days between the (summer) solstice and the aphelion and between the (winter) solstice and the perihelion. See Table 12.5.

Table 12.5
Dates of Different Sun Positions

	Approx. Date	Approx Day No.
Perihelion	2...4 January	2
Vernal Equinox	20...21 March	80
Solstice (Summer, N. Hemisphere)	21 June	172
Aphelion	4...5 July	184
Autumnal Equinox	22...23 September	262
Solstice (Winter, N. Hemisphere)	21...22 December	355

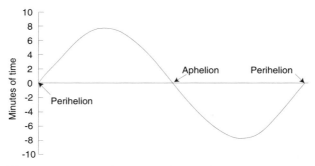

Figure 12.17   That part of the equation of time resulting from the ellipticity of Earth's orbit has a value (in minutes of time) equal to $4C$, where C is the equation of center—that is, the difference between the true anomaly and the mean anomaly. Observe that this function has a zero at both perihelion and aphelion.

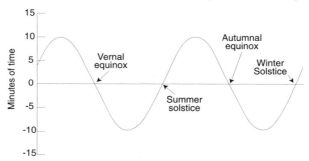

Figure 12.18   The part of the equation of time resulting from the obliquity of Earth's orbit has a value (in minutes of time) equal to $4(\Lambda - \Re)$—that is, 4 times the difference between the solar longitude and its right ascension. Zeros occur at the equinoxes and solstices.

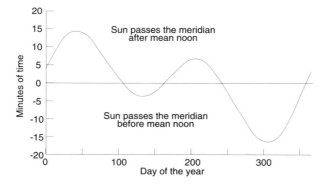

Figure 12.19   The observed equation of time is the combination of the effects owing to ellipticity and obliquity of Earth's orbit.

12.38

References

Duffie, John A., and William A. Beckman, Solar energy thermal processes, *John Wiley*, **1974**.

Welford, W. T., and R. Winston, The optics of non-imaging concentrators, *Academic Press*, **1978**.

Wilkins, E. S. et al., Solar gel ponds, *Science 217*, 982, Sept. 10 **1982**.

Further Reading

Dohrn-van Rossum, Gerhard, History of the hour: clocks and modern temporal orders, *University of Chicago Press*, **1996**.

PROBLEMS

12.1   A time traveler finds himself in an unknown place on Earth at an unknown time of the year. Because at night the sky is always cloudy, he cannot see the stars, but he can accurately determine the sunrise time and also the length of a shadow at noon. The sun rises at 0520 local time. At noon, a vertical mast casts a shadow 1.5 times longer than its height. What is the date and what is the latitude of the place? Is this determination unambiguous?

12.2   An astronaut had to make an emergency de-orbit and landed on an island in the middle of the ocean. He is completely lost but has an accurate digital watch and a copy of *Fundamentals of Renewable Energy Processes* that NASA always supplies.[†] He times carefully the length of the day and discovers that the time between sunrise and sunset is 10:49:12. He knows the date is Jan. 1, 1997. He can now figure his latitude. Can you?

12.3   A building in Palo Alto, CA (latitude 37.4 N) has windows facing SSE. During what period of the year does direct light from the sun enter the window at sunrise? Assume no obstructions, good weather, and a vanishing solar diameter.

What is the time of sunrise on the first day of the period? And on the last day? What is the insolation on the SSE-facing wall at noon at the equinoxes?

12.4   Consider an ideal focusing concentrator. Increasing the concentration causes the receiver temperature to increase—up to a point.

Beyond certain maximum concentration, the temperature remains constant. What is the maximum concentration that can be used on Mars for a 2-D and for a 3-D case? Some data:

Radius of the orbit of Mars is 1.6 AU.
1 AU is 150 million km.
The angular diameter of the moon, as seen from Earth, is 0.5 degrees.

12.5   Consider the arbitrary distribution function

$$\frac{dP}{df} = f - \frac{1}{2}f^2$$

Determine for what value of f is this function a maximum.
Plot dP/df as a function of f for the interval in which $dP/df > 0$.
Now define a new variable, $\lambda \equiv c/f$, where c is any constant.

[†] Just kidding!

Determine for what value of f is the distribution function $dPd\lambda$ a maximum.

Plot $|dP/d\lambda|$ as a function of f.

12.6   An expedition to Mars is being planned. Let us make a preliminary estimate of the energy requirements for the first days after the expedition lands.

Landing date is November 15, 2007 which is Mars day 118. At the landing site (17.00° N, 122° E) it will be just after local sunrise. The five-person landing team has all of the remaining daylight hours to set up the equipment to survive the cold night.

Prior to the manned landing, robots will have setup a water extraction plant that removes the liquid from hydrated rocks by exposing them to concentrated sunlight. Assume an adequate (but not generous) supply of water.

Power will be generated by photovoltaics and will be stored in the form of hydrogen and oxygen obtained from water electrolysis.

The photovoltaics are blankets of flexible material with 16.5% efficiency at 1 (Mars) sun. No concentrators will be used. These blankets will be laid horizontally on the Mars surface.

The electrolyzers operate at 90% efficiency.

MARS DATA (relative to Earth)

Mean radius of orbit	1.52
Gravity	0.38
Planetary radius	0.53
Length of day	1.029
Length of year	1.881
Density	0.719

The inclination of the plane of the Martian equator referred to the plane of its orbit is 25.20°.

The inclination of the plane of the Martian orbit referred to the ecliptic is 1.85°.

The average daytime Martian temperature is 300 K (just a little *higher* than the average daytime Earth temperature of 295 K). The Martian night, however, is cold! Average temperature is 170 K (versus 275 K, for Earth).

The vernal equinox occurs on Mars day 213.[†]

Define a Martian hour, h_m, as 1/24 of the annual average of the period between consecutive sunrises.

[†] Not really. Although the other data are accurate, this date was pulled out of a hat.

1. How long does the sunlit period last on the day of arrival?
2. Determine the available insolation on a horizontal surface (watts m^{-2}, averaged over a Martian day—that is, over a 24 h$_m$ period).
3. Estimate the O$_2$ consumption of the five astronauts. Consider that they are on a strictly regulated diet of 2500 kilocalories per Martian day. Assume that all this is metabolized as glucose. Use 16 MJ per kg of glucose as combustion enthalpy.
4. How much energy will be required to produce the necessary amount of oxygen from water by electrolysis?
5. What area of solar cell blanket must be dedicated for the production of oxygen?
6. Assume that the mean temperatures of Mars are the actual temperatures at the astronaut's settlement. Assume further that the temperature of the Martian air falls instantaneously from its daytime 300 K to its nighttime 175 K and vice-versa.

   The astronauts are housed in a hemispherical plastic bubble 10 m in diameter. The wall material is rated at R-12 (in the American system) as far as its thermal insulation is concerned. No heat is lost through the floor.

   The interior of the bubble is kept at a constant 300 K. During the night, stored hydrogen has to be burned to provide heat. The inside wall of the bubble is at 300 K, while the outside is at 175 K. Assume that the effective emissivity of the outside surface is 0.5.

   How much hydrogen is needed per day? Express this in solar cell blanket area.

12.7   How long was the shadow of a 10-m tall tree in Palo Alto, CA (37.4 N, 125 W) on March 20, 1991 at 0200 PM (PST)? Desired accuracy is ± 20 cm.

12.8   You are on a wind swept plain on the planet Earth but you know neither your position nor the date. There are no hills and you can see the horizon clearly. This allows you to time the sunset accurately, but unfortunately your watch reads Greenwich time. Take a straight pole and plant it exactly vertically in the ground (use a plumb line). You have no meter- or yardstick but you assign arbitrarily 1 unit to the length of the pole above ground. Observe the shadow throughout the day: at its shortest (at 08:57 on your watch) the shadow is 2.50 units long. Sun sets at 13:53. From these data alone, determine the date, latitude, and longitude unambiguously (i.e., decide if you are in the Northern or the Southern hemisphere).

12.9   Calculate the azimuth of a vertical surface that results in the maximum annual average relative insolation under the conditions below:

The surface is situated at latitude of 40° N in a region in which there is always a dense early morning fog (insolation = zero) up to 10:00 and then the rest of the day is perfectly clear.

Relative insolation is defined here as the ratio of the insolation on the surface in question to that on a *horizontal* surface at the equator on the same day of the year.

12.10   At what time of day is the sun due east in Palo Alto?

The latitude of Palo Alto is 37.4° north; the longitude is 122° west of Greenwich.

Do this for two days of the year: August 11 and November 15.

12.11   What is the insolation (W m^{-2}) on a surface facing true east and with 25° elevation erected at a point 45° N at 10:00 local time on April 1, 1990?

Assume that the insolation on a surface facing the sun is 1000 W m^{-2}.

12.12   The average food intake of an adult human is, say, 2000 kilocalories per day. Assume that all this energy is transformed into a constant uniform heat output.

Ten adults are confined to a room 5×5×2 meters. The room is window-less and totally insulated with R-11 fiberglass blankets (walls, door, ceiling, and floor). Outside the room the temperature is uniformly at 0 C. The air inside the room is not renewed. Assume that temperature steady state is achieved before the prisoners suffocate. What is the room temperature?

12.13   Here is a way in which a person lost in a desert island can determine her/his latitude with fair precision even though the nights are always cloudy (Polaris cannot be seen).

Use a vertical stick and observe the shadow. It will be shortest at local noon. From day to day, the shortest noon shadow will change in length. It will be shortest on the day of the summer solstice: this will tell you the solar declination, δ.

On any day, comparing the length of the horizontal shadow with the length of the stick will let you estimate the solar zenithal angle, χ, quite accurately, even if no yard- (or meter-) stick is on hand.

Knowing both δ and χ it is possible to determine the latitude. Or is it?

Develop a simple expression that gives you the latitude as a function of the solar declination and the noon zenithal angle. No trigonometric functions, please! Is there any ambiguity?

12.14   As determined by a civilian-type Trimble GPS receiver, the latitude of my house in Palo Alto, CA, is 34.44623° N.

At what time of the year are consecutive sunrises exactly 24 hours apart?

Call Δt the time difference between consecutive sunrises. At what time of the year is Δt a maximum?

What is the value of this maximum Δt? Express this in seconds.

12.15   An explorer in the Arctic needs to construct a cache in which to store materials than cannot stand temperatures lower than -10 C. The surface air over the ice is known to stay at -50 C for long periods of time.

The cache will be built by digging a hole in the ice until the bottom is 0.5 meters from the water. The roof of the cache is a flat surface flush with the ice surface. It will be insulated with a double fiberglass blanket, each layer rated at R-16 (in the American system).

The heat conductivity of ice is 1.3 W m^{-1}K^{-1}.

Assume that the area of the cache is so large that the heat exchange through the vertical walls can be neglected.

Estimate the temperature inside the cache when the outside temperature has been a steady -50 C for a time sufficiently long for the system to have reached steady state. Assume the temperature inside the cache is uniform.

12.16   What is the solar azimuth at sunset on the day of the summer solstice at 58° north?

12.17   A battery of silicon photocells mounted on a plane surface operates with an efficiency of 16.7% under all conditions encountered in this problem. It is installed at a location 45° north. The time is 10:00 on April 1, 1995.

When the battery faces the sun directly, it delivers 870 W to a resistive load. How much power does it deliver if the surface is set at an elevation of 25° and faces true east?

12.18   Consider a mechanical heat pump whose coefficient of performance is exactly half of the ideal one. It uses an amount W of mechanical energy to pump Q_C units of energy from an environment at -10 C into a room at 25 C where an amount $Q_A = Q_C + W$ of energy is deposited. How many joules of heat are delivered to the room for each joule of mechanical energy used?

12.19   You have landed on an unknown planet and, for some obscure reason, you must determine both your latitude and the angle, ι, of inclination of the planet's spin axis referred to the its orbital plane.

To accomplish such a determination, all you have is a ruler and plenty of time. Erect a vertical pole exactly 5 m tall and observe the length of the shadow at noon as it varies throughout the year.

The shortest length is 1.34 m and the longest is 18.66 m.

12.20   The smallest zenithal angle of the sun was, on January 1, 2000, 32.3°. At that moment, the sun was to your south. What is your latitude?

12.21   You are 30° north. The day is September 15 of a nonleap year. It is 12:44 true solar time. A 10 meter long rod tilted due west is planted in the ground making a 30° angle with the vertical.

Calculate the position of the sun.

What is the length of the shadow of the rod?

12.22   On which day of the year does the sun appear vertically overhead at the three locations below. Determine the time (in standard time) in which this phenomenon occurs. Disregard the Equation of Time. Also assume that the time zones are those dictated by purely geographic considerations, not by political ones.

Locations:

Palo Alto, California (USA): 37° 29' N, 122° 10' W.

Macapá, Amapá (Brazil): 0.00, 51° 07' W.

Brasília, Federal District (Brazil): 15° 53' S, 47° 51' W.

12.23   If you consult the URL <http://mach.usno.navy.mil/> and follow the instructions, you will find that for San Francisco, CA (W122.4, N37.8), on February 19, 2002, the sunrise occurred at 06:55 and the sun crossed the meridian at 12:24.

1. Using the information in the textbook, verify the meridian crossing time.
2. Still using the formulas in the textbook, calculate the sunrise time. If there is any discrepancy, indicate what causes it.

12.24   Aurinko (pronounced OW-rin-ko, where the stressed syllable, OW, has the same sound as in "how") is a (fictitious) unmanned airplane designed to serve as a repeater for radio signals replacing expensive satellites. It is equipped with 14 electric motors, each of which can deliver 1.5 kW. Its cruising speed is 40 km/hr (slightly faster than a man can run in a 100- or 200-meter dash). It operates at 30 km altitude.

Wingspan: 75.3 m.

62,120 solar cells.

Maximum electric output of the cells: 32 kW when full sunlight is normal to the cells.

1. When Aurinko is in its orbit (30 km above sea level) how far is its geometric horizon?[†] The geometric horizon differs from the radio horizon because the latter is somewhat extended by atmospheric refraction.
2. What is the area on the ground that is reached by the direct rays from Aurinko. Disregard atmospheric refraction.
3. Assume that the airplane orbits a point 37.8° north. On the day in which the sunlight hours are the least, how long is the sunlit period as seen from the airplane at 30 km altitude? Disregard atmospheric refraction.
4. Assuming that the solar cells mounted on top of the wing of the airplane are always horizontal, what is the insolation averaged over the sunlit period on the day of the previous question? Since the airplane is above most of the atmosphere, the solar constant can be taken as 1200 W/m^2.
5. Assume that the total power consumption of the airplane while in orbit is 10 kW. (This includes propulsion, house keeping, and communications.) The electric energy obtained from the photovoltaic array is in part directly used by the load and the excess is stored to be used during the period when the array output is less than the load demand. The storage system has a turn-around efficiency of $\eta_{turnaround}$—that is, only a fraction, $\eta_{turnaround}$, of the energy fed into the system can be retrieved later.

Assume, for simplicity, that $\eta_{turnaround} = 1.0$. Assume also that the efficiency of the photovoltaic collectors is 20% independently of the sun power density.

The solar array covers all the wing surface except for a rim of 20 cm. More clearly, there is a space of 20 cm between the leading edge and the array and the same space at the trailing edge and the wing tips. Consider rectangular (nontapered) wings.

6. What must the chord of the wing be (chord = distance between leading and trailing edges)?

12.25   In the solar spectrum, what fraction of the total power density is absorbed by silicon? Hint: Use Table 12.1.

12.26   This is an experiment technically easy to carry out. Unfortunately, it takes 365 days to complete. Hence, we are going to invert the procedure and, from the theory developed in the textbook, we will calculate what the results of the experiment would be. If you have not been exposed to it, you may be surprise with what you get.

[†] This is not a space orbit around a planet. It is an orbit (as the term is used in aviation) around a point 30 km above sea level.

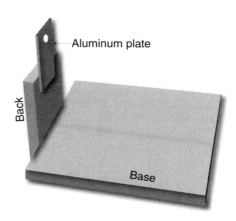

To set up the experiment, you would have to build the simple device illustrated in the figure. It consists of a wooden base large enough to support a standard piece of paper. A vertical piece of wood ("back") is attached to the base and mounted on it is a thin aluminum rectangle with a small hole in it. The aluminum must be thin enough (say, 1 mm or less) so that the sun can shine through the hole even when it is far from perpendicular to the plate.

In the model we used, the hole was 129 mm above the base, but this is not a critical dimension. Orient the device so that the hole faces equatorward. The noon sun will cast a shadow, but shining through the hole will cause a little dot of light to appear on the base. What we want to do is to follow the path this dot of light will trace out. The exact time in which the observations are made is important. You have to start at the astronomical noon on one of the following days (in which the equation of time, EOT, is zero): April 15, June 14, September 2, or December 25.

To follow the dot of light, place a sheet of paper on the base. Immobilize it by using some adhesive tape. With a pen, mark the center of the dot of light at the moment the sun culminates. Since on the suggested starting dates the equation of time is zero, the sun will culminate (i.e., cross the local meridian) at 1200 Standard Time corrected by the longitude displacement. This amounts to 4 minutes of time for each degree of longitude away from the longitude of the meridian at the center of the time zone. The center of the time zone is (usually) at exact multiples of 15° of longitude. For instance, the Pacific Time Zone in North America is centered on 120° W. If you happen to be in Palo Alto (125° W, your time offset (on days when the EOT is zero) is $15° \times (120° - 125°) \times 4 = -20$ minutes. Thus, the sun will culminate at 1220 PST.

The next observation must be made 240 hours (or an exact multiple of 24 hours) late—that is, it must, if you are in Palo Alto, be made exactly at 1220 PST. After one year, an interesting pattern will emerge.

12.47

The assignment in this problem is to calculate and plot the position of the light dot every 10 days over a complete year. Start on April 15 and do not forget the equation of time. Assume that you are in Palo Alto, CA: 125° W 37.4° N.

12.27   The time interval between two consecutive full moons is called **a lunar month**, or a **lunation**, and lasts, on average, 29.53 days. Explain why this is not the **length of the lunar orbital period**—that is, the time it takes for the moon to complete one full orbit around the Earth.

Calculate the length of the lunar orbital period.

12.28   A supraluminar (faster than light) probe was sent to reconnoiter an Earth-like extra-solar planet whose characteristics are summarized in the table below. All units used are terrestrial units, except as noted below. It was determined that the planet had a very tenuous atmosphere, totally transparent to the solar radiation. As the probe materialized in the neighborhood of the alien sun, it measured the light power density which was 11 kW/m². The measurement was made when the probe was at exactly 50 million km from the sun. It was also determined that the planet was in an essentially circular orbit.

"Day" is defined as the time period between consecutive solar culminations (noons). It differs from the terrestrial day.

"Year" is the length of the orbital period—the time period between successive vernal equinoxes. It differs from the terrestrial year.

The "date" is designated by the day number counting from Day Number 1—the day of the vernal equinox.

Orbital radius	$132{\times}10^6$	km
Orbital period	28.98×10^6	sec
Orbital inclination	26.7	degrees
Planetary radius	5800	km
Spin period	90,000	sec

The direction of the spin is the same as the direction of the orbital motion.

The probe will land at at point with a latitude of $+45°$.

1. At the probe landing site, what is the length, in seconds, of the longest and of the shortest daylight period of the year?
2. What are the dates of the solstices?
3. What is the daily average insolation on a horizontal surface on day 50?

12.48

Chapter 13
Biomass

13.1 Introduction

Wikipedia, the free online encyclopedia, defines **biomass** as "all non-fossil material of biological origin." This is an acceptable definition even though we will focus almost exclusively on vegetable matter that can lead to the production of useful energy.

Any biomass-based energy process begins with the capture of sunlight and production of a chemical compound. This complicated step, called **photosynthesis**, leads basically to glucose. Subsequent biochemical transformations result in the creation of a very large number of compounds, some of very great commercial value. However, to the engineer interested in extracting fuels from biomass, the products of importance are glucose and its polymers—starch and cellulose—other sugars (some of which form hemi-cellulose), and lignin. At best, photosynthesis proceeds with efficiencies of less than 8%. By the time the final product is available for consumption, large chunks of energy have been spent in cultivation, fertilizing, harvesting, transporting the raw biomass, removing the excess water and extracting the desired fuel. The overall efficiency is usually a fraction of 1%. This serves as a prime example of practical energy processes that, although extremely inefficient, are of commercial interests mainly because the economic and ecological aspects are favorable.

13.2 The Composition of Biomass

A plant can be considered as a structure that supports specialized organs. The structure consists of wood and the specialized organs include the leaves that perform photosynthesis and the roots that collect water and nutrients. Materials are transported from one site to another by sap. Fruits perform the function of sexual reproduction. Energy in a plant is stored in roots and tubers, in the sap and in the fruit. Structural parts of the plant also represent an accumulation of energy.

Plant tissue consists of 50% to 95% water.

Leaves are the plant's main photosynthetic organ. They contain, among other substances, proteins and much of the minerals taken up. Good plantation management frequently involves returning leaves to the ground as mulch and fertilizer.

Leaves can be transformed into **biogas** (a mixture of methane and carbon dioxide) by anaerobic digestion leaving a residue of sludge, valuable as fertilizer because it contains minerals.

In wood, the structural part of plants, one finds few minerals (typically, wood ashes represent less than 1% of the dry mass that was burned). Wood is made of cellulose, hemicellulose, and lignin in variable proportions (say, 50%, 30%, and 20%, respectively). Cellulose, a polyhexose, and hemicellulose, a polypentose, are carbohydrates (see the subsection on organic chemistry). Lignin has a phenolic structure highly resistant to microorganisms.

Energy is stored by the plant in the form of carbohydrates, hydrocarbons, and esters.

Carbohydrates consist of sugars and their polymers such as starches, cellulose, and hemicellulose. Sugars can be stored as such in sap (sugar cane), tubers (sugar beets), and fruits.

Starches are mostly found in fruit (grain), tubers (potatoes), and roots (manioc).

The hydrocarbons found in plants are generally polyisoprenes (terpenes)—that is, polymers of the alkyne hydrocarbon, isoprene, C_5H_8. They are found in some euphorbia such as the rubber tree.

Vegetable oils are esters, chemically quite different from mineral oils that are hydrocarbons. Usually, the oil from the pulp of the fruit is of a different nature from that obtained from the seed or kernel. The olive is one exception, its fruit and seed yield the same type of oil.

One oil of interest to the energy engineer is that from the castor bean: it is one of the rare oils that is insoluble in gasoline and can, therefore, be used to lubricate gasoline pumps. Palm oil, originally used by Rudolf Diesel in his engine, is of major interest as fuel.

13.2.1 A Little Bit of Organic Chemistry

Hydrogen is the most abundant element in the universe, but carbon is the most versatile. By itself it forms graphite and diamond and those fantastic and useful molecules, buckyballs and nanotubes.

13.2.1.1 Hydrocarbons

The many hydrocarbons are among the simplest carbon compounds and, of these, the simplest is methane, CH_4. A whole slew of methane-like compounds exist having the general form C_nH_{2n+2}. They constitute the family of **alkanes** known also as **paraffin** or **saturated** hydrocarbons. Other series exist such C_nH_{2n} (**alkenes** or **olefins**), C_nH_{2n-2} (**alkynes**), and so on all the way to C_nH_{2n-6}, exemplified by the important **aromatic** hydrocarbon, benzene. Organic chemistry starts with these compounds. It is instructive to investigate some of their functional derivatives.

13.2.1.2 Oxidation Stages of Hydrocarbons

Let us examine the successive stages of oxidation of a methane-series hydrocarbon. We chose propane because it is the simplest compound in which the carbons are not all at the end of the chain as in methane and ethane.

The first oxidation step consists of replacing a hydrogen atom by the **functional group**, OH (hydroxyl). The result is an **alcohol**.

If the carbon to which the OH is attached is at the end of the chain, we have a **1-alcohol** (1-propanol, in this case); if it is the carbon in the middle of the chain, we have a **2-alcohol** (2-propanol).

More than one hydrogen can be replaced. If three hydrogens in propane are replaced by OHs, we have **propanotriol** better known as glycerine, one of the most important alcohols in biochemistry.

The next step consists of the loss of two hydrogens: one in the OH and one previously attached to the carbon atom to which the OH was bonded. The remaining oxygen is now doubly bonded to the carbon. If the carbon was at the end of the chain, a functional group O=C–H is formed, characteristic of **aldehydes**. If the carbon was in the middle of the chain, a **ketone** is formed.

The third step consists of substituting the H of the C=O–H group by an OH, creating a **carboxyl** group, O=C–O–H, characteristic of a **carboxylic acid**. Notice that acids are derived from aldehydes, not from ketones.

The final step in the oxidation of either an acid or a ketone is the formation of carbon dioxide and water.

The sequence is illustrated in Figure 13.1.

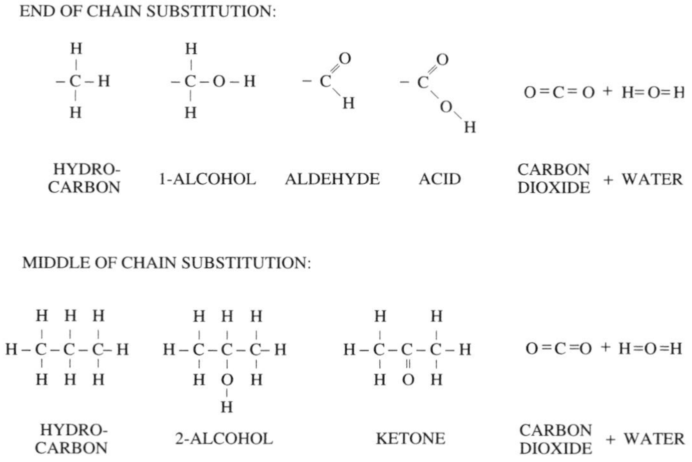

Figure 13.1   Oxidation sequence of a methane-series hydrocarbon.

13.2.1.3 Esters

From the biomass energy point of view, the important chemicals are oils and carbohydrates, especially the latter.

Oils are esters—the result of a reaction of an acid (frequently a carboxylic acid) and an alcohol liberating a molecule of water. Technically, most esters are waxes except the esters of glycerine that are oils or fats.

Perhaps the simplest ester is methyl formate, a combination of methyl alcohol (methanol) and formic acid, both derived from methane. See Figure 13.2.

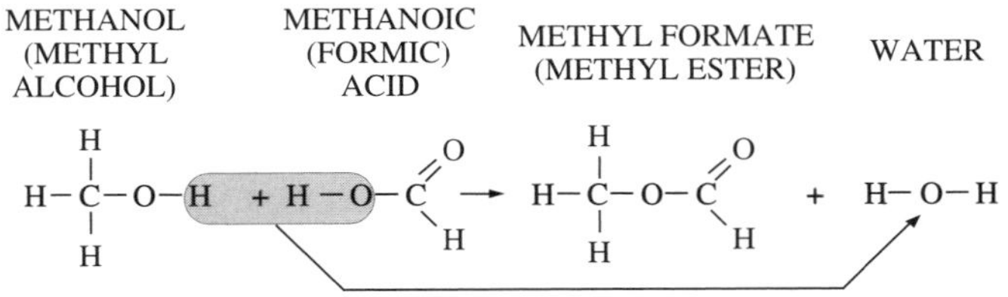

Figure 13.2   An ester (wax, oil, or fat) is formed when an alcohol combines with a carboxylic acid. The oxygen in the water comes from the acid.

The general formula for an ester is $R_{ac}COOR_{al}$, where R_{ac} is the acid group and R_{al} is the alcohol group.

Esters are responsible for many scents of fruits and flowers and for the aroma of wine. For example, pineapple, banana, and wintergreen smells are due to, respectively, ethyl butyrate, amyl[†] acetate, and methyl salicylate.

Palm oil, a potentially important fuel, is the result of the combination of glycerine and palmitic acid (the acid derived from $C_{16}H_{34}$).

13.2.1.4 Carbohydrates

Carbohydrates are the most common components of biomass. They are sugars or condensation polymers of simple sugars.

A condensation polymer is one in which, when the monomers combine, a small molecule, such as H_2O, is eliminated in a manner similar to that indicated in Figure 13.2: a hydroxyl from one monomer and a hydrogen from the next combine to form water. The reverse—depolymerization—reaction requires the addition of water and is called **hydrolysis**.

Simple sugars, known as **monosaccharides** are compounds of carbon, hydrogen, and oxygen containing from 3 to 8 carbon atoms. 5-carbon sugars are called **pentoses**, 6-carbon sugars, **hexoses**, and so on.

Mono- and disaccharides are fermentable—that is, can biologically be transformed directly into ethanol. If the plant contains free sugar, its juices

---

[†] Amyl is the same as pentyl.

can be fermented directly as is the case of sugar cane. More complex carbohydrates must first be hydrolyzed into simple sugars before fermentation.

Carbohydrate hydrolysis can be accomplished by weak acids or by enzymes. Yields obtained with inexpensive acids, such as sulfuric, are low (some 50% of the potential glucose is made available). Enzymes are expensive but lead to larger yields. Either process is energy intensive.

It may be possible to use microorganisms to perform both hydrolysis and fermentation in a single step.

Hexoses have the empirical formula $(CH_2O)_6$. There are 16 different hexoses with this same formula. To distinguish one from another, it is necessary to examine their structural arrangement. Hexoses can exist in either a linear chain or in a cyclic structure, the latter being more common and the only one found in aqueous solutions. To simplify the explanation that follows, we will use the linear formula.

A simple way to arrange the atoms that constitute a hexose would be:

$$
\begin{array}{ccccccc}
\text{H} & \text{H} & \text{H} & \text{H} & \text{H} & \text{H} \\
| & | & | & | & | & | \\
-\text{C}- & \text{C}- & \text{C}- & \text{C}- & \text{C}- & \text{C}- \\
| & | & | & | & | & | \\
\text{O} & \text{O} & \text{O} & \text{O} & \text{O} & \text{O} \\
| & | & | & | & | & | \\
\text{H} & \text{H} & \text{H} & \text{H} & \text{H} & \text{H} \\
* & \#
\end{array}
$$

The preceding structure satisfies the empirical formula but leaves two dangling valences at the end of the chain. To correct this, one can remove the H marked with an asterisk, *, and place it at the other end. The formula then represents correctly the (linear) structure of glucose:

$$
\begin{array}{ccccccc}
\text{H} & \text{H} & \text{H} & \text{H} & \text{H} & \text{H} \\
| & | & | & | & | & | \\
\text{C}- & \text{C}- & \text{C}- & \text{C}- & \text{C}- & \text{C}-\text{H} \\
|| & | & | & | & | & | \\
\text{O} & \text{O} & \text{O} & \text{O} & \text{O} & \text{O} \\
 & | & | & | & | & | \\
 & \text{H} & \text{H} & \text{H} & \text{H} & \text{H}
\end{array}
$$

Notice that glucose has an O=C–H group: it is an aldehyde, while another common hexose called **fructose** is a ketone because the H that was moved is the one marked with a # and is not at the end of the chain:

$$
\begin{array}{ccccccc}
\text{H} & & \text{H} & \text{H} & \text{H} & \text{H} \\
| & & | & | & | & | \\
\text{H}-\text{C}- & \text{C}- & \text{C}- & \text{C}- & \text{C}- & \text{C}-\text{H} \\
| & || & | & | & | & | \\
\text{O} & \text{O} & \text{O} & \text{O} & \text{O} & \text{O} \\
| & & | & | & | & | \\
\text{H} & & \text{H} & \text{H} & \text{H} & \text{H}
\end{array}
$$

13.5

Sucrose is a dimer of the two preceding sugars. It consists of a glucose and a fructose molecule attached to one another with loss of a water molecule.

13.3   Biomass as Fuel

There are numerous ways to use biomass as fuel:
1. Biomass can be burned as harvested. Wood, saw dust, corn cobs, rice husks, and other agricultural residue can be burned directly in appropriate furnaces. The heat of combustion of most dry biomass lies in the 15 to 19 MJ/kg range.
2. The utility of wood as a fuel can, traditionally, be improved by transforming it into charcoal.
3. Wood can be gasified to drive vehicles and fuel industries.
4. Methanol can be made from wood.
5. Palm oil can be used in diesel engines.
6, The sap of some plants is so rich in hydrocarbons that it also can be used directly as diesel fuel.
7. Sugars and starches can be fermented into ethanol.
8. Biomass residues can be digested into a methane-rich biogas.

No matter which technology is used, biomass has, over fossil fuels, the singular advantage of not contributing to the increase of CO_2 in the atmosphere. Some pessimistic but plausible scenarios foresee, for the not too far future, a CO_2-driven change of the planetary mean temperature which could lead either to the melting of the polar ices causing the flooding of coastal cities such as New York and Los Angeles, or else, to such a change in ocean currents that the Gulf Stream might be diverted bringing extreme cold weather to Europe.

Regardless of whether a near-future climate catastrophe will occur or not, it is undeniable that the CO_2 concentration in the atmosphere is growing alarmingly. The American daily consumption of gasoline alone is now up to 7.3 million barrels per day. This amounts to 1.2×10^9 liters or about 0.7×10^9 kg per day and means that the exhaust of American cars spews into the atmosphere 610,000 tons of carbon every day in the form of carbon dioxide or 220 million tons of carbon per year. This is such a large number that it is difficult to visualize: imagine adding yearly to our atmosphere an amount of carbon massing the same as over 5000 of our largest battleships! Our carbon consumption in the form of gasoline is equivalent to burning 35 million giant trees every year. This is due to gasoline alone. If all the oil used up is considered, the above numbers become 3 to 4 times larger. There is also a vast amount of natural gas and coal being used (56% of all the electricity in the United States of America is derived from coal). All these fuels generate unbelievable amounts of carbon dioxide.

When Earth was young and devoid of life, its atmosphere consisted mostly of nitrogen and carbon dioxide, with no free oxygen. Early life was completely anaerobic (more accurately, anoxygenic). But bacteria began to learn how to use sunlight to extract free hydrogen from the water in which they lived. The available hydrogen then became the energy source to drive further chemistry to build carbohydrates and proteins and, of course, DNA (or perhaps, RNA). The only trouble with this process is that when water is split, in addition to hydrogen, one also gets an extremely reactive and poisonous gas—oxygen, a gas in which the bacteria and their descendants, the modern plants, had little interest. Oxygen was thus released into the atmosphere as a pollutant, while carbon dioxide was removed and, in part, retained and fossilized creating the fossil fuels we now use. With the passing of time, most of the CO_2 was scrubbed off the air and stored underground. Life, being adaptable, learned not only to tolerate oxygen but actually to benefit from its availability. Thus, animals were created.

At present we are busily burning fossil fuels in a serious attempt to restore the atmosphere to its pristine anoxygenic days.

When fuel produced from biomass is burned, all the carbon released is carbon that was removed from the air the previous growing season, not carbon that has been stored underground these last 600 million years. Thus, the net amount of atmospheric carbon dioxide does not change. An additional advantage of biomass-derived fuel is that it has a low sulfur content, thus reducing the acid rain problem.

13.3.1 Wood Gasifiers

The fuel shortage during World War II motivated the development of gasifiers (using charcoal or wood) for fueling automobiles and tractors.

Gasifiers operate by destructive distillation of a solid fuel using part of this very fuel to generate the necessary heat.

Recently, advanced gasifiers have made their appearance. Their output is a mixture of hydrogen, carbon monoxide, carbon dioxide, and nitrogen (from the air used in the partial combustion process). When a gasifier is fed oxygen instead of air, its product gases can generate high flame temperatures making these devices useful in the steel industry.

The syngas produced by the gasifiers can be fed directly to oil furnaces provided with special burners. This simplifies the conversion of oil-burning installations to wood-burning ones without requiring a completely new furnace.

A good wood gasifier will deliver to the furnace 85% of the combustion energy of the wood. Part of this energy is the chemical energy of the generated fuels and part is the heat content of the hot gases. If the gases are allowed to cool prior to burning, only 65% of the wood energy is recovered.

Wood contains little sulfur and is, from this standpoint, a clean fuel. Its burning does, however, generate tars that tend to pollute the exhaust gases and to clog passages in the equipment. To avoid this problem, most modern gasifiers are of the downdraft type. This causes the recirculation of the generated gases through the hottest region of the flame cracking the tars and yielding a surprisingly clean burning fuel. Usually, the gases are circulated through a cyclone to remove the particulates.

Advanced wood gasifiers were developed in Brazil where the shortage of oil has spurred the search for alternate fuels (see the subsection on ethanol). Most of the wood is from eucalyptus trees that grow quickly even in poor, dry land. In the US, a variety of hybrid willow bush that grows over 3 m per year is being investigated as a fuel source for big power plants. If future limitations on CO_2 emissions are legislated, wood will be in a clearly advantageous position since there is no *net* CO_2 produced by its combustion. Willow shrubs (that do not look like the usual willow trees) will, like eucalyptus, grow in marginal land. Edwin H. White is the dean of research of the College of Environmental Science and Forestry of the State University of New York where the experimental work is being done. The original idea of using willow shrubs is from Sweden.

The syngas produced by gasifiers can be used as feedstock for a number of chemicals employing processes outlined in Chapter 11. The production of methanol from wood-generated syngas is especially attractive.

13.3.2 Ethanol

13.3.2.1 Ethanol Production

Countries with large territories and small oil resources may profit from the use of ethanol to satisfy part of their fuel requirements. Owing to its high antiknock properties, ethanol is more efficient than gasoline as an automotive fuel because higher compression ratios can be tolerated (11:1 compared with 9:1 for gasoline). This partially compensates the smaller volumetric energy density of ethanol.

Ethanol can be derived from a number of vegetal products: sugars, starches, and cellulose. Only the sugar-derived ethanol has been produced in scales vast enough to be used as a major automotive fuel. However, other sources of ethanol do hold some promise and deserve further investigation.

It is important to keep in mind that, in addition to the raw material (sucrose), a substantial amount of energy is needed in the ethanol production process especially for distillation. If a plant is rich in sugar but poor in some additional fuel, one must consider a separate source of fuel, such as wood in the beets/eucalyptus combination. The advantage of sugar cane is that is produces both the sugar in its sap and the fuel from its bagasse (the squeezed-out stalks).

At present, 75 tons of **raw** sugar cane are produced annually per hectare in Brazil. The cane delivered to the processing plant is called **burned and cropped (b&c)** and represents 77% of the mass of the raw cane. The reason for this reduction is that the stalks are separated from the leaves (which are burned and whose ashes are left in the field as fertilizer) and from the roots that remain in the ground to sprout for the next crop. Average cane production is, therefore, 58 tons of b&c per hectare per year.[†]

Each ton of b&c yields 740 kg of juice (135 kg of sucrose and 605 kg of water) and 260 kg of moist bagasse (130 kg of dry bagasse). Since the higher heating value of sucrose is 16.5 MJ/kg and that of the bagasse is 19.2 MJ/kg, the total heating value of a ton of b&c is 4.7 GJ of which 2.2 GJ come from the sucrose and 2.5 from the bagasse.

Per hectare per year, the biomass produced corresponds to 0.27 TJ. This is equivalent to 0.86 W per square meter. Assuming an average insolation of 225 W per square meter, the photosynthetic efficiency of sugar cane is 0.38%.

The 135 kg of sucrose found in 1 ton of b&c are transformed into 70 liters of ethanol with a combustion energy of 1.7 GJ. The practical sucrose-ethanol conversion efficiency is, therefore, 76% (compare with the theoretical 97%).

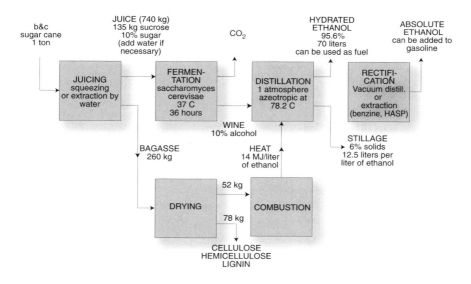

Figure 13.3   Flow chart for the production of ethanol (typical Brazilian practice).

[†] Hawaiian and continental US sugar cane plantations are substantially more productive thanks to intensive fertilization. Energy plantations must minimize the use of fertilizers because these are energy consumers themselves.

One hectare of sugar cane yields 4060 liters of ethanol per year (without any additional energy input because the bagasse produced exceeds the amount needed to distill the final product). This, however does not include the energy used in tilling, transportation, and so on. Thus, the solar energy-to-ethanol conversion efficiency is 0.13%. This is much less than that of photovoltaic converters. Nevertheless, ethanol production for fuel has been demonstrated as economically attractive.

Of the 130 kg of dry bagasse produced per ton of b&c, only 52 are needed for distillation, the rest is available as a by-product valuable as chemical feedstock. The excess cellulose finds many applications as, for example, a raw material for production of paper. Hemicellulose can easily (compared with cellulose) be hydrolyzed and yields a series of 5-carbon sugars (pentoses) from which products such as furfural can be obtained. Furfural is used in the separation of alkanes from other hydrocarbons in petroleum (for production of lubricating oils) and in the preparation of furfuric resins used in sand casting. Lignin can be the raw material for production of some phenolic compounds. It also can be transformed into metallurgical coke.

A large Brazilian ethanol plant can produce 1.2 million liters/day, energetically equivalent to some 5000 barrels of oil per day. It draws sugar cane from 110,000 hectares and has a daily output of 8600 tons of cellulose, 5200 tons of hemicellulose and 1900 tons of coke as by-products.

Industrial ethanol production starts with the separation of the juice from the bagasse. This can be done mechanically by squeezing or by extraction with water. The juice, suitably diluted to permit fermentation is placed into vats to which a yeast culture (*Saccharomyces cerevisae*) has been added.[†] The liquid is maintained at 37 C and, within a short time, carbon dioxide begins to evolve. The process lasts some 36 hours, after which the wine containing about 10% alcohol is transferred to the distillery. The carbon dioxide is, usually, collected and compressed for sale.

Ethanol is miscible with water—that is, the two substances can be dissolved in one another in any proportion without settling out. They form an azeotrope with the boiling point of 78.2 C at normal pressure (pure ethanol boils at 78.5 C). Water and alcohol co-distill with a fixed composition of 95.6% alcohol. Atmospheric pressure distillation yields this **hydrated** alcohol that contains too much water to be added to gasoline to make gasohol. Nevertheless, it is perfectly adequate as a fuel when used all by itself.

To obtain **anhydrous** alcohol, the hydrated form must be **rectified**. This can be done by "vacuum" distillation (at 95 mm Hg, the azeotrope consists of 99.5% ethanol), or as is currently done, by extracting the water.

[†] Another alcohol producing microorganism is the bacterium *Zymomona mobilis*, which can ferment more concentrated sugar solutions than *Saccharomyces*.

Modern water extraction methods may save distillation energy. It takes little energy to concentrate the wine into 80% ethanol; however, to go above this value, the energy increases markedly. It is important to investigate if it makes sense to truncate the distillation at, say, the 80% level and remove the rest of the water by extraction. In fact, there is a substance sold under the trademark HASP that selectively absorbs water from a water/ethanol mixture. HASP can then be dried at a modest cost of energy.

Present Brazilian practices require some 14 MJ of heat to distill one liter of ethanol. Although this represents 57% of the energy of the final product, the distillation energy comes from a low utility energy source: the bagasse.

Distillation yields two outputs: concentrated ethanol and the leftover stillage containing much of the nutrients the sugar cane extracted from the soil. Stillage represents a problem: for each liter of alcohol, 12.5 liters of stillage are produced which, owing to the danger of pollution, cannot be dumped into public waters. Many suggestions for stillage disposal are being investigated. Among these, are its use as fertilizer, which requires close proximity of the distillery to the plantation because the stillage is highly diluted (6% solids) and cannot be economically transported long distances.

Ethanol is a good substitute for gasoline but not for diesel or fuel oil, two products that, in Brazil, continue to be derived from (partly) imported petroleum. Since it is impossible to produce diesel and fuel oil without producing gasoline, Brazil finds itself in the peculiar position of having to export gasoline that, thanks to its ethanol program, it does not use. The fuel program can only be considered a success when biomass-derived chemicals fully replace all petrochemicals. As pointed out previously, wood gasifiers are beginning to replace fuel oil and, possibly, palm oil could replace diesel oil.

13.3.2.2 Fermentation

The production of ethanol discussed above is based on the fermentation of glucose that proceeds as described below.

The first step in the fermentation of glucose is its enzymatic splitting into two identical, 3-carbon, glyceraldehyde molecules:

$$
\begin{array}{ccc}
\text{H} & \text{H} & \text{H} \\
| & | & | \\
\text{C}- & \text{C}- & \text{C}-\text{H} \\
|| & | & | \\
\text{O} & \text{O} & \text{O} \\
& | & | \\
& \text{H} & \text{H}
\end{array}
$$

The glyceraldehyde reacts with water and is oxidized into glyceric acid by an enzyme that removes and binds two H atoms:

$$
\begin{array}{ccccc}
\underset{\substack{\text{GLYCER-}\\\text{ALDEHYDE}}}{\overset{\displaystyle \underset{\substack{\overset{|}{C}-\overset{|}{C}-\overset{|}{C}-H}{\underset{\substack{||\;\;|\;\;|}{O\;\;O\;\;O}}}}{\underset{|\;\;|}{H\;\;H}}{H\;\;H\;\;H}}{}}
& + H_2O &\rightarrow& & + H_2
\end{array}
$$

| GLYCER-ALDEHYDE | WATER | GLYCERIC ACID | HYDROGEN bound to enzyme |

Glyceric acid loses one water molecule and is transformed into pyruvic acid that decomposes into carbon dioxide and acetaldehyde. Thus, the 6-carbon glucose is transformed into the 2-carbon acetaldehyde, which is the number of carbons in ethanol:

PYRUVIC ACID $\rightarrow$ ACET-ALDEHYDE $+ CO_2$

Finally, the acetaldehyde interacts with the hydrogen-carrying enzyme and is reduced to ethanol, regenerating the enzyme:

ACET-ALDEHYDE + HYDROGEN bound to enzyme $\rightarrow$ ETHANOL

The sequence above is one of several ways in which organisms use glucose. It is the metabolic path used by *Saccharomyces cerevisae* (*sakcharon* = sugar + *mykes* = fungus + *cerevisae* = of beer), the yeast commonly used in the fermentation of sugars into alcohol. The utilization of glucose by muscles follows exactly the same path up to the production of pyruvic acid. In muscles, this acid is directly reduced by the hydrogen-carrying enzyme (skipping the acetaldehyde step) and producing lactic acid:

PYRUVIC ACID + HYDROGEN bound to enzyme $\rightarrow$ LACTIC ACID

The overall reaction for the fermentation of glucose is

$$(CH_2O)_6 \rightarrow 2\ CO_2 + 2\ C_2H_5OH. \tag{1}$$

The heat of combustion of glucose is 15.6 MJ/kg and, since its molecular mass is 180 daltons, its heat of combustion per kilomole is 2.81 GJ. Ethanol has a heat of combustion of 29.7 MJ/kg and a molecular mass of 46 daltons. This amounts to 1.37 GJ/kilomole. When burned, the 2 kilomoles of ethanol in the reaction above produce 2.73 GJ of heat. Consequently, the transformation of glucose into ethanol proceeds with the surprisingly high ideal efficiency of $2.73/2.81 = 0.975$.

13.3.3 Dissociated Alcohols

Alcohols, both methanol and ethanol, can have their efficiency as automotive fuel boosted by being dissociated prior to burning. The endothermic dissociation can be driven by the waste exhaust heat.

Three different factors contribute to the efficiency increase:

1. The dissociated products have larger heats of combustion than the original alcohol;
2. the dissociated products have large resistance to "knocking" and, thus, tolerate higher compression ratios leading to better thermodynamic efficiency; and
3. leaner mixtures can be used because hydrogen, the major component of the dissociation, has a wide flammability range.

For practical reasons the dissociation of alcohol has to be accomplished in the vehicle just before the use of the fuel so that only liquids have to be transported in the vehicle. Alcohols can be dissociated to hydrogen and carbon monoxide or, if water is introduced for the shift reaction, to hydrogen and carbon dioxide. The direct dissociation of methanol occurs at 250 to 350 C, while steam reforming can be performed at some 200 C. Equilibrium tends toward almost complete dissociation. Undesirable side reactions constitute a problem requiring careful choice of catalysts.

The system involves a catalytic converter that includes a vaporizer, a superheater, a reactor, and a gas cooler. Methanol is evaporated by the heat from the engine coolant and leaves the vaporizer at 80 C. It enters the superheater where engine exhaust heat raises its temperature to 250 C. The hot alcohol vapor is then fed to the reactor where additional exhaust heat drives the dissociation in the presence of a catalyst. To improve volumetric efficiency, the mixture of hydrogen and carbon dioxide (accompanied by small amounts of impurities) is cooled to 100 C before returning to the engine.

Test results with a 65 kW Chevrolet Citation (1980) engine showed efficiency increases of up to 48% over gasoline. The compression ratio of the engine was changed from its original 8.3:1 to 14:1.

13.3.4 Anaerobic Digestion

Fermentation transforms a limited number of substances (sugars) into an ethanol/water mixture from which the alcohol has to be recovered by means of an energy intensive distillation. Anaerobic digestion, in contrast, transforms many different vegetable and animal substances into methane, a gas that, essentially insoluble in water, evolves naturally and can be collected with a minimal expenditure of energy. Fermentation is well understood, whereas digestion has been studied to a lesser extent because, up to recently, important industrial applications were few.

Most of the energy in the raw material appears in the methane produced; only little is used in creating microbial cells. Digestion is, therefore, an efficient way to refine biomass.

Consider the digestion of glucose:

$$(CH_2O)_6 = 3\ CH_4(g) + 3CO_2(g) \tag{2}$$

One kilomole of glucose has a higher heat of combustion of 2.81 GJ. The 3 kilomoles of methane have a heating value of $3 \times 0.89 = 2.67$ GJ. The reaction is exothermic, but little energy is lost: 95% of it appears in the methane. The enthalpy change of the reaction is -0.14 GJ per kilomole of glucose. On the other hand, the free energy change is much larger: -0.418 GJ per kilomole of glucose. Since $\Delta S = (\Delta H - \Delta G)/T = 927$ kJ/K per kilomole of glucose, it can be seen that the reaction leads to a substantial increase in entropy; the thermodynamic driving force for the reaction is, consequently, large.

Anaerobic digestion proceeds in three distinct steps:

1. A group of different reactions mediated by several types of fermentative bacteria degrade various substances into fragments of lower molecular mass (polysaccharides into sugars, proteins into peptides and amino acids, fats into glycerine and fatty acids, nucleic acids into nitrogen heterocycles, ribose, and inorganic phosphates).

2. Further degradation is promoted by acetogenic bacteria that convert the alcohols and higher acids into acetic acid, hydrogen, and carbon dioxide.

3. The acetic acid, hydrogen, and carbon dioxide produced in Steps 1 and

2 are used by methanogenic archea[†] to produce methane and carbon dioxide from the acid and methane and water from the hydrogen and carbon dioxide.

Table 13.1 shows some of the reactions that take place in each of the above steps.

The result of the digestion process is a methane/carbon dioxide mixture called **biogas**, typically containing 65% of methane.

Steps 1 and 2 are mediated both by **obligate** anaerobes and by **facultative** anaerobic bacteria that can operate in the absence of oxygen but are not poisoned by this gas. Step 3 is mediated by strictly anaerobic archea which do not survive the presence of oxygen. Digestion is the result of the cooperation of many microorganisms—a complete ecological system.

In a batch mode, the rate limiting step is that involving methanogens— Steps 1 and 2 proceed rapidly while Step 3 proceeds at a much slower rate.

Digestion can be easily demonstrated in the laboratory by mixing organic matter (manure, for instance) with water and placing it in a sealed jar as suggested in Figure 13.4. The center opening is for loading the mixture, the one on the left is for the removal of samples of the liquid (mostly for pH determination) and the one on the right is the biogas outlet. The gas produced by the digester can be washed by bubbling it through a water filled beaker after which it is collected for chromatographic analysis.

To expedite the process, the mixture should be inoculated with sludge from another digester (the local sewage plant is a convenient source). The jar should be kept at about 37 C. After a few days, the pH of the mixture drops markedly indicating that the acid forming step is in progress.

Initially, little gas evolves, but sampling indicates that oxygen is being consumed. Soon, gas starts bubbling out of the liquid and the chromatograph will show it to be mostly carbon dioxide. In some 10 days, methane starts appearing in the output and its relative concentration builds up rapidly to the final level.

[†] Modern taxonomy classifies all life forms into one of three domains: Eubacteria (true bacteria in which the genetic material is not collected into a nucleus), archea, and eucarya (organisms whose cells have a nucleus). Some of the archea have unusual behavior, thriving at surprisingly high temperatures (above 100 C) they are **hyperthermophiles**, while other microorganisms are psychrophilic, mesophilic, or thermophilic; see further on. Enzymes from hyperthermophiles, being stable at high temperatures, are being adapted to industrial uses where operation at elevated temperatures are necessary. Notice that in the above classification of living beings there is no place for viruses, which are, according to this view, not considered alive.

Table 13.1

Estimated Free Energy Changes

of Some Reactions in Aerobic Digestion

(after D. L. Klass)

Reaction	Free energy (MJ/Kmole)
Fermentative bacteria	
$C_6H_{10}O_5 + H_2O \rightarrow (CH_2O)_6$	-17.7
$(CH_2O)_6 \rightarrow 3CH_3CO_2^- + 3H^+$	-311
$(CH_2O)_6 + 2H_2O \rightarrow CH_3CH_2CO_2^- + H^+ + 3CO_2 + 5H_2$	-192
$(CH_2O)_6 \rightarrow CH_3CH_2CH_2CO_2^- + H^+ + 2CO_2 + 2H_2$	-264
$(CH_2O)_6 + 6H_2O \rightarrow 6CO_2 + 12H_2$	-25.9
Acetogenic bacteria	
$(CH_2O)_6 + 2H_2O \rightarrow 2CH_3CO_2^- + 2H^+ + 2CO_2 + 4H_2$	-216
$CH_3CH_2CO_2^- + H^+ + 2H_2O \rightarrow CH_3CO_2^- + H^+ + CO_2 + 3H_2$	$+71.7$
$CH_3CH_2CH_2CO_2^- + H^+ + 2H_2O \rightarrow 2CH_3CO_2^- + 2H^+ + 2H_2$	$+48.3$
$CH_3CH_2OH + H_2O \rightarrow CH_3CO_2^- + H^+ + 2H_2$	$+9.7$
$2CO_2 + 4H_2 \rightarrow CH_3CO_2^- + H^+ + 2H_2O$	-94.9
Methanogenic archea	
$CH_3CO_2^- + H^+ \rightarrow CH_4 + CO_2$	-35.8
$CO_2 + 4H_2 \rightarrow CH_4 + 2H_2O$	-131
$HCO_3^- + H^+ + 4H_2 \rightarrow CH_4 + 3H_2O$	-136

Conditions: 25 C, pH = 7.

Figure 13.4   A laboratory digester.

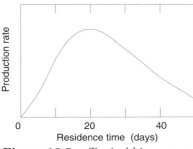

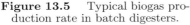

Figure 13.5   Typical biogas production rate in batch digesters.

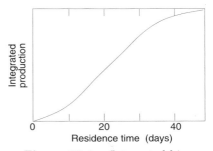

Figure 13.6   Integrated biogas production in batch digesters.

The rate of gas production—that is, the digestion rate, varies as shown in Figure 13.5. The integrated production is shown in Figure 13.6. Practical digesters operate either in the batch or the continuous mode. Batch digesters are loaded with an input slurry and then sealed. The different microorganisms must then grow and the three digestion phases will occur in sequence. Steady state is never achieved and long residence times (30 to 60 days) are required.

Continuous or semicontinuous digesters receive the slurry at one end of the tank while the spent sludge is withdrawn from the other. The material slowly moves through regions of the tank where the different reactions occur in nearly steady state conditions. A shorter residence time is thus achieved (10 to 20 days). Long residence times translate into small **loading rates**— that is, into the need for large tanks for a given gas production rate. This increases the investment cost.

The influent to the digester consists of a watery slurry containing chopped pieces of organic matter. Part of the solids, whether in suspension or in solution, can (ideally) be digested and part cannot. The former is called **volatile solids**, VS, and the latter, **fixed solids**, FS.

Methane yield, Y, is the amount of methane produced from 1 kg of VS. Depending on the influent, the ideal yield, Y_0, corresponding to a total digestion of the VS, ranges between 21 and 29 MJ of methane per kg of VS.

Digesters are able to use only a fraction, R, of the VS; their yield is $Y = RY_0$. R depends on the digester design, on the nature of the influent and on the residence time. It may range from as low as 25% for grass clippings to as high as 50% for activated sewage sludge.

Actual gas production depends on the **allowable loading rate**, L, measured in kg (VS) per cubic meter of the digester per day.

The power density of the digester is

$$P = LRY_0 \qquad \text{joules/m}^3\text{per day.} \tag{3}$$

For representative values, $Y_0 = 25$ MJ(CH$_4$)/kg(VS) and $R = 0.4$,

$$P = 10L \qquad \text{MJ(CH}_4)/\text{m}^3 \text{per day.} \qquad (4)$$

or

$$P \approx 100L \qquad \text{W(CH}_4)/\text{m}^3. \qquad (5)$$

Batch digesters can handle loading rates up to 1.5 kg (VS) m^{-3}day^{-1} and can produce methane at a rate of 150 W/m^3 of digester. Continuous digesters, owing to their loading rates of up to 6 kg (VS) m^{-3}day^{-1}, produce some 600 W/m^3. Larger power densities lead to more economical methane production. One way to increase power density is to decrease the residence time. This can be accomplished in several ways, including:

1. operation at higher temperature,
2. agitation,
3. microorganism immobilization,
4. strain selection,
5. material addition, and
6, pre-treatment of the influent.

There are three temperature ranges that favor digestion:

1. **psychrophilic** range centered around 5 C,
2. **mesophilic** range centered around 37 C, and
3. **thermophilic** range centered around 55 C.

The fastest digestion occurs at the thermophilic range. However, only large installations can be economically operated at a high temperature owing to excessive heat losses inherent in small digesters (because of their large surface-to-volume ratio).

Methane and carbon dioxide are only slightly soluble in water; microbubbles of these gases tend to surround the methanogenic organisms that produce them reducing their contact with the liquid and retarding the reaction. Agitation will dislodge these bubbles, as will lowering the pressure and increasing the temperature. Agitation can, sometimes, be economically achieved by rebubbling the produced gas through the slurry.

The greatest gain in digestion speed is achieved by microorganism immobilization. The ecosystem that promotes digestion is slow in forming and every time the residue is removed from the equipment, so are the microorganisms and new cultures have to be established. This is true for both batch and continuous digesters. One way to ensure the retention of the microflora is to use a **biological filter** in which the digester is filled with inert chips on which a bacterial slime is formed. The residue is removed while the chips remain in the equipment.

For this arrangement to operate properly, it is necessary that the residue be sufficiently fluid, a condition difficult to achieve with many of the raw materials used. In fact, a problem in the operation of a digester is the clogging caused by the thick sludge formed. Thus, adequate pretreatment of the raw material may be essential for efficient operation.

To improve digestion rates one can add to the influent certain materials such as enzymes, growth factors, fertilizers, and so on. Since proper carbon-to-nitrogen (C/N) and carbon-to-phosphorus (C/P) ratios are necessary for efficient digestion, it may be important to add chemicals for the correction of these ratios when certain raw materials are to be processed. Good C/N ratios hover around 20:1 and C/P ratios, around 80:1.

If the biogas is to be shipped over long distance, it may be important to upgrade it to pipeline quality (by removing much of the CO_2) using any of the methods discussed in Chapter 10.

Biogas is produced spontaneously by garbage dumps and by wastewater treatment plants. The gas from garbage dumps is, most often, simply allowed to escape into the atmosphere. The methane in it is a potent greenhouse gas and it may be accompanied by objectionable impurities such as hydrogen chloride. In some instances, wells are driven into garbage dumps, the biogass is collected and purified, and then injected in normal natural gas pipelines. Wastewater treatment plants frequently collect the gas and flare it making no use of the energy it contains.

It is estimated that methane, corresponding, energetically, to some 5000 barrels of oil, is released daily by wastewater treatment plants. A modest amount of energy but one that should not be wasted. Some municipalities feed this gas to fuel cells to produce electricity and usable heat. This eliminates the need for emergency generators.

Currently, the main application of anaerobic digestion is in sewage treatment. Sewage and other liquid organic waste, from which, in a **preliminary treatment**, large solid objects have been removed, are piped to settling tanks (**primary treatment**) where a great deal of the solids precipitate out forming a **sludge** that is sent to an anaerobic digester. The output of the latter is methane gas and a type of sludge that can be sold to farmers as fertilizer (provided heavy metals and other toxic substances are not present). The liquid left over from this phase is, sometimes, sent to a **secondary treatment** tank where *aerobic* bacteria remove much of the pathogens still left in the system. The action in this tank can be expedited by supplying the oxygen the bacteria consume. This can be done either by aeration (at the cost of some energy) or by growing microalgae that photosynthetically produce the gas. The output of this stage is safe to release into the environment (again, provided it is not contaminated by toxic substances). Unfortunately this effluent contains nitrates and phosphates that may promote algal bloom (**eutrophication**). Further improvement of the effluent can be accomplished in a **tertiary treatment**, which may yield pure drinking water. Many technologies can be used at this last stage including reverse osmosis. Alternatively, the effluent can be sent to artificially created wet lands, where nature will take care of the purifying.

Sewage treatment does not necessarily include all of the preceding stages. It can be truncated, frequently right after the primary treatment. Tertiary treatment is not too commonly used. Sometimes, extreme truncation is the practice: the sewage is simply not treated at all.

The use of algae for aeration suggests that these could be harvested for further anaerobic digestion thus producing additional methane. For this to be possible, selected algae must be use to make the harvesting practical. They must be relatively large so they can be filtered out. The difficulty is in assuring that the correct species remain in the system, otherwise they may be naturally replaced by species unsuitable for harvesting.

One possibility for producing additional methane and more fertilizer and for removing toxic substances is to allow certain vascular aquatic plants to grow in the sewage. Among the most attractive of these is the water hyacinth (Eichhornia crassipes). This is a floating plant of extraordinary productivity. Under favorable circumstances it will produce 600 kg of *dry* biomass per hectare per day (Yount). This enormous vegetative capacity has transformed E. crassipes into a vegetable pest. It has invaded lakes and rivers in Africa and Asia and even the United States interfering with the movement of boats, with fishing, and with hydroelectric plants. In addition, E. crassipes can harbor in its leaves the snail carrying the fluke that causes the serious disease, schistosomiasis. The problem was big enough for a specialized journal, the *Hyacinth Control Journal*, to be created. Only in its country of origin is the water hyacinth in (partial) equilibrium with its local predator—the Brazilian manatee, *Trichechus manatus manatus*, known locally as "peixe boi" (cow fish). One very useful characteristic of E. crassipes is its capacity for removing heavy metals and phenols from polluted waters.

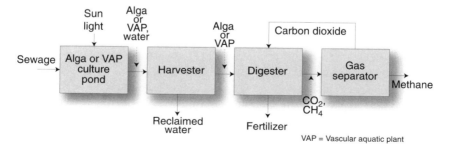

Figure 13.7   Block diagram of a combination sewage treatment–fertilize–methane plant.

Wolverton has demonstrated that the plant can remove, daily, some 0.3 kg of heavy metals and over 50 kg of phenol per hectare. To maintain this removal capacity, periodic harvest (every 5 weeks?) must be carried out. The collected material can be used for CH_4 production but the sludge should not be used as fertilizer owing to the accumulation of toxics.

13.4 Photosynthesis

Photosynthesis is an extremely complicated process best left to plant physiologists and biochemists. Here, with an apology to these experts, we will make a somewhat unscientific attempt to look at a plant leaf as a system to be examined from the point of view of the engineer. This, although far from accurate, will, it is hoped, lead to some insight into the behavior of plants and of some of the mechanisms involved.

Formally, photosynthesis is a process through which carbon dioxide and water are transformed into carbohydrates and oxygen according to

$$nCO_2 \; + \; nH_2O \rightarrow (CH_2O)_n + nO_2 \tag{6}$$

It is well known that this is a complex process. It is not a simple dissociation of the carbon dioxide with subsequent attachment of the freed carbon to a water molecule, as it might appear from an inspection of the chemical equation, above. It can be shown, by means of radioactive tracers, that the oxygen liberated comes from the water, not from the carbon dioxide. Figure 13.8 represents schematically the photosynthesizing unit in a plant leaf. Photochemical reactions occur in the compartment labeled **chloroplast**. This compartment must have a transparent wall (labeled **upper epidermis** in the figure) to allow penetration of light. Water and carbon dioxide must be conveyed to the chloroplast.

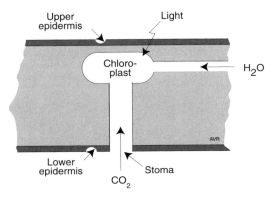

Figure 13.8   Schematic representation of the photosynthesizing unit in a plant leaf.

In this schematic, water is brought in through a duct, while carbon dioxide enters from the atmosphere via a channel whose opening is on the **lower epidermis**.

This opening is called a **stoma** (plural: either stomas or stomata; from the Greek word for "mouth"). A waxy coating renders both epidermis impermeable to water. Thus, losses of water occur mainly through the stomata.

Carbon dioxide from the atmosphere moves towards the chloroplast by diffusion. Its flux, ϕ, is, therefore, proportional to the difference between the concentration, $[CO_2]_a$, of the carbon dioxide in the atmosphere and the concentration, $[CO_2]_c$, of this gas at the chloroplast.

The flux is also proportional to the **stomatal conductance**, V, a quantity that has the dimensions of velocity:

$$\phi = V\left([CO_2]_a - [CO_2]_c\right) \tag{7}$$

The carbon dioxide sink is, of course, the photosynthetic reaction in the chloroplast. Under steady state conditions, the rate of carbon dioxide uptake per unit leaf area is equal to the incoming flux. This rate is proportional to the light power density, P (W m^{-2}), the dioxide concentration, $[CO_2]_c$ (kmoles m^{-3}), and to the reaction rate, r (m^{3}J^{-1}). Thus,

$$\phi = r[CO_2]_c P \tag{8}$$

Eliminating $[CO_2]_c$ from Equations 7 and 8 , one obtains

$$\phi = \frac{r[CO_2]_a P}{1 + (r/V)P} \tag{9}$$

ϕ is measured in kmoles of CO$_2$ uptake per second per square meter of leaf area.

Under conditions of low-light power densities—that is, when $P << V/r$, the carbon dioxide uptake rate is proportional to the concentration of this gas in the atmosphere and to the light power density:

$$\phi = r[CO_2]_a P \tag{10}$$

When the power density is high—that is, when $P >> V/r$, the uptake rate becomes

$$\phi = V[CO_2]_a \tag{11}$$

At high light power densities, the carbon dioxide uptake rate is independent of P, but is still proportional to $[CO_2]_a$.

The reaction rate, r, is a basic characteristic of photosynthesis, and one would expect it to be the same for all advanced plants. Measurements made by plant physiologists confirm this expectation for cases in which there is a normal concentration of oxygen in the air (Ehleringer and Bjorkman, 1977).

r can be determined by measuring the CO$_2$ uptake rate as a function of the light power density in the region in which Equation 10 is valid.

We have, then,

$$r = \frac{1}{[CO_2]_a} \frac{\phi}{P}. \tag{12}$$

From the Ehleringer and Bjorkman data, it can be seen that the slope of the ϕ versus P plot in the linear region (Figure 13.11), is 178 nmoles/J $= 178 \times 10^{-12}$ kilomoles/J. The measurements were made at RTP with a carbon dioxide concentration of 330 ppm—that is, with a CO_2 pressure of 330 μatm. At RTP the volume, $\forall$, occupied by 1 kilomole of any perfect gas is

$$\forall = \frac{RT}{p} = \frac{8314 \times 298}{1.0133 \times 10^5} = 24.45 \ \text{m}^3/\text{kmole}, \tag{13}$$

hence, 330 ppm correspond to

$$[CO_2] = \frac{330 \times 10^{-6}}{24.45} = 13.5 \times 10^{-6} \ \text{kmoles/m}^3. \tag{14}$$

Thus,

$$r = \frac{178 \times 10^{-12} \ \text{kmoles/J}}{13.5 \times 10^{-6} \ \text{kmoles/m}^3} = 13.2 \times 10^{-6} \ \text{m}^3/\text{J}, \tag{15}$$

and

$$\phi = \frac{13.2 \times 10^{-6}[CO_2]_a P}{1 + 13.2 \times 10^{-6}P/V} = \frac{178 \times 10^{-12}P}{1 + 13.2 \times 10^{-6}P/V}. \tag{16}$$

The second part of the equation is for a CO_2 concentration of 330 ppm. This equation should (roughly) predict the carbon dioxide uptake rate as a function of the sunlight power density for any plant provided the stomatal velocity, V, be known. Assigning arbitrarily to V a value of 0.004 m/s, independent of the light power density, we would get the plot of Figure 13.9.

At this point we are able to predict how the photosynthetic efficiency depends on light power density. The useful output of photosynthesis is the energy content of the created biomass, which is about 440 MJ per kilomole of carbon. This is more than the enthalpy of formation of carbon dioxide (393.5 MJ/kmole, in absolute value). The additional energy comes from the hydrogen always present in plant tissue.

The uptake of 1 kmole of CO_2 corresponds, thus, to the **fixation** of 440 MJ. Hence, the rate of fixation is 440 ϕ MWm^{-2} and the photosynthetic efficiency is

$$\eta_\nu \simeq 4.4 \times 10^8 \phi/P. \tag{17}$$

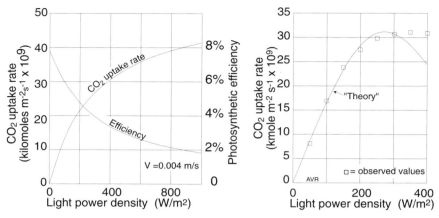

Figure 13.9 A very simple model of photosynthesis predicts a reduction of efficiency with rising light power density.

Figure 13.10 Using a light power density-dependent V a better match between theory and observation is obtained.

If this is true, then the largest possible photosynthetic efficiency is

$$4.4 \times 10^8 \times 178 \times 10^{-12} = 0.078. \tag{18}$$

In order to test the simple model developed above, we need measured data on carbon dioxide uptake rate as a function of light power density.

Consider the data of Bjorkman and Berry (1973) for *Atriplex rosea* (Common name: Redscale). We find that although the general shape of the ϕ vs P curve of predicted and measured values is roughly the same, the numerical values are quite different. It would be astounding if it were otherwise because we took the value, $V = 0.04$ m/s, out of a hat. In addition, we assumed, without justification, that V is independent of P. By experimentation, we find that if we make $V = 0.0395 \exp(-0.00664P)$,[†] we do get a better match between our "theory" and observation, at least in the lower light power density region. Se Figure 13.10

There is indeed a mechanism to regulate the stomatal conductance. The lips of the stomata consist of special cells that, when full of water, become turgid (swell up). When dry, these cells flatten out and close the opening. It is not impossible that the size of the stomatal opening is controlled, at least indirectly, by the light intensity on the leaf because the more light the greater the desiccation. It is not unreasonable that V decreases as P increases. *Increasing the illumination may cause a reduction in the stomatal velocity accentuating the predicted decrease in efficiency.*

[†] With a medieval twist of mind, we chose this exponential dependence because it is well known that nature, which abhors a vacuum, actually loves exponentials.

If we return to the Bjorkman and Berry work we find that their data were collected not only for A. rosea but also for another plant of the same genus but of a different species: A triangularis (common name: Spearscale).

From Figure 13.11, it can be seen that A. rosea is substantially more efficient than A. triangularis under the conditions of the experiment ($T = 25$ C and $[CO_2] = 300$ ppm). Note that the CO_2 uptake rate for either plant peaks at light power densities well below those of full sunlight (1000 Wm^{-2}). Clearly, plant leaves are optimized for the average light power density the plant receives (one leaf shading another), not for full sun light.

To understand the reason for the difference in photosynthetic performance of the two species of Atriplex, one must become a little bit more familiar with the photosynthesis mechanism.

In most plants, photosynthesis is carried out by two pigments—**chlorophyll A and chlorophyll B**. These, when purified and in solution, are green in color—that is, they reflect green light that, therefore, cannot be utilized by the plant and, consequently would be wasted. In living plants, the situation may be less extreme but leaves are still green and thus fail to use a band of the solar spectrum in which the power density is substantial. Figure 13.12 shows the absorption spectra for the two chlorophylls.

In all chlorophyll plants, transformation of CO_2 into carbohydrates is performed by a chain of chemical reactions called the **Calvin-Benson cycle**. The essential mechanism involves the combination of CO_2 with a 5-carbon sugar called RuDP (ribulose-1,5-diphosphate) and the formation of two molecules of a 3-carbon substance called PGA (phosphoglyceric acid). The PGA is then converted into carbohydrates and, part of it, into more RuDP to perpetuate the cycle.

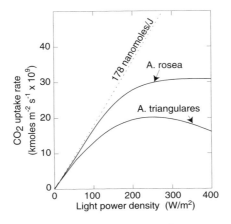

Figure 13.11 CO_2 uptake rate by two different varieties of *Atriplex*.

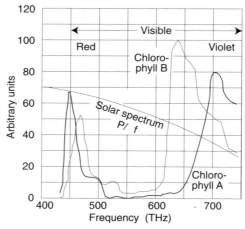

Figure 13.12   Absorption spectra of chlorophylls A and B.

The difference between A. rosea and A. triangularis resides not in the basic cycle but rather in the manner in which CO_2 is delivered to the cycle. In A. triangularis, atmospheric CO_2 is taken directly to the RuDP, while in A. rosea, there is an intervening ancillary "pumping" mechanism that delivers the CO_2 to the RuDP in a manner somewhat parallel to the delivery of oxygen to animal cells by the flowing blood. See Figure 13.13.

The "hemoglobin" of plants is PEP (phospho-enol-pyruvate) that has great affinity for CO_2. The end products of the PEP-CO_2 reaction are malic and aspartic acids, each containing 4 carbons per molecule. In the presence of appropriate enzymes, the two acids release their CO_2 to the RuDP and are eventually transformed back to PEP, closing the pumping cycle.

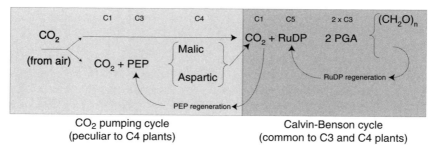

Figure 13.13   The difference between a C3 and a C4 plant is that the latter has an additional mechanism to extract CO_2 from the air and deliver it to the Calvin-Benson cycle. The symbols C3, C4, and so on indicate the number of carbon atoms in the molecule.

Plants that use PEP and consequently create a 4-carbon acid as the first product in the photosynthesis chain are called **C4** plants, while those whose first photosynthesis product is a 3-carbon acid, are called **C3** plants. As a general rough rule almost all trees and most shrubs are C3 plants, while many tropical savanna grasses and sedges are C4.

Carbon dioxide is much more reactive with PEP than with RuDP. In addition, when the CO_2/O_2 ratio becomes small, the presence of oxygen interferes with the functioning of the RuDP and photosynthesis is halted. Thus, under low CO_2 concentrations, C3 plants interrupt their photosynthesis while C4 plants continue to operate.

Low CO_2 concentrations at the chloroplast occur when the stomatal conductance is insufficient owing to the desiccation of the lipcells in the stoma. This saves water (which evaporates through the stomata) but also hinders the flow of CO_2. Nevertheless, the plant metabolism continues even though photosynthesis may be halted. The plant continues its respiration, burning carbohydrates and evolving CO_2, an activity exactly opposite to photosynthesis. This phenomenon can be detected in any plant in the dark. C3 plants, because they interrupt their photosynthesis under high light power densities, are also net CO_2 *producers* under such conditions. This is called **photorespiration**, and cannot be detected in C4 plants.

An interesting experiment demonstrates the difference between C3 and C4 plants. A C3 and a C4 plant are placed under a bell jar and are properly illuminated. Both may prosper as long as there is an adequate CO_2 supply. Soon, however, this gas is used up and the C3 plant will go into photorespiration consuming its own tissues, shriveling up and producing the CO_2 that the C4 plant scavenges to grow.

When the CO_2 supply is adequate, C4 plants lose their advantage because of the no longer necessary "pumping" mechanism that uses up energy. Figure 13.14 shows that, at high CO_2 concentrations, A. triangularis is more efficient than A. rosea (data from Bjorkman and Berry).

The advantages of the C4 plants over the C3 are maximum under conditions of high-light power density, high temperatures, and limited water supply. These are the conditions found in semi-arid tropical regions. Sugar cane, a tropical plant, for instance, is a C4 plant. C4 plants also outperform C3 ones when CO_2 concentrations are low. During the last glacial maximum (some 20,000 years ago), the CO_2 pressure in air fell to 190–200 μatm. This caused grasses to replace forests in many regions. See Street-Perrott et al.

Water plants, of course, do not have to save water. One would expect that their stomata do not have regulating valves (a water-saving adaptation). Indeed, leaves of water plants wilt quite rapidly when plucked. In such plants, the stomatal conductance will be independent of light level and their CO_2 uptake rate will not fall off with increasing light power density.

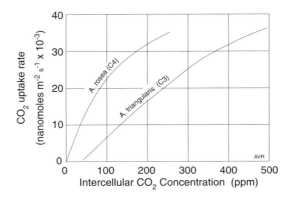

Figure 13.14   At high CO_2 concentrations, C3 plants
are more efficient than C4.

Thus, vascular aquatic plants are potentially a more efficient biomass producer than land plants. This is confirmed by the high productivity of such water plants as the water hyacinth (*Eichhornia crassipes*).

References

Finegold, J., and T. McKinnon, Dissociated methanol test results, *Alcohol Fuels Program Technical Review*, U.S. Dept Energy (SERI), Summer **1982**.

Klass, Donald L., Methane from anaerobic fermentation, *Science*, **223**, 1021, Mar. 9, **1984.**

Street-Perrott, F. Alayne, Y. Huang, R. A. Perrott, G. Eglinton, P. Baker, L. Ben Khelifa, D. D. Harkness, and D. O. Olago, Impact of lower atmospheric carbon dioxide on tropical mountain ecosystems, *Science.* **278** 1422, Nov. 21, **1997.**

Wolverton, B. C. (with coauthors), A Series of Nasa Technical Memoranda, TM-X-727720 through 27, TM-X-727729 through 31, *Nasa National Space Technology Laboratories, Bay St. Louis, MS 39520,* **1975**.

Yount, J. L., Report of the 35th annual meeting, *Florida Anti-mosquito Association* p. 83, **1964**.

PROBLEMS

13.1 A plant leaf takes up 0.05 μmoles of CO_2 per minute per cm^2 of leaf area when exposed to sunlight with a power density of 50 W/m^2 in an atmosphere containing 330 ppm of CO_2. When the light power density is raised to 600 W/m^2, the uptake is 0.36 μmoles min^{-1}cm^{-2}. Assume that in the above range, the stomata do not change their openings.

What is the expected uptake at 100 W/m^2 in the same atmosphere and the same temperature as above? And at 1000 W/m^2?

Measurements reveal that the uptake at 1000 W/m^2 is only 0.40 μmoles min^{-1}cm^{-2}. This reduction must be the result of a partial closing of the stomata. What is the ratio of the stomatal area at 1000 W/m^2 to that at 100 W/m^2?

What is the expected uptake at 100 W/m^2 if the CO_2 concentration is increased to 400 ppm?

13.2 An automobile can be fueled by dissociated alcohol. The energy necessary for such dissociation can come from waste exhaust heat. In the presence of catalysts, the process proceeds rapidly at temperatures around 350 C.

Consider liquid methanol that is catalytically converted to hydrogen and carbon monoxide. Compare the lower heats of combustion of methanol with those of the products. Do you gain anything from the dissociation?

Compare the entropy of gaseous methanol with that of the products. Does this favor the reaction?

Assume that a gasoline engine with a 9:1 compression ratio is fueled by:

1. gasoline,
2. methanol, and
3. dissociated methanol.

Assuming that the three fuels lead to identical engine behavior, compare the energy per liter of the fuel compared with that of gasoline.

The compression ratio is now changed to the maximum compatible with each fuel:

gasoline: 9:1,
methanol: 12:1,
dissociated methanol: 16:1.

Remembering that the efficiency of a spark-ignition is

$$\eta = 1 - r^{1-\gamma},$$

where r is the compression ratio and γ is the ratio of the specific heats (use 1.4), compare the new energy per liter ratios.

The gasoline and methanol molecules are complex and this leads to a low value of γ. With hydrogen and carbon monoxide, γ is much higher. Change the above calculations using 1.2 for gasoline and methanol, and 1.7 for the dissociated methanol.

13.3   Under proper conditions, water hyacinths (Eichhornia crassipes), a floating plant, will grow at such a rate that their dry biomass increases 5% per day. The total water content of these plants is high (94%). Nevertheless, 400 kg of dry matter can be harvested daily from one hectare of plantation.

Consider a plantation with one hectare area consisting of a long canal (folded upon itself). At the starting point (seeding end), the canal is 0.5 m wide. It expands gradually enough to just accommodate the growing plants that are slowly swept along by the current. Assume a current of constant speed such that the plants take 60 days to float from the seeding end to the harvesting end.

1. How wide must the canal be at the harvesting end?
2. How long must the canal be?
3. How much energy is harvested per day (express this in GJ and in barrels of oil)?
4. How many kilograms of wet plants must be used daily as seed?
5. Assume that 50% of the biomass energy is converted into methane through a digestion process. Estimate the methane yield in cubic meters per day.
6. If the methane is burned in a gas turbine or in a diesel engine with 20% efficiency, what is the average electric power that can be generated?
7. Assuming that the average depth of the canal is 1 m, what is the amount of water that has to flow in daily provided there is no loss by evaporation and infiltration?
8. Do you have any good ideas of how to insure that the water velocity is kept reasonably constant notwithstanding the expanding canal?

13.4   Sugar cane is submitted to an illumination of 500 W/m^2. Assuming a stomatal velocity of 6 mm/s, what is the photosynthetic efficiency (defined as the ratio of the heat of combustion of the dry biomass generated to the incident solar energy)?

13.5   Here is a typical task that an energy consultant might tackle:

The operator of a large alcohol distillery wants to know if it makes economic sense to use the leftover bagasse as a further source of ethanol. In the traditional process, the amount of bagasse obtained from 1 ton of burned and cropped sugar cane is larger than the amount that has to be burned to drive the distillation process. The excess is either sold or used to generate electricity for the plant. The question is how much additional alcohol can be obtained by hydrolyzing all the polysaccharides (cellulose and hemicellulose) in the leftover bagasse. We will make the following simplifying assumptions:

a. The hydrolysis will yield 600 grams of sugars (glucose and pentoses) per kilogram of polysaccharides.
b. The hydrolysis requires no energy (not true!).
c. The glucose-to-ethanol and the pentose-to-ethanol yields are the same as the sucrose-to-ethanol yields of the traditional process.
d. The data for this problem are those discussed in Section 13.3.2 of the textbook.

Calculate the additional amount of alcohol that can be obtained from 1 ton of burned and cropped sugar cane. Comment.

13.6   A digester consists of a cylindrical stainless steel tank with a diameter, d, and a height, $2d$. The metal is 3 mm thick.

An R-5 (American system) fiberglass blanket completely covers the tank.

The contents of the digester are agitated so that they are, essentially, at a uniform temperature of 37 C. To simplify this problem, assume that the influent (the material fed in) is preheated to 37 C.

Stainless steel has a thermal conductivity of $\lambda = 60$ W m^{-1} K^{-1}.

We desire a net production rate of 1 kW of methane. The digester must produce, in addition, enough methane to fire a heater that keeps the material in it at a constant temperature (the digestion process, itself, generates negligible heat). The efficiency of the heater is 70%.

Loading rate is $L = 4$ kg of volatile solids per cubic meter of digester per day. Assume that 1 kg of volatile solids produces 25 MJ of methane and that 40% of all the volatile solids in the influent are digested.

The outside temperature is such that the external walls of the digester are at a uniform 20 C.

Estimate the diameter of the digester.

13.7   A hypothetical plant has perfectly horizontal leaves. The carbon dioxide uptake rate, ϕ, depends linearly (in the usual manner) on the solar light power density, P, provided $P \leq 150$ W/m^2. Above this value, ϕ is constant, independently of P.

Assume the insolation at normal incidence is 1000 W/m^2 during all daylight hours. The latitude is $45° N$

1. What is the amount of carbon fixed by each square meter of leaf area during the winter solstice day?

13.31

2. Estimate the number of kilograms of dry biomass the plant produces on the winter solstice day per hectare of leaves.

13.8   A hemispherical, perfectly transparent container has a 5-m radius. The bottom part is covered with moist soil on which a bush with horizontal leaves is planted. The leaves are arranged in such a way that they do not shade one another. The volume occupied by the plant by the plant is 0.75 m^3 and its leaf area is 4 m^2.

The "air" inside the container has the following composition:

O_2	52 kg
N_2	208 kg
A	2.6 kg
H_2O	26 kg
CO_2	0.78 kg

The temperature inside the container is a uniform 298 K.

The soil is moist enough to supply all the water needed by the plant. However no water is ever exchanged directly between soil and air.

The plant has the usual value of r (13.2×10^{-6} m^3/J) and has a stomatal conductance of 10 mm/s.

Argon has a molecular mass of 40 daltons.

The plant has a carbon dioxide uptake rate proportional to the illumination power density, P, for values of $P < 200$ W/m^2 and independent of P for greater values. The light has the same spectral distribution as the sun.

1. What is the air pressure inside the container?
2. What is the *total* carbon dioxide uptake rate (kmoles/s and mg/s) under the above circumstances, when the illumination power density is 150 W/m^2?
3. Make a rough estimate of how long will it take to reduce the CO_2 concentration to the level of normal (outside) air.
4. Assume that the leaf area of the plant does not change. Calculate more accurately the time required to reduce the carbon dioxide concentration to the normal value.
5. What is the composition of the air when the carbon dioxide concentration reaches its normal level?

Chapter 14
Photovoltaic Converters

14.1 Introduction

Economics are crucial to the success of any energy utilization system, and photovoltaic converters are no exception. At present, development of such converters is in the stage in which prices are still coming down rapidly. It is, nevertheless, difficult to compare photovoltaic with other systems solely on the basis of investment cost. While photovoltaics have low operating cost, since they consume no fuel, their peak power can be only realized on a clear day, with the converter facing the sun. The average power will be less than half the peak power for sun tracking systems (owing to nighttime) and less than one-fourth of peak power for nontracking systems.[†]

Because of the intermittence of the sunlight, storage systems or standby power generators are frequently required, substantially increasing investment costs. Exceptions include photovoltaics used to supply peak loads that coincide with periods of maximum insolation. Also, photovoltaics used directly to power irrigation systems need no storage provisions.

An important consideration in any type of power plant is the ratio of the *average* power delivered to the peak installed power. This is related to the plant utilization factor. In Chapter 1, we saw that, in the United States., nuclear power plants operate with an utilization factor of over 87%—that is, over 0.87 watts are (on average) actually delivered for each watt of installed plant capacity. Maintenance shut-downs are the main limit to full utilization of the installed power capacity. In the case of wind and solar power, the main limitation comes from the unreliability and intermittence of the wind or the sun, respectively. In Chapter 12, we listed representative average insolations for a number of US localities. Remember that peak insolation is 1000 W/m^2. It can be seen from that list that the average is between 1/4 and 1/6 of the peak. One would think that for photovoltaic plants, this would be the expected ratio of average to peak power. However, practical experience is always somewhat disappointing when compared to expectations. Two German utility-scale photovoltaic installations for which data are available report their average-to-peak ratio as only 0.11. The Cal State Campus at Hayward, in sunny California, being blessed by a better climate, estimates an average-to-peak ratio of 0.16. Hence, one can roughly estimate that a photovoltaic system must have 6 times bigger peak capacity than a nuclear plant if we want both to deliver the same average power.

[†] Clearly, these limitations do not apply to systems that collect sunlight out in space.

The efficiency of photovoltaic systems is low compared with that of traditional thermal or hydroelectric plants. It may be over 20% for sophisticated crystalline silicon systems and some 5% for some inexpensive thin film ones. However, efficiency is not of primary interest in many photovoltaic installations. The cost per peak watt may be the important characteristic. At $0.20 per peak watt (if this figure is ever attained), even low efficiency photovoltaics would be attractive. Remember that conventional hydroelectric and fossil fueled plants cost around $1.00/W, while nuclear ones may cost over $5.00/W. But, again, remember that these other energy sources, unlike photovoltaics, can operate continuously and thus, on average, produce much more energy.

At the low price mentioned, it is possible that many buildings will have their external surfaces covered with photovoltaics. In some cases, the average energy generated may well exceed the needs of the building. It will, however, be generated only on sunny days, not on rainy ones nor will there be any generation at night. Consequently, adequate storage facilities must be available, especially in case of residences, where demand during the day may be small while at night, requirements are higher.

If the building is **off-grid**, as some rural properties are—that is, if it has to be entirely self-sufficient, expensive batteries or some other storage scheme is needed. On the other hand, if the building is connected to the power grid, storage can be provided by the local utility company in what is called a **utility-tie** system. The excess energy generated by the customer can be sold to the utility for a price below that charged by the utility to the customer.[†] The price differential would pay for the storage and distribution. A dual metering system can be used, one meter measures the outgoing power from the custumer to the utility and the other the power from the utility to the customer. A **utility-intertied inverter** must be used so that the frequency and the phase of the customer-generated electricity matches the grid.

In areas of unreliable energy supply, subjected to frequent blackouts, a hybrid of the off-grid and the utility-tie system may be useful.

Such **building-integrated photovoltaic** (BIPV) systems will become progressively more popular as the price of solar collectors decreases. The land area of the building, the structure on which to mount the solar collectors (roof and external walls), the very roof, and the connection to the grid are all investments made even if no BIPV is used and thus should not be charged to the BIPV cost but must be included in the cost of centrally generated PV systems.

[†] The Public Utilities Regulatory Policy Act of 1978 fixes the so-called **avoided cost** that is the minimum amount that a utility has to pay an independent power producer. Frequently, this is only the cost of the fuel saved.

BIPV seems more appropriate for individual residences, which have a lot of roof area per inhabitant, and for any one- or two-floor structures such as those used by some factory or office complexes. Apartment buildings, with their much larger population density, would be at a disadvantage. Of course, shade trees would be contraindicated. Roofs would have to resist hailstorms and, in California, be immune to the pelting by avocados falling from overhanging trees.

In evaluating the performance of any particular grid-intertied building integrated photovoltaic installation, the first thing to do is to select the proper degree of optimism. At the current state of the art (2005) uncertainties are large and improvements are still coming fast: it is possible, by using reasonable assumptions, to prove either the effectiveness of the system or its complete undesirability.

Let us consider a typical one family residence in California. It will have, perhaps, a 200 m² roof area. If electricity is used for cooking but not for heating, it will use an average of 50 kWh per day at a peak of less than 10 kW. The expected average yearly insolation is 250 W/m². If relatively inexpensive thin film photovoltaic units are used, one can count on 5% efficiency, which means, under optimal conditions, each square meter of roof will generate 12.5 W on average or 2500 W for the whole roof. Almost certainly both the tilt and the orientation of the roof will not be optimum. Let us say, one can count on 10 W/m² or 2000 W total (on a day-and-night) average. The peak production will be 4 times larger or 8 kW. This does not quite satisfy the 10 kW peak requirement but the excess can come from the grid.

2 kW average correspond to 48 kWh per day, very close to the desired value of 50 kWh. Thus, the energy balance is acceptable.

At peak insolation, the photovoltaics will generate 40 W/m². If the cost is $1.00/W (peak), the collectors will cost $40/m² or $8000 for the roof. We now must make a number of additional assumptions:

1. Labor cost: $3000.
2. Ancillary equipment (controls, inverter, etc): $2000.
3. Cost of capital: $15%/year on a 10-year loan.
4. Longevity of equipment: > 10 years.

Under the assumption above, the initial investment is of $13,000, and the yearly cost is $1950. During one year, the system will generate 48 × 365 = 17,500 $/kWh.

(continued)

(continued)

Hence the cost of the generated electricity is 1950/17,500 = 0.111 $/kWh, which, by coincidence, is almost precisely the price of electricity in Northern California. Thus, at first glance, during the first 10 years, it makes no economic difference if the electricity comes from the utility or from the BIPV system. If there are no maintenance or repair costs, then, after 10 years, the electricity is free for as long as the equipment holds out.

Of course, there are additional costs. For example, although the average power generated is equal to the average power used, the custumer will sell his surplus to the utility at a price much lower than what he will pay for the energy received from the utility.

This example illustrates the difficulty of making a reasonable assessment of the economic possibilities of a system still in development. For instance, if the efficiency of the photovoltaic blankets is much higher than the value of 5% assumed and/or the cost per peak watt is much lower than the $1.00/W we used, then the system will suddenly be extremely attractive. One cannot dismiss the possibility that in the near future, efficiencies of 10% and costs of $0.50/W will become reality.

BIPV systems have not yet made a significant contribution to power generation, but there are a number of **utility-scale** photovoltaic plants in operation. Japan leads the way in total generating capacity, but all installations are relatively small, none reaching the 1 MW peak level. In this latter category, the champion is Germany with over 20 MW installed, followed by the United States with over 15 MW. When one considers that one single typical nuclear plant generates over 1 GW, about the same as a large thermal installation and a large hydroelectric plant such as Itaipu in Brazil has a 12.5 GW capacity, it is apparent that utility-scale photovoltaic have, so far, made only very modest contributions, much smaller than even that of wind turbines of which over 2 GW are operating in the United States.

Table 14.1 lists all known utility-scale photovoltaic plants with more than 1 MW peak output operating worldwide in mid-2004. The data were extracted from http://www.pvresources.com whose operators do an excellent job in keeping the information up to date. An important future contribution to this list will be the large (5 MW peak) installation that will come online in Prescott, Arizona, in 2005.

14.4

Table 14.1
Utility-Scale Photovoltaic Plants
Larger Than 1 MW Peak Output (2004)

Place	Country	Peak Power (MW)
Hemau	Germany	4
Rancho Seco, CA	USA	3.9
Tucson, AZ	USA	3.78
Serre	Italy	3.3
Haarlemmermeer-NieuWland	Netherlands	2.3
Munich	Germany	2.1
Neustadt	Germany	2
Prescott, AZ	USA	2
Passau	Germany	1.75
Passau	Germany	1.7
Markstetten/Oberofalz	Germany	1.6
Relzow	Germany	1.5
Saarbrucken	Germany	1.4
Twenty-Nine Palms, CA	USA	1.3
Tudela	Spain	1.18
Santa Rita, CA	USA	1.18
Meerane	Germany	1.1
Toledo	Spain	1.1
Hayward, CA	USA	1.05
Fürth	Germany	1
Herne	Germany	1
Herne	Germany	1
Vasto	Italy	1
Amersfoort-Nieuwland	Netherlands	1
Davis, CA	USA	1
Farmingdale, NY	USA	1
San Diego, CA	USA	1

Another area in which reality does not follow expectations is in plant cost estimates. A cost of \$5.00/W is being quoted for utility-class photovoltaic installations, yet the very modern Hayward plant (not yet completed) seems to be coming in at \$6.8/W.

Photovoltaics are nearly essential for powering satellites and other space craft that do not stray too far away from the sun. Craft that venture into deeper space—Jupiter and beyond—must be powered by much more expensive radioisotope thermal generators (RPG). See the chapter on thermoelectrics. Photovoltaics are also the power source of choice for many remotely located devices and, of course, for a large number of consumer electronics.

Broadly, three distinct techniques are used in building solar cells:

1. Crystalline material, most frequently silicon. These are expensive to produce but yield good efficiencies ($> 20\%$) if they are made of single crystals. Somewhat cheaper, but less efficient, are polycrystalline units.

2. Amorphous thin films (Si, GaAs, $CuInSe_2$, TiO_2, etc.) have efficiencies of some 7% but are much less expensive. They can be made into flexible sheets.

3. Organic polymers, still in early development, could easily become the best overall solution. They promise to be low cost, light weight, and flexible. One can imagine photovoltaic blankets cheap enough to serve as roofing materials replacing present day shingles, tarpaper, or tiles. Their manufacturing methods will probably be less toxic than those of inorganic materials.

14.2   Theoretical Efficiency

It is hoped that the reader is familiar with the general behavior of semiconductors: the existence of two energy bands separated by an energy gap, W_g joules wide. Electrons in the filled lower (valence) band cannot (in general) conduct electricity, while electrons in the upper (conduction) band can, but this band is normally empty. This is the case, at least at absolute zero temperature.

If an incoming photon has an energy, hf larger than W_g, an electron from the valence band can be kicked into the conduction band leaving behind a hole. If this incoming photon has insufficient energy, it may not interact with the material which, consequently, is transparent to such radiation. When $hf > W_g$, the photon is absorbed, hopefully creating **excitons**—bound electron-hole pairs, and the material is opaque to the radiation in question. A p-n junction will separate some excitons into their component electron and hole, making the electron available as an external current.

The exact boundary between transparency and opacity depends on the type of material considered. Table 14.2 displays the data for some semiconductors. Diamonds, a form of carbon that crystallizes in the same manner as silicon and germanium being highly resistant to heat and radiation, are a promising material for transistors that have to operate in hostile environments.

The few readers who need to refresh their basic knowledge of semiconductors may take advantage of Appendix B in this chapter, which offers a very elementary description of the band structure and of the formation of p-n junctions.

Table 14.2
Light Absorption Limits for Some Semiconductors

Material	ν_0 (THz)	λ (nm)	W_g (eV)	Region in which transition from transparent to opaque occurs
α-Sn	19.3	15500	0.08	Far infrared
Ge	162	1850	0.67	Infrared
Si	265	1130	1.10	Infrared
GaAs	326	920	1.35	Near infrared
GaP	540	555	2.24	Visible
C	1300	230	5.40	Ultraviolet

A structure that allows light to create electron-hole pairs and then separates them by means of a built-in *p-n* junction constitutes a **photovoltaic diode** or, simply, a **photodiode**.

Photodiodes exposed to monochromatic light can, theoretically, achieve 100% efficiency in converting radiation to electric energy. However, in the majority of practical cases, photodiodes are exposed to broadband radiation—that is, to a stream of photons of different energies. When exposed to such radiation, the efficiency is limited by two mechanisms:

1. Photons with less than the "band-gap energy" (the energy necessary to disrupt covalent bonds) are incapable of creating electron-hole pairs and either pass right through the material or interact with it producing only heat. Thus, only part of the spectrum is usable.
2. Photons with more than the band-gap energy create electrons and holes with energies that exceed the mean thermal energy of these carriers. This excess is quickly dissipated as heat and is lost as far as electric energy is concerned. Thus, only part of the photon energy is usable.

In all cases, whether we are considering ideal or practical devices, their efficiency is defined as the ratio of the power, P_L, delivered to the load to the power, P_{in}, of the incident radiation,

$$\eta \equiv \frac{P_L}{P_{in}}. \tag{1}$$

The characteristics of broad-band radiation can be described by specifying the power density, ΔP, of the radiation in a given frequency interval, Δf, as was done for solar radiation in Table 12.1 (Chapter 12). Alternatively, taking the $\Delta P / \Delta f$ ratio to the limit, one writes an equation expressing the dependence of $\partial P / \partial f$ on f. The total incident power density is, then,

$$P_{in} = \int_0^\infty \frac{\partial P}{\partial f} df. \tag{2}$$

In the case of black body, $\partial P / \partial f$ is given by **Planck's equation**,

$$\frac{\partial P}{\partial f} = A \frac{f^3}{e^{\frac{hf}{kT}} - 1} \tag{3}$$

where A is a constant having the units of W m^{-2}Hz^{-4}. Hence,

$$P_{in} = A \int_0^\infty \frac{f^3}{e^{\frac{hf}{kT}} - 1} df. \tag{4}$$

Let $x \equiv \frac{hf}{kT}$, then

$$df = \frac{kT}{h} dx \quad \text{and} \quad f^3 = \left(\frac{kT}{h}\right)^3 x^3. \tag{5}$$

$$P_{in} = A \left(\frac{kT}{h}\right)^4 \int_0^\infty \frac{x^3}{e^x - 1} dx \tag{6}$$

The definite integral, $\int_0^\infty \frac{x^3}{e^x-1} dx$ has the value $\pi^2/15$, therefore

$$P_{in} = A \left(\frac{kT}{h}\right)^4 \frac{\pi^4}{15} = aT^4, \tag{7}$$

where a (W m^{-2} K^{-4}) is also a constant.

When the temperature of a black body radiator increases, not only does the total power, P, increase (Equation 7), but, in addition, the peak radiation is shifted to higher frequencies as can be seen from Figure 14.1. There is a simple relationship between the frequency, f_{peak}, and the temperature, T.

The proportionality between the light power density and the fourth power of the temperature is related to the **Stefan-Boltzmann** law.

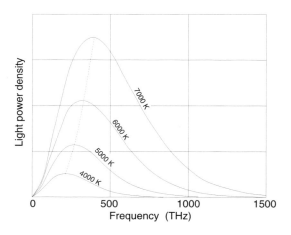

Figure 14.1   The peak of the p vs f curve of a black body moves toward higher frequencies as the temperature increases.

From Equation 3, we see that the shape of the distribution curve is determined by the factor, $\dfrac{f^3}{e^{\frac{hf}{kT}}-1}$. The peak occurs when

$$\frac{d}{df}\left(\frac{f^3}{e^{\frac{hf}{kT}}-1}\right)=0. \tag{8}$$

Making the $x\equiv\dfrac{hf}{kT}$ substitution and taking the derivative, we obtain the equation

$$(3-x)\exp x-3=0, \tag{9}$$

whose numerical solution is $x=2.821$. From the definition of x,

$$f_{peak}=\frac{k}{h}xT=59.06\times10^9 T. \tag{10}$$

The relation between f_{peak} and T is the **Wien's displacement law**.

It is useful to relate the total flux, ϕ, of photons that, given a specified spectral distribution, corresponds to a power density, P_{in}.

Consider a small frequency interval, Δf, centered on the frequency f. Since each photon has energy hf, the power density of radiation in this interval is

$$\Delta P=\Delta\phi\,hf\ \text{W/m}^2, \tag{11}$$

where $\Delta\phi$ is the photon flux (photons $\text{m}^{-2}\text{s}^{-1}$) in the interval under consideration. In the limit, when $\Delta f\to0$ (and dividing both sides by df),

$$\frac{d\phi}{df}=\frac{1}{hf}\frac{\partial P}{\partial f}, \tag{12}$$

and

$$\phi=\frac{1}{h}\int_0^\infty\frac{1}{f}\frac{\partial P}{\partial f}df. \tag{13}$$

Particularizing for the black body case and, once more, letting $x\equiv hf/kT$,

$$\phi=\frac{A}{h}\int_0^\infty\frac{1}{f}\frac{f^3}{e^{\frac{hf}{kT}}-1}df=\frac{A}{h}\int_0^\infty\frac{f^2}{e^{\frac{hf}{kT}}-1}df, \tag{14}$$

$$\phi=\frac{A}{h}\left(\frac{kT}{h}\right)^3\int_0^\infty\frac{x^2}{e^x-1}dx=2.404\frac{A}{h}\left(\frac{kT}{h}\right)^3. \tag{15}$$

because the definite integral, in this case, has the value 2.404.

Still for black body radiation, we can find the ratio of the light power density to the corresponding photon flux. From Equations 7 and 15,

$$\frac{P}{\phi}=\frac{A\left(\frac{kT}{h}\right)^4\frac{\pi^4}{15}}{2.404\frac{A}{h}\left(\frac{kT}{h}\right)^3}=37.28\times10^{-24}T. \tag{16}$$

Not surprisingly, the ratio of total power to total photon flux increases proportionally to the temperature because, as we saw when we derived Wien's displacement law, the higher the temperature the more energy does the average photon have.

Example 14.1

What is the photon flux when light radiated from a 6000 K black body has a power density of 1000 W/m^2?

From Equation 16,

$$\phi = \frac{P}{37.28 \times 10^{-24}T} = \frac{1000}{37.28 \times 10^{-24} \times 6000}$$
$$= 4.47 \times 10^{21} \quad \text{photons m}^{-2}\text{s}^{-1}. \tag{17}$$

For the ideal case, the efficiency of the device is, of course,

$$\eta_{ideal} = \frac{P_{L_{ideal}}}{P_{in}}. \tag{18}$$

We now need to know $P_{L_{ideal}}$.

If broad-band radiation falls on a semiconductor with a band-gap energy, $W_g = hf_g$, the photons with frequency $f < f_g$ will not create excess carriers. A fraction

$$G_L = \frac{1}{P}\int_0^{f_g} \frac{\partial P}{\partial f}\,df, \tag{19}$$

of the total radiation power density, P_{in}, will be lost.

Let ϕ_g be the total flux of photons with $f > f_g$. Each photon creates a single electron-hole pair with energy hf. However, as stated, the energy in excess of W_g will be randomized and will appear as heat and each photon contributes only W_g joules to the electric output. The useful electric energy (the energy, P_L, delivered to a load) will be,

$$P_L = \phi_g W_g \ \text{ W/m}^2. \tag{20}$$

The flux of photons with energy larger than hf_g is (adapting Equation 13),

$$\phi_g = \frac{1}{h}\int_{f_g}^{\infty} \frac{1}{f}\frac{\partial P}{\partial f}\,df. \tag{21}$$

The useful power is

$$P_L = hf_g\phi_g = f_g\int_{f_g}^{\infty} \frac{1}{f}\frac{\partial P}{\partial f}\,df, \tag{22}$$

and the efficiency is

$$\eta_{ideal} = \frac{P_L}{P_{in}} = f_g \frac{\int_{f_g}^{\infty} \frac{1}{f} \frac{\partial P}{\partial f} df}{\int_{0}^{\infty} \frac{\partial P}{\partial f} df}. \tag{23}$$

Observe that η_{ideal} depends only on the spectral distribution and on the band-gap of the semiconductor. It completely ignores the manner in which the device operates. Unlike the efficiency of real photodiodes, η_{ideal} does not depend on the level of illumination. Again, for the black body case,

$$\phi_g = \frac{A}{h} \int_{f_g}^{\infty} \frac{f^2}{e^{\frac{hf}{kT}} - 1} df = \frac{A}{h} \left(\frac{kT}{h}\right)^3 \int_{X}^{\infty} \frac{x^2}{e^x - 1} dx, \tag{24}$$

where $X = hf_g/kT = qV_g/kT$.

It should be obvious that the ratio $\sigma \equiv \phi_g/\phi$ depends only on the nature of the radiation considered, not on its intensity. The ratio is

$$\sigma \equiv \frac{\phi_g}{\phi} = \frac{\int_{X}^{\infty} \frac{x^2}{e^x - 1} dx}{\int_{0}^{\infty} \frac{x^2}{e^x - 1} dx} = \frac{\int_{X}^{\infty} \frac{x^2}{e^x - 1} dx}{2.404} = 0.416 \int_{X}^{\infty} \frac{x^2}{e^x - 1} dx. \tag{25}$$

For 6000-K black body radiation the ratio is a fixed 0.558 if $W_g = 1.1$ eV. The ideal efficiency of a photodiode is then

$$\eta_{ideal} = \frac{15}{\pi^4} \left(\frac{h}{k}\right)^4 \frac{f_g}{T^4} \int_{f_g}^{\infty} \frac{f^2}{e^{\frac{hf}{kT}} - 1} df. \tag{26}$$

It is more convenient to work with the band-gap voltage, V_g, instead of the corresponding frequency, $f_g = \frac{q}{h} W_g$,

$$\eta_{ideal} = \frac{15}{\pi^4} \left(\frac{h}{k}\right)^4 \frac{q V_g}{h T^4} \int_{\frac{qV_g}{h}}^{\infty} \frac{f^2}{e^{\frac{hf}{kT}} - 1} df. \tag{27}$$

Letting $x \equiv \frac{hf}{kT}$ as before,

$$\eta_{ideal} = \frac{15}{\pi^4} \left(\frac{h}{k}\right)^4 \frac{q}{h} \left(\frac{kT}{h}\right)^3 \frac{V_g}{T^4} \int_{\frac{qV_g}{kT}}^{\infty} \frac{x^2}{e^x - 1} dx = \frac{15}{\pi^4} \frac{q}{k} \frac{V_g}{T} \int_{\frac{qV_g}{kT}}^{\infty} \frac{x^2}{e^x - 1} dx$$

$$= 1780 \frac{V_g}{T} \int_{\frac{qV_g}{kT}}^{\infty} \frac{x^2}{e^x - 1} dx. \tag{28}$$

Note that the lower limit of the integral is now that value of x that corresponds to f_g.

There is no analytical solution to the preceding integral, but it can either be solved numerically or the table in Appendix A to this chapter can be used to determine the value of the definite integral (which is, of course, a simple number, function of the lower limit of the integral).

Example 14.2

What is the flux of photons that have more energy than that of the silicon band-gap (1.1 eV, i.e., $V_g = 1.1$ V) when light radiated from a 6000-K black body has a power density of 1000 W/m^2?

Equation 25 gives us the ratio, σ of ϕ_g to ϕ.

For the particular combination of this example ($V_g = 1.1$ V and $T = 6000$ K), the ratio is 0.558, and from Example 14.1, $\phi = 4.47 \times 10^{21}$ photons m^{-2} s^{-1}.

Consequently,

$$\phi_g = \sigma\phi = 0.558 \times 4.47 \times 10^{21} = 2.49 \times 10^{21} \text{ photons m}^{-2}\text{ s}^{-1}. \quad (29)$$

Example 14.3

What is the ideal efficiency of the photocell under the circumstance of the previous example?

Using Equation 28,

$$\eta_{ideal} = 1780 \frac{1.1}{6000} \int_{2.125}^{\infty} \frac{x^2}{e^x - 1} dx. \quad (30)$$

The lower limit of the integral is $X = hf_g/kT = qV_g/kT = 2.125$.

The value of the definite integral is 1.341 (by interpolation in the table in Appendix A to this chapter), hence,

$$\eta_{ideal} = 1780 \frac{1.1}{6000} 1.341 = 0.438. \quad (31)$$

Figure 14.2 shows how the ideal efficiency of a photodiode depends on its band-gap energy when exposed to a black body at 5800 K (about the temperature of the sun). Since the solar spectrum is not exactly that of a black body, the dependence is somewhat different from that shown in the figure. Also, the exact spectral distribution of sunlight in space differs from that on the ground owing to atmospheric absorption.

Efficiencies greater than these **black body spectrum efficiencies** can be achieved. Three techniques are discussed in the next three sections.

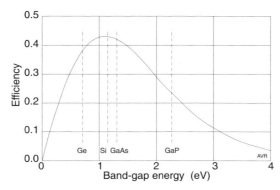

Figure 14.2   Dependence of the efficiency of a photodiode
on its band-gap energy. Black body at 5800 K.

14.3 Carrier Multiplication

In the derivation of the ideal efficiency of photocells in the preceding
section, we made the assumption that photons with energy greater than
that of the band-gap will yield a single **exciton** (electron-hole pair). The
excess energy of the photon will appear as excess kinetic energy of the
electron and of the hole and will be quickly be dissipated as heat: it is
thermalized. This assumption is almost precisely true for photocells made
of bulk material.

Why does a photon with energy $> 2W_g$ not yield additional excitons?
One reason is that a photon is indivisible, it cannot split into two parts
and act on two different covalent bonds. One single exciton is generated
from the interaction. However, the energetic carriers created could cause
ionization of an atom by impact and thus create another exciton. This
impact ionization occurs with low efficiency ($< 1\%$, in bulk Si) and makes
a minute contribution to the number of excitons (and thus to the number
of carriers) created.

Schaller and Klimov have recently (2004) demonstrated that carrier
multiplication by impact ionization can proceed with nearly 100% efficiency
if PbSe is in the form of nanocrystals (of about 5 nm size).

The excess photon energy, $W_{ph} - W_g$, is partitioned between the created
electron and hole according to their respective masses. Since in PbSe these
particles have nearly the same mass, each particle will receive approximately
the same excess energy: $(W_{ph} - W_g)/2$, which, for impact ionization, must
exceed W_g.

So, it is not surprising that impact ionization in lead selenide will occur
when $W_{ph} > 3W_g$. This establishes, for this particular case, the **threshold
energy** necessary for impact ionization.

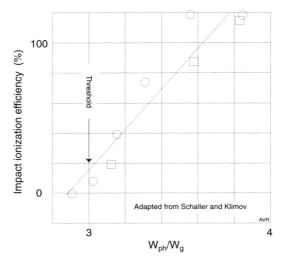

Figure 14.3   Impact ionization efficiency in PbSe as a function of the excess photon energy.

The process efficiency increases (initially) as the excess energy of the photon increases beyond the threshold value. See Figure 14.3, adapted from the paper by Schaller and Klimov. We fitted a straight line through the data to more clearly show the trend. Disregard the difference between circular and square data points—they represent two different measurement techniques.

Calculations show that, for lead selenide cells, the ideal efficiency of photodiodes can exceed 60%. To be sure, no such cells were tested in the Schaller and Klimov study, but Greenham et al. demonstrated solar cells based on nanocrystal back in 1996.

PbSe is a material of great interest for solar cells because it can have its W_g adjusted from about 0.3 to 1.3 eV. It absorbs radiation strongly from near infrared all the way into the ultraviolet. Band-gap adjustment is discussed further in Subsection 14.4.1.

To better understand carrier multiplication, do Problem 14.26.

14.4   Spectrally Selective Beam Splitting

The broad-band solar spectrum can be split into slices of radiation whose frequencies match the properties of the semiconductors used in the photodiodes. Thus, for silicon, a narrow band beginning at 265 THz (in the infrared) would best match its 1.1 eV band-gap energy. Other spectral slices would be directed either to other photodiodes or to an absorber to generate heat.

14.4.1 Cascaded Cells

Conceptually, the simplest beam splitting system is achieved when one superposes two photodiodes with different band-gaps. Assume that Diode #1, with the larger band-gap, W_{g1}, is on top (i.e., it is the one that receives the direct unfiltered light). It will absorb photons with energy above W_{g1}, but is transparent to those with energies below this threshold. If the diode underneath (Diode #2) has a lower band-gap, W_{g2}, it will absorb part of the photons that passed through Diode #1. Such an arrangement, called **cascade** can, for obvious reasons, yield efficiencies larger than those of single devices.

For instance, if Diode #1 has a band-gap of 1.8 eV and Diode #2 one of 1.1 eV, the ideal efficiency of the combination is 0.56 when exposed to sunlight in space. Compare this with the best possible single diode efficiency of 0.43.

One problem that arises when one attempts to cascade diodes is that of interconnection. Normally, the nonilluminated face of the device is metallized to act as a current collector. This makes it opaque and, consequently, useless in a cascade configuration.

One solution is to build the two (or more) diodes from a single semiconductor crystal with the individual junctions in series. To do this, the materials of the two diodes must have (nearly) the same lattice constant, but, of course, different band-gaps. By varying the stoichiometry of quaternary compounds such as AlInGaAS, it is possible to tailor the band-gap to the desired value while keeping the lattice constant about the same.

Walukiewicz, among others, is working (2003) on an interesting solution to the cascaded cells problem. For years, LED emitted red light (GaAsP) or, at most, green (GaP). The quest for blue LEDs lasted for a number of years until, in the 1980s, Akasaki and Nakamura developed the technique for growing thin films of gallium nitride whose large band-gap energy (3.4 eV) caused the generation of near ultraviolet light. Doping with indium allowed the changing of W_g so as to obtain blue light. Although the films had an enormous amount of defects, the material displayed great tolerance to imperfection. GaAsP and GaP LEDs with the same density of defects would emit no light.

The Berkeley team and its collaborators grew indium gallium nitride crystals of high quality and was able to show that by varying the relative amount of indium, the band-gap could be tuned from 3.4 eV (pure GaN) all the way down to 0.7 eV (pure InN). This 5:1 range covers most of the solar spectrum.

Cascaded cells can thus be built choosing the best W_g combination such as 1.7/1.1 eV. At present the best available combination is 1.83/1.43 eV whose efficiency is lower than that of the 1.7/1.1 combination.

The difference in the lattice constant of the different layers of cascaded cells—that is, the mismatch of the geometry of the crystal—introduces strains that, at best, create imperfections detrimental to the performance of the device and can even cause cracks in the material. Here it is where the great tolerance to defects favors the various $In_xGa_{1-x}N$ layers, which, presumably, can work well even under substantial strain.

The $In_xGa_{1-x}N$ can be doped with silicon to produce n-type material but, at present, there are difficulties in finding an adequate dopant for the needed p-type material.

It should be pointed out here that the $In_xAl_{1-x}N$ system has an even wider (0.7 to 6.2 eV) range than the $In_xGa_{1-x}N$ system.

14.4.2   Filtered Cells

Spectrally selective beam splitters can be of the absorption or of the interference filter type. For instance, concentrated sunlight can be filtered through cells containing cobalt sulfate solution that will absorb part of the energy while transmitting the rest to underlying silicon diodes.

Hamdy and Osborn measured the transmittance of 5-cm thick cobalt sulfate filters with different concentrations. For a concentration of 1 gram per liter, the transmittances at different light frequencies were as shown in Table 14.3.[†] The absorbed energy is used to generate useful heat. This kind of filter has a window in the 350 to 550 THz range and thus passes radiation that matches silicon characteristics reasonably well. The band-gap energy of silicon corresponds to 265 THz.

Table 14.3

Transmittance of a Cobalt Sulfate Filter (1 g/liter)

f (THz)	T
200	0.020
273	0.29
333	0.15
375	0.88
429	0.93
500	0.90
600	0.30
750	0.92

[†]   The transmittance value for 750 THz is not a typo. The cobalt sulfate filter is quite transparent to near UV.

Another class of spectrally selective filters are the **interference** filters. These filters consist of a number of layers of materials with different refractive indices. At each layer interface, radiation is partly reflected and partly transmitted. At some frequencies all reflected components add up while the transmitted ones cancel one another. In these regions the energy is reflected. At other regions the energy is transmitted.

Hamdy and Osborn constructed a 12-layer filter consisting of six pairs of SiO_2 and TiO_2 layers each having a $\lambda/4$ thickness.

Such a filter had a band-reflect region between 312 and 385 THz with very steep skirts and some 95% reflectance. This reflected spectral slice is suitable for work with silicon. There was considerable ripple in the transmitted region (both above 385 THz and below 312 THz) but the average transmittance hovered around 80%.

14.4.3 Holographic Concentrators

Perhaps the most attractive beam splitting technique is the holographic concentrator. A holographic plate is prepared in such a way that it acts as an extremely dispersive cylindrical lens. Sunlight is concentrated by factors of 50 to 100 and dispersed into a rainbow as indicated in Figure 14.4. Two (or more) photodiodes are mounted perpendicular to the holographic plate and positioned in such a way as to be exposed to that part of the spectrum at which they are most efficient.

The far infrared, of little interest for photovoltaic conversion, is directed to a region where the corresponding heat can be dissipated without affecting the photodiode, which, as discussed later, must operate at low temperature to avoid loss in efficiency.

One should remember here the definition of efficiency of a hologram: it is the ratio of the light power diffracted to the incident light power. Most of the undiffracted light just passes through the hologram. This efficiency is frequency dependent and, thus, a hologram has a given bandwidth.

For photovoltaic applications, a holographic plate must have high efficiency (some 90%) and a large bandwidth, because the solar spectrum has significant energy over a larger than 2:1 frequency range.

Northeast Photosciences has demonstrated holographic photovoltaic systems with over 30% efficiency.

14.5 Thermo-photovoltaic Cells

Another scheme to increase the efficiency of solar converters recirculates the photons that, owing to the transparency of the photodiodes, failed to be absorbed. To accomplish this, the body that emits the radiation must be in the immediate vicinity of the diodes.

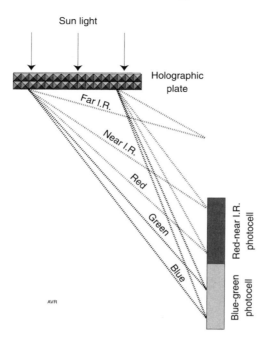

Figure 14.4   A holographic plate serves as both a concentrator and a spectrum splitter.

Consider the configuration illustrated in Figure 14.5 in which a radiator (heated by the sun through concentrators or by flames from a combustion unit) illuminates a photodiode with a band-gap, W_g.

Photons with energy above W_g interact with the semiconductor (solid lines) and generate an electric current. Those with less than W_g pass through the semiconductor and, on being reflected by the mirror, are returned to the radiator (dashed lines).

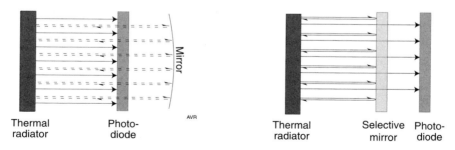

Figure 14.5   Two configurations of thermo-photovoltaic converters.

The mirror can be the back electrode of the diode. In devices developed by Richard Swanson at Stanford, the mirror is a layer of silver separated from the diode by a small thickness of oxide. The electrical connection between the mirror and the diode is assured by a polka dot pattern of contacts that obscure only 1% of the mirror. The graphite radiator works at temperatures between 2200 and 2300 K. The semiconductor is silicon.

Another solution uses a selective mirror placed between the radiator and the semiconductor. This mirror reflects photons of energy below W_g but transmits those with more energy. Such mirrors are difficult to realize.

To improve the efficiency, it is important to reduce the heat losses from the radiator. Since one loss mechanism is convection, it is attractive to evacuate the space between radiator and diode. This, however, permits the material of the radiator to distill and coat the diode with a light-absorbing layer that reduces the efficiency of the device. It is then necessary to choose radiator materials that have low vapor pressures at the operating temperature. Sometimes, it is practical to accept some convection losses in order to reduce radiator sublimation by operating in an inert gas atmosphere. The Swanson device mentioned above operated in an argon atmosphere at about 50 kPa.

The temperature of the radiator determines the choice of the band-gap for the semiconductor.

Thermo-photovoltaic cells have been proposed as a simple heat engine directly converting the radiation of a flame into electricity.

14.6   The Ideal and the Practical

So far, we have discussed the ideal photocell without specifying any particular implementation or technology. A simple ideal cell (one in which there is no carrier multiplication) is transparent to part of the spectrum and totally opaque to the rest. Each of the absorbed photons gives rise to a single electron-hole pair, and every pair drifts safely to the built-in p-n junction, which then separates the electron from the hole. The carriers thus created are quickly thermalized and the available output energy is exactly W_g joules per photon absorbed.

The efficiency of simple ideal cells when exposed to monochromatic light whose photons have an energy just above W_g joules is 100%. When exposed to broad-band radiation, the efficiency depends only on the nature of the incident radiation and the value of the band-gap energy. This is the **simple ideal spectrum efficiency** of the cell.

As discussed in preceding sections, it is possible to exceed the simple ideal efficiencies by:

1. Making use of photons that, owing to their insufficient energy, failed to be absorbed by the cell. That is, for instance, the case of beam splitting system or of thermo-photovoltaic arrangements.
2. Making use of the excess energy which most carriers have at the moment of their generation. This is the case of the carrier multiplication systems.

Practical photocells fail to reach the the ideal efficiency owing to a number of loss mechanisms:

1. Some of the incident photons are reflected away from the cell instead of being absorbed or may be absorbed by obstructions such as current collecting electrodes.
2. If the thickness of the photoactive material is insufficient, not all photons with energy above W_g are absorbed—the material may not be opaque enough.
3. Not all electron-hole pairs created live long enough to drift to the p-n junction. If their lifetime is small or if they are created too far from the junction, these pairs will recombine and their energy is lost. Electron-hole pairs may drift toward the surface of the device where the recombination rate is high and thus will not make it to the junction.
4. Carriers separated by the p-n junction will lose some of their energy while on the way to the output electrodes owing to the resistance of the connections. This constitutes the **internal resistance** of the cell.
5. Mismatch between photocell and load may hinder the full utilization of the generated power.

14.7 The Photodiode

So far, we have discussed photodiodes making no reference to the exact mechanisms that make them work. In this section we will consider p-n junctions and their use in converting light into electric energy.

For a more detailed discussion of p-n junctions, see da Rosa, A.V. Those who require a more basic introduction to the nature of semiconductors can read Appendix B to this chapter.

Consider a single semiconductor crystal consisting of two regions in close juxtaposition—an **n-region** into which a minute amount of certain foreign atoms capable of easily releasing an electron has been introduced, and a **p-region** containing atoms capable of binding an electron. These foreign atoms or **dopants** are, respectively, called **donors** and **acceptors**. For semiconductors such as germanium and silicon, donors are phosphorus, arsenic, and antimony, and acceptors are boron, aluminum, gallium, and indium.

14.20

A donor that has released an electron becomes positively charged, but it remains firmly attached to the crystal and cannot move under the influence of an electric field—it constitutes an **immobile charge**. On the other hand, the released electron is free and being able to **drift** in an electric field becomes a negative **carrier** (of electricity). Correspondingly, acceptors, having acquired an extra electron, become negatively charged but are also immobile, whereas the hole can drift and become a positive carrier. A hole, although it is simply an absence of an electron where normally an electron should be (see Appendix B), can be treated as a particle with positive mass and a positive charge equal, in absolute value, to that of the electron.

The n-side of the junction contains positive donors and an abundance of free electrons, while the p-side contains negative acceptors and positive holes. Free electrons, more abundant in the n-side, diffuse toward the p-side, whereas holes from the p-side migrate to the n-side.

If these particles were uncharged, the diffusion process would only stop when their concentration became uniform across the crystal. Being charged, their separation causes an electric field to be established and a compensating drift current makes carriers flow back in a direction opposite to that of the diffusion current.

The drift current is driven by a **contact potential** created by the following mechanism: The migrating electrons not only transport negative charges to the p-side but, also, leave uncompensated positively charged donors in the n-side. The holes coming in from the p-side contribute to the accumulation of positive charges in the n-side and leave uncompensated negatively charged acceptors in the p-side. Thus the n-side becomes positive and the p-side negative.[†]

In an unbiased p-n junction (one to which no external voltage is applied), the overall current is zero, not because diffusion and drift currents are themselves zero but because these currents exactly balance one another. In other words, the equilibrium in a junction is **dynamic**, not static.

The equilibrium consists of four currents:

$j_{n_D} = -qD_n dn/dx$ (diffusion current of electrons from the n-side to the p-side);

$j_{n_E} = q\mu_n nE$ (drift current of electrons from the p-side to the n-side);

$j_{p_D} = -qD_p dp/dx$ (diffusion current of holes from the p-side to the n-side);

$j_{p_E} = q\mu_p pE$ (drift current of holes from the n-side to the p-side).

[†] Contact potential is not an exclusive property of semiconductors. Two different metals, when joined, will also develop a contact potential between them.

The symbols used in this section include:

n, concentration of electrons—number of free electrons per unit volume;

p, concentration of holes;

N_a, concentration of acceptors;

N_d, concentration of donors;

n_i, intrinsic carrier concentration;

D_n, diffusion constant of electrons;

D_p, diffusion constant of holes;

μ_n, mobility of electrons;

μ_n, mobility of holes.

The **intrinsic carrier concentration** is a temperature dependent quantity that characterizes a given semiconductor. It is

$$n_i = BT^{3/2} \exp\left(\frac{W_g}{kT}\right), \tag{32}$$

where B can be calculated from the knowledge of the **effective masses** of electrons and holes in the semiconductor in question. Usually, one does not find a tabulation of different values of B. What one finds are tabulations of n_i for a given semiconductor at a specified temperature. From that, B is calculated, and values on n_i at different temperatures can be found. The usual values of n_i, at 300 K, are 10×10^{10} cm^{-3} for silicon and 2.5×10^{13} cm^{-3} for germanium.

The total current, j, is the sum of the electron and the hole currents and must be equal to zero because there is no external connection to the device:

$$j = j_n + j_p = 0. \tag{33}$$

This equation can be satisfied by having either $j_n = -j_p \neq 0$ or $j_n = j_p = 0$. If either current were different from zero, holes or electrons would accumulate in one side of the junction—a situation that could not be sustained for any appreciable length of time. Hence, each of the currents must, individually, be zero. In other words, the drift of holes must (under equilibrium conditions) be exactly equal to the diffusion of holes. The same must be true for electrons.

To find the magnitude of the contact potential one solves one of the above equations. Take, for instance, the hole current:

$$j_p = q\left(\mu_p p E - D_p \frac{dp}{dx}\right) = 0 \tag{34}$$

and, since $E = -dV/dx$,

$$\mu_p p \frac{dV}{dX} + D_p \frac{dp}{dx} = 0, \tag{35}$$

$$\frac{dp}{p} = -\frac{\mu_p}{D_p} dV = -\frac{q}{kT} dV. \tag{36}$$

Integrating this equation from deep into the p-side to deep into the n-side, one obtains

$$\ln \frac{p_p}{p_n} = [V(p_n) - V(p_p)] \frac{q}{kT}. \tag{37}$$

In the absence of an external current, there is no voltage drop across the undisturbed crystal (that part of the crystal far away from the junction). Thus, $V(p_n) - V(p_p)$ is the contact potential, V_C:

$$V_C = \frac{kT}{q} \ln \frac{p_p}{p_n} \approx \frac{kT}{q} \ln \frac{N_a N_d}{n_i^2}. \tag{38}$$

The approximation in Equation 38 is valid only when $p_n = n_i^2/N_d$.

Example 14.4

The doping concentrations of a particular silicon junction are $N_a = N_d = 10^{16}$ cm^{-3}. The operating temperature is 300 K. What is the contact potential?

The values above are representative doping levels for many silicon devices. Since in each cubic centimeter of silicon one finds a total of 5×10^{22} silicon atoms, we can see that the impurity levels are very small (in relative terms): 1 impurity atom for each 5 million silicon atoms. From the chemical point of view, we are dealing here with super pure silicon! Since the diode is at 300 K, the intrinsic carrier concentration is $n_i = 10^{10}$ cm^{-3}.

Using Equation 38,

$$V_C = \frac{kT}{q} \ln \frac{N_a N_d}{n_i^2}$$

$$= \left(\frac{1.38 \times 10^{-23} \times 300}{1.6 \times 10^{-19}} \right) \ln \left(\frac{10^{16} \times 10^{16}}{10^{20}} \right) = 0.72 \text{ V}.$$

Photocells and tunnel diodes, among others, use heavily doped semiconductors and, consequently, their contact potential is somewhat higher.

With doping concentrations of $N_a = N_d = 10^{19}$ cm^{-3}, the contact potential is $V_C = 1.08$ V.

14.23

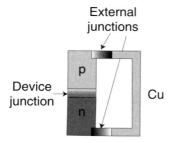

Figure 14.6   Two external junctions are formed when a photodiode is shorted out by a metallic wire.

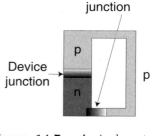

Figure 14.7   A single external junction is formed when a diode is shorted out by its own material.

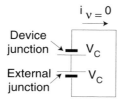

Figure 14.8   No current circulates in a shorted junction at even temperature and no illumination.

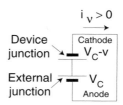

Figure 14.9   When light reduces the potential across the device junction, an external current circulates.

The contact potential cannot be measured by applying a voltmeter to a p-n junction because the potentials around a close circuit will exactly cancel out (provided the circuit is at uniform temperature).[†] If the potentials did not cancel out, then a p-n junction would drive a current through some external load delivering energy. This would constitute a heat engine delivering a useful output under a zero temperature differential—a thermodynamic impossibility.

Another way to see that the potentials in a close circuit must cancel out is to consider that any connection between the free ends of a p-n junction must involve at least one additional external junction whose voltage opposes that of the original one, as explained next. Figure 14.6 shows a p-n junction shorted out by a copper wire. In addition to the device junction, there are two external junctions, one between the p-material and the copper wire and one between the copper wire and the n-material. A conceptually simpler way to short out the device is shown in Figure 14.7 where the external wire is made of the same material as the p-side of the diode. Now there is a single external junction between the p-wire and the n-material. Clearly the

---

[†] If there is a temperature difference along the circuit, then thermoelectric effects will appear, as discussed in Chapter 5.

two p-n junctions (the device junction and the external junction) have the same contact potential and oppose one another. Electrically, the situation is as depicted in Figure 14.8. Obviously, the external current is zero.

To understand what happens when the device is exposed to light, examine Figure 14.10, which depicts the potential across the unilluminated photo diode. The values shown correspond to a Si diode with equal doping in the two sides (10^{18} dopant atoms per cm^3). The contact potential at ambient temperature is 0.958 V, the n-region being positive with respect to the p-region. The potential in the device is uniform except in a narrow transition region only 2.5×10^{-6} cm wide where there is a strong electric field of 380000 V/cm.

Light creates exitons that will recombine after a short **lifetime**, τ, unless an electric field separates the electrons from the corresponding holes. Owing to the narrowness of the transition region, not too many electron-hole pairs are created in the region itself—most are created in the neighborhood of the region and, because they move at random, they have a chance of stumbling into the high electric field where the separation occurs: the electrons created in the negatively charged p-region will drift to the n-side causing it to become less positive. By the same token, holes created in the positively charged n-region will drift to the p-region causing the latter to become less negative. Thus, light has the effect of *reducing* the contact potential which becomes, let's say, $V_c - v$, as depicted in Figure 14.9. Since, as we are going to explain, the external junction is not light sensitive and, hence, still has a contact potential, V_c, there is no longer an exact cancellation of the two contact potentials and a net external voltage, v, will appear constituting the output of the photodiode. A current, I_ν, will flow through the short in the direction indicated.

For the electron-hole pairs created in the field-free region to reach the transition region, they must survive long enough—that is, their lifetime, τ, must be sufficiently large. If the lifetime is small, these pairs will recombine contributing nothing to the output of the photodiode. The larger τ, the larger the efficiency of the device.

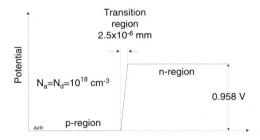

Figure 14.10   The electric field of a p-n junction is confined to a very narrow transition region.

Thus, it is of paramount importance that the material from which the diode is made be as free from defects as possible because defects reduce τ. On the other hand, the external junction is, intentionally, made to have extremely low τ and thus is quite insensitive to light.[†]

The conventional direction of the external current is from the p-side of the device (the cathode) to the n-side (the anode). Its magnitude is

or
$$I_\nu = q\phi_g A \text{ A,} \tag{39}$$
$$J_\nu = q\phi_g \text{ A/m}^2, \tag{39a}$$

where q is the charge of the electron, ϕ_g is the flux of photons with energy larger than the band-gap energy, W_g, of the semiconductor, and A is the active area of the junction, This assumes 100% quantum efficiency— each photon creates one electron-hole pair capable of reaching the potential barrier of the junction.

The short-circuit current is proportional to the photon flux and can be used for its measurement. There is a minute temperature dependence of I_ν, but only if the incident light is not monochromatic. This results from the small change in the band-gap energy with temperature. The higher the temperature, the smaller this energy and, consequently, the larger the fraction of the spectrum that has sufficient energy to disturb covalent bonds and, thus, to generate current.

A short-circuited photodiode can be represented by a shorted current source as suggested by Figure 14.11. If however, the short is replaced by a load, R_L, then a voltage, $V_L = I_L R_L$, will appear across it. This voltage will drive a current, I_D, through the diode. See Figure 14.12.

Clearly,
$$I_L = I_\nu - I_D \tag{40}$$

where I_D is a function of the voltage, V_L, across the diode. We need to know the functional relationship between I_D and V_L.

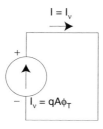

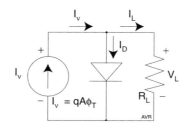

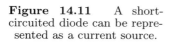

Figure 14.11   A short-circuited diode can be represented as a current source.

Figure 14.12   Model of an ideal photodiode with a resistive load.

[†] The very act of soldering a wire to the diode will create enough imperfections as to reduce the τ at this junction to negligible values.

We saw that in an unbiased diode, there is a balance between the diffusion and the drift currents—they exactly cancel one another. When an external voltage (bias) is applied, the potential barrier at the junction is altered. If the bias reinforces the built-in barrier, only a minute current (the **reverse saturation current**), I_0, flows. When the bias is forward— that is, the potential barrier is lowered, a substantial current flows. This forward current occurs when a positive voltage is applied to the p-side of the device—that is, the positive (conventional) direction of the forward current is into the p-side. The diode behaves in a markedly asymmetric fashion with respect to the applied voltage.

It is shown in any elementary text on electronics that the current through a p-n diode is

$$I_D = I_0 \left[\exp\left(\frac{qV}{kT}\right) - 1 \right]. \tag{41}$$

Figure 14.13b illustrates the V-I characteristics of a p-n diode for small applied biases.

If $|V|$ is much lager than kT/q, and $V < 0$, the exponential term becomes negligible compared with 1 and I_D **saturates** at a value $I = -I_0$, i.e., the current becomes independent of the applied voltage. Since at room temperature, $kT/q = 0.026$ V, the saturation occurs for voltages as small as -0.1 V.

If the applied voltage is positive and much larger than kT/q, then the exponential term dominates and the current increases exponentially with voltage. A positive bias of as little as 0.5 V causes the current through the diode to grow to $2 \times 10^8 I_0$. It can be seen that the reverse saturation current, I_0 is a quantity extremely small compared with the diode currents under modest forward biases.

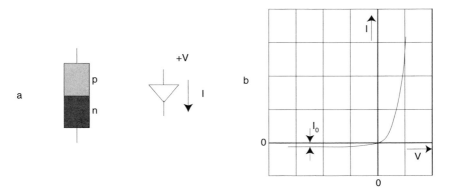

Figure 14.13   The V-I characteristics of a p-n diode.

Applying Equation 41 to Equation 40,

$$I_L = I_\nu - I_0 \left[\exp\left(\frac{qV}{kT} \right) - 1 \right]. \tag{42}$$

The current, I_D, that flows through the diode diverts some current from the load. Thus, it is clear that the smaller, I_D, the more current flows through the load, and, presumably, the more efficient the diode. Notice that I_D is proportional to I_0 (Equation 41). This explains the great effort made to produce photodiodes with the smallest possible reverse saturation current, I_0.

Inverting Equation 42,

$$V = \frac{kT}{q} \ln \left[\frac{I_\nu - I_L}{I_0} + 1 \right]. \tag{43}$$

A real diode has an internal resistance, R_s, that causes the voltage across the load to drop to

$$V_L = \frac{kT}{q} \ln \left[\frac{I_\nu - I_L}{I_0} + 1 \right] - I_L\, R_s. \tag{44}$$

To a first approximation, a photodiode acts as a current generator. It is a low voltage, high current device—several must be connected in series to generate practical voltage levels. There are difficulties in such connections.

The current through the series is the current generated by the weakest diode. A partial shadowing of a solar panel consisting of many diodes in series can drastically reduce the overall output.

One possible solution consists of building a larger number of minute diodes forming series connected groups, which then are paralleled to produce an useful current. Partial shadowing will then disable a limited number of diodes.

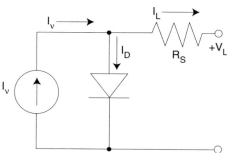

Figure 14.14   Circuit model of a real photodiode.

The open-circuit voltage of the photodiode is found by setting $I_L = 0$ in Equation 44:

$$V_{oc} = \frac{kT}{q} \ln\left(\frac{I_\nu}{I_o} + 1\right) \approx \frac{kT}{q} \ln\left(\frac{I_\nu}{I_0}\right) \qquad (45)$$

The voltage depends on the I_ν/I_0 ratio. Both currents are proportional to the cross-section of the junction; their ratio, therefore, does not depend on the cross-section. It depends on the photon flux (the light-power density) and on the characteristics of the diode. Typically, this ratio is of the order of 10^7 at one sun (1000 W/m^2) in silicon. At 300 K, this corresponds to an open-circuit voltage of 0.42 V.

Owing to its logarithmic dependence, the open-circuit voltage varies slowly with changes in light power-density and is not a convenient photometric measure.

The V-I characteristic of a photodiode can be obtained by plotting Equation 44 as shown in Figure 14.15. The currents were normalized by dividing by I_0. Three different illumination levels are displayed, corresponding to I_ν/I_0 of 10^6, 5×10^6, and 10^7.

For the highest level of illumination in the figure, in addition to the characteristics for diodes with no series resistance, we have also plotted the curve for a diode with normalized resistance of $R_s = 7.5 \times 10^{-9}/I_0$ ohms.

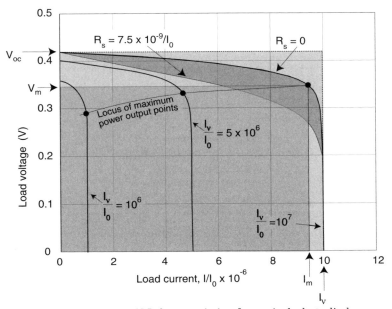

Figure 14.15 V-I characteristic of a typical photodiode.

The power a photodiode delivers to a load depends on, among other things, the value of the load resistance. The optimum operating point occurs at a specific voltage, V_m, and a corresponding current, I_m (see derivation further on). These points are indicated by the small black circles in Figure 14.15.

The area under the curves has the dimension of power. The large darker gray rectangle represents the $V_m \times I_m$ product (the maximum power the diode can deliver when an ideally matched load is used), whereas the lighter gray rectangle represents the $V_{oc} \times I_\nu$ product. The ratio between these two areas (or the two powers) is called the **fill factor**, F_f. Both gray areas are for the case of $I_\nu/I_0 = 10^7$, which is the highest level of illumination appearing in the figure. V_m and I_m are for a diode with no series resistance.

$$F_f = \frac{V_m \times I_m}{V_{oc} \times I_\nu}. \tag{46}$$

The gray region with curved boundaries corresponds to the difference in area between the characteristics of a diode with no series resistance and one with a given series resistance—this area represents the losses associated with this internal resistance. Clearly, the smaller the series resistance (i.e., the smaller the losses), the larger the fill factor. This factor indicates how closely a given diode approaches the power $V_{oc} \times I_\nu$,

It is important to inquire how the maximum deliverable power from a photodiode depends on a number of different parameters such as level of illumination, temperature, etc. This can, of course, be done by numerical experimentation using the expressions for V_m and I_m. However, since the calculation of V_m involves an equation (Equation 50) that cannot be solved analytically, this procedure does not yield a simple formula from which the effects of the different parameters can be assessed.

On the other hand the $V_{oc} \times I_\nu$ product can be calculated from a simple equation (Equation 48). For this reason, in the considerations below, we inferred the behavior of $P_{L_{max}}$ from that of the $V_{oc} \times I_\nu$ product. This would be perfectly correct if these two quantities were strictly proportional to one another. However, this is not quite the case.

Table 14.4 displays values of the open-circuit voltage, V_{oc}, of the maximum-power voltage, V_m, of the maximum-power current, I_m, and of the fill factor F_f for a hypothetical resistanceless photodiode having a reverse saturation current of 10^{-7} A/m^2. It can be observed that, although the ratio is not constant, it varies relatively little when the illumination level changes by 3 orders of magnitude. Consequently, studying the behavior of the $V_{oc} \times I_\nu$ product leads to a simplified insight on the behavior of $P_{L_{max}}$.

Table 14.4
Fill Factor as Function
of the Level of Illumination

I_ν (A)	V_{oc} (V)	V_m (V)	I_m (A)	Fill factor $\frac{V_m \times I_m}{V_{oc} \times I_\nu}$
0.1	0.358	0.293	0.092	0.752
1	0.417	0.348	0.931	0.777
10	0.477	0.404	9.40	0.796
100	0.537	0.461	94.7	0.813

There are four important observations regarding the efficient operation of photovoltaic devices:

1. All else being the same, the efficiency of a photodiode increases with an increase in the light-power density.

To gain a qualitative insight on the effect of the input light-power density, P_{in}, on the efficiency, consider the following facts:
The *open-circuit* voltage, V_{oc}, is, as seen before,

$$V_{oc} = \frac{kT}{q} \ln \left(\frac{I_\nu}{I_0} \right). \tag{47}$$

The $I_\nu V_{oc}$ product is

$$I_\nu V_{oc} = I_\nu \frac{kT}{q} \ln \left(\frac{I_\nu}{I_0} \right). \tag{48}$$

Since I_ν is proportional to the input light-power density, P_{in}, and V_{oc} grows logarithmically with I_ν, and, hence, with P_{in}, it follows that the $I_\nu V_{oc}$ product grows faster than linearly with P_{in}.

Accepting that the power output of the photodiode follows the behavior of the $I_\nu V_{oc}$ product, then, plausibly, an increase in P_{in} causes a more than proportional increase in P_{out}—that is, the efficiency increases with P_{in}.

One can reach the same conclusion more rigorously as shown below.
The power delivered by the diode is

$$P = VI = VI_\nu - VI_0 \left[\exp(qV/kT) - 1 \right]. \tag{49}$$

Maximum power flows when the voltage is $V = V_m$ such that $dP/dV = 0$.

This leads to a transcendental equation that must be solved numerically:

$$\left(1 + \frac{qV_m}{kT} \right) \exp \left(\frac{qV_m}{kT} \right) = J_\nu/J_0 + 1 \tag{50}$$

Here, we have switched from currents, to current densities.

An empirical formula that closely approximates the solution of the above equation is

$$V_m = V_A \ln \left(\frac{P_{in}}{V_B J_0} \right), \tag{51}$$

where

$$V_A = 2.2885 \times 10^{-2} - 139.9 \times 10^{-6} \ln J_0 - 2.5734 \times 10^{-6} (\ln J_0)^2, \tag{52}$$

$$V_B = 4.7253 - 0.8939 \ln J_0. \tag{53}$$

From an inspection of Figure 14.15, it can be seen that the optimum operating point must lie near the knee of the V-I characteristics as indicated by the solid circles in the figure.

From Equation 42,

$$J_m = J_\nu - J_0 \left[\exp \left(\frac{q V_m}{kT} \right) - 1 \right], \tag{54}$$

where J_m is the current density when $V = V_m$.

The maximum power (per unit active area) the diode delivers to a load is

$$P_m = V_m J_m, \tag{55}$$

and the maximum efficiency (when the load is properly matched to the photodiode) is

$$\eta_m = \frac{V_m J_m}{P_{in}}. \tag{56}$$

In Section 14.2, we define a device-independent **ideal efficiency**, η_{ideal}, a function of only the spectral distribution and the band-gap of the semiconductor, W_g. The **ideal-diode efficiency**, η_m, defined above, considers the case of an ideal photodiode which has a non-zero reverse saturation current, I_0. Designers strive to make I_0 as small as possible but it cannot be made vanishingly small (see Section 14.8). If I_0 were taken as zero, then the diode would be modeled as a simple current generator (with no shunting diode) and no open-circuit voltage could be defined because current generators cannot deliver a current to an infinite resistance. On the other hand, from the manner in which a p-n diode works, it can be seen that the output voltage is limited to the contact potential, V_C. One could then argue that as $I_0 \to 0$, $J_m \to J_\nu$, and $V_m \to V_C$ and the power delivered to the load (per unit active area) would be $P_L = I_\nu V_C$. The value of V_C can be obtained from Equation 38. Thus,

$$P_{L_{max}} = J_\nu \frac{kT}{q} \ln \frac{N_a N_d}{n_i^2} \tag{57}$$

But $J_\nu \propto P_{in}$. This relationship is demonstrated in Example 14.5, later on. Let γ be the constant of proportionality between J_ν and P_{in}, then

$$P_{L_{max}} = \gamma P_{in} \frac{kT}{q} \ln \frac{N_a N_d}{n_i^2}, \tag{58}$$

and the efficiency would be

$$\eta_m = \gamma \frac{kT}{q} \ln \frac{N_a N_d}{n_i^2}. \tag{59}$$

The largest value that either N_a or N_d can have, in silicon, is around 10^{19} atoms/cm^3. Remember that silicon has an atomic concentration of about 5×10^{22} atoms per cm^3. This is roughly the limit of solubility of most dopants in silicon. A larger concentration of such dopants will cause some of them to settle out. Using $\gamma = 0.399$ (again, please refer to Example 14.5), we get for the case of a 6000-K black body radiation,

$$\eta_m = 0.399 \left(\frac{1.38 \times 10^{-23} \times 300}{1.6 \times 10^{-19}} \right) \ln \left(\frac{10^{19} \times 10^{19}}{10^{20}} \right) = 0.43. \tag{60}$$

Although this result is almost precisely the ideal efficiency of a silicon photocell exposed to 6000-K black body radiation, its precision is spurious. One reason is that zero reverse saturation currents cannot be realized even theoretically. In addition, there is nothing magical about the maximum level of doping which is limited by the solubility of dopants in silicon and can, by no stretch of the imagination, be directly related to the ideal efficiency of a photocell.

2.   *The efficiency of a photodiode decreases with an increase in the reverse saturation current.*

We have shown that the behavior of the $I_\nu V_{oc}$ product is a good indicator of the behavior of the $I_m V_m$ product and, consequently, of the efficiency of the photo diode. Referring back to Equation 48, it can be seen that an increase in I_0 results in a decrease in $I_\nu V_{oc}$ and, thus, plausibly, in the efficiency. Also, the numerical experimentation in Example 14.5 confirms this conclusion.

Good photodiodes are designed to minimize I_0. This, as explained in Section 14.8, can be accomplished by

1.   using highly doped semiconductors and
2.   striving for the largest possible minority carrier lifetimes.

Example 14.5

How does the efficiency of an ideal silicon photodiode vary with light-power density and with the reverse saturation current density, J_0, when exposed to the radiation from a 6000-K black body?

To answer the above question one must solve a number of equations numerically. A spread sheet is an excellent platform do to so.

Start by picking a value for P_{in} and one for J_0. Calculate the corresponding efficiency using

$$\eta_m = \frac{V_m J_m}{P_{in}}. \tag{61}$$

Increment J_0 and recalculate η_m. After all desired values of J_0 have been used, select a new P_{in} and repeat the whole procedure.

To calculate η_m, we need to have the values of V_m and I_m as a function of the selected parameters P_{in} and J_0.

<u>Calculate V_m</u>

Given these parameters, calculate V_m from

$$V_m = V_A \ln \left(\frac{P_{in}}{V_B J_0} \right), \tag{51}$$

where

$$V_A = 2.2885 \times 10^{-2} - 139.9 \times 10^{-6} \ln J_0 - 2.5734 \times 10^{-6} (\ln J_0)^2, \tag{52}$$

$$V_B = 4.7253 - 0.8939 \ln J_0. \tag{53}$$

<u>Calculate I_m</u>

Use

$$J_m = J_\nu - J_0 \left[\exp \left(\frac{q V_m}{kT} \right) - 1 \right].$$

To do this we need J_ν as a function of P_{in}.
From Equation 25,

$$\sigma = 0.416 \int\limits_{\frac{q V_g}{kT} = 2.1256}^{\infty} \frac{x^2}{e^x - 1} dx = 0.416 \times 1.3405 = 0.5577. \tag{62}$$

The value of the definite integral was from Appendix A.
Still from Equation 25,

$$\phi_g = \sigma \phi = 0,5577 \phi. \tag{63}$$

(continued)

(continued)

From Equation 16,

$$\phi = \frac{P_{in}}{223.7 \times 10^{-21}}, \tag{64}$$

$$\phi_g = 0.5577 \frac{P_{in}}{223.7 \times 10^{-21}}, \tag{65}$$

$$J_\nu = q\phi_g = 1.6 \times 10^{-19} \times 0.5577 \frac{P_{in}}{223.7 \times 10^{-21}} = 0.399 P_{in}, \tag{66}$$

$$J_m = 0.399 P_{in} - J_0 \left[\exp\left(\frac{qV_m}{kT}\right) - 1 \right]. \tag{67}$$

Using the formulas above, we calculated the ideal efficiency of silicon photodiodes exposed to the radiation from a 6000-K black body (roughly, the radiation from the sun).

This was done for various light-power densities and different values of the reverse saturation current density. The results are tabulated in Table 14.5 and plotted in Figure 14.16.

Table 14.5

$J_0(\text{nA/m}^2) \rightarrow$ $P_{in}(\text{W/m}^2) \downarrow$	10 η_m	20 η_m	50 η_m	100 η_m	200 η_m	500 η_m	1000 η_m
10	0.1637	0.1570	0.1482	0.1415	0.1349	0.1261	0.1195
20	0.1704	0.1637	0.1548	0.1482	0.1415	0.1327	0.1261
50	0.1793	0.1726	0.1637	0.1570	0.1503	0.1415	0.1349
100	0.1860	0.1793	0.1704	0.1637	0.1570	0.1482	0.1415
200	0.1928	0.1860	0.1771	0.1704	0.1637	0.1548	0.1482
500	0.2017	0.1950	0.1860	0.1793	0.1726	0.1637	0.1570
1000	0.2085	0.2017	0.1928	0.1860	0.1793	0.1704	0.1637
2000	0.2153	0.2085	0.1996	0.1928	0.1860	0.1771	0.1704
5000	0.2243	0.2175	0.2085	0.2017	0.1950	0.1860	0.1793
10000	0.2311	0.2243	0.2153	0.2085	0.2017	0.1928	0.1860
20000	0.2379	0.2311	0.2221	0.2153	0.2085	0.1996	0.1928
50000	0.2469	0.2401	0.2311	0.2243	0.2175	0.2085	0.2017
100000	0.2538	0.2469	0.2379	0.2311	0.2243	0.2153	0.2085

It can be seen from Figure 14.16 that there is a logarithmic dependence of the efficiency on the level of illumination and that the efficiency decreases when the reverse saturation current increases.

(continued)

(continued)

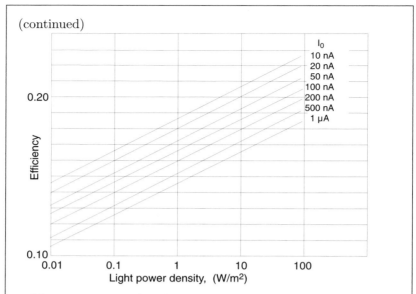

Figure 14.16   The efficiency of lossless silicon photodiodes exposed to 6000-K black body radiation increases logarithmically with light-power density and decreases with increasing reverse saturation current.

3.   *The efficiency of a photodiode decreases with an increase in the operating temperature.*

Equation 61 shows that the efficiency is proportional to V_m and, since V_m increases when V_{oc} increases, it is sufficient to investigate the temperature dependence of the latter.

Referring to Equation 47, it is not immediately clear what happens to V_{oc} when the temperature changes. The kT/q factor causes the voltage to rise proportionally to T, but the $\ln(I_\nu/I_0)$ factor causes the temperature to influence the voltage in the opposite direction through its effect on I_0. A more careful analysis has to be carried out.

Using information from the next section, we can write

$$V_{oc} = \frac{kT}{q} \ln\left(\frac{J_\nu}{J_0}\right) = \frac{kT}{q} \ln\left(\frac{I_\nu}{qAn_i^2 \left[\sqrt{\frac{D_p}{\tau_p}\frac{1}{N_d}} + \sqrt{\frac{D_n}{\tau_n}\frac{1}{N_a}}\right]}\right). \tag{68}$$

Here, we inserted the expression for J_0,

$$J_0 = qn_i^2 \left[\sqrt{\frac{D_p}{\tau_p}\frac{1}{N_d}} + \sqrt{\frac{D_n}{\tau_n}\frac{1}{N_a}}\right], \tag{69}$$

where n_i is the **intrinsic concentration** of holes and electrons and N_d and N_a are, respectively, the donor and acceptor concentrations, D and τ are, respectively, the **diffusion constant** and the **lifetime** of the minority carriers.

The lifetime of minority carriers is a function of the quality of the material and of the heat treatment it has undergone. It increases with temperature (see Ross and Madigan). The diffusion constant behavior depends on the doping level. In purer materials it decreases with temperature and in heavily doped ones (as is the case of photodiodes) it may even increase with temperature. At any rate, the combined effect of temperature on D and τ is relatively small and, for this general analysis, will be disregarded. This allows us to write,

$$J_0 = \Lambda n_i^2, \tag{70}$$

where Λ is

$$\Lambda \equiv q \left[\sqrt{\frac{D_p}{\tau_p}} \frac{1}{N_d} + \sqrt{\frac{D_n}{\tau_n}} \frac{1}{N_a} \right]. \tag{71}$$

The intrinsic carrier concentration—the number of either electrons or of holes per unit volume when the semiconductor is perfectly (and impossibly) pure—is given by

$$n_i = BT^{3/2} \exp\left(-\frac{qV_g}{2kT} \right), \tag{72}$$

where V_g is the band-gap voltage and B is a constant that varies from material to material.

The expression for the open-circuit voltage becomes,

$$V_{oc} = \frac{kT}{q} \ln\left(\frac{J_\nu \exp \frac{qV_g}{kT}}{\Lambda B^2 T^3} \right), \tag{73}$$

in which all quantities, other than T are (completely or nearly) temperature independent. Remember that J_ν has only a minuscule dependence on T.

From the expression above,

$$\frac{dV_{oc}}{dT} = \frac{k}{q} \left\{ \ln\left[\frac{J_\nu \exp\left(\frac{qV_g}{kT} \right)}{\Lambda B^2 T^3} \right] - 3 - \frac{qV_g}{kT} \right\}$$

$$= \frac{k}{q} \left\{ \ln\left[\frac{J_\nu}{J_0(T)} \right] - 3 - \frac{qV_g}{kT} \right\}$$

$$= \frac{k}{q} \left\{ \ln\left[\frac{300^3 J_\nu}{T^3 J_0} \right] - 3 - \frac{qV_g}{kT} \right\}. \tag{74}$$

Here, $J_0(T)$ is the temperature dependent reverse saturation current density and it is equal to $(T/300)^3 J_0$ where J_0 is the reverse saturation current density at 300 K.

It can be seen that dV_{oc}/dT is negative (V_{oc} goes down with increasing temperature) as long as

$$\ln\left[\frac{J_\nu}{J_0(T)}\right] < 3 + \frac{qV_g}{kT}, \tag{75}$$

or

$$\frac{J_\nu}{J_0(T)} < \exp\left(3 + \frac{qV_g}{kT}\right) \sim 200 \times 10^{12}. \tag{76}$$

We used $qV_g/kT \sim 30$ as a representative minimum value.

We also recognize that for an excellent photo diode exposed to 1 sum (1000 W/m²) the J_ν/J_0 ratio, at 300 K, is $\sim 10^8$. For the inequality above to become non-valid, J_ν would have to reach the value of 2 million A/m², a patent impossibility. Thus, under all plausible circumstances, V_{oc} decreases with an increase in temperature.

The theoretical dependence of dV_{oc}/dT on the illumination level is plotted for three different semiconductors in Figure 14.17. The plot is based on Equation 74.

It is apparent that

1. all temperature coefficients of the open-circuit voltage are negative; that is, in all cases, an increase in T results in a decrease in V_{oc} and, consequently, in the power output and efficiency;
2. the larger the band-gap, the larger (in absolute value) the temperature coefficient of the open-circuit voltage;
3. the larger the illumination, the smaller the temperature coefficient.

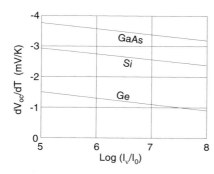

Figure 14.17 Temperature coefficient of the open-circuit voltage.

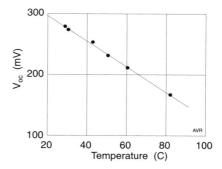

Figure 14.18 Observed temperature variation of the open-circuit voltage.

14.38

Actual measured open-circuit voltages of a silicon photodiode are displayed in Figure 14.18 (see Rappaport, 1959). The observed temperature coefficients are about 2 mV/K, roughly what is predicted by Equation 74.

Unfortunately, there is no information about the temperature and the illumination level used in Rappaport's measurements so a more precise comparison between prediction and observation cannot be made.

Large efficiencies are obtained with large concentration ratios (i.e., large light power densities). However, increasing the concentration tends to raise the operating temperature of the diode degrading the performance. As a result, in order to profit from high concentration, it is necessary to provide adequate cooling. Silicon with its larger thermal conductivity, has, here, an advantage over gallium arsenide.

Sometimes it is possible to operate diodes under a thin layer of water that removes the heat.

4. For maximum efficiency, a load must "match" the photodiode.

If good efficiency is desired, photodiodes must be operated near I_m. This means that for each level of illumination the optimum load must have a different resistance.

Consider a photodiode operating at the largest expected light-power density, $P_{in_{max}}$. A properly matched load is one that causes the diode to operate at the maximum output power, i.e., with a load voltage, V_m, defined by Equation 50. The corresponding load current, I_m, can be obtained from Equation 54. It should be noticed that, owing to the almost vertical characteristic of the diode at high currents, $I_m \approx I_\nu$.

The load must be adjusted to

$$R_L = \frac{V_{m_{max}}}{I_{m_{max}}}, \tag{77}$$

and the power delivered is

$$P_{L_{max}} = I_{m_{max}}^2 R_L \approx I_{\nu_{max}}^2 R_L. \tag{78}$$

Expose, now, the photodiode to a light power density $P_{in} = \lambda P_{in_{max}}$ where λ is a quantity smaller than 1, but maintain the same load resistance, R_L. The short circuit current, I_ν, is strictly proportional to P_{in} and is now,

$$I_\nu = \lambda I_{\nu max}. \tag{79}$$

The power delivered to the load is

$$P_L = I_L^2 R_L \approx \lambda^2 I_{\nu_{max}} R_L, \tag{80}$$

because $I_L \approx I_\nu$.

In other words, with a constant load resistance the electric power delivered falls with the square of the ratio of the sun power densities. Thus, if the maximum expected illumination is, say, 1000 W/m^2 and the load is adjusted to match these conditions, then if the illumination falls to 100 W/m^2, the dc output falls all the way to 1% of the former value. In contrast, if the load resistance is changed to cause operation at the "matched" condition, the dc power will only fall to 10%.

Since non-suntracking solar systems are subject to widely fluctuating illumination throughout a day, it is important to use devices (**load followers**) that automatically adjust the load characteristics to match the requirements of the photodiodes.

14.8 The Reverse Saturation Current

The reverse saturation current in a diode is the sum of the saturation current owing to holes, I_{p0}, and that owing to electrons, I_{n0}. Simple p-n diode theory leads to the expression

$$I_0 = I_{p0} + I_{n0} = qA\left(\frac{D_p}{L_p}p_n + \frac{D_n}{L_n}n_p\right). \tag{81}$$

Notice that the ratio of the diffusion constant, D, to the diffusion length, L, has the dimension of velocity so that, when multiplied by the carrier concentration (p_n or n_p), it yields a diffusion flux. Multiplying all by qA (where A is the area of the junction) converts the flux into current.

L and D are related to one another by the lifetime, τ, of the minority carriers:

$$L = \sqrt{D\tau} \tag{82}$$

Consequently

$$I_0 = qA\left[\sqrt{\frac{D_d}{\tau_p}}p_n + \sqrt{\frac{D_n}{\tau_n}}n_p\right] \tag{83}$$

and, since

$$p_n = \frac{n_i^2}{N_d}, \tag{84}$$

and

$$n_p = \frac{n_i^2}{N_a}, \tag{85}$$

it follows that

$$I_0 = qAn_i^2\left[\sqrt{\frac{D_p}{\tau_p}}\frac{1}{N_d} + \sqrt{\frac{D_n}{\tau_n}}\frac{1}{N_a}\right], \tag{86}$$

where n_i is the intrinsic concentration of holes and electrons and N_d and N_a are the donor and acceptor concentrations, respectively.

Table 14.6
Diffusion Constants in Silicon

	$N = 10^{16}$ cm^{-3}	$N = 10^{18}$ cm^{-3}	
D_n	27	8	cm^2/s
D_p	12	5	cm^2/s

In the derivation of I_0, several simplifying assumptions were made.[†]

An examination of Equation 86 reveals that the larger the doping (N_d and N_a) the smaller the saturation current and, thus, the larger the efficiency. However, the doping level also influences the values of the diffusion constant and of the lifetime of the minority carriers. The diffusion constant tends to decrease with increasing doping, thus helping in the reduction of I_0. Typically, for silicon at 300 K, we have the values shown in Table 14.6.

In Table 14.6, N represents either N_a or N_d.

The lifetime of the minority carriers goes down with increasing doping and thus partially counteracts the improvement in saturation current resulting from the other effects. There is no fundamental physical relationship between lifetime and doping level.

As the technology progresses, better lifetimes are achieved with a given impurity concentration. Empirically, it has been found that

$$\tau = \frac{\tau_0}{1 + N/N_0}, \tag{87}$$

where N is the doping level and τ_0 and N_0 are parameters that depend on the type of carrier. For holes, for example, τ_0 is given as 400 μs and N_0 as 7×10^{15} cm^{-3}. These data are somewhat old and may be outdated.

[†] Among the most important simplifications are the following:
1. All the recombination of minority carriers was assumed to occur in the bulk of the material, none at the surface. The surface recombination velocity, s, was taken as zero or, at least, much smaller than D/L.
2. The width, w, of the diode, normal to the junction, was taken as much larger than the diffusion length ($w \gg L$).
3. It was assumed that no generation and no recombination of electron-hole pairs occurred in the transition region of the junction.

If assumption 1 is removed and assumption 2 is kept ($s \gg D/L$)—that is, if recombination is dominated by surface effects, then I_0 is given by

$$I_0 = qAn_i^2 \left[\sqrt{\frac{D_p}{\tau_p}} \frac{1}{N_d} \coth \frac{w_n}{L_p} + \sqrt{\frac{D_n}{\tau_n}} \frac{1}{N_a} \coth \frac{w_p}{L_n} \right].$$

If, on the other hand, 2 is removed and 1 is kept then I_0 is given by

$$I_0 = qAn_i^2 \left[\sqrt{\frac{D_p}{\tau_p}} \frac{1}{N_d} \tanh \frac{w_n}{L_p} + \sqrt{\frac{D_n}{\tau_n}} \frac{1}{N_a} \tanh \frac{w_p}{L_n} \right].$$

If both assumptions a and b are removed, the the expression for I_0 becomes rather complicated (c.f. Rittner).

If N exceeds some 10^{17} cm^{-3},

$$\tau_p N_d \approx \tau_n N_a \approx 2.8 \times 10^{12} \text{ sec cm}^{-3}. \tag{88}$$

Introducing this empirical relationship into our simplified formula for the reverse saturation current,

$$\frac{I_0}{A} = 10^{-5} \times \left[\sqrt{\frac{D_p}{N_d}} + \sqrt{\frac{D_N}{N_a}} \right] \text{ A cm}^{-2}, \tag{89}$$

for silicon at 300 K.

Using the above information, we can calculate I_0/A for the case of a lightly doped diode ($N_a = N_d = 10^{16}$ cm^{-3}) and for a heavily doped one ($N_a = N_d = 10^{18}$ cm^{-3}). We get, respectively, 0.9 pA cm^{-2} and 0.05 pA cm^{-2}. The larger current is just a rough estimate because our approximation is not valid for such low doping. These extremely low saturation currents lead, of course, to high open-circuit voltages: 0.637 V and 0.712 V, respectively, for silicon at 1 sun.

Experimentally observed voltages at 1 sun have reached 0.68 V and, at 500 suns, 0.800 V . This corresponds to 28% efficiency.

14.9 Practical Efficiency

As is invariably the case, the practical efficiency of photodiodes departs substantially from the ideal one. The reasons for this include:

1. *Surface reflection.*   Surface reflection is caused by the abrupt change in refractive index when the rays pass from air to the diode material. A more gradual (even though stepwise) change can be achieved by coating the diodes with a transparent medium having an index of refraction somewhere between that of air and that of silicon. This is identical to the "bluing" technique used in photographic lenses.
2. *Opacity of current collectors.*   Electrodes have to be attached to the diodes. On the back side, the electrode can be a metallized surface, This, of course, does not work with cascaded cells. On the front side, the electrodes should introduce the least possible interference with the illumination.
3. *Poor utilization of the available surface.*   Some photodiodes are produced from circular wafers of the type employed in the semiconductor industry. When arrayed side-by-side, considerable empty space is left.
4. *Thinness of the cells.*   Excessively thin diodes fail to interact with all the available light. This is more a problem with silicon than with gallium arsenide owing to the latter's greater opacity. Silicon cells must be over 100 μm thick versus only a few μm for GaAs. However, gallium is quite expensive (roughly 1/4 of the price of gold) and, in

addition, it is difficult to create single crystals with more than a few square centimeters area. This limits crystalline GaAs to space use (where its high resistance to radiation damage becomes important) or as photocells in concentrator systems where the cost is transferred from the cell itself to that of the concentrator. In this application, GaAs's greater insensitivity to heat is an asset.

5. *Series resistance of the cells.* The material from which the diodes are made is not a good conductor of electricity. The path between the region where the current is generated and where it is collected must be minimized. On the other hand, these collectors, as explained before, should not obscure the source of light. To achieve a compromise between these two requirements, a grid of thin metal strips is built on the diode as illustrated in Figure 14.19. An ingenious solution for the problem, developed by Peter Borden, is shown in Figure 14.20. Grooves, about 50 micrometers apart, are created on the surface of the semiconductor. One face of these grooves is metallized and constitutes the current collector; the other face receives the illumination.

Light is caused to shine slantwise on the cell in a direction perpendicular to the active face, fully illuminating it while leaving the contact face in shadow. Thus, the collectors, although large, do not interfere with the incoming light. By choosing an appropriate crystallographic orientation, it is possible to create grooves whose cross section is not that of an isosceles triangle—that is, grooves that have the active face larger than the current collecting one. This reduces the required amount of tilt.

6. *Lifetime of the minority carriers.* Some electron-hole pairs are created too far from the potential barrier to survive the length of time it would take to reach it by diffusion. The longer the minority carrier lifetime, the larger the number of electrons and holes that can be separated by the barrier and, consequently, the larger the diode efficiency. This subject was discussed in more detail towards the end of the preceding section.

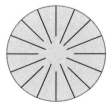
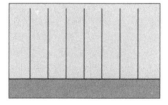

Figure 14.19   Typical current collector strips on photodiodes.

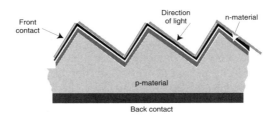

Figure 14.20   Configuration used to reduce the
series resistance of photodiodes.

14.10   Solar-Power Satellite

Science, science fiction, and magic—three areas of human endeavor separated by fuzzy and changing boundaries. Magic today may be science tomorrow. Arthur C. Clarke expressed this in his third Law: "Any sufficiently advanced technology is indistinguishable from magic."

Serious research, advanced enough to resemble science fiction, if not magic, may occasionally lead to practical results: landing on the moon. Often, it leads to a dead end, sometimes because of impossible technological obstacles, more frequently, because of impossible economics. Nevertheless, valuable knowledge can be generated.

The NASA's Solar Power Satellite (SPS) project was a study that ran into insurmountable financial barriers, even though it could have led to the generation of low cost electricity. Its dependence on the economy of scales was such that to make it economical, an investment of trillions of dollars would be required.

Critics pointed out valid difficulties in the project. It was shown that the energy needed to fabricate the photocells exceeded the total energy the system would provide over its lifetime. True enough if the photocells were produced by the Czochralski method then universally used by the semiconductor industry. Yet, modern polymers may, one day, yield photovoltaic systems much lighter (and easier to deploy) than the single crystal system envisioned and at attractive energy pay-back ratios.

It also looked impossible to launch into orbit the huge masses required. Again true using current technology. But another science fiction idea could come to the rescue: the space elevator—the dangling of a rope from a geostationary satellite, its other end anchored to the surface of earth.

(continued)

(continued)

"Climbers" would crawl up and down this rope delivering their cargo to outer space at a cost much less than by using rockets. What made this proposal a piece of science fiction is that it is easy to show that even the strongest steel cannot sustain its own weight from geo-stationary orbit. But then, at the time the SPS was being investi-gated, no one had ever hear of carbon nanotubes (discovered in 1991 by Sumio Iijima). It now seems that ribbons made of a composite of polymers and nanotubes can do much more than hold their own weight.

All told, the SPS study did not lead to any major practical ap-plication, but one can learn from it. We will attempt to do so in this section.

A drawback of photovoltaic power generation is the intermittent and unreliable nature of insolation. To compensate, one needs either large stor-age systems or a source of backup power. Both solutions substantially increase the required capital investment. A further disadvantage is that load centers tend to be concentrated in regions where good insolation is unavailable. This translates into the need of long transmission lines.

Transmission lines can be made quite short if the power is beamed by means of microwaves directly to the neighborhood of consuming cen-ters. Solar collectors in Arizona could feed New York or Chicago through microwaves generated on the ground and reflected by satellites. Power man-agement would be simplified by simply switching the beam from one user to another.

To get around the unreliability of insolation, the photodiodes must be placed in space, in a geostationary orbit. It is then possible to achieve almost constant exposure to the sun.[†]

Originally proposed by Peter Glazer of the Arthur D. Little Corpo-ration, the SPS concept was investigated by NASA but met considerable opposition.

The proposed configuration would use satellites, each one capable of delivering 5 GW to the power grid on earth.

The development of the SPS involved four major elements:

1. Energy conversion in space
2. Energy transmission to earth
3. Space transportation
4. Space construction

[†] A short eclipse would still occur near local midnight during the equinoxes. The duty cycle of a solar-power satellite (in geostationary orbit) is about 99%.

14.10.1   Beam from Space

Power generated in geostationary orbit must, somehow, be sent to the consumer on earth. One of the surprising results brought out by the SPS study is that a microwave beam can transmit energy more efficiently than a physical transmission line of comparable length. In fact, the calculated efficiency from the input of the transmitting antenna on the satellite to the receiving antenna output on ground is 74%, more than double of the efficiency realizable with a metallic transmission line. In addition, the power a beam can carry is much larger than the carrying capacity of the largest existing transmission line (3.25 GW transmission lines between Itaipu and São Paulo, Brazil).

There are numerous constraints in designing such a microwave beam. The frequency must be such as to minimize ionospheric and atmospheric absorption and it must be within one of the bands allocated to industrial heating. This led to the choice of the "microwave oven" frequency of 2.45 GHz.

There is also a constraint regarding the maximum power density of the beam as it transverses the ionosphere. A maximum level of 230 W/m^2 was deemed acceptable. Ionospheric nonlinearities would presumably cause power densities much above this limit to generate harmonics of the 2.45 GHz signal capable of interfering with other radio services.

Concerns with interference also prompted the requirement that the beam power level be quite low outside the designated receiving area. Because it is physically impossible to abruptly truncate an electromagnetic beam, the one coming down from the satellite was to have a gaussian radial distribution.

Another major concern had to do with the proper aiming of the beam. An ingenious solution was found for this problem. See the subsection on the radiation system.

14.10.2   Solar Energy to DC Conversion

A thermo-mechanical-electric conversion system using Rankine engines was considered but was abandoned in favor of photovoltaics. Two different photovoltaic materials were studies: silicon and gallium aluminum arsenide, leading to two different structures as shown in Figure 14.21.

The total surface area of both configurations was nearly the same (5.2 km by 10.4 km for the silicon version and 5.25 km by 10.5 km for the gallium arsenide one), a total of some 55 km^2. Of this, 96% is sun-collecting area so that, at 1360 W/m^2, a total of 72 GW would be collected under the most favorable conditions.

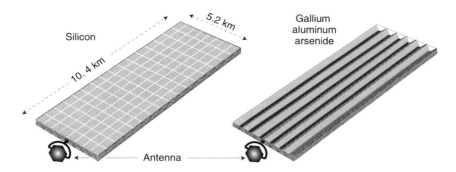

Figure 14.21   Two proposed configurations for the 5-GW power satellite. The one on the left uses silicon cells, the one on the right, gallium aluminum arsenide.

The silicon cells were to be exposed to unconcentrated sunlight, while the gallium arsenide cells would work with a concentration ratio of 2, achieved by means of a simple through-type reflector. This is possible because the gallium arsenide cells can be operated efficiently at a higher temperature (125 C) than the silicon ones (36 C). At these temperatures cell efficiency would be 18.2% for the GaAs and 16.5% for the Si. Owing to a 8.5% reflection loss in the concentrator, both cases would deliver about the same 10.5 GW of dc power.

A secular degradation of the SPS was expected owing, in part, to the progressively decreasing efficiency of the cells in space. The decay can be reduced by *in situ* annealing of the silicon cells carried out periodically by means of lasers mounted on a roving structure. GaAs cells, operating at a higher temperature, would be self-annealing. Decay caused by micrometeorite damage was expected to amount to only 1% in 30 years.

Of the 10.5 GW, only 8.2 GW would reach the microwave generators, the rest being lost in the feeding conductors. These losses may seem excessive. Consider, however, that there is a limit to the dc voltage transmitted. The microwave generators can use, at most, 40,000 volts. Also, a much higher voltage would cause break down in the (very) tenuous atmosphere at geostationary heights. 10.5 GW at 40 kV correspond to 262,000 amperes! It does not take a large resistance to cause huge I^2R loses.

14.10.3 Microwave Generation

Several microwave generators were considered, including

1. magnetrons,
2. transistors,
3. amplitrons, and
4. klystrons.

Magnetrons would have difficulty in delivering the required spectral purity necessary to avoid interference with other services. Remember that, even if spurious radiation is kept 60 dB below the carriers, this amounts to nearly 6 kW of undesired signal.

For use in the SPS, transistors needed considerable more development. Given enough time, it might be possible to created appropriate transistors.

Amplitrons would have the following advantages:

1. low mass: 0.4 kg/kW compared with 0.7 kg/kW for klystrons;
2. high efficiency: 88% (possibly, 90%) compared with 85% for klystrons;
3. almost infinite cathode life: amplitrons operate by secondary emission and could use platinum cathodes.

The advantages of klystrons are

1. per tube, klystrons have an order of magnitude more output than amplitrons;
2. klystrons use high anode voltage (40 kV versus 20 kV for amplitrons). This reduces the currents and consequently the mass of the dc power distribution system, which represents a substantial portion of the mass of the satellite;
3. klystrons have much higher gains than amplitrons. The resulting lower rf excitation power facilitates phase control essential for forming and aiming the beam.

NASA was inclined to use high efficiency klystrons (85%) delivering 7 GW to the transmitting antenna. Since the power output of each klystron is 50 kW, a total of 140,000 tubes would be needed. Assuming a mean lifetime between failure of 5000 hours, one would expect a failure every 2 minutes! This points out the necessity of using tubes with extremely long life and of having automated procedures for diagnosing and repairing defects.

14.10.4 Radiation System

The microwave power is transferred to the antenna (the small hexagonal structure in Figure 14.21) through a rotary joint. The latter is necessary to keep the antenna aimed at earth while the collector points toward the sun. Thermal considerations limit the power density of the transmitting array to 22 kW/m^2. Based on this figure, an area of 300,000 m^2 of transmitting antenna would be required—a circle with 600 m diameter. But the power density across the antenna aperture cannot be constant because this would generate a beam whose shape does not satisfy the ground safety requirements—some regions of the array must have larger power densities than others (still observing the maximum of 22 kW/m^2). This results in a transmitting antenna 1 km in diameter.

The antenna was a planar phased array consisting of a large number of radiators, grouped in subarrays, each of which must be exactly at the same distance from the ground target. Since the transmitted wavelength is 12 cm (2.45 GHz), the mechanical alignment of the individual radiators would have to be correct within less than 1 centimeter over the whole 80 hectares of antenna in order to form a beam properly directed to the ground station. This is impossible to achieve. Of course, it is not the geometric distance that counts, it is the electric distance. Hence, the unavoidable mechanical misalignment can be compensated by changing the phase of individual subarrays. To this end, a transmitter on the ground sends up a "pilot" radio signal at a frequency slightly different from the one beamed down. If all radiators, now acting as receivers, were in their correct position, the phase of the received pilot signal would be the same from subarray to subarray. Any mispositioning will appear as a phase difference (relative to a reference phase received at the center of the antenna). The phase of the transmitted signal is altered by exactly the negative of the received phase error. This insures that the phase of all radiators in the plane of the antenna has now the correct value to focus on the spot on the ground whence the pilot signal came from. An accidental loss of phase control defocuses the beam, spreading it out harmlessly over a wide area.

As mentioned before, the surprisingly good efficiency of 74% can be achieved in the transmission link. Thus, 5.15 GW would be available at the output of the ground antenna, and, of these, 5 GW would be delivered to the grid.

The beam width of the satellite antenna was such that, if the illuminated area on the ground were at the equator, it would have the shape of a circle 10 km in diameter.[†] This is equivalent to an average power density of a little over 80 W/m². However, the beam is not uniform: it has a gaussian shape with a peak power density of 230 W/m². This shape specification is important to keep the radiation level outside the collecting area on the ground low enough to avoid health hazards and interference.

14.10.5 Receiving Array

The ground receiving system consisted of a number of rectennas, i.e., antennas equipped with their individual rectifiers. This scheme avoids the necessity to adjust the phase of the antenna current so that their output can be added up.

[†] At 40° latitude, the footprint is a 10- by 14-km ellipse.

The proposed antennas were half-wave dipoles. The effective area of such dipoles is

$$A = 1.64 \, \frac{\lambda^2}{4\pi,} \tag{90}$$

where the factor, 1.64, is the gain of a dipole relative to an isotropic radiator. Since the wavelength is 12.2 cm, each antenna sees an area of 20 cm^2. The total number of antennas in the 10 km diameter circle is 50 billion! To build this many antennas in one year, 1500 antennas have to be built per second, a clearly challenging task.

The antenna problem is complicated by the fact that the rectifier attached to it is, necessarily, a nonlinear device. Strong harmonic generation will take place and this must be kept from being reradiated. Each antenna must be equipped with an appropriate filter. The antenna, filter rectifier combination must cost a small fraction of a dollar otherwise the cost of the ground collecting system would be prohibitive.

The dc output from all antennas was to be added up and fed to inverters for conversion to 60 Hz ac and distribution to customers.

14.10.6 Attitude and Orbital Control

The many factors that perturb both the orbit and the attitude of the SPS include

1. solar and lunar gravitational pull,
2. lack of spherical symmetry in the geogravitational field,
3. solar radiation pressure,
4. microwave recoil,
5. rotary junction friction,
6. aerodynamic drag,
7. interaction with the geomagnetic field, and
8. gravity gradient torques.

Some of the orbital and attitudinal corrections were to be made by argon ion rockets. As much as 50,000 kg of argon would be needed annually for this purpose.

14.10.7 Space Transportation and Space Construction

Both size and mass of the solar power satellites present a major challenge to the astronautical engineer. Clearly, the satellites would have to be assembled in space. It is estimated that each would require 850 man-years of space labor even using automatic assembly machines. Large crews living long months in orbit would be needed.

The transport vehicle for the materials would have to be one order of magnitude larger than the space shuttle. For each satellite, some 400 launches would be needed. If the construction rate were 1 satellite per year, this would translate into more than one launch per day. To be economical some 50 satellites must be placed in orbit; the 1 per year rate would require a sustained effort for half a century.

14.10.8 Future of Space Solar Power Projects

Early enthusiasm for space-based solar power projects was dampened when more realistic cost estimates were made. The DOE estimated the cost of research and construction of one demonstration satellite with its ground-based rectenna at over $100 billion. Each additional unit would cost 11.5 billion. In 1981, the National Research Council pushed the cost estimate to some $3 trillion and the time of completion to 50 years. Many groups opposed the whole idea. The ground solar people did not want such a grand project sucking in most of the funds for solar energy research. Fusion proponents also hated the idea of a major competitor. To counter this, SPS advocates maintain that fusion power is the power of the future and always will be.

The cost of launching the components into geosynchronous orbit and of assembling a solar power satellite in space are staggering. This has led some to propose the construction of the satellite on the moon using lunar-mined materials. A launch from the moon requires substantially less energy than from earth. Alternatively, the solar collectors and microwave generators could be permanently based on the moon if an economical way of beaming the energy to earth can be found.

Appendix A

Values of two definite integrals
used in the calculation of photodiode performance.

X	$\int_X^\infty \dfrac{x^2}{e^x - 1} dx$	$\int_X^\infty \dfrac{x^3}{e^x - 1} dx$	X	$\int_X^\infty \dfrac{x^2}{e^x - 1} dx$	$\int_X^\infty \dfrac{x^3}{e^x - 1} dx$
0.0	2.4041	6.4935			
0.1	2.3993	6.4932	2.6	1.0656	4.5094
0.2	2.3855	6.4911	2.7	1.0122	4.3679
0.3	2.3636	6.4855	2.8	0.9605	4.2259
0.4	2.3344	6.4753	2.9	0.9106	4.0838
0.5	2.2988	6.4593	3.0	0.8626	3.9420
0.6	2.2576	6.4366	3.1	0.8163	3.8010
0.7	2.2115	6.4066	3.2	0.7719	3.6611
0.8	2.1612	6.3689	3.3	0.7293	3.5226
0.9	2.1073	6.3230	3.4	0.6884	3.3859
1.0	2.0504	6.2690	3.5	0.6494	3.2513
1.1	1.9911	6.2067	3.6	0.6121	3.1189
1.2	1.9299	6.1363	3.7	0.5766	2.9892
1.3	1.8672	6.0579	3.8	0.5427	2.8622
1.4	1.8034	5.9719	3.9	0.5105	2.7381
1.5	1.7390	5.8785	4.0	0.4798	2.6171
1.6	1.6743	5.7782	4.1	0.4507	2.4993
1.7	1.6096	5.6715	4.2	0.4231	2.3848
1.8	1.5452	5.5588	4.3	0.3970	2.2737
1.9	1.4813	5.4408	4.4	0.3722	2.1660
2.0	1.4182	5.3178	4.5	0.3488	2.0619
2.1	1.3561	5.1906	4.6	0.3267	1.9613
2.2	1.2952	5.0596	4.7	0.3058	1.8642
2.3	1.2356	4.9254	4.8	0.2861	1.7706
2.4	1.1773	4.7886	4.9	0.2675	1.6806
2.5	1.1206	4.6498	5.0	0.2501	1.5941

A reasonable approximation of the integral $\displaystyle\int_X^\infty \frac{x^2}{e^x - 1} dx$ is given by

$$\int_X^\infty \frac{x^2}{e^x - 1} dx \approx 2.4164 - 0.086332X - 0.37357X^2$$
$$+ 0.099828X^3 - 0.0078158X^4. \qquad (91)$$

Appendix B

A Semiconductor Primer

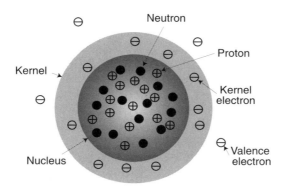

A silicon atom consists of a **nucleus** (dark sphere), containing 14 **protons** (heavily rimmed circles with a "+" sign) and 14 **neutrons** (solid black), surrounded by 14 **electrons** (circles with a "−" sign). The nucleus carries a charge of +14 but the atom (unless it is ionized) has no net charge owing to the 14 electrons swarming around it.

Observe that 10 of the 14 electrons are tightly bound to the nucleus and are difficult to remove. However, four electrons (called **valence electrons**) can be easily removed (ionizing the atom) and are, therefore, able to take part in chemical reactions. Silicon is, consequently, **tetravalent**.

It proves convenient to represent a silicon atom as consisting of a **kernel** surrounded by four valence electrons. The kernel, because it has fourteen positive charges from the nucleus and eight tightly bound electrons, has a net charge of +4. The atom, as a whole, is, of course, neutral because the charge of the four valence electrons cancels that of the kernel.

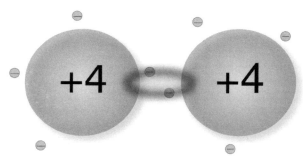

Two silicon atoms can be bound one to another by exchanging valence electrons. Such a bond is called a **covalent bond**.

Because it has four valence electrons, each silicon atom can make *four* covalent bonds attaching itself to four neighboring atoms. This might lead to a lattice structure as depicted below. Indeed, it does so in the case of carbon which crystallizes in a two-dimensional fashion when it forms graphite. Silicon (and carbon in the diamond form) has a three-dimensional crystal difficult to depict in a flat drawing. For simplicity, we will continue to use the flat picture.

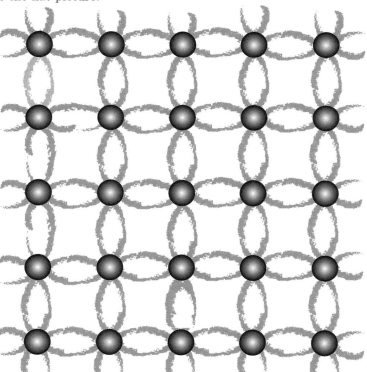

14.54

At 0 K, all valence electrons are engaged in covalent bonds and are, therefore, unavailable as **carriers**—that is as transporters of electric charge. No current can flow through the crystal—it is an **insulator**.

However, if a bond is disrupted (by thermal agitation of the lattice or through the impact of a photon or a high speed free electron), then one of the valence electrons is ejected from the bond and becomes free to carry electricity, leaving behind an incomplete bond, one in which a **hole** exists into which an electron from a neighboring bond can fall. This causes the hole to move to a new place. Thus, the disruption of a bond creates a *pair* of carriers—an electron and a hole imparting some degree of conductivity to the material.

It is clear from our picture, that, in this particular case, the number of free (**conduction**) electrons is exactly equal to the number of holes. Such materials are called **intrinsic**.

One can now understand why semiconductors have (most of the time) an electric conductivity that increases when the temperature increases: the warmer the material the greater the number of carriers.

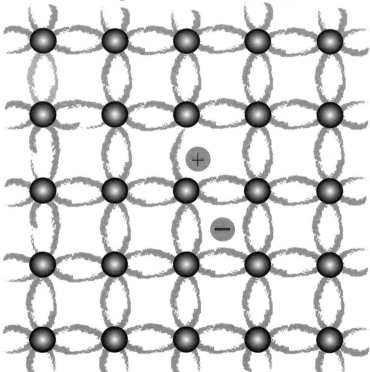

14.55

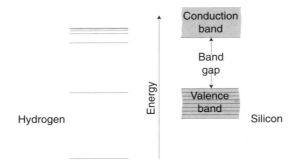

Another way of looking at this situation is to realize that, just as in isolated atoms, electrons in a crystal can only occupy certain discrete energy levels. In the case of solids, electrons are confined to certain **bands** of energy levels. In semiconductors, such as silicon, one band, called the **valence band** is completely occupied and electrons cannot move. The next band (**conduction band**) is completely empty. The energy difference between the top of the valence band and the bottom of the conduction band is called the **band-gap**. In silicon, the band-gap is 1.1 eV.

While the free electron is a negative carrier of charge, the hole can be treated as a positive carrier. In the presence of an electric field their motion is in opposite directions, yet the resulting current is in the same direction because of the opposite charges they carry.

A simple analogy to the motion of electrons and holes is that of the motion of VW beetles and the corresponding "holes."

In a garage in San Francisco with a floor fully occupied by Volkswagen beetles (in neutral and with no brakes on, so they can roll freely), the cars cannot move even if an earthquake tilts the garage floor.

If a car is hoisted to the next floor, then, with a tilted garage, the car will roll to the left. Simultaneously some cars in the lower floor will also move, in effect, causing the "hole" in the row of cars to move *up*.

The energy necessary to lift a car from one floor to the next is equivalent to the band-gap energy in semiconductors.

14.56

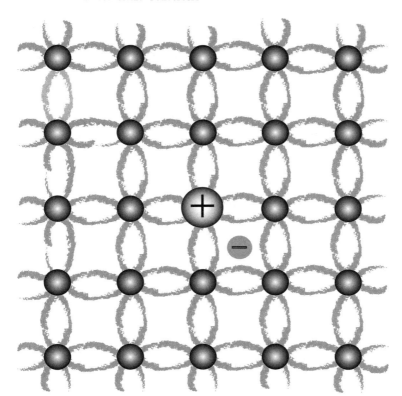

So far, we have dealt with absolutely pure silicon. In fact, all silicon contains some impurity which may dramatically alter the properties of the material. In the picture above, a phosphorus atom has replaced one silicon atom. Since phosphorus has a valence of 5, it has enough electrons to complete the four covalent bonds that anchor the atom in the lattice. In addition, there is one electron left over, which then acts as a carrier. Electron-hole pairs (not shown in the picture) are still being created by the thermal agitation of the lattice but the number of electrons now exceeds that of holes—the material is *n*-**silicon**, one in which the dominant carrier is negative. The phosphorus kernel has a +5 charge and the crystal site has only 4 covalent bond electrons. Thus, the site has a +1 positive charge, which being immobile (because it is tied to the lattice) does *not* constitute a carrier. It is called a **donor**.

If instead of a pentavalent atom such as phosphorus, the impurity is a trivalent one such aluminum, then there is an insufficiency of electrons and only three of the covalent bonds are satisfied, leaving one incomplete—that is, leaving a free hole—the material is *p*-**silicon**, one in which the dominant carrier is a positive hole. Similarly to the case of phosphorus, the aluminum atom represents a −1 immobile charge—it is an **acceptor**.

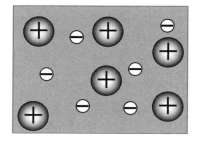
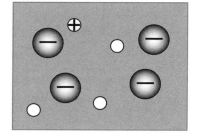

From now on, for clarity, we will omit most of the background of silicon atoms and covalent bonds from our drawings and show only the carriers and their corresponding donors or acceptors. It must be emphasized that the donors and acceptors are not carriers because they cannot move.

In the drawing, the left hand one represents an n-material and the right hand one, a p-material. Both are electrically neutral, having equal number of positive and negative charges.

The introduction of certain impurities into the mass of silicon is called **doping**. Typically, the amount of doping is small, ranging from 1 dopant atom for every 10 thousand silicon atoms (extremely heavy doping) to 1 dopant atom for every 100 million silicon atoms (very light doping).

Things start becoming really interesting when a crystal has a p-region juxtaposed to an n-region as shown above.

Electrons, more abundant in the n-side, tend to diffuse to the p-side, while holes tend to diffuse into the n-side. The donors and acceptors, of course, cannot move. The net effect of these diffusions is that the n-side becomes positive (in the illustration, there are 8 positive and only 3 negative charges). By the same token, the p-side becomes negative (7 negative and 2 positive charges). Thus, a **contact potential** is created. which in silicon at room temperature can be around 1 V, depending on the doping.

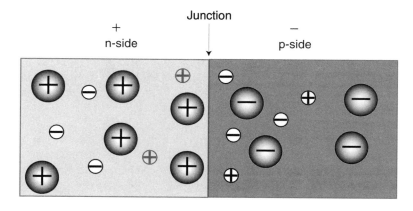

14.58

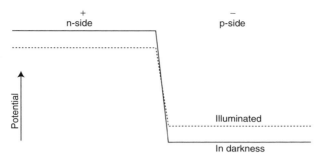

The potential across most of the n-side and p-side of the device is constant (no electric field). All the field is concentrated across a narrow transition region. Owing to the narrowness of this region (a few tens of a nanometer), the electric fields developed can be enormous—tens of millions of volts/meter.

When light shines near a p-n junction, electron-hole pairs may be created on either side. If they are very far from the transition region, they will recombine after a few microseconds. If, however, they are near, they may drift toward the region of high electric field. In this case, an electron created in the p-side may fall to the n-side, while a hole created on the n-side may fall to the p-side. In either case, these charges counteract the contact potential. Thus, the effect of light on a p-n junction is the lowering of the contact potential.

References

da Rosa, Aldo V., *Fundamentals of Electronics*, Optimization Software, Inc., (**1989**).

Greenham, N. C., X. Peng, and A. P. Alivisatos, *Phys.Rev.* B **54**, 17628 (**1996**).

Hamdy, Adel M. and D. E. Osborn, The potential for increasing the efficiency of solar cells in hybrid photovoltaic/thermal concentrating systems by using beam splitting. *Solar and Wind Technology*, **7**, No 23, 147–153, (**1990**).

Ludman, Jacques E., Holographic solar concentrator, *Applied Optics*, **212**, 3057, Sept. 1, **1982**.

Ludman, Jacques E., Approximate bandwidth and diffraction efficiency in thick holograms, *Rome Air Development Center*, **1981**.

Ludman, Jacques E., J. L. Sampson, R. A. Bradbury, J. G. Martín, J. R. Riccobono, G. Sliker, and E. Rallis, Photovoltaic systems based on spectrally selective holographic concentrators, SPIE, 1667 *Practical Holography VI*, 182, **1992**.

Ludman, Jacques E., Juanita Riccobono, Nadya Reinhand, Irina Semenova, José Martín, William Tai, Xiao-Li Li, and Geof Syphers, Holographic solar concentrator for terrestrial photovoltaics, *First WCPEC*, Dec 5–9, **1994**, Hawaii, IEEE.

Rappaport, Paul, The photovoltaic effect and its utilization, *RCA Rev.*, **20**, 373–397, Sept. **1959**.

Rittner, Edmund S., An improved theory for Silicon p-n junctions in solar cells, *J. Energy*, **1**, 9, Jan. **1977**.

Ross, B, and J. R. Madigan, Thermal generation of recombination centers in silicon, *Phys. Rev.*, **108** 1428-1433, Dec. 15, **1957**.

Schaller, R. D., and V. I. Klimov, High efficiency carrier multiplication in PbSe nanocrystals: Implications for solar energy conversion, *Phys. Rev. Lett.* **92**, 186601-4 (**2004**).

Wysocki, Joseph J., and Paul Rappaport, Effect of temperature on photovoltaic solar energy conversion, *J. Appl. Phys.*, **31** 571–578, Mar. **1960**.

PROBLEMS

14.1   What is the theoretical efficiency of cascaded photodiodes made of two semiconductors, one with a band-gap energy of 1 eV and the other with 2 eV when exposed to sunlight?

14.2   As any science fiction reader knows, there are many parallel universes each one with different physical laws. In the parallel universe we are discussing here, the black body radiation at a given temperature, T, follows a simple law (P is the power density (W m^{-2}) and f is in Hz.):

$\frac{\partial P}{\partial f}$ is zero at $f = 0$,

$\frac{\partial P}{\partial f}$ grows linearly with f to a value of 1 W m^{-2}THz^{-1} at 500 THz,

from 500 THz it decreases linearly to zero at 1000 THz.

1. What is the total power density of the radiation?
2. What is the value of the band-gap of the photodiode material that results in the maximum theoretical efficiency of the photodiode exposed to the above radiation?

14.3   Under circumstances in which there is substantial recombination of carriers in the transition region of a diode, the v-i characteristic becomes

$$I = I_\nu - I_R \left[\exp \left(\frac{qV}{2kT} \right) - 1 \right] - I_0 \left[\exp \left(\frac{qV}{kT} \right) - 1 \right].$$

In solving this problem use the more complicated equation above rather than the equation give in the text which is

$$I = I_\nu - I_0 \left[\exp \left(\frac{qV}{kT} \right) - 1 \right].$$

A silicon diode has 1 cm^2 of effective area. Its reverse saturation current, I_0, is 400 pA, and the current, I_R, is 4 μA. These values are for $T = 300$ K. Assume that this is the temperature at which the diode operates. Assume also 100% quantum efficiency—that is, that each photon with energy above 1.1 eV produces one electron-hole pair. Finally, assume no series resistance.

At one sun, the power density of light is 1000 W/m^2, and this corresponds to a flux of 2.25×10^{21} photons s^{-1} m^{-2} (counting only photons with energy above 1.1 eV).

1. What is the open-circuit voltage of this diode at 1 sun?
2. At what voltage does the diode deliver maximum power to the load?
3. What is the maximum power the diode delivers?
4. What load resistance draws maximum power from the diode?
5. What is the efficiency of the diode?

14.61

Now assume that a concentrator is used so that the diode will receive 1000 suns. This would cause the operating temperature to rise and would impair the efficiency. Assume, however, that an adequate cooling system is used so that the temperature remains at 300 K. Answer the questions above using 100% concentrator efficiency.

In fact, the concentrator is only 50% efficient. Is there still some advantage in using this diode with the concentrator?

14.4   Assume that you are dealing with perfect black-body radiation ($T = 6000$ K). We want to examine the theoretical limits of a photodiode. Assume no light losses by surface reflection and by parasitic absorption in the diode material. Consider silicon (band-gap energy, W_g, of 1.1 eV).

Clearly, photons with energy less than W_g will not interact with the diode because it is transparent to such radiation.

1. What percentage of the power of the black-body radiation is associated with photons of less than 1.1. eV?
2. If all the energy of the remaining photons were transformed into electric energy, what would the efficiency of the photodiode be?
3. Would germanium ($W_g = 0.67$ eV) be more or less efficient?
4. Using Table 11.1 in the text, determine the percentage of the solar energy absorbed by silicon.
5. A photon with 1.1 eV will just have enough energy to produce one electron-hole pair, and, under ideal conditions, the resulting electron would be delivered to the load under 1.1 V of potential difference. On the other hand, a photon of, say, 2 eV, will create pairs with 0.9 eV excess energy. This excess will be in the form of kinetic energy and will rapidly be thermalized, and, again, only 1.1 eV will be available to the load. Thus, all photons with more than W_g will, at best, contribute only W_g units of energy to the load.

   Calculate what fraction of the black-body radiation is available to a load connected to an ideal silicon photodiode.
6. The short circuit current of a diode under a certain illumination level is 10^7 times the diode reverse saturation current. What is the relative efficiency of this diode compared with that of the ideal diode above?

14.5   Consider radiation with the normalized spectral power density distribution given by:

$$\frac{\partial P}{\partial f} = 0 \text{ for } f < f_1 \text{ and } f > f_2,$$

$$\frac{\partial P}{\partial f} = 1 \text{ for } f_1 < f < f_2$$

where $f_1 = 100$ THz and $f_2 = 1000$ THz.

14.62

1. What is the theoretical efficiency of a photodiode having a band-gap energy of $W_g = h f_1$?
2. What band-gap energy maximizes the efficiency of the diode?
3. If the band-gap energy is $h \times 500$ THz, and if the material is totally transparent to radiation with photons of less energy than W_g, what fraction of the total radiation power goes through the diode and is available on its back side?
4. If behind the first diode, there is a second one with $W_g = h \times 100$ THz, what is the efficiency of the two cascaded diodes taken together?

14.6   Two photodiodes, each with an effective area of 10 cm^2, are exposed to bichromatic radiation having power densities of 500 W/m^2, in narrow bands one around 430 THz and the other, around 600 THz.

One diode has a band-gap energy of 1 eV, the other has 2 eV. When the diode is reverse biased (in the dark), the saturation current is 10 nA.

The diodes operate at 300 K.

1. What are the short-circuit photo currents?
2. What is the open-circuit voltage of each diode?
3. What is the maximum theoretical efficiency of each diode?
4. What is the maximum power each diode can deliver to a load (assume no series resistance in the diodes)?

14.7   An ideal photodiode is made of a material with a band-gap energy of 2.35 eV. It operates at 300 K and is illuminated by monochromatic light with wavelength of 400 nm. What is its maximum efficiency?

14.8   What is the short-circuit current delivered by a 10 cm by 10 cm photocell (100% quantum efficiency, $R_s = 0$) illuminated by monochromatic light of 400 nm wavelength with a power density of 1000 W/m^2.

14.9   Under different illumination, the cell of Problem 14.8 delivers 5 A into a short circuit. The reverse saturation current is 10 nA/m^2. What is the open-circuit voltage at 300 K?

14.10   The optical system of a solar photovoltaic system consists of a circular f:1.2† lens with a focal length, F, of 3 m.

When aimed directly at the sun (from the surface of planet Earth), what is the diameter, D_i, of the solar image formed in the focal plane? Assume a 90% efficiency of the optical system and a perfectly clear atmosphere at noon.

---

† The f-number is, as in any photographic camera, the ratio of the focal length to the diameter of the lens.

1. What is the concentration ratio, C?
2. What is the total light power, P, that falls on a photovoltaic cell exposed to the solar image?
3. What is the power density, p?

   Assume:
   3.1 no conduction losses;
   3.2 no convection losses;
   3.3 the silicon photovoltaic cell is circular and has a diameter equal to that of the solar image. The cell intercepts all the light of the image;
   3.4 the efficiency of the photocell is equal to 60% of the maximum theoretical value for a black body radiator at 5800 K.
   3.5 all the power the photocell generates is delivered to an external load;
   3.6 the effective heat emissivity, ϵ, is 0.4.

4. What is the power delivered to the load?
5. What is the temperature of the photocell?
6. What is the temperature of the photocell if the load is disconnected? If you do your calculations correctly, you will find that the concentrated sunlight will drive the cell to intolerably high temperatures. Silicon cells should operate at temperatures of 500 K or less.
7. How much heat must be removed by a coolant to keep the cell at 500 K when no electric power is being extracted?

   The coolant will exit the cell at 480 K and drives a steam engine that rejects the heat at 80 C and that realizes 60% of the Carnot efficiency.
8. How much power does the heat engine deliver?
9. What is the overall efficiency of the photocell cum steam engine system?

14.11   *Treat the photodiode of this problem as an ideal structure. Assume 100% quantum efficiency.*
    A photodiode has an area of 1 by 1 cm and is illuminated by monochromatic light with a wavelength of 780 nm and with a power density of 1000 W/m². At 300 K, the open circuit voltage is 0.683 V.

1. What is its reverse saturation current, I_0?
2. What is the load resistance that allows maximum power transfer?
3. What is the efficiency of this cell with the load above?

14.12   The power density of monochromatic laser light (586 nm) is to be monitored by a 1×1 mm silicon photodiode. The quantity observed is the short-circuit current generated by the silicon. Treat the diode as a perfect ideal device.

14.64

1. What current do you expect if the light level is 230 W/m²?
2. How does the temperature of the semiconductor affect this current? Of course, the temperature has to be lower than that which will destroy the device (some 150 C, for silicon).
3. Instead of being shorted out, the diode is now connected to a load compatible with maximum electric output. Estimate the load voltage.

14.13   A silicon photocell being tested measures 4 by 4 cm. Throughout the tests the temperature of the device is kept at 300 K. Assume the cell has no significant series resistance. Assume 100% quantum efficiency. The band-gap energy of silicon is 1.1 eV.

Initially, the cell is kept in the dark. When a current of 100 μA is forced through it in the direction of good conduction, the voltage across the diode is 0.466 V.

Estimate the open circuit voltage developed by the cell when exposed to bichromatic infrared radiation of 412 nm and 1300 nm wavelength. The power density at the shorter wavelength is 87 W/m² while, at the longer, it is 93 W/m².

14.14   What is the ideal efficiency of a photocell made from a semiconducting material with a band-gap energy, $W_g = 2$ eV, when illuminated by radiation with the normalized spectral power distribution given below:

f	$\partial P/\partial f$
<200 THz	0
200 to 300 THz	$0.01f-2$
300 to 400 THz	$4-0.01f$
>400 THz	0

In the table above, the frequency, f, is in THz.
Repeat for a semiconductor with $W_g = 1$ eV.

14.15   What is the theoretical efficiency of a photocell with a 2.5 V band-gap when exposed to 100 W/m² solar radiation through a filter having the following transmittance characteristics:

Pass without attenuation all wavelengths between 600 and 1000 nm. Reject all else.

14.16   A photodiode is exposed to radiation of uniform spectral power density ($\frac{\partial P}{\partial f}$ = constant) covering the range from 300 to 500 THz. Outside this range there is no radiation. The total power density is 2000 W/m².

1. Assuming 100% quantum efficiency, what is the short-circuit photo current of a diode having an active area of 1 by 1 cm?
2. When exposed to the radiation in Part a of this problem, the open-

circuit voltage delivered by the diode is 0.498 V. A 1.0 V voltage is applied to the diode (which is now in darkness) in the reverse conduction direction (i.e., in the direction in which it acts almost as an open circuit). What current circulates through the device? The temperature of the diode is 300 K.

14.17   The sun radiates (roughly) like a 6000-K black body. When the power density of such radiation is 1000 W/m^2—"one sun"—the total photon flux is 4.46×10^{21} photons per m^2 per second. Almost exactly half of these photons have energy equal or larger than 1.1 eV (the band-gap energy, W_g, of silicon).

Consider a small silicon photodiode with a 10 by 10 cm area. When 2 V of reversed bias is applied, the resulting current is 30 nA. This is, of course, the reverse saturation current, I_0.

When the photodiode is short-circuited and exposed to black body radiation with a power density of 1000 W/m^2, a short-circuit current, I_ν, circulates.

1. Assuming 100% quantum efficiency (each photon creates one electron-hole pair and all pairs are separated by the *p-n* junction of the diode), what is the value of this current?
2. What is the open-circuit voltage of the photodiode at 300 K under the above illumination?
3. Observe that the *v-i* characteristics of a photodiode are very steep at the high current end. In other words, the best operating current is only slightly less that the short-circuit current. This knowledge will facilitate answering the question below.

   Under an illumination of 1000 W/m^2, at 300 K, what is the maximum power the photodiode can deliver to a load. What is the efficiency? Do this by trial and error and be satisfied with 3 significant figures in your answer. Consider an ideal diode with no internal resistance.
4. What is the load resistance that leads to maximum efficiency?
5. Now repeat the power, efficiency, and load resistance calculations for an illumination of 10000 W/m^2.
6. What happens to the efficiency and the optimal load resistance when the power density of the illumination on a photodiode increases?

14.18   Everything else being the same, the efficiency of a photodiode rises

1. when the operating temperature rises,
2. when the operating temperature falls,
3. when the light power density rises,
4. when the light power density falls.

14.66

14.19   A photodiode with a band-gap energy of $W_g = 1.4$ eV is exposed to monochromatic radiation (500 THz) with a power density $P = 500$ w/m^2. The active area of the device is 10 by 10 cm.

   Treat it as an ideal device in series with an internal resistance of 2 mΩ.

   All measurements were made at 298 K.

   The open-circuit voltage is 0.555 V.

1. Estimate the short-circuit current.
2. How much power does the diode deliver to 200 mΩ load?
3. What is the efficiency of the device when feeding the 200 milliohm load?

14.20   *Suggestion: To solve this problem, use a spread sheet and tabulate all the pertinent values for different hour angles (from sunrise to sunset). 5° intervals in α will be adequate.*

   A flat array of silicon photodiodes is set up at 32° N. The array faces south and is mounted at an elevation angle that maximizes the yearlong energy collection, assuming perfectly transparent air.

1. What is the elevation angle of the array?
2. On April 15, 2002, how does the insolation on the array vary throughout the day? Plot the insolation, P, versus the time of day, t, in hours.
3. What is the average insolation on the collector.
4. Assuming ideal silicon photodiodes with a reverse saturation current density of 10 nA/m^2, what is the average power delivered during the day (from sunrise to sunset) if a perfect load follower is used, i.e., if the load is perfectly matched at all the different instantaneous values of insolation? What is the average overall efficiency?
5. Estimate the average power collected if the array is connected to a load whose resistance maximizes the efficiency at noon. In other words, the average power when no load-follower is used.

14.21   To simplify mathematical manipulation, we will postulate a very simple (and unrealistic) spectral power distribution:

$$\frac{\partial P}{\partial f} = \begin{cases} A \text{ for } 300 \text{ THz} < f < 500 \text{ THz}; \\ 0, \text{ otherwise.} \end{cases}$$

1. If $A = 10^{-12}$ W m^{-2} Hz^{-4}, what is the power density of the radiation?
2. Assuming 100% quantum efficiency, what is the short-circuit current density, J_ν, in an ideal photodiode having a band-gap energy smaller than the energy of 300 THz photon?
3. At 300 K, assuming a reverse saturation current density of $J_0 = 10^{-7}$ A/m^2, what is the open-circuit voltage of the photocell?

14.67

4. At what load voltage, V_m, does this photocell deliver its maximum power output?
5. What is the current density delivered by the photocell when maximum power is being transferred to the load?
6. What is the efficiency of the photocell?
7. What is the load resistance under the above conditions?
8. Repeat all the above for a light power density of 2 W/m^2.
9. What would be the efficiency at these low light levels if the load resistance had the optimum value for 200 W/m^2?

14.22 It is hoped that high efficiency cascaded photocells can be produced at a low cost. This consists of a sandwich of two cells of different band-gaps. The bottom cell (the one with the smaller band-gap) can be made using $CuIn_xGa_{1-x}Se_2$, known as CIGS. This material has a band-gap of about 1 eV and has been demonstrated as yielding cells with 15% efficiency.

The question here is what band-gap of the top cell yields the largest efficiency for the combined cascaded cells. Assume radiation from a black body at 6000 K. Assume no losses; that is, consider only the theoretical efficiency.

14.23 The V-I characteristic of a photocell is describable by a rather complex mathematical formula, which can be handled with a computer but is too complicated for an in-class exam. To simplify handling, we are adopting, rather arbitrarily, a simplified characteristic consisting of two straight lines as shown in the figure above. The position of the point, C, of maximum output varies with the I_ν/I_0 ratio. Empirically,

$$V_C = V_{OC}\left(0.7 + 0.0082\ln\frac{I_\nu}{I_0}\right) \tag{1}$$

and

$$I_C = I_\nu\left(0.824 + 0.0065\ln\frac{I_\nu}{I_{0.}}\right) \tag{2}$$

Consider now silicon photodiodes operating at 298 K. These diodes form a panel, 1 m^2 in area, situated in Palo Alto (latitude 37.4° N, longitude 125° W). The panel faces true south and has an elevation of 35°. In practice, the panel would consist of many diodes in a series/parallel connection. In the problem, here, assume that the panel has a single enormous photodiode.

Calculate the insolation on the surface at a 1130 PST and at 1600 PST on October 27. Assume clear meteorological conditions.

Assume that the true solar time is equal to the PST.

1. Calculate the insolation on the collector at the two moments mentioned.
2. What are the short-circuit currents (I_ν) under the two illuminations? Consider the sun as a 6000-K black body.

14.68

3. When exposed to the higher of the two insolations, the open-circuit voltage of the photodiode is 0.44 V. What is the power delivered to a load at 1130 and at 1600? The resistance of the load is, in each case, that which maximizes the power output for that case. What are the load resistances? What are the efficiencies?

4. Suppose that at 1600 the load resistance used was the same as that which optimized the 1130 output. What are the power in the load and the efficiency?

5. Let the load resistance be the same at both 1130 and 1600 but, unlike Question 4, not necessarily the resistance that optimizes the output at 1130. The idea is to operate the panel at slightly lower efficiency at 1130 and at somewhat higher efficiency than that of Question 4 at 1600 in the hope that the overall efficiency can be improved.

What is the value of this common load resistance?

14.24

1. What is the ideal (theoretical) efficiency of a gallium phosphide photocell exposed to the radiation of a 6000-K black body? For your information: the corresponding efficiency for silicon is 43.8%.

2. What is the efficiency of an ideal silicon photocell when illuminated by monochromatic light with a frequency of 266 THz?

3. What is the efficiency of an ideal silicon photocell when illuminated by monochromatic light with a frequency of 541.6 THz?

4. A real silicon photocell measuring 10 by 10 cm is exposed to 6000-K black body radiation with a power density of 1000.0 W/m². The temperature of the cell is 310 K. The measured open-circuit voltage is 0.493 V. When short-circuited, the measured current is 3.900 A. The power that the cell delivers to a load depends, of course, on the exact resistance of this load. By properly adjusting the load, the power can maximized. What is this maximum power?

14.25   Consider a solar cell made of semiconducting nanocrystals with a band-gap energy of $W_g = 0.67$ eV. What is the theoretical efficiency when the solar cell is exposed to the radiation of a 6000-K black body? Assume that photons with less than 3.3 W_g create each one single electron/hole pair and that those with more than 3.3 W_g create 2 electron/hole pairs each owing to impact ionization.

14.26   The Solar Power Satellites proposed by NASA would operate at 2.45 GHz. The power density of the beam at ionospheric heights (400 km) was to be 230 W/m². The collector on the ground was designed to use dipole antennas with individual rectifiers of the Schottky barrier type. These dipoles were dubbed **rectennas**.

The satellites would have been geostationary (they would be on a 24-h equatorial orbit with zero inclination and zero eccentricity).

1. Calculate the orbital radius of the satellites.
2. Calculate the microwave power density on the ground at a point directly below the satellite (the subsatellite point). Assume no absorption of the radiation by the atmosphere.
3. The total power delivered to the load is 5 GW. The rectenna system has 70% efficiency. Assume uniform power density over the illuminated area. What is the area that the ground antenna farm must cover?
4. A dipole antenna abstracts energy from an area given in the text. How many dipoles must the antenna farm use?
5. Assuming (very unrealistically) that the only part of each rectenna that has any mass is the dipole itself, and assuming that the half wave dipole is made of extremely thin aluminum wire, only 0.1 mm in diameter, what is the total mass of aluminum used in the antenna farm?
6. How many watts must each dipole deliver to the load?
7. If the impedance of the rectenna is 70 ohms, how many volts does each dipole deliver?

14.27   The Solar Power Satellite radiates 6 GW at 2.45 GHz. The transmitting antenna is mounted 10 km from the center of gravity of the satellite. What is the torque produced by the radiation?

14.28   Compare the amount of energy required to launch a mass, m, from the surface of the earth to the energy necessary to launch the same mass from the surface of the moon. "Launch" here means placing the mass in question an infinite distance from the point of origin. Consult the *Handbook of Chemistry and Physics* (CRC) for the pertinent astronomical data.

Part IV

Wind and Water

Chapter 15

Wind Energy

15.1 History

The use of wind energy dates back to ancient times when it was employed to propel sailboats. Extensive application of wind turbines seems to have originated in Persia where it was used for grinding wheat. The Arab conquest spread this technology throughout the Islamic world and China.

In Europe, the wind turbine made its appearance in the eleventh century. Two centuries later it had become an important tool, especially in Holland.

The development of the American West was aided by wind-driven pumps and sawmills.

The first significant wind turbine designed specifically for the generation of electricity was built by Charles Brush in Cleveland, Ohio. It operated for 12 years, from 1888 to 1900 supplying the needs of his mansion. Charles Brush was a mining engineer who made a fortune with the installation of arc lights to illuminate cities throughout the United States. His wind turbine was of the then familiar multivane type (it sported 144 blades) and, owing to its large solidity (see Section 15.10), rotated rather slowly and required gears and transmission belts to speed up the rotation by a factor of 50 so as to match the specifications of the electric generator.

The wind turbine itself had a diameter of 18.3 meters and its hub was mounted 16.8 meters above ground.

The tower was mounted on a vertical metal pivot so that it could orient itself to face the wind. The whole contraption massed some 40 tons.

Owing to the intermittent nature of the wind, electric energy had to be stored—in this case in 400 storage cells.

Although the wind is free, the investment and maintenance of the plant caused the cost of electricity to be much higher than that produced by steam plants. Consequently, the operation was discontinued in 1900 and from then on the Brush mansion was supplied by the Cleveland utility.

In 1939, construction of a large wind generator was started in Vermont. This was the famous Smith-Putnam machine, erected on a hill called Grandpa's Knob. It was a propeller-type device with a rated power of 1.3 MW at a wind speed of 15 m/s. Rotor diameter was 53 m. The machine started operation in 1941, feeding energy synchronously directly into the power network. Owing to blade failure, in March 1945, operation was discontinued. It ought to be mentioned that the blade failure had been predicted but during World War II there was no opportunity to redesign the propeller hub.

After World War II, the low cost of oil discouraged much of the alternate energy research and wind turbines were no exception. The 1973 oil crises re-spurred interest in wind power as attested by the rapid growth in federal funding. This led to the establishment of **wind farms** that were more successful in generating tax incentives than electric energy. Early machines used in such farms proved disappointing in performance and expensive to maintain. Nevertheless, the experience accumulated led to an approximately 5-fold reduction in the cost of wind-generated electricity. In the beginning of 1980, the cost of 1 kWh was around 25 cents; in 1996 it was, in some installations, down to 5 cents. To be sure, the determination of energy costs is, at best, an unreliable art. Depending on the assumptions made and the accounting models used, the costs may vary considerably. The calculated cost of the kWh depends on a number of factors including

1. The cost of investment—that is, the cost of the installed kW. This number was around $1000/kW in 1997 and does not appear to have changed much since. The largest windpower plant outside the United States is the 50 MW plant (84 turbines each with a 600 kW capacity) that the French company, Cabinet Germa, installed in Dakhla on the Atlantic coast of Morocco. The project, that started operations in 2000, cost $60 million, or $1200/kW.

   These investment costs are comparable with those of fossil-fueled and hydroelectric plants. However, the latter types of plants operate with utilization factors of at least 50% whereas wind power plants operate with a factor of some 20%. The utilization factor compares the amount of energy produced over, say, a year with that that would be produced if the plant operated at full power 24 hours per day. The intermittent and variable nature of the wind is the cause of the low factors achieved. Thus, for a one-on-one comparison, the cost of wind power plants should be multiplied by $0.5/0.2 = 2.5$.

   The one time investment cost must be translated to yearly costs by including the cost of borrowing the necessary funds. See Section 1.12 in Chapter 1. The cost of the kWh produced is extremely sensitive to the cost of the money (which may include interest, taxes, insurance, etc.). This is illustrated in Problem 15.17 at the end of this chapter.
2. Fuel costs, which are, of course, zero for wind and hydroelectric plants.
3. Operating and maintenance costs.
4. Decommissioning costs.
5. Land costs.

Even though the real cost of wind-generated energy may be uncertain, what is certain is that it has come down dramatically these last 15 years. In 1997, the selling of wind-generated electricity under a scheme called "green pricing" started becoming popular. In such a scheme, the consumers commit themselves to buying electricity for at least one year in monthly blocks of 100 kWh at typically 2.5 cents/kWh *more* than the going rate. Ecologically minded consumers can thus volunteer to support nonpolluting energy sources.

The relative cheapness of oil in the 1990s resulted in another ebbing in development funds. Nevertheless, the installed wind-generating capacity in the United States in January 2004 exceeded 6.3 GW (over 2 GW, in California). The installed capacity in Europe is larger than in North America. It had passed the 20 GW mark in January 2003. It should be pointed out that the installed wind turbine capacity in the United States corresponded (2004) to only some 0.6% of the total electricity-generating capacity.

The energy produced by wind farms is high-utility electricity whereas the energy from fuel is low-utility heat. Thus, to be fair, wind energy should be multiplied by 3 when comparison is made with fossil fuel use to reflect the roughly 30% conversion efficiency from the latter to electricity.

Notwithstanding the small present day contribution of wind energy to the total energy picture, there is merit in pursuing aggressively the technology, especially in view of its favorable ecological aspects:

1. Clearly, wind power emits absolutely no CO_2, by far the major pollutant when fuels (other than hydrogen or biomass) are burned.
2. The operation of wind turbines leaves behind no dangerous residues as do nuclear plants.
3. Decommissioning costs of wind turbines are much smaller than those of many other types of power plants especially compared with those of nuclear generators.
4. Land occupied by wind farms can find other uses such as agriculture.

Some groups are opposed to wind turbines because of the danger they constitute to the birds that fly near the wind farms.

The optimal size of a wind turbine has been the subject of disputes these last few decades. Government sponsored research in both the United States and Germany favored large (several MW) machines, while private developers opted for much smaller ones. Large machines fitted in well with the ingrained habits of the power generation industry accustomed to the advantages of economy of scale. Such advantages, however, do not seem to apply to wind turbines. Consider an extremely oversimplified reasoning:

For a given wind regimen, the amount of energy that can be abstracted from the wind is proportional to the swept area of the turbine. The area swept out by a rotor with 100 m diameter is the same as that of 100 machines with 10 m diameter. The mass of the plant (in a first-order scaling) varies with the *cube* of the diameter. The aggregate mass of the 100 small machines is only 10% of the mass of the large one. Hence, for the same amount of energy produced, the total equipment mass varies inversely with the diameter. Since costs tend to grow with mass, many small turbines ought to be more economical than one large one. This reasoning would suggest that the best solution is to use an infinite number of infinitely small turbines. Taken to this extreme, the conclusion is patently absurd.

Other factors play an important role in the economy (or diseconomy) of scale for wind turbines, complicating the situation to the point that a plausible model of how energy cost varies with turbine size becomes difficult to construct. Larger machines could arguably be more efficient and might simplify maintenance. They also would require less ancillary equipment (for instance, one single large transformer instead of many small ones) and possible less land area. Small machines profit from mass production economies, from modularity (allowing an easy expansion of the capacity of a wind farm) and from a greater immunity to breakdown (the breakdown of a few turbines affects only a fraction of the total wind farm capacity).

One of the largest early wind turbines tested was the German "Growian"[†] (100-m diameter, 3 MW) and the Boeing Mod-5B (98-m diameter, 3.2 MW).

The development of large wind machines seems to have been temporarily abandoned. On the other hand, the size of the turbines in large practical wind farms has been growing. In 1996, sizes between 500 kW and 750 kW were being favored; larger ones are now being installed. Thus, the trend toward larger turbines continues.

15.2   Wind Turbine Configurations

Several wind turbine configurations have been proposed, including:

1. drag-type,
2. lift-type (with vertical or horizontal axes),
3. Magnus effect wind plants,
4. Vortex wind plants.

Essentially all present day wind turbines are of the lift type and, over 90% of these are of the horizontal axis type. Magnus effect and vortex plants have never played a serious practical role.

15.2.1   Drag-Type Wind Turbines

In a drag-type turbine, the wind exerts a force in the direction it is blowing—that is, it simply pushes on a surface as it does in a sailboat sailing before the wind. Clearly, the surface on which the wind impinges cannot move faster than the wind itself.

[†] The German love for acronyms has given us words like *Stuka* and *Flak*. *Growian* stands for "Grosse Wind Energie Anlage."

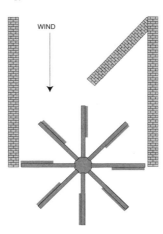

Figure 15.1   Top view of an ancient Persian wind turbine.

The ancient Persian wind turbine was a drag-type machine. Figure 15.1 is a sketch of such a mill seen from above. It consisted of a vertical axis to which horizontal radial arms were attached. Near the extremities of these arms, a vertical curtain was installed and this was the surface on which the wind exerted its useful force. Two walls channeled the wind, forcing it to blow on only one side of the device, thus creating a torque. Notice that one wall forms a funnel concentrating the collected wind.

The bucket wind turbine, sketched in Figure 15.2, is another vertical-axis drag-type device. It rotates because the convex surface offers less wind drag than the concave one. This device can be cheaply built by amateurs using an oil barrel cut along its vertical axis. It operates inefficiently.

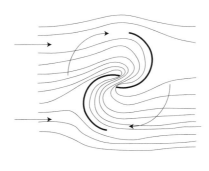

Figure 15.2   A 2-bucket
wind turbine.

Figure 15.3   Air flow in a
Savonius rotor.

Improved performance can be obtained by staggering the buckets as shown in Figure 15.3 so that a gap is left between them. The air is accelerated as it passes the gap reducing the front drag of the convex bucket. It is then blown on the reverse side of the bucket aiding in the creation of torque. This type of device is called a **Savonius** rotor and actually uses a certain amount of lift (in addition) to drag.

Savonius turbines cannot compete in efficiency with pure lift-type machines, but they are easy to build and find application as sensors in anemometers and eolergometers and as starters for vertical-axis lift-type machines.

15.2.2 Lift-Type Wind Turbines

In a lift-type machine, the wind generates a force perpendicular to the direction it is blowing. The familiar propeller wind turbines are of the horizontal-axis, lift-type. All lift-type turbines are analogous to sailboats sailing cross wind. The sailboat (or the blade of the turbine) can move substantially faster than the wind itself. Figure 15.4 shows such turbines.

Notice that the propeller-driven shaft that delivers the collected energy is high above ground level. This usually forces one of two solutions: either the electric generator is placed on top of the tower next to the propeller, or a long shaft, with associated gears, is used to bring the power to a ground-level generator. The first solution, although requiring reinforced towers, is the preferred one because of the cost and difficulties of transmitting large mechanical power over long shafts. Mounting the generator on top of the tower increases the mass of that part of the system that has to swivel around when the wind changes direction.

Some wind turbines have the propeller upstream from the generator and some downstream. It has been found that the upstream placement reduces the noise produced by the machine.

A propeller wind turbine that employs a ground-level generator but avoids the use of a long shaft is the suction-type wind turbine. It resembles a conventional wind turbine but the rotating blades act as a centrifugal pump. The blades are hollow and have a perforation at the tip so that air is expelled by centrifugal action creating a partial vacuum near the hub. A long pipe connects the hub to an auxiliary turbine located at ground level. The inrushing air drives this turbine. The system does not seem promising enough to justify further development.

One wind turbine configuration not only allows placing the generator on the ground but also avoids the necessity of reorienting the machine every time the wind changes direction—it is the vertical-axis lift-type wind turbine. The one illustrated in the center of Figure 15.4 is a design that was proposed by McDonnell-Douglas and was called "Gyromill." It would have been capable of generating 120 kW, but it was never commercialized.

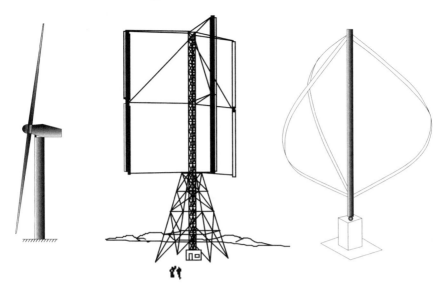

Figure 15.4   From left to right: a horizontal axis (propeller) type turbine, and two vertical axis machines—a "Gyromill" and a Darrieus.

One obvious disadvantage of the gyromill is the centrifugal force that causes the wings to bend outward, placing considerable stress on them. An elegant way to avoid centrifugal stresses is to form the wings in the shape assumed by a rotating rope loosely attached to the top and bottom of the rotating shaft. This leads to the familiar "egg beater" shape and, of course, causes the wing to work only in tension.

The shape of such a rotating rope is called a **troposkein** and resembles closely a **catenary**. There is, however, a difference. The catenary is "the shape assumed by a perfectly flexible inextensible cord of uniform density and cross-section hanging freely from two fixed points." Each unit length of the cord is subject to the same (gravitational) force. In the case of the troposkein, the force acting on each section of the cord depends on the distance of the section from the axis of rotation.

The troposkein wing (right-hand drawing of Figure 15.4) was first suggested by a French engineer called Darrieus after whom this type of wind turbine is named.

15.2.3 Magnus Effect Wind Machines

Magnus effect machines have been proposed but look unpromising. This effect, discussed in Section 15.14, is the one responsible for, among other things, the "curve" in baseball.

When a pitcher throws a curve, he causes the ball to spin creating an asymmetry: one side of the ball moves faster with respect to the air than the other and, consequently, generates the "lift" that modifies the trajectory of the ball. An identical effect occurs when a vertical spinning cylinder is exposed to the wind. The resulting force, normal to the wind direction, has been employed to move sailboats and wind machines.

15.2.4   Vortex Wind Machines

Finally, it is possible to abstract energy from the wind by making it enter tangentially through a vertical slit into a vertical hollow cylinder. As a result, the air inside is forced to gyrate and the resulting centrifugal force causes a radial pressure gradient to appear. The center of this air column, being at lower than atmospheric pressure, sucks outside air through openings at the bottom of the cylinder. The inrushing air drives a turbine coupled to a generator. The spinning air exits through the open top of the cylinder forming a vortex continuously swept away by the wind. This type of machine has been proposed by Gruman.

15.3   Eolergometry

Later on in this chapter, we are going to show that the power of a wind turbine is proportional to the cube of the wind velocity:

$$P_D = \frac{16}{27}\frac{1}{2}\rho v^3 A \eta \tag{1}$$

where $\frac{1}{2}\rho v^3$ is the **power density** in the wind, $\frac{16}{27}\frac{1}{2}\rho v^3$ is the **available power density** from the wind, A is the **swept area** of the wind turbine and η is the **efficiency** of the wind turbine.

The mean power output from the wind turbine over a period from 0 to T is proportional to the cube of the **mean cubic wind velocity**, $<v>$:

$$<v> \equiv \left(\frac{1}{T} \int_0^T v^3 dt \right)^{1/3}. \tag{2}$$

Owing to this cubic dependence on velocity, local wind conditions exert a strong influence on the useful energy generated by wind turbines, and site selection plays a critical role in overall planning.

Anemometers—instruments that measure or record wind velocity—can be used in wind surveys. Anemometric records have to be converted to eolergometric data—that is, data on wind power density. The mean cubic velocity, $<v>$, must be calculated from velocity measurements taken at frequent intervals.

The usual anemometric averages, $\bar{v}$—that is, the arithmetical average of v—are not particularly suitable for siting wind turbines. Consider a wind that blows constantly at a speed of 10 m/s (average speed of $\bar{v} = 10$ m/s). It carries an amount of energy proportional to $v^3 = 1000$. A wind that blows at 50 m/s 20% of the time and remains calm the rest of the time also has a $\bar{v}$ of 10 m/s, yet the energy it carries is proportional to $0.2 \times 50^3 = 25,000$ or 25 times more than in the previous case. In the first case, $<v> = 10$ m/s, while in the second, it was 29.2 m/s.

The quantity, $<v>$, is not normally measured directly by meteorologists although it can, as stated above, be derived from anemograms. Instruments that measure the cubic mean velocity, or else the total available energy density over a prescribed period of time are called **eolergometers** from *Aeolus* (god of winds) + *ergon* (work) + *metrein* (to measure). Such instruments are useful in surveys of the availability of wind energy.

Eolergometric surveys are complicated by the variability of the wind energy density from point to point (as a function of local topography) and by the necessity of obtaining vertical wind energy profiles. It is important that surveys be conducted over a long period of time—one year at least—so as to collect information on the seasonal behavior.

Clearly, values of $\bar{v}$ are easier to obtain than those of $<v>$ and, consequently, there is the temptation to guess the latter from the former. However, the ratio $\gamma \equiv <v>$ is a function of the temporal statistics of the wind velocity and is strongly site dependent. For perfectly steady winds, it would, of course, be 1. For the extreme case of the example in one of the preceding paragraphs, it is 2.92. Since the wind power density is proportional to the cube of the velocity, uncertainties in this ratio lead to large errors in the estimated available energy. Nevertheless, it has been suggested that a ratio of 1.24 can be used to make a (wild) guess of the energy availability at California sites.

When no information on the variation of wind velocity with height above ground is available, one can use the scaling formula

$$v(h) = v(h_0) \left(\frac{h}{h_0} \right)^{1/7}. \tag{3}$$

15.4 Availability of Wind Energy

In Chapter 1, we saw that 30% of the 173,000 TW of solar radiation incident on Earth is reflected back into space as the planetary albedo. Of the 121,000 TW that reach the surface, 3% (3600 TW) are converted into wind energy and 35% of this is dissipated in the lower 1 km of the atmosphere. This corresponds to 1200 TW. Since humanity at present uses only some 7 TW, it would appear that wind energy alone would be ample to satisfy all of our needs.

This kind of estimate can lead to extremely overoptimistic expectations. For one thing, it is difficult to imagine wind turbines covering all the ocean expanses. If we restrict wind turbines to the total of land areas, we would be talking about 400 TW. Again, it would be impossible to cover all the land area. Say that we would be willing to go as far as 10% of it, which is more than the percentage of land area dedicated to agriculture. We are now down to 40 TW. But, owing to the cubic dependence of power on wind speed, it is easy to see that much of this wind energy is associated with destructive hurricane-like winds, which actually generate no energy, since any reasonable wind turbine must shut itself down under such conditions.

The difficulty is that wind energy is very dilute. At a 10 m height, the wind power density may be some 300 W/m^2 at good sites and, at 50 m height, it can reach some 700 W/m^2, as it does in the San Gregorio Pass in California and in Livingston, Montana. Notice that these are not values of *available* wind power densities, which are 59% (16/27) of the preceding values. See Subsection 15.6.5.

Any sizable plant requires a large collecting area which means many turbines spread over a large area of land.[†]

Winds tend to be extremely variable so that wind plants must be associated with energy storage facilities (usually in small individual installations) or must debit into the power distribution network (in large wind farms).

In some areas of the world where the wind is quite constant in both speed and direction, wind turbines would operate with greater efficiency. All along the northeastern coast of Brazil, trade winds blow almost uninterruptedly from a northeasterly direction with steady speeds of some 13 knots (about 7 m/s) which corresponds to 220 W/m^2 of wind power density or 130 W/m^2 of *available* power density. This constant wind direction would allow the construction of fixed wind concentrators capable of substantially increasing the wind capture area.

15.5 Wind Turbine Characteristics

To minimize the cost of the produced electricity, the rated power of the generator must bear an appropriate relationship to the swept area of the wind turbine. The rated power of a generator is the maximum power it can deliver under steady conditions. Usually, the power actually generated is substantially lower than the rated. The ratio of the rated generator power to the swept area of the turbine is called **specific rated capacity** or **rotor loading**. Modern machines use rotor loadings of 300 to 500 W/m^2.

[†] Of course, land used for wind farms may be still available for agriculture.

The ideal wind turbine would be tailored to the wind conditions of each site. With steady unvarying winds ($\gamma = 1$) the rated power of the generator should be the same as that of the turbine. However, when $\gamma > 1$, which is invariably the case, the most economical combination has to be determined by considering many different factors. See, for example, Problem 15.3.

The Boeing Model 2 wind turbine was of the horizontal-axis (propeller) type. The rotor had a diameter of 91.5 m with variable pitch in the outer 14 m of each blade. This allowed the control of start-up and shutdown as well as the adjustment of its rotating speed. The performance of this wind turbine is shown in Figure 15.5. Note that with wind speeds below 3.9 m/s, the propeller does not rotate; at 6.3 m/s the machine reaches 17.5 rpm and its output is synchronized with the power grid.

The power generated increases rapidly up to wind velocities of 12.5 m/s and then remains constant up to 26.8 m/s.

Above this speed, the machine shuts itself down for safety reasons. It can, however, withstand winds up to some 56 m/s. With this arrangement, at high wind speeds, the turbine extracts but a small fraction of the available energy. The Model 2 generated 2.5 MW of electricity with any wind in the 12.5 m/s to 25 m/s range notwithstanding there being 8 times more energy at the higher speed. At even higher wind speeds, the machine shuts itself down to avoid destructive stresses, and, thus, delivers no energy just when there is a largest amount of power in the wind.

Typically, the power delivered by a given wind plant depends on the wind velocity in a manner similar to the one displayed in Figure 15.5.

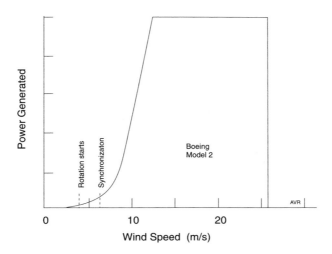

Figure 15.5   Power output of the Boeing Model 2 wind turbine as a function of wind velocity.

15.11

Wind turbines frequently deliver their energy to a utility-operated net and must do so with alternating current of the correct frequency. There are two general solutions to the synchronization problem:

1. Maintaining the rotation of the turbine at a constant rate (by changing the blade pitch, for instance).
2. Allowing the turbine to rotate at the speed dictated by load and wind velocity. In this case dc is generated and electronically "inverted" to ac. Such **variable-speed** machines are somewhat more expensive but are more efficient and have a longer life.

15.6   Principles of Aerodynamics

The symbol, P, in this chapter stands for both *power* and *power density*—that is, power per unit area, depending on the context. The lower case, p, is reserved for *pressure*.

The following subscripts are used:

$P_W = \dfrac{1}{2}\rho v^3$ "Power density in the wind." This is the amount of energy transported across a unit area in unit time.

$P_A = \dfrac{16}{27}\dfrac{1}{2}\rho v^3$ "Available power density." This is the theoretical maximum amount of power that can be extracted from the wind.

$P_D = \dfrac{16}{27}\dfrac{1}{2}\rho v^3\,A\eta$ "Power delivered." This is the power that a wind turbine delivers to its load.

15.6.1   Flux

The flux of a fluid is defined as the number of molecules that cross a unit area (normal to the flow) in unit time. It can be seen that, if n is the concentration of the molecules (number per unit volume) and v is the bulk velocity of the flow, then the flux, ϕ, is

$$\phi = nv \quad \mathrm{m^{-2}s^{-1}}. \tag{4}$$

The total flow across an area, A, is, consequently

$$\Phi = \phi A \quad \mathrm{s^{-1}}. \tag{5}$$

15.6.2 Power in the Wind

If the mean mass of the gas molecules is m, then the mean energy of a molecule owing to its bulk drift (not owing to its thermal motion) is $\frac{1}{2}mv^2$. The amount of energy being transported across a unit area in unit time is the power density of the wind:

$$P_W = \frac{1}{2}mv^2\phi = \frac{1}{2}mnv^3 = \frac{1}{2}\rho v^3 \quad \text{W m}^{-2}. \tag{6}$$

Notice that the power density is proportional to the cube of the wind velocity. The quantity, ρ, is the gas density—that is, the mass per unit volume:

$$\rho = mn \quad \text{kg m}^{-3}. \tag{7}$$

At standard temperature and pressure (STP),[†] the density of air is

$$\rho = \frac{0.2 \times 32 + 0.8 \times 28}{22.4} = 1.29 \quad \text{kg m}^{-3}. \tag{8}$$

The numerator is the average molecular mass of air containing 20% O_2 and 80% N_2, by volume. The denominator is the number of cubic meters per kilomole at STP:

From the perfect gas law, at STP,

$$V = \frac{RT}{p} = \frac{8314 \times 273.3}{1.013 \times 10^5} = 22.4 \quad \text{m}^3. \tag{9}$$

15.6.3 Dynamic Pressure

Since 1 m³ of gas contains n molecules and each molecule carries $\frac{1}{2}mv^2$ joules of energy owing to its bulk motion, the total energy density—that is, the total energy per unit volume is

$$W_d = \frac{1}{2}nmv^2 = \frac{1}{2}\rho v^2 \quad \text{J m}^{-3} \text{ or N m}^{-2}. \tag{10}$$

Energy per unit volume has the dimensions of force per unit area—that is, of pressure. Thus W_d is referred to as **dynamic pressure**.

15.6.4 Wind Pressure

Wind exerts pressure on any surface exposed to it. Consider the (unrealistic) flow pattern depicted in Figure 15.6. The assumption is that any

---

[†] STP corresponds to one atmosphere and 0 celsius.

molecule striking the surface is reflected and moves back against the wind without interfering with the incoming molecules.

Under such a simplistic assumption, each molecule transfers to the surface a momentum, $2mv$, because its velocity change is $2v$ (it impacted with a velocity, v, and was reflected with a velocity, $-v$). Since the flux is nv, the rate of momentum transfer per unit area, i.e., the generated pressure, is $2mv \times nv = 2\rho v^2$. The assumption is valid only at very low gas concentrations, when, indeed, a molecule bouncing back may miss the incoming ones.

In a more realistic flow, the reflected molecules will disturb the incoming flow, which would then roughly resemble the pattern shown in Figure 15.7. This leads to a pressure smaller than that from the ideal flow case, a pressure that depends on the shape of the object. To treat this complicated problem, aerodynamicists assume that the real pressure is equal to the dynamic pressure multiplied by an experimentally determined correction factor, C_D, called the **drag coefficient**.

$$p = \frac{1}{2}\rho v^2 C_D. \tag{11}$$

The drag coefficient depends on the shape of the object and, to a certain extent, on its size and on the flow velocity. This means, of course, that the pressure exerted by the wind on a surface is not strictly proportional to v^2 as suggested by Equation 11.

The drag coefficient of a large flat plate at low subsonic velocities is usually taken as $C_D = 1.28$.

15.6.5 Available Power

Electrical engineers are familiar with the concept of available power. If a source (see Figure 15.8) has an open circuit voltage, V, and an internal resistance, R_s, the maximum power it can deliver to a load is $V^2/4R_s$. This occurs when $R_L = R_s$.

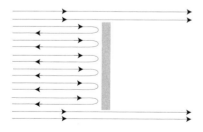

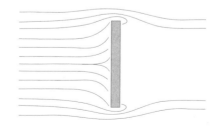

Figure 15.6   A simplistic flow pattern.

Figure 15.7   A more realistic flow pattern.

Figure 15.8   An electric source and its load.

The same question arises when power is to be extracted from the wind. If the surface that interacts with the wind is stationary, it extracts no power because there is no motion. If the surface is allowed to drift downwind without any resistance, then it again will extract no power because the wind will exert no force on it. Clearly, there must be a velocity such that maximum power is extracted from the wind.

The power density, P, extracted from the wind is the product of the pressure, p, on the surface and the velocity, w, with which the surface drifts downwind. The wind pressure is

$$p = \frac{1}{2}\rho C_D(v - w)^2, \tag{12}$$

hence

$$P = pw = \frac{1}{2}\rho C_D(v - w)^2 w. \tag{13}$$

Setting $\partial P/\partial w$ to zero, an extremum of P is found. This is a maximum and occurs for $w = v/3$ independently of the value of C_D. Thus,

$$P_{max} = \frac{2}{27}\rho C_D v^3 \quad \text{W m}^{-2}. \tag{14}$$

The ratio of maximum extractable power to power in the wind is

$$\frac{P_{max}}{P_w} = \frac{\frac{2}{27}\rho C_D v^3}{\frac{1}{2}\rho v^3} = \frac{4}{27}C_D. \tag{15}$$

The largest possible value of C_D is that predicted by our simplistic formula that, by stating that $p = 2\rho v^2$, implies that $C_D = 4$. Thus, at best, it is possible to extract 16/27 or 59.3% of the "power in the wind." This is the **available power density** from the wind:

$$P_A = \frac{16}{27}\frac{1}{2}\rho v^3. \tag{16}$$

15.6.6 Efficiency of a Wind Turbine

The efficiency of a wind turbine is the ratio of the power, P_D, delivered to the load, to some reference power. There is a certain amount of arbitrariness in this definition. Some authors choose the "power in the wind" ($\frac{1}{2}\rho v^3$) as reference. In this text we will use the available power (P_A):

$$\eta = \frac{P_D}{P_A}. \tag{17}$$

In a well-designed wind turbine, the efficiency can reach 0.7.

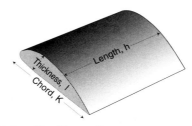

Figure 15.9   Significant dimensions in an airfoil.

15.7 Airfoils

Airplane wings, helicopter rotors, empennage surfaces, and propeller blades are examples of aerodynamic surfaces (**airfoils**), which must generate a great deal of lift with a correspondingly small drag. The performance of an airfoil depends greatly on the shape of its cross section.

Figure 15.10 shows a section through an airfoil. The line (A.A') represents the trace of an arbitrary **reference plane**. Notice that the region above this plane differs from the one below it—the airfoil is **asymmetric**. In **symmetric** airfoils, the reference plane is the plane of symmetry, and the region above it is a mirror image of that below. When air flows relative to the airfoil along the x-axis in the figure, a force is exerted on the foil. Such force is usually decomposed into a **lift** component (normal to the velocity) and a **drag** component (parallel to the velocity). The corresponding pressures are indicated by the vectors, p_L and p_D, in the figure. The angle between the wind direction and the reference line is called the **angle of attack**, α.

For each shape of the airfoil, p_L and p_D are determined experimentally in wind tunnels under specified conditions. The observed pressures are related to the dynamic pressure, $\frac{1}{2}\rho v^2$, by proportionality constants, C_L and C_D, called, respectively, the **lift coefficient** and the **drag coefficient**:

$$p_L = \frac{1}{2}\rho v^2 C_L, \tag{18}$$

$$p_D = \frac{1}{2}\rho v^2 C_D. \tag{19}$$

These coefficients are functions of the angle of attack[†] and can be found in tabulations, many of which were prepared by NACA (the forerunner of NASA) in the United States and by Göttingen in Germany. Since, presumably, airplanes move only forward, the tabulations are usually made for only a small range of angles of attack near zero. However, for some airfoils, data are available for all 360° of α as in Figure 15.11 which shows the dependence of C_L on α for the airfoil known as Göttingen-420.

[†] The coefficients depend also, although more weakly, on the Reynolds number, R, and the aspect ratio, Λ, as we are going to show.

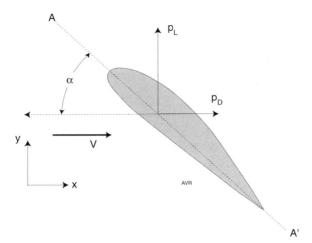

Figure 15.10   Pressure on an airfoil.

Notice that the airfoil under discussion generates lift even with negative angles of attack (as long as they remain small). When there is a positive angle of attack, one can intuitively understand the creation of lift even for a flat surface—after all the airflow is hitting the surface from below and the drag it exerts has a lift component. However, when the angle of attack is zero or slightly negative, the lift must be due to more complicated mechanisms. Observations show that the air pressure immediately above the airfoil is smaller than the pressure immediately below it and this is the obvious cause of the lift. The problem is to explain how such a pressure difference comes about.

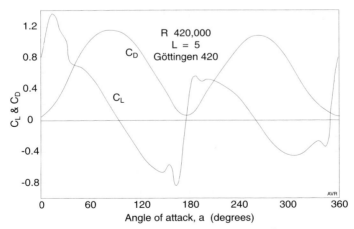

Figure 15.11   The lift and drag coefficients of the Göttingen-420 airfoil.

15.17

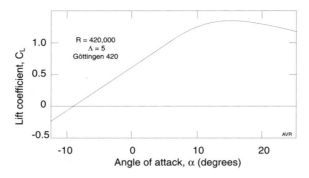

Figure 15.12   The lift coefficient of the Göttingen-420 airfoil.

Further observations show also that

1. the air flow tends to follow the curvature of the top of the airfoil instead of simply being deflected away from it;
2. the air flow velocity is substantially increased lowering the pressure, according to Bernoulli's principle.

We need some explanation for these observations.

The Coanda[†] Effect is the cause of the airflow's tendency to follow the shape of the airfoil. It is extremely easy to demonstrate this phenomenon. Open a faucet and allow a thin stream of water to fall from it. Now take a curved surface—a common drinking glass will do—and let the stream hit the side of the glass at a glancing angle. The water will run along the side and then, making a sharp turn, will flow along the bottom instead of simply falling vertically down. The water tends to follow the glass surface just as the air tends to follow the airfoil surface.

The bulging part of the airfoil restricts the air flow (as in the strangulation in a venturi) causing an acceleration of the flow. The transit time of air molecules along the path over the airfoil from leading to trailing edge is not the same as that for the flow under the airfoil. Such synchronism, frequently invoked in explanations of wing lift, does, in fact, not occur.

In the airfoil of Figure 15.12, the lift is linearly related to α up to some $10°$. At higher angles the airfoil **stalls**—that is, a further increase in α actually reduces the lift. Near zero angle of attack, this airfoil develops a lift over 16 times larger than its drag.

Because the lift and drag coefficients are not strictly independent of the air velocity or the dimensions of the wing, it is impossible to scale any experimental results exactly. However, the data are valid for different sizes and speeds as long as the **Reynolds number** is preserved.

[†] Henri-Marie Coanda, Romanian scientist (1885–1972) described this effect in 1930.

15.8 Reynolds Number

The size of most airplanes exceeds, by far, that of available wind tunnels causing engineers to do their tests and measurements on reduced scale models. The dimensions of such models are an accurate constant fraction of the original. One thing, however, can usually not be scaled down proportionally: the size of the molecules in air. For this reason, measurement on models may not be converted with precision to expected forces on the real plane. In fact, the forces observed on an object moving in a fluid are not only the **dynamic** forces, F_d (proportional to $\frac{1}{2}\rho v^2$), but include also **viscous** forces, F_v.

When an airfoil moves through still air, molecules of gas in immediate contact with the surface are forced, through friction, to move with the velocity, v_x (assuming movement in the x-direction), of the foil, while those at a large distance do not move at all. A velocity gradient, ∇v_x, is established in the y-direction.

The resulting velocity shear causes the viscous force, F_v, to appear. This force is proportional to the wing area, A, to the velocity gradient, $\partial v_x/\partial y$, and to the coefficient of viscosity, μ, a property of the medium through which the wing moves.

$$F_v = \mu \frac{\partial v_x}{\partial y} A. \tag{20}$$

For accurate scaling, it is necessary to preserve the ratio of the dynamic to the viscous forces. Define a quantity, R, called **Reynolds number** as

$$R \propto \frac{F_d}{F_v} = \frac{\frac{1}{2}\rho v_x^2 A}{\mu \partial v_x/\partial y A} \propto \frac{\rho}{\mu} \cdot \frac{v_x^2}{\partial v_x/\partial y} \tag{21}$$

This is too complicated. Make two simplifying assumptions:

1. $\frac{\partial v_x}{\partial y}$ is independent of y, and
2. the air is disturbed only to a distance, K, above the wing (where K is the **chord length**).

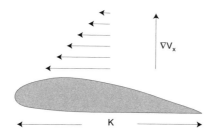

Figure 15.13   Wind shear over a wing.

Under such circumstances,

$$\frac{\partial v_x}{\partial y} = \frac{v}{K},$$ (22)

and

$$R = \frac{\rho v^2}{\mu v/K} = \frac{\rho}{\mu} v K.$$ (23)

The value of μ for air is 1.84×10^{-5} kg m^{-1} s^{-1} and is, contrary to one's gut feeling, independent of pressure and density (see the box at the end of this section). However, the ratio, μ/ρ, called the **kinematic viscosity** increases when the pressure decreases. Fluids, at low pressure exhibit great kinematic viscosity and this explains why vacuum pumps need large diameter pipes. For STP conditions, this ratio for air is $1/70,000$ m^2/s because $\rho = 1.29$ kg m^{-3}.

Since for any given angle of attack, the coefficients (C_L and C_D) are functions of R, measurements made with models cannot be extrapolated to life-size wings unless the Reynolds number is the same. Fortunately the coefficients depend only weakly on R as illustrated in Figure 15.14, where C_D and $C_{L_{max}}$ are plotted versus R for the NACA 0012 symmetric airfoil. $C_{L_{max}}$ is the largest value that C_L can reach as a function of the angle of attack.

For first order calculations, one can ignore the effects of varying Reynolds number. In more precise calculations, the correct Reynolds number must be used especially because, as we shall see, the wing of a vertical axis wind turbine perceives a variable wind velocity throughout one cycle of its rotation.

In general, things tend to improve with larger Reynolds numbers (because the lift-over-drag ratio usually increases). This means that, on these grounds, larger wind turbines tend to be more efficient than smaller ones.

A 3-meter chord wing moving at 360 km/h has a Reynolds number of

$$R = 7 \times 10^4 \times 100 \times 3 = 21 \times 10^6.$$ (24)

To measure the characteristics of this wing using a model with a 0.3 m chord under the same Reynolds number, one would need a wind velocity of 3600 m/s. However, the result would not be valid because the speed in question is supersonic. This explains why much of the wind tunnel data correspond to modest Reynolds numbers. The Göttingen-420 data were measured at $R = 420,000$.

There is a class of wind tunnels, the NACA **variable-density tunnels**, in which R is increased by increasing the static air pressure. This raises ρ but does not affect μ. Thus, large Reynolds numbers can be achieved with small models at moderate wind speeds.

15.20

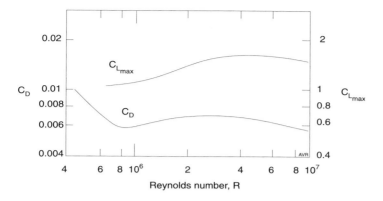

Figure 15.14   Effect of the Reynolds number
on the lift and drag coefficients.

Why is μ independent of pressure or density? When a molecule of gas suffers a (isotropic) collision, the next collision will, statistically, occur at a distance one mean free path, ℓ, away. This locates the collision on the surface of a sphere with radius, ℓ, centered on the site of the previous collision. The projected area of this sphere is proportional to ℓ^2. One can, therefore, expect that

$$\mu \propto \nu n \ell^2. \tag{25}$$

where ν is the **collision frequency** and n is the **concentration** of the molecules. But ℓ is inversely proportional to the concentration while the collision frequency is directly proportional to it. Thus,

$$\mu \propto n \times n \times \frac{1}{n^2}, \tag{26}$$

i.e., μ is independent of n.

15.9   Aspect Ratio

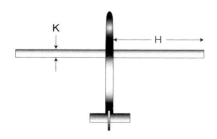

In a rectangular wing, the ratio between the length, H, and the chord, K, is called the **aspect ratio**, $\mathcal{R}$.

$$\mathcal{R} \equiv \frac{H}{K} = \frac{H^2}{KL} = \frac{H^2}{A}, \tag{27}$$

where A is the area of the wing.

Hence, the aspect ratio can be defined as the ratio of the square of the wing length to the wing area. This definition must be used in the case of tapered (nonrectangular) wings, which have a variable chord length.

The drag experienced by a body moving through a fluid is caused by a number of different mechanisms. An ideal infinitely smooth and infinitely long wing would experience only a **pressure drag**. However, real wings are not perfectly smooth and the air tends to adhere to the surface causing viscous shearing forces to appear and generating a **skin-friction drag**.

Wing lift is the result of the pressure being larger under the wing than over it. At the tip, there is a "short circuit" between the underside and the top and air circulates around the tip forming a vortex. Energy is used to impart the circular motion to the air in the vortex and this energy must come from the forward motion of the wing. It manifests itself as an additional drag force called the **induced drag.** The induced drag can be lessened by

1. increasing the number of wings (or the number of wing tips),
2. increasing the aspect ratio of the wing,
3. tapering the wing so that the chord is smaller near the tip, and
4. placing a vertical obstacle to the flow around the wing tip. Sometimes additional fuel tanks are mounted there.

Each wing tip generates its own vortex. Energetically, it is more economical to have many small vortices instead of one large one because the losses are proportional to the square of the vortex size. It is easy to see that the sum of the squares is smaller than the square of the sum. Biplanes (having four wing tips) have less induced drag than monoplanes, everything else being the same. Soaring birds reduce the induced drag by spreading the feathers so that, near the end of the wing, there are many tips.

Obviously the smaller the chord at the tip of the wing, the smaller the induced drag. In a rectangular wing of a given area, a larger length, H, results in a smaller chord, K. The wing has a larger aspect ratio and, consequently, a smaller induced drag.

Tapered wings have smaller wing-tip chords than rectangular wings of the same area and, again, have smaller induced drag.

Gliders have long, slender wings to maximize the aspect ratio (minimizing the induced drag). In fast moving airplanes, the **parasitic drag**[†] is dominant, making the induced drag unimportant: fast planes can tolerate small aspect ratios.

When a wing is tested, the total drag measured includes the induced drag, hence it is customary to indicate the aspect ratio, $Æ\!R$, of the test section when aerodynamic coefficients are tabulated.

[†] Parasitic drag is caused by parts of the machine that offer resistance to the flow of air but do not generate lift.

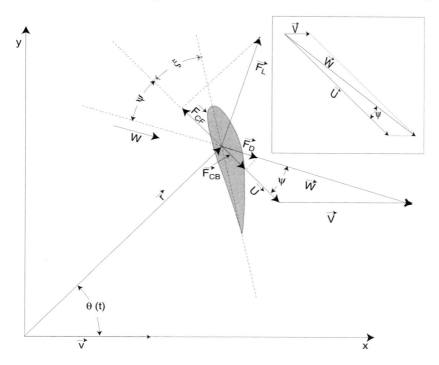

Figure 15.15   Angles and forces on a wing. Usually, $\vec{U}$ is much larger than $\vec{V}$, but for clarity in the drawing (but not in the inset), they were taken as approximately equal. This exaggerates the magnitude of the angle, ψ. The inset shows the wind vectors.

15.10 Wind Turbine Analysis

> As an example of wind turbine analysis, we have chosen a gyromill because, even though this type of turbine is not in use, its analysis is much simpler than that of, say, propeller-type machines. It, nevertheless, brings out important conclusions that, with suitable modifications, are also applicable to other lift-type devices.
>
> For those interested in an introductory analysis of propeller turbines, we recommend reading the book by Duncan. See References.

Consider a vertical-axis wind turbine of either the Darrieus or the McDonnell-Douglas type. Consider also a right-handed orthogonal coordinate system with the z-axis coinciding with the vertical axis of the machine. The system is so oriented that the horizontal component of the wind velocity, V, is parallel to the x-axis.

We will assume that the airfoil, whose cross section lies in the x-y plane, has a chord that makes an angle, ξ, with the normal to the radius vector. This is the **setup** angle chosen by the manufacturer of the wind turbine. It may be adjustable and it may even vary during one rotation. In this analysis, it is taken as a constant.

The radius vector makes an angle, $\theta(t)$, with the x-axis. Figure 15.15 indicates the different angles and vectors involved in this derivation. The inset shows how the wind velocity, V, and the velocity, U (owing to the rotation), combine to produce an **induced** velocity, W, which is the actual air velocity perceived by the wing. U is, of course, the velocity the wing would perceive if there were no wind.

Refer to Figure 15.15. The angle of attack is $\alpha = \psi + \xi$.

$$\vec{\omega} = \vec{k}\,\omega, \tag{28}$$

$$\vec{r} = r(\vec{i}\,\cos\theta + \vec{j}\,\sin\theta), \tag{29}$$

$$\vec{U} = -\vec{\omega} \times \vec{r} = -\begin{pmatrix} \vec{i} & \vec{j} & \vec{k} \\ 0 & 0 & \omega \\ r\,\cos\theta & r\,\sin\theta & 0 \end{pmatrix} = U\,(\vec{i}\,\sin\theta - \vec{j}\,\cos\theta), \tag{30}$$

$$\vec{W} = \vec{U} + \vec{V} = \vec{i}(V + U\,\sin\theta) - \vec{j}\,U\,\cos\theta, \tag{31}$$

$$\begin{aligned} W &= \sqrt{V^2 + U^2\sin^2\theta + 2UV\sin\theta + U^2\cos^2\theta} \\ &= \sqrt{V^2 + U^2 + 2UV\sin\theta} \equiv \Gamma V, \end{aligned} \tag{32}$$

where

$$\Gamma \equiv \sqrt{1 + \frac{U^2}{V^2} + 2\frac{U}{V}\sin\theta}, \tag{33}$$

$$\begin{aligned} \vec{U}\cdot\vec{W} &= (U\,\sin\theta)(V + U\,\sin\theta) + U^2\cos^2\theta \\ &= U^2 + UV\,\sin\theta = UW\,\cos\psi, \end{aligned} \tag{34}$$

from which

$$\cos\psi = \frac{U^2 + UV\,\sin\theta}{UW} = \frac{U + V\,\sin\theta}{\sqrt{V^2 + U^2 + 2UV\,\sin\theta}} = \frac{U/V + \sin\theta}{\Gamma}. \tag{35}$$

For a given wind speed, V, and angular velocity, ω, of the wind turbine, the ratio, U/V, is constant. The quantity, Γ, and the angle, ψ, vary with the angular position of the wing, thus they vary throughout the revolution. Consequently, the angle of attack also varies. Clearly, if there is no wind, α is constant. If there is a high U/V ratio (if the wind speed is much smaller than that of the rotating wing), then α varies only a little throughout the revolution.

Given a U/V ratio and a wind velocity, it is possible to calculate both α and W for any wing position, θ. For a given W and α, the wing will generate a lift

$$F_L = \frac{1}{2}\rho W^2 A_p C_L, \tag{36}$$

and a drag

$$F_D = \frac{1}{2}\rho W^2 A_p C_D. \tag{37}$$

In the preceding equations, A_p is the area of the wing and F stands for the force on the wing.

Note that $\vec{F}_L$ is normal to $\vec{W}$ in the x-y plane and that $\vec{F}_D$ is parallel to $\vec{W}$. The lift force, $\vec{F}_L$, has a component, $\vec{F}_{CF}$, normal to the radius vector, causing a forward torque. The drag force, $\vec{F}_D$, has a component, $\vec{F}_{CB}$, also normal to the radius vector, causing a retarding torque.

The resulting torque is

$$\Upsilon = r(F_{CF} - F_{CB}). \tag{38}$$

From Figure 15.15,

$$F_{CF} - F_{CB} = F_L \sin\psi - F_D \cos\psi = \frac{1}{2}\rho W^2 A_p(C_L \sin\psi - C_D \cos\psi). \tag{39}$$

Thus,

$$\Upsilon = \frac{1}{2}\rho W^2 A_p r(C_L \sin\psi - C_D \cos\psi)$$

$$= \frac{1}{2}\rho V^2 A_p r\left[\Gamma^2(C_L \sin\psi - C_D \cos\psi)\right]. \tag{40}$$

The average torque taken over a complete revolution is

$$<\Upsilon> = \frac{1}{2\pi}\int_0^{2\pi}\Upsilon(\theta)\,d\theta. \tag{41}$$

In the expression for Υ, only the part in brackets is a function of θ. Let us define a quantity, D:

$$D \equiv \Gamma^2(C_L \sin\psi - C_D \cos\psi), \tag{42}$$

$$<D> = \frac{1}{2\pi}\int_0^{2\pi} D\,d\theta, \tag{43}$$

$$<\Upsilon> = \frac{1}{2}\rho V^2 A_p r <D>. \tag{44}$$

The power delivered by the turbine to its load is

$$P_D = \omega<\Upsilon>N. \tag{45}$$

Here, N is the number of wings on the wind turbine. The swept area is (see Figure 15.16)

$$A_v = 2rH, \tag{46}$$

and the area of each wing is

$$A_p = KH, \tag{47}$$

where H is the (vertical) length of the wing and K is the chord (assumed uniform).

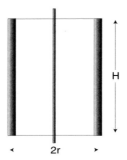

Figure 15.16   The aspect ratio of a wind turbine.

A **solidity**, S, is defined:

$$S = \frac{N A_p}{A_v} = N \frac{K}{2r} \tag{48}$$

The **available** power from the wind is

$$P_A = \frac{16}{27} \frac{1}{2} \rho V^3 A_v, \tag{49}$$

$$\eta = \frac{P_D}{P_A} = \frac{\frac{1}{2}\rho V^2 N A_p r \omega <D>}{\frac{1}{2}\rho V^3 A_v \frac{16}{27}} = \frac{27}{16} \frac{U}{V} <D>S. \tag{50}$$

The efficiency formula derived above is correct only to first order. It ignores parasitic losses owing to friction and to the generation of vortices; it disregards the reduction in wind velocity caused by the wind turbine itself; it fails to take into account the interference of one wing blade on the next. In fact, Equation 50 predicts that with large enough solidities, the efficiency can exceed unity. We will discuss this question a little later.

Notice that the $U/V<D>$ product is a function of the parameter U/V. $<D>$ must be obtained from numerical analysis looking up values of C_L and C_D for the various α that appear during one revolution.

To gain an idea of the shape of the $U/V<D>$ vs U/V graph, consider the situation when $U = 0$. Clearly, $U/V = 0$ and since D cannot be infinite, $U/V<D>$ must also be 0.[†]

[†] When $U = 0$, Γ is unity (Equation 33) and, from Equation 35, $\cos \psi = \sin \theta$ and $\sin \psi = \cos \theta$. This makes
$$D = C_L \cos \theta - C_D \sin \theta, \tag{51}$$
and consequently, provided C_L and C_D are constant,
$$<D> = 0, \tag{52}$$
because the mean value of $\sin \theta$ and of $\cos \theta$ is zero. Since the torque is proportional to $<D>$, this type of wind turbine has no torque when stalled: it has zero starting torque and requires a special starting arrangement (such as a small Savonius on the same shaft). Actually, C_L and C_D do depend on Θ, and Equation 51 holds only approximately.

When $U \to \infty$, $W \to U$ and $\psi \to 0$. From Equation 42,

$$D = \Gamma^2(C_L \sin \psi - C_D \cos \psi) \to -\Gamma^2 C_D,$$

thus, for large values of U/V, $D < 0$ and, consequently, $U/V<D> < 0$.

This means that, at high rpm, the wind turbine has a negative torque and tends to slow down. One can, therefore, expect that the efficiency has a maximum at some value of U/V in the range $0 < U/V < \infty$.

As an example, we have computed $U/V<D>$ for various values of U/V and a number of setup angles, ξ. The airfoil used was the Gö-420. The results are shown in Figure 15.17. It can be seen that the optimum setup angle (the one that leads to the highest $U/V<D>$ is $-6°$. Symmetric airfoils work best with $\xi = 0$.

For $\xi = -6°$, the airfoil reaches a $U/V<D>$ of 4.38 (nondimensional) at a U/V of 6.5. Thus, in this particular case, the efficiency formula yields:

$$\eta_{max} = 7.39S. \tag{53}$$

Were one to believe the formula above, efficiencies greater than 1 could be reached by using solidities, S, larger than 0.135. Clearly, there must be some value of solidity above which the formula breaks down. In Figure 15.18, the efficiency of a wind turbine is plotted versus solidity. The linear dependence predicted by Equation 53 is represented by the dashed line with the 7.39 slope of our example. Using a more complicated aerodynamical model, Sandia obtained the results shown in the solid line. It can be seen that increasing the solidity beyond about 0.1 does not greatly affect the efficiency. One can distinguish two regions in the efficiency versus solidity curve: one in which, as predicted by our simple derivation, the efficiency is proportional to the solidity, and one in which the efficiency is (roughly) independent of the solidity.

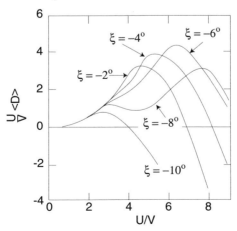

Figure 15.17   Performance of the Gö-420 airfoil in a vertical-axis turbine.

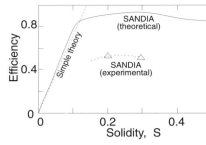

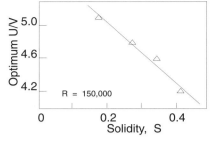

Figure 15.18   Effect of solidity on efficiency of a vertical-axis wind turbine.

Figure 15.19   Dependence of the optimum U/V on solidity. Experimental data from the Sandia 2-m diameter Darrieus turbine.

Triangles in Figure 15.18 indicate values of efficiency measured by Sandia using small models of the wind turbines. Measured efficiencies are about half of the calculated ones. This discrepancy is discussed further on.

The main reason for the behavior depicted in Figure 15.18 is that our simple theory failed to account for the interference of one wing with the next. The larger the solidity, the farther the disturbance trails behind the wing and the more serious the interference thus counteracting the efficiency gain from a larger S. Consequently, the optimum U/V shrinks as S increases.

Figure 15.19 shows the experimentally determined effect of solidity on the optimum U/V. If the straight line were extrapolated, one would conclude that for $S = 1$, the optimum U/V would be about 0.7.

In the range of solidities that have only a small effect on the efficiency, increasing S results in a wind turbine that rotates more slowly (because of the smaller optimum U/V) and has more torque (because the efficiency—and consequently the power—is the same). Increasing S has the effect of "gearing down" the wind turbine. Since the cost of a wind turbine is roughly proportional to its mass, and hence to its solidity, one should prefer machines with S in the lower end of the range in which it does not affect the efficiency. That is why, for large machines, propellers are preferred to vanes. Nevertheless, for small wind turbines, the simplicity of vane construction may compensate for the larger amount of material required.

Although the equations derived above will help in the understanding of the basic wind turbine processes, they fall far short from yielding accurate performance predictions. Numerous refinements are needed:

1. Frictional losses in bearings must be taken into account.
2. The rotating wings create vortices that represent useless transformation of wind energy into whirling motion of the air. The effect of such vortices has to be considered.
3. The wind, having delivered part of its energy to the machine, must nec-

essarily slow down. Thus, the average wind velocity seen by the blades is less than the free stream velocity and the power is, correspondingly less than that predicted by the formulas.

Single streamtube models are based on an average wind slowdown. These models ignore the nonuniformity of the wind velocity in the cross section of the wind turbine. By contrast, if the wind slowdown is considered in detail, then we have a **multiple streamtube** model.

Figure 15.20 shows how the ratio of the streamtube to freestream velocity varies with position in the Sandia 2-m diameter Darrieus turbine. When the calculations are based on the multistreamtube model, the predicted performance approaches reality as can be seen in Figure 15.21 in which the measured efficiency (squares) is compared with the values predicted using both the single and the multi-streamtube models. Even the multi-streamtube model overestimates the performance.

4. The accuracy of the prediction is improved by using the appropriate Reynolds number, a quantity that actually varies throughout the revolution. Figure 15.22 shows the effect of the Reynolds number on the performance of the wind turbine. As expected, the measurements, made at Sandia, show that the larger R, the better the performance.

Clearly, a refined wind turbine performance model is too complicated to be treated in this book.

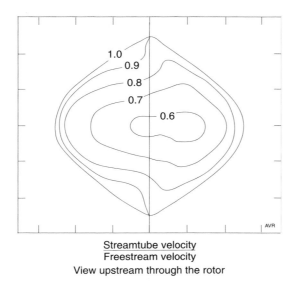

Streamtube velocity
Freestream velocity
View upstream through the rotor

Figure 15.20   Streamtube velocity through the rotor of a Sandia 2-m Darrieus turbine.

15.29

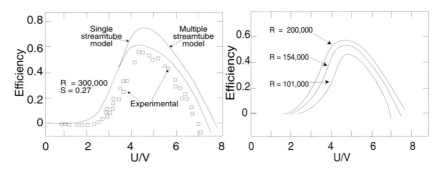

Figure 15.21   Calculated efficiencies compared with observed values. Data from Sandia.

Figure 15.22   Influence of the
Reynolds number on the performance. Data from Sandia.

15.11   Aspect Ratio (of a wind turbine)

In Figure 15.16 (previous section), the aspect ratio, $A\!R_{turb}$, of a Gyromill (McDonnell-Douglas vertical-axis wind turbine) was defined:

$$A\!R_{turb} = \frac{H}{2r} = \frac{A_v}{4r^2} = \frac{H^2}{A_v}. \tag{54}$$

This is, of course, different from the aspect ratio of the wing, which is

$$A\!R_w = \frac{H}{K}, \tag{55}$$

where K is the chord length of the wing. Since the solidity is $S = NK/2r$,
the wing aspect ratio can be rewritten as $A\!R_w = HN/2rS$. However, the
aspect ratio of the wind turbine is $A\!R_{turb} = H/2r$, hence,

$$A\!R_{turb} = \frac{S}{N} A\!R_w. \tag{56}$$

For constant N and constant S (constant first-order efficiency), the
aspect ratio of the wind turbine is proportional to that of the wing. Wingtip
drag decreases with increasing wing aspect ratio, consequently, wind turbine
efficiencies actually tend to go up with increasing $A\!R_{turb}$.

From Equation 48, it can be seen that if both the swept area, A_v
(power), and the solidity S (efficiency), are kept constant, then, to a first
approximation, the mass of the wings remains constant because their area,
A_p, is constant. However, larger aspect ratios require higher towers. The
mass of the tower increases with height faster than linearly because higher
towers must have a larger cross-section. On the other hand, the struts that
support the wings become smaller with increasing $A\!R_{turb}$, again nonlinearly.

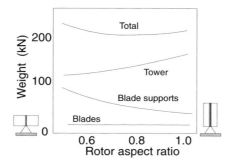

Figure 15.23   Influence of the aspect ratio on the mass of a wind
turbine. McDonnell-Douglas 120 kW Gyromill.

Thus, we have a situation like the one depicted in Figure 15.23 which
displays data for the McDonnell-Douglas "gyromill." There is a minimum
in total mass of the wind turbine at a $Æ\!R_{turb}$ of, roughly, 0.8.

15.12 Centrifugal Force

Consider a section of a wing with mass M. Its weight is Mg (g is the
acceleration of gravity). The centrifugal force acting on the section is

$$F_c = \frac{MU^2}{r} = M\omega^2 r. \tag{57}$$

The ratio of centrifugal force to weight is

$$\frac{F_c}{W} = \frac{\omega^2 r}{g}. \tag{58}$$

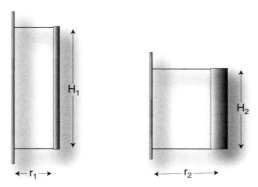

Figure 15.24   Two wind turbines with the same area, but
with different aspect ratios.

In a 10-m diameter wind turbine rotating at 100 rpm ($\omega = 10.5$ rad/s), a section of the wing will experience a centrifugal force 55 times larger than its weight. This illustrates the need for careful design to minimize the centrifugal effects. For instance, what aspect ratio minimizes these effects?

Compare two wind turbines with the same swept area. If they have the same solidity and use the same airfoil, they will deliver the same power when operated at the same U/V ratio. If they have different aspect ratios then they must have:

$$r_1 H_1 = r_2 H_2. \tag{59}$$

The centrifugal force per unit length is

$$F_c = \frac{MU^2}{r}, \tag{60}$$

where M is the mass of the wing per unit length. The ratio of the centrifugal forces on the two wind turbines is

$$\frac{F_{c_1}}{F_{c_2}} = \frac{M_1 r_2}{M_2 r_1}. \tag{61}$$

If the two wings have a solid cross section, then the masses (per unit length) are proportional to the square of the chord, K. In general,

$$M = bK^a, \tag{62}$$

where a is an exponent that depends on the type of construction. From this

$$K = \left(\frac{M}{b}\right)^{1/a}. \tag{63}$$

Since the wind turbines have the same solidity, $NK/2r$,

$$\frac{N_1}{2r_1}\left(\frac{M_1}{b}\right)^{1/a} = \frac{N_2}{2r_2}\left(\frac{M_2}{b}\right)^{1/a}, \tag{64}$$

$$\frac{M_1}{M_2} = \left(\frac{N_2 r_1}{N_1 r_2}\right)^a, \tag{65}$$

$$\frac{F_{c_1}}{F_{c_2}} = \left(\frac{N_2}{N_1}\right)^a \left(\frac{r_1}{r_2}\right)^{a-1}, \tag{66}$$

and, if the two wind turbines have the same number of wings,

$$\frac{F_{c_1}}{F_{c_2}} = \left(\frac{r_1}{r_2}\right)^{a-1}. \tag{67}$$

From the preceding formula, it becomes clear that for any $a > 1$, the larger the radius, the larger the centrifugal force. Since it is difficult to achieve low a's, this favors large aspect ratios.

The centrifugal force in a "gyromill" poses a difficult problem for the wind turbine designer. However, as explained in Subsection 15.2.2, this problem disappears in the case of the Darrieus (egg-beater) configuration.

15.13 Performance Calculation

Consider a vertical-axis wind turbine with the characteristics given in the table below.

Table 15.1
Wind Turbine Characteristics

Number of wings	$N = 3$
Height (length of wings)	$H = 16$ m
Radius	$r = 10$ m
Solidity	$S = 0.27$
Site	sea level
Wing performance	see Figure 15.25 and Table 15.2.

Assume that the performance data are valid for the actual Reynolds number. The wind turbine drives a load whose torque obeys the relationship

$$\Upsilon_L = 2000\omega. \tag{68}$$

The performance of the wind turbine is presented in the form of a torque vs angular velocity plot. A different plot must be constructed for each wind velocity of interest.

Take, for instance, $V = 10$ m/s. The questions to be answered are:

1. What is the operating rpm?
2. What is the power delivered to the load?
3. What is the actual (average) Reynolds number?
4. What is the efficiency of the wind turbine?

Using the technique described in Section 15.10, a plot of efficiency versus U/V is constructed. Such a plot (similar to those in Figure 15.17) is shown in Figure 15.25. The corresponding numerical data appear in Table 15.2.

Now, let us construct the torque versus angular velocity plot:

$$\omega = \frac{U}{r} = \frac{U}{V} \times \frac{V}{r} = \frac{U}{V}. \tag{69}$$

Observe that in this particular example, the value of V/r is 1.

$$A_v = 2rH = 320 \text{ m}^2, \tag{70}$$

$$\rho = 1.29 \text{ kgm}^{-3} \text{ (sea level)}, \tag{71}$$

$$P_D = \frac{16}{27}\frac{1}{2}\rho V^3 A_v \eta = 122{,}000 \ \eta, \tag{72}$$

$$\Upsilon = \frac{P_D}{\omega} = 122{,}000\frac{\eta}{U/V}. \tag{73}$$

Notice that, Υ, above, is the total torque of the wind turbine, not the torque per wing, as before.

Using the performance table (Table 15.2), it is easy to calculate η for any U/V (or for any ω). The plot of Υ versus ω is shown in Figure 15.26. In the same figure, the Υ_L versus ω characteristic of the load is plotted. It can be seen that two different values of ω yield wind turbine torques that match that of the load. In point A, if there is a slight increase in wind velocity, the wind turbine will speed up, its torque will become larger and the turbine will accelerate, increasing the torque even further. It is an unstable point. In point B, an increase in wind velocity will reduce the wind turbine torque causing it to slow down back to its stable operating point.

At the stable point, B, the angular velocity in our example is 5.4 rad/s, equivalent to 0.86 rps or 52 rpm. The power generated will be

$$P = \Upsilon\omega = 2000 \ \omega^2 = 58{,}300 \text{ W}. \tag{74}$$

Table 15.2
Efficiency Versus U/V for the Wing in the Example

U/V	Efficiency
2.0	0.00
2.2	0.02
2.4	0.05
2.6	0.08
2.8	0.12
3.0	0.15
3.5	0.33
4.0	0.49
4.5	0.56
5.0	0.53
5.5	0.44
6.0	0.33
6.5	0.17
7.0	0.00

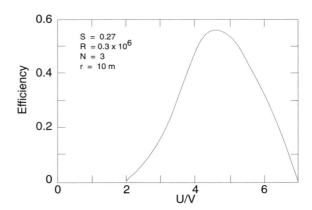

Figure 15.25   Efficiency versus U/V for the wing in the example.

The efficiency that corresponds to $U/V = \omega = 5.4$ is 0.46.
The solidity is

$$S = \frac{NK}{2r} \qquad \therefore \qquad K = \frac{2rS}{N} = 1.8 \text{ m}. \tag{75}$$

The Reynolds number is

$$R = 70{,}000 \, WK \approx 70{,}000 \, UK = 6.8 \times 10^6. \tag{76}$$

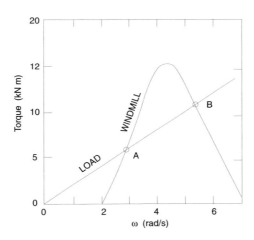

15.14 Magnus Effect

In the introduction to this chapter, we made reference to the use of rotating cylinders to generate lift useful in driving wind turbines. The lift is caused by the **Magnus** effect, used by baseball pitchers when throwing a "curve."

Consider a rotating cylinder exposed to a stream of air.

Sufficiently far away from the cylinder, the air is undisturbed—that is, it moves with the velocity of the wind.

Immediately in contact with the cylinder, at point a (see Figure 15.26), the air flows from right to left in a direction opposite to that of the wind because the (rough) cylinder surface exerts friction on the air and accelerates it in the direction of its own motion. Its velocity is equal to the linear velocity of the cylinder surface. A gradient of velocity is established as illustrated in the figure.

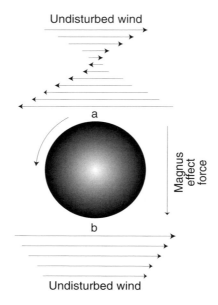

Figure 15.26 A rotating cylinder induces a velocity gradient in the wind from which a lateral force results.

On the opposite side, b, the cylinder impels the air from left to right causing it to flow faster than the stream. Again a velocity gradient may be established.

The average stream velocity is larger on the b-side than on the a-side of the cylinder.

According to Bernoulli's law, the high velocity side, b, experiences a smaller pressure than the low velocity side, a. Consequently, a force from a to b is exerted on the cylinder. This force, the result of the aerodynamic reaction on a rotating body, is called the **Magnus effect**. It is proportional to $\vec{\Omega} \times \vec{V}$, where $\vec{\Omega}$ is the angular velocity of the cylinder and $\vec{V}$ is the velocity of the wind.

References

Duncan, W. J., A. S. Thom, and A. D, Young, *An Elementary Treatise on the Mechanics of Fluids*, The English Language Book Society, **1960**.

Gipe, Paul, *Wind energy comes of age*, John Wiley, **1995**.

PROBLEMS

15.1   At a given sea-level location, the wind statistics, taken over a period of one year and measured at an anemometer height of 10 m above ground, are as follows:

Number of hours	Velocity (m/s)
90	25
600	20
1600	15
2200	10
2700	5
remaining time	calm

The velocity is to be assumed constant in each indicated range (to simplify the problem).

Although the wind varies with the 1/7th power of height, assume that the velocity the windmill sees is that at its center.

The windmill characteristics are:

Efficiency	70%
Cost	150 $/m^2
Weight	100 kg/m^2

The areas mentioned above are the swept area of the windmill.

If h is the height of the tower and M is the mass of the windmill on top of the tower, then the cost, C_T, of the tower is

$$C_T = 0.05 \ h \ M.$$

How big must the swept area of the windmill be so that the average delivered power is 10 kW? How big is the peak power delivered—that is, the rated power of the generator? Be sure to place the windmill at the most economic height. How tall will the tower be?

Neglecting operating costs and assuming an 18% yearly cost of the capital invested, what is the cost of the MWh?

If the windmill were installed in La Paz, Bolivia, a city located at an altitude of 4000 m, what would the average power be, assuming that the winds had the velocity given in the table above? The scale height of the atmosphere is 8000 meters (i.e., the air pressure falls exponentially with height with a characteristic length of 8000 m).

15.2   A utilities company has a hydroelectric power plant equipped with generators totaling 1 GW capacity. The utilization factor used to be exactly 50%—that is, the plant used to deliver every year exactly half of the energy the generators could produce. In other words, the river that feeds the plant reservoir was able to sustain exactly the above amount of energy. During the wet season, the reservoir filled but never overflowed.

Assume that the plant head is an average of 80 m and that the plant (turbines and generators) has an efficiency of 97%.

What is the mean rate of flow of the river (in m^3/s)?

With the industrial development of the region, the utilities company wants to increase the plant utilization factor to 51%, but, of course, there is not enough water for this. So, they decided to use windmills to pump water up from the level of the hydraulic turbine outlet to the reservoir (up 80 m).

A careful survey reveals that the wind regime is the one given in the table below:

$<v>$ (m/s)	Θ
5	0.15
7	0.45
10	0.30
12	0.10

$<v>$ is the mean cubic wind velocity and Θ is the percentage of time during which a given value of $<v>$ is observed. The generator can be dimensioned to deliver full power when $<v>= 12$ m/s or when $<v>$ is smaller. If the generator is chosen so that it delivers its rated full power for, say, $<v>$ m/s, then a control mechanism will restrict the windmill to deliver this power even if $<v>$ exceeds the 10 m/s value.

Knowing that the cost of the windmill is \$10 per m^2 of swept area, that of the generator is \$0.05 per W of rated output, the efficiency of the windmill is 0.7 and that of the generator is 0.95, calculate which is the most economic limiting wind velocity.

What is the swept area of the windmill that will allow increasing the plant factor to 51% ? (The pumps are 95% efficient.) What is the cost of the MWh generated by the windmills, assuming an annual cost of investment of 20% and neither maintenance nor operating costs.

The windmills are, essentially, at sea level.

15.3   A windmill is installed at a sea-level site where the wind has the following statistics:

v (m/s)	% of time
0	30
3	30
9	30
12	8
15	2

The velocity in the table above should be the mean cubic velocity of the wind, however, to simplify the problem, assume that the wind actually blows at a constant 3 m/s 30% of the time, a constant 9 m/s another 30% of the time and so on.

The windmill characteristics are:

Efficiency (including generator)	0.8
Windmill cost	200 $/m² of swept area
Generator cost	200 $/kW of rated power.

Rated power is the maximum continuous power that the generator is supposed to deliver without overheating. The duty cycle is 1—that is, the windmill operates continuously (when there is wind) throughout the year. Consider only investment costs. These amount to 20% of the investment, per year.

The system can be designed so that the generator will deliver full (rated) power when the wind speed is 15 m/s. The design can be changed so that rated power is delivered when the wind speed is 12 m/s. In this latter case, if the wind exceeds 12 m/s, the windmill is shut down. It also can be designed for rated power at 9 m/s, and so on.

We want a windmill that delivers a maximum of 1 MW. It has to be designed so that the cost of the generated electricity over a whole year is minimized. What is the required swept area? What is the cost of electricity?

15.4   *For this problem, you need a programmable calculator or a computer.*

Consider an airfoil for which
$$C_L = 0.15\alpha,$$
$$C_D = 0.015 + 0.015|\alpha|,$$
for $-15° < \alpha < 15°$ where α is the angle of attack. A wing with this airfoil is used in a vertical-axis windmill having a radius of 10 m. The setup angle is 0.

Tabulate and plot α as well as the quantity D (see text) as a function of θ, for $U/V = 6$. Use increments of θ of 30° (i.e., calculate 12 values).

Considerations of symmetry facilitate the work. Be careful with the correct signs of angles. It is easy to be trapped in a sign error. Find the mean value of $<D>$.

15.5   In the region of Aeolia, on the island of Anemos, the wind has a most peculiar behavior. At precisely 0600, there is a short interval with absolutely no wind. Local peasants set their digital watches by this lull. Wind velocity then builds up linearly with time, reaching exactly 8 m/s at

2200. It then decays, again linearly, to the morning lull.

A vertical-axis windmill with 30 m high wings was installed in that region. The aspect ratio of the machine is 0.8 and its overall efficiency is 0.5. This includes the efficiency of the generator.

What is the average power generated? What is the peak power? Assuming a storage system with 100% turnaround efficiency, how much energy must be stored so that the system can deliver the average power continuously? During what hours of the day does the windmill charge the storage system?

Notice that the load always gets energy from the storage system. This is to simplify the solution of the problem. In practice, it would be better if the windmill fed the load directly and only the excess energy were stored.

15.6   A vertical-axis windmill with a rectangular swept area has an efficiency whose dependence on the U/V ratio (over the range of interest) is given, approximately, by

$$\eta = 0.5 - \frac{1}{18} \left(\frac{U}{V} - 5 \right)^2 .$$

The swept area is 10 m^2 and the aspect ratio is 0.8.

The wind velocity is 40 km/h.

What is the maximum torque the windmill can deliver? What is the number of rotations per minute at this torque? What is the power delivered at this torque? What is the radius of the windmill and what is the height of the wings? If the windmill drives a load whose torque is given by

$$\Upsilon_L = \frac{1200}{\omega}$$
                                                                          Nm

where ω is the angular velocity, what is the power delivered to the load? What are the rpm when this power is being delivered?

15.7   An engineering firm has been asked to make a preliminary study of the possibilities of economically generating electricity from the winds in northeastern Brazil. As a first step, a quick and very rough estimate is required. Assume that the efficiency can be expressed by the ultra-simplified formula of Equation 48 in the text. The results will be grossly overoptimistic because we fail to take into account a large number of loss mechanisms and we assume that the turbine will always operate at the best value of U/VD, which, for the airfoil in question is 4.38. Our first cut will lead to unrealistic results but will yield a ballpark idea of the quantities involved.

In the region under consideration, the trade winds blow with amazing constancy, at 14 knots. Assume that this means 14 knots at a 3 m

15.40

anemometer height. The wind turbines to be employed are to have a rated power of 1 MWe (MWe = MW of electricity).

For the first cut at the problem we will make the following assumptions:

a. The configuration will be that of the McDonnell-Douglas gyromill, using three wings.
b. Wind turbine efficiency is 80%.
c. The wings use the Göttingen-420 airfoil (see drawing at the end of the problem). We will accept the simple efficiency formula derived in this chapter.
d. The wind turbine aspect ratio will be 0.8.
e. The wings will be hollow aluminum blades. Their mass will be 25% of the mass of a solid aluminum wing with the same external shape. Incidentally, aluminum is not a good choice for wind turbine wings because it is subject to fatigue. Composites are better.
f. The total wind turbine mass will be 3 times the mass of the three wings taken together.
g. Estimated wind turbine cost is $1.00/kg. Notice that the cost of aluminum was $1200/kg in 1852, but the price is now down to $0.40/kg.
h. The cost of investment is 12% per year.

Estimate:

1. The swept area.
2. The wing chord.
3. The wing mass.
4. The Reynolds number when the turbine is operating at its rated power. Assume that this occurs at the optimum U/V ratio.
5. The rpm at the optimum U/V ratio.
6. The torque under the above conditions.
7. The tension on each of the horizontal supporting beams (two beams per wing).
8. The investment cost per rated kW. Assuming no maintenance or operating cost, what is the cost of the generated kWh? Use a utilization factor of 25%.

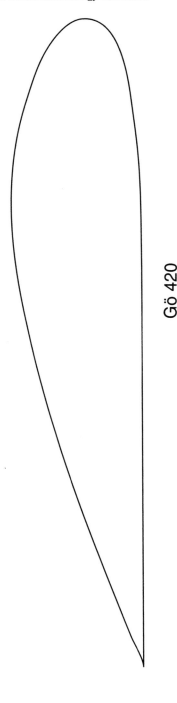

Gö 420

15.42

15.8 A sailboat has a drag, F, given by

$$F = aW^2,$$

where F is in newtons, W is the velocity of the boat with respect to the water (in m/s), and $a = 80$ kg/m.

The sail of the boat has a 10 m^2 area and a drag coefficient of 1.2 when sailing before the wind (i.e., with a tailwind).

Wind speed is 40 km/h.

When sailing before the wind, what is the velocity of the boat? How much power does the wind transfer to the boat? What fraction of the available wind power is abstracted by the sail?

15.9 A vertical-axis windmill of the gyromill configuration extracts (as useful power) 50% of the available wind power. The windmill has a rectangular swept area with a height, H, of 100 m and an aspect ratio of 0.8.

The lower tips of the wings are 10 m above ground. At this height, the wind velocity is 15 m/s. It is known that the wind increases in velocity with height according to the 1/7th power law.

Assuming that the windmill is at sea level, what power does it generate?

15.10 *Solve equations by trial and error. Use a computer.*

Can a wind-driven boat sail directly into the wind? Let's find out (forgetting second order effects).

As a boat moves through the water with a velocity, W (relative to the water), a drag force, F_w, is developed. Let $F_w = 10W^2$.

The boat is equipped with a windmill having a swept area of 100 m^2 and an overall efficiency of 50%. The power generated by the windmill is used to drive a propeller, which operates with 80% efficiency and creates a propulsive force, F_p, that drives the boat.

The windmill is oriented so that it always faces the *induced* wind, i.e., the combination of V and W.

The wind exerts a force, F_{WM}, on the windmill. This force can be estimated by assuming a $C_D = 1.1$ and taking the swept area as the effective area facing the wind.

Wind velocity, V, is 10 m/s. What is the velocity, W_S, of the boat in the water? Plot W as a function of the angle, ϕ, between the direction of the wind and that of the boat. In a tailwind, $\phi = 0$ and in a headwind, $\phi = 180°$.

The boat has a large keel, so that the sideways drift caused by the "sail" effect of the windmill is negligible.

15.11   *Solve equations by trial and error. Use a computer.*

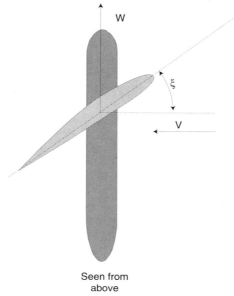

Seen from
above

A vehicle is mounted on a rail so that it can move in a single direction only. The motion is opposed by a drag force, F_W,

$$F_W = 100W + 10W^2$$

where W is the velocity of the vehicle along the rail. Notice that the drag force above does not include any aerodynamic effect of the "sail" that propels the vehicle. Any drag on the sail has to be considered in addition to the vehicle drag.

The wind that propels the vehicle is perpendicular to $\vec{W}$ and has a velocity, $V = 10$ m/s. It comes from the starboard side (the right side of the vehicle).

The "sail" is actually an airfoil with an area of 10.34 m^2. It is mounted vertically, i.e., its chord is horizontal and its length is vertical. The reference line of the airfoil makes an angle, ξ, with the normal to $\vec{W}$. See the figure.

The airfoil has the following characteristics:

$$C_L = 0.15\alpha,$$

$$C_D = 0.015 + 0.015|\alpha|.$$

α is the angle of attack expressed in degrees. The two formulas above are valid only for $-15° < \alpha < 15°$. If α exceeds 15°, the wing stalls and the lift falls to essentially zero.

15.44

Conditions are STP.

Calculate the velocity, W, with which the vehicle moves (after attaining a steady velocity) as a function of the setup angle, ξ. Plot both W and α as a function of ξ.

15.12   An electric generator, rated at 360 kW at 300 rpm and having 98.7% efficiency at any reasonable speed, produces power proportionally to the square of the number of rpm with which it is driven.

This generator is driven by a windmill that, under given wind conditions, has a torque of 18,000 Nm at 200 rpm, but produces no torque at both 20 and 300 rpm. Assume that the torque varies linearly with the number of rpm between 20 and 200 and between 200 and 300 rpm.

What is the electric power generated?

15.13   A car is equipped with an electric motor capable of delivering a maximum of 10 kW of mechanical power to its wheels. It is on a horizontal surface. The rolling friction (owing mostly to tire deformation) is 50 N regardless of speed. There is no drag component proportional to the velocity. Frontal area is 2 m^2 and the aerodynamic drag coefficient is $C_D = 0.3$.

Assume calm air at 300 K and at 1 atmosphere. What is the cruising speed of the car at full power?

With the motor uncoupled and a tailwind of 70 km/h, what is the car's steady state velocity. The drag coefficient is $C_D = 1$ when the wind blows from behind.

15.14   A building is 300 m tall and 50 m wide. Its C_D is 1. What is the force that the wind exerts on it if it blows at a speed of 10 m/s at a height of 5 m and if its velocity varies with $h^{1/7}$ (h being height above ground).

15.15   A windmill has a torque versus angular velocity characteristic that can be described by two straight lines passing through the points:

50 rpm, torque = 0;
100 rpm, torque = 1200 Nm;
300 rpm, torque = 0.

1. If the load absorbs power according to $P_{load} = 1000\omega$, what is the power taken up by this load? What is the torque of the windmill? What is the angular velocity, ω?
2. If you can adjust the torque characteristics of the load at will, what is the maximum power that you can extract from the windmill? What is the corresponding angular velocity?

15.16   A windmill with a swept area of 1000 m^2 operates with 56% efficiency under STP conditions.

At the location of the windmill, there is no wind between 1800 and 0600. At 0600, the wind starts up and its velocity increases linearly with time from zero to a value that causes the 24-h average velocity to be 20 m/s. At 1800, the wind stops abruptly.

What is the maximum energy the windmill can generate in one year?

15.17   US Windpower operates the Altamont Wind Farm near Livermore, CA. They report a utilization factor of 15% for their 1990 operation. Utilization factor is the ratio of the energy generated in one year, compared with the maximum a plant would generate if operated constantly at the rated power. Thus, a wind turbine rated at 1 kW would produce an average of 150 W throughout the year. Considering the intermittent nature of the wind, a 15% utilization factor is good. Hydroelectric plants tend to operate with a 50% factor.

1. It is hoped that the cost of the kWh of electricity will be as low as 5 cents. Assuming that the operating costs per year amount to 15% of the total cost of the wind turbine and that the company has to repay the bank at a yearly rate of 12% of the capital borrowed, what cost of the wind turbine cannot be exceeded if the operation is to break even? For your information, the cost of a fossil fuel or a hydroelectric plant runs at about 1000 dollars per installed kW.
2. If, however, the wind turbine actually costs $1000 per installed kW, then, to break even, what is the yearly rate of repayment to the bank?

15.18   Many swimmers specialize in both freestyle (crawl) and butterfly. The two strokes use almost the same arm motion, but the crawl uses an alternate stroke whereas the butterfly uses a simultaneous one. The kicks are different but are, essentially, equally effective. Invariably, swimmers go faster using the crawls.

From information obtained in this course, give a first order explanation of why this is so.

15.19   *Solve equations by trial and error. Use a computer.*

An air foil of uniform chord is mounted vertically on a rail so that it can move in a single direction only. It is as if you had cut off a wing of an airplane and stood it up with its longer dimension in the vertical. The situation is similar to the one depicted in the figure of Problem 15.11.

There is no friction in the motion of the air foil. The only drag on the system is that produced by the drag coefficient of the airfoil.

The *setup angle* is the angle between the airfoil reference plane and the normal to the direction of motion. In other words, if the rail is north-south, the airfoil faces east when the setup angle is zero and faces north when the set-up angle is 90°.

Consider the case when the set-up angle is zero and the wind blows from the east (wind is abeam). Since the wind generates a lift, the airfoil will move forward. What speed does it attain?

Conditions are STP. Area of the airfoil is 10 m². Wind speed is 10 m s^{-1}.

The coefficients of lift and drag are given by

$$C_L = 0.6 + 0.066\alpha - 0.001\alpha^2 - 3.8 \times 10^{-5}\alpha^3.$$

$$C_D = 1.2 - 1.1\cos\alpha.$$

15.20   Consider a vertical rectangular (empty) area with an aspect ratio of 0.5 (taller than wide) facing a steady wind. The lower boundary of this area is 20 m above ground and the higher is 200 m above ground. The wind velocity at 10 m is 20 m/s and it varies with height according to the $h^{1/7}$ law.

Calculate:

1. the (linear) average velocity of the wind over this area,
2. the cubic mean velocity of the wind over this area,
3. the available wind power density over this area,
4. the mean dynamic pressure over this area.

Assume now that the area is solid and has a C_D of 1.5. Calculate:

5. the mean pressure over this area,
6. the torque exerted on the root of a vertical mast on which the area is mounted.

15.21   The retarding force, F_D, on a car can be represented by

$$F_D = a_0 + a_1 V + a_2 V^2.$$

To simplify the math, assume that $a_1 = 0$.

A new electric car is being tested by driving it on a perfectly horizontal road on a windless day. The test consists of driving the vehicle at constant speed and measuring the energy used up from the battery. Exactly 15 kWh of energy is used in each case.

When the car is driven at a constant 100 km/h, the distance covered is 200 km. When the speed is reduced to 60 km/h, the distance is 362.5 km.

If the effective frontal area of the car is 2.0 m², what is the coefficient of aerodynamic drag of the vehicle?

15.22   The army of Lower Slobovia needs an inexpensive platform for mounting a reconnaissance camcorder that can be hoisted to some height between 200 and 300 m. The proposed solution is a kite that consists of a Göttingen-420 airfoil with 10 m^2 of area tethered by means of a 300 m long cable. To diminish the radar signature the cable is a long monocrystal fiber having enormous tensile strength so that it is thin enough to be invisible, offers no resistance to airflow, and has negligible weight.

In the theater of operation, the wind speed is a steady 15 m/s at an anemometer height of 12 m, blowing from a 67.5° direction. It is known that this speed grows with height exactly according to the 1/7–power law.

The wing loading (i.e., the total weight of the kite per unit area) is 14.9 kg m^{-2}.

The airfoil has the following characteristics:

$$C_L = 0.5 + 0.056\alpha,$$

$$C_D = 0.05 + 0.012|\alpha|,$$

where α is the attack angle in degrees.

The above values for the lift and drag coefficients are valid in the range $-10° < \alpha < 15°$.

The tethering mechanism is such that the airfoil operates with an angle of attack of 0.

The battlefield is essentially at sea level.

The questions are:

1. Assume that the kite is launched by somehow lifting it to an appropriate height above ground. What is the hovering height?
2. What modifications must be made so that the kite can be launched from ground (i.e., from the height of 12 m). No fair changing the wing loading. Qualitative suggestions with a minimum of calculation are acceptable.

15.23   A car massing 1000 kg has an effective frontal area of 2 m^2. It is driven on a windless day on a flat, horizontal highway (sea level) at the steady speed of 110 km/h. When shifted to neutral, the car will, of course, decelerate and in 6.7 seconds its speed is down to 100 km/h.

From these data, estimate (very roughly) the coefficient of aerodynamic drag of the car. What assumptions and/or simplifications did you have to make to reach such estimate?

Is this estimate of C_D an upper or a lower limit? In other words, do you expect the real C_D to be larger or smaller than the one you estimated?

15.24   In early airplanes, airspeed indicators consisted of a surface exposed to the wind. The surface was attached to a hinge (see figure) and a spring (not shown) torqued the surface so that, in absence of air flow, it would hit a stop and, thus, assume a position with $\theta = 90°$.

The wind caused the surface to move changing θ. This angle, seen by the pilot, was an indication of the air speed.

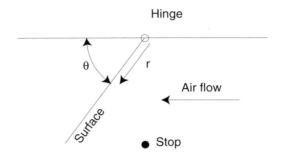

In the present problem, the surface has dimensions L (parallel to the axis of the hinge) by D (perpendicular to L). $L = 10$ cm, $D = 10$ cm.

The spring exerts a torque

$$\Upsilon_{spring} = \frac{0.1}{\sin \theta} \text{ N m.}$$

The coefficient of drag of the surface is 1.28. Air density is 1.29 kg/m^3. Calculate the angle, θ, for wind velocities, V, of 0, 10, 20, and 50 m/s.

15.25   An EV (electric vehicle) is tested on a horizontal road. The power, P, delivered by the motors is measured in each run, which consists of a 2 km stretch covered at constant ground velocity, V. Wind velocity, W, may be different in each run.

Here are the test results:

Run	Wind Direction	Wind Speed (m/s) W	Car Speed (km/h) V	Power (kW) P
1	—	0	90	17.3
2	Headwind	10	90	26.6
3	Headwind	20	90	39.1
4	—	0	36	2.1
5	Tailwind	35	90	4.3

How much power is required to drive this car, at 72 km/h, into a 30 m/s head wind?

15.49

15.26   Here are some data you may need:

Quantity	Earth	Mars	Units
Radius	6.366×10^6	3.374×10^6	m
Density	5517	4577	kg/m^3
Surface air pressure	1.00	0.008	atmos.
Air composition	20% O$_2$, 80% N$_2$	100% CO$_2$	
Gravitational constant	6.672×10^{-11}		N m^{2}kg^{-2}

A parachute designed to deliver a 105 kg load to mars is tested on Earth when the air temperature is 298 K and the air pressure is 1.00 atmospheres. It is found that it hits the surface with a speed of 10 m/s.

Assume that mass of the parachute itself is negligible. Assume the drag coefficient of the parachute is independent of the density, pressure, and temperature of the air.

If we want to have a similar parachute deliver the same load to Mars, what must be its area be? Compare with the area of the test parachute used on Earth.

15.27   An EV experiences an aerodynamic drag of 320 N when operated at sea level (1 atmosphere) and 30 C.

What is the drag when operated at the same speed at La Paz, Bolivia (4000 m altitude, air pressure 0.6 atmospheres), and at a temperature of -15 C?

15.28   A trimaran is equipped with a mast on which a flat rigid surface has been installed to act as a sail. This surface is kept normal to the induced wind direction. The boat is 25 km from the shore, which is due north of it. A 36 km/h wind, V, blows from south to north. How long will it take to reach the shore if it sails straight downwind? Ignore any force the wind exerts on the boat except that on the sail.

The area, A, of the sail is 10 m^2.

The coefficient of drag of a flat surface is $C_D = 1.28$.

The air density is $\rho = 1.2$ kg/m^3.

The water exerts a drag force on the trimaran given by

$$F_{water} = 0.5 \times W^2,$$

where, W, is the velocity of the boat relative to the water (there are no ocean currents).

15.29   Two identical wind turbines are operated at two locations with the following wind characteristics:

Location 1

Percent of time	Wind speed
50	10 m/s
30	20 m/s
20	25 m/s

Location 2

Percent of time	Wind speed
50	15 m/s
50	21 m/s

Which wind turbine generates more energy? What is the ratio of energy generated by the two wind turbines?

15.30   What is the air density of the planet in Problem 1.22 if the temperature is 450 C and the atmospheric pressure is 0.2 MPa?

15.31   One may wonder how an apparently weak effect (the reduction of pressure on top of an airfoil caused by the slightly faster flow of air) can lift an airplane.

Consider a Cessna 172 (a small 4-seater). It masses 1200 kg and has a total wing area of 14.5 m². In horizontal flight at sea level, what is the ratio of the average air pressure under the wing to the pressure above the wing?

15.32   A car has the following characteristics:

Mass, m, = 1200 kg.
Frontal area, A, = 2.2 m².
Coefficient of drag, C_D = 0.33.

The experiment takes place under STP conditions.

When placed on a ramp with a $\theta = 1.7°$ angle, the car (gears in neutral, no brakes) will, of course, start moving and will accelerate to a speed of 1 m/s. This speed is maintained independently of the length of the ramp. In other words, it will reach a **terminal velocity** of 1 m/s.

When a steeper ramp is used ($\theta = 2.2°$), the terminal speed is 3 m/s.

Now place the car on a horizontal surface under no wind conditions. Accelerate the car to 111.60 km/h and set the gears to neutral. The car will coast and start decelerating. After a short time, Δt, the car will have reached the speed of 104.4 km/h.

What is the value of Δt?

15.33   The observed efficiency of a "gyromill" type wind turbine is
$\eta = 0,$ for $U/V \leq 2,$
$\eta = 0.280\,(U/V - 2),$ for $2 \leq U/V \leq 5,$

15.51

$$\eta = -0.420U/V + 2.940, \quad \text{for } U/V > 5.$$

The turbine has 2 blades or wings each 30 m long and the radius of the device is 9 m.

When operating at sea level under a uniform 15-m/s wind what power does it deliver to a load whose torque is 50,000 Nm independently of the rotational speed? What is the rotation rate of the turbine (in rpm)?

15.34   A standard basketball has a radius of 120 mm and a mass of 560 grams. Its coefficient of drag, C_D, is 0.3 (a wild guess), independently of air speed.

Such a ball is dropped from an airplane flying horizontally at 12 km altitude over the ocean. What is the velocity of the ball at the moment of impact on the water?

Make reasonable assumptions.

15.35   This is a terrible way to sail a ship, but leads to a simple problem.

In the absence of wind relative to the boat, a boat's engine power, P_{Eng}, of 20,680 W is needed to maintain a speed of 15 knots (1 knot is 1852 meters per hour). The efficiency of the propeller is 80%. Assume that water drag is proportional to the water speed squared.

Under similar conditions, only 45 W are needed to make the boat move at 1 m/s.

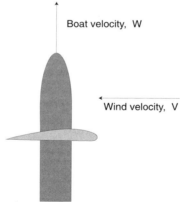

This very boat, is now equipped with an 10 m^2 airfoil, mounted vertically and oriented perpendicularly to the boat's axis. See figure.

The coefficient of lift of the airfoil is

$$C_L = (0.05\alpha + 0.5)$$

and is valid for $-10 < \alpha < 10$. In these two equations, α is in degrees.

15.52

The airfoil exerts a lift that, it is to be hoped, propels the boat due north when a 15 knot wind blows from the east.

What is the speed of the boat?

15.36   A sail plane (a motorless glider) is at a 500 m altitude and is allowed to glide down undisturbed. The atmosphere is perfectly still (no wind, no thermals [vertical wind]). Air temperature is 0 C, and air pressure is 1 atmosphere.

The wings have a 20 m^2 area and their lift coefficient is $C_L = 0.5$ and their drag coefficient is $C_D = 0.05$. Assume, to simplify the problem, that the rest of the sail plane (fuselage, empennage, etc.) produces no lift and no drag. The whole machine (with equipment and pilot) masses 600 kg.

Naturally, as the sail plane moves forward, it loses some altitude. The *glide ratio* is defined as the ratio of distance moved forward to the altitude lost.

1. What is the glide ratio of this sail plane?
2. What is the forward speed of the plane?
3. To keep the plane flying as described, a certain amount of power is required. Where does this power come from and how much is it?

15.37   The drag force on a car can be expressed as power series in v, the velocity of the car (assuming no external wind):

$$F_D = a_0 + a_1 v + a_2 v^2. \qquad (77)$$

For simplicity, assume $a_0 = 0$.

A car drives 50 km on a horizontal road (at sea level) at a steady speed of 60 km/h. Careful measurements show that a total of 1.19×10^7 J were used. Next, the car drives another 50 km at a speed of 120 km/h and uses 3.10×10^7 J. The frontal area of the car is 2.0 m^2. What is the coefficient of drag, C_D, of the car?

15.38

Percentage of time	m/s
10	Calm
20	5
40	10
30	15

The wind statistics (over a whole year) at a given site are as shown in the table.

When the wind has a speed of 15 m/s, the wind turbine delivers 750 kW. What is the number of kWh generated in a one year period?

15.39

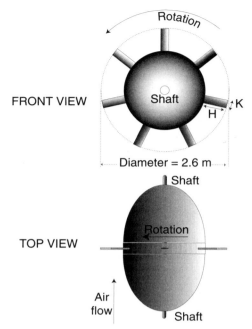

Consider a turbine consisting of seven blades each shaped as a NACA W1 symmetric airfoil with a 32 cm chord, K, and sticking out 52 cm, H, above a fairing. These blades have their reference plane aligned with the plane of rotation of the turbine. The diameter is 2.6 m. The device spins at 1050 rpm in such a direction that, if the turbine rotates in calm air, the angle of attack is zero.

To an acceptable approximation, the coefficients of the airfoil are:

$$C_L = 0.08\alpha - 0.0001\alpha^3$$

For $|\alpha| < 11°$:

$$C_D = 0.0062 \exp(0.2|\alpha|)$$

For $11° < |\alpha| < 21°$:

$$C_D = -0.415 + 0.0564|\alpha| - 0.001|\alpha|^2.$$

In the formulas above, α is in degrees.

The air stream that drives the turbine flows vertically up in the drawing (it flows parallel to the turbine shaft) and has a density 3 times that of the air at STP. It's velocity is 28.6 m/s.

How much power does the turbine deliver to the shaft?

15.54

Chapter 16
Ocean Engines

16.1 Introduction

In Chapter 4, we discussed the utilization of the thermal energy of the oceans. Other forms of ocean energy are also available, such as mechanical energy from waves, currents and tides, and chemical energy from salinity gradients. A summary of the oceanic energy resources is presented by Isaacs and Schmitt, 1980.

16.2 Wave Energy

OTECs took advantage of the ocean acting as an immense collector and storer of solar radiation thus delivering a steady flow of low grade thermal energy. A similar role is also played by the ocean in relation to the wind energy which is transformed into waves far steadier than the air currents that created them. Nevertheless, waves are neither steady enough nor concentrated enough to constitute a highly attractive energy source notwithstanding their large total power: about 2 TW can possibly be captured according to the World Energy Council.

16.2.1 About Ocean Waves[†]

The following specialized terminology and symbology is required for the discussion of ocean [surface] waves:

Duration – Length of time the wind blows.

Fetch – Distance over which the wind blows.

d – the **depth** of the water.

g – the **acceleration of gravity**, 9.81 m/s^2.

h – the **height** of the wave—that is, the vertical distance between the through and the crest of a wave.

T – the **period**—that is, the time interval between two successive wave crests at a fixed point.

v – the **phase velocity** of the wave—that is, the ratio between the wavelength and the period.

v_g – the **group velocity** of the wave—that is, the velocity of wave energy propagation.

[†] A much more extensive discussion of ocean waves can be found in *Introduction to Physical Oceanography*, Chapter 16, by Robert H. Stewart, Texas A&M University accessible at http://oceanworld.tamu.edu/resources/ocng_text book/.

λ – the **wavelength**—that is, the horizontal distance between two successive wave crests, measured along the direction of propagation.

16.2.1.1 The Velocity of Ocean Waves

There is little net horizontal motion of water in a surface ocean wave. A floating object drifts in the direction of the wave with about 1% of the wave velocity. A given elementary cell of water will move in a vertical circle, surging forward near the crest of the wave but receding by an almost equal amount at the trough. Near the surface, the diameter of the circle is equal to the wave height. As the depth increases, the diameters diminish— the motion becomes negligible at depths much larger than one wavelength. Thus, in deep waters $(d \gg \lambda)$ the wave does not interact with the sea bottom and its behavior is independent of depth. The wave velocity is, then, a function of the wavelength,

Deep water, $(d \gg \lambda)$:
$$v = \sqrt{g} \left(\frac{\lambda}{2\pi} \right)^{1/2}. \qquad (1)$$

Any system in which the wave velocity depends on wavelength is called **dispersive**. Hence the deep ocean is dispersive. However, when the water is sufficiently shallow, the circular motion of the water is perturbed by the seafloor and the wave looses some of its energy. When $\lambda \gg d$, the velocity no longer varies with λ (the system is no longer dispersive) but depends now on the depth, d.

Shallow water, $(\lambda \gg d)$:
$$v = \sqrt{g}\, d^{1/2}. \qquad (2)$$

For intermediate depths, there is a transitional behavior of the wave velocity. If the water is very shallow (at $d \approx \lambda/7$), the velocity of the crest of the wave is too fast compared to that of the trough and the wave breaks.

We saw that in shallow waters, the wave velocity diminishes as the depth is reduced. This is the reason why waves tend to come in parallel to a beach. In waves coming in at a sharp angle, the part closer to the shore (presumably, in shallower water) slows down allowing the more distant parts to catch up. If one constructs an undersea mound in the shape of a spherical segment, it will act as a lens focusing the waves into a small region.

In a sloping beach, the wave farther out moves faster than one nearer shore, so the wavelength diminishes as the waves come in. The period, however, is not affected.

All the preceding refers to the phase velocity of the wave. The group velocity—the velocity of energy propagation—differs from the phase velocity in a dispersive medium. Using the two limiting approximations (deep and shallow water), we find that

Deep water, $(d \ll \lambda)$: $$v_g = \frac{v}{2}$$ (3)

Shallow water, $(\lambda \gg d)$: $$v_g = v$$ (4)

16.2.1.2 Wave Height

The height of the wave depends on both the fetch and the duration of the wind. Figure 16.1 shows the empirically determined relationship involved. The fetch data assume that the wind has been blowing for a long time and the duration data assume that the wind has been blowing over a sufficiently long fetch. Both sets of curves seem to approach some final asymptotic value for each wind velocity.

The largest storm[†] wave ever observed (1933) had a height of 34 meters and a wavelength of 342 m. Its period was 14.8 s. This leads to a phase velocity of 23.1 m/s and a group velocity of about 11.5 m/s.

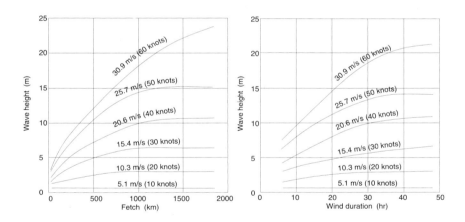

Figure 16.1   Wave height as a function of the fetch (left) and of the duration of the wind (right). A second order regression was fitted through observed data points.

[†] Tsunamis can be much larger than storm waves. A tsunami in Lituya Bay, Alaska, on July 9, 1958 reached a height of 524 m.

16.2.1.3 Energy and Power

The energy density of an ocean wave is (approximately)

$$W = g\rho \left(\frac{h}{2}\right)^2 \quad \text{J/m}^2, \tag{5}$$

where ρ is the density of the water (roughly 1000 kg/m^3). The power associated with a wave is

$$P = v_g W \quad \text{W/m}. \tag{6}$$

Consider the power of the 1933 record wave (see above).

$$W = 9.8 \times 1000 \times \left(\frac{34}{2}\right)^2 = 2.8 \times 10^6 \quad \text{J/m}^2. \tag{7}$$

Since the group velocity of this wave was 11.5 m/s, the power density was $11.5 \times 2.8 \times 10^6 = 32.5 \times 10^6$ W/m.

16.2.2 Wave Energy Converters

Perhaps the major difficulty with wave energy converters is the great range of powers in ocean waves. A moderate sea may have waves with a power of some 50 kW/m, while a big storm may generate powers of 10 MW/m. This 200:1 range is difficult of accommodate. Machines capable of economically using 50 kW/m seas tend to be too frail to resist large storms.

Renewable energy resources are frequently unsteady (variable wind velocities, fluctuating insolation, changing sea states, etc.) and of low power densities. A good wind farm site may operate with 400 W/m^2 rotor loading. Average insolation even in sunny Arizona barely exceeds some 250 W/m^2. Wave energy may be argued to have a more attractive mean power density. Indeed, one can find a number of sites throughout the world where ocean wave power density is above 50 kW/m. Assuming that a power plant uses, say, 10 m of land perpendicular to the shore, such sites would provide some 5 kW/m^2, an order of magnitude more than wind and sun. Clearly, this kind of comparison is only marginally meaningful, but may still be used as a point in favor of the use of ocean waves.

Another advantage of ocean wave power systems is that they can be integrated into coastal structures, including sea walls and jetties, constructed to combat erosion caused by the action of the sea. Such integration may be economically interesting. However, wave energy converters do not necessarily have to be built on the shore line—they may also be placed offshore, even though this may created substantial mooring or anchoring problems.

16.2.2.1   Offshore Wave Energy Converters[†]

Offshore wave energy converters usually belong to one of four categories:

1. Heaving buoys
2. Hinged contour
3. Overtopping
4. Oscillating water column. This category of wave-energy converters can be used both offshore or on shoreline. We will discuss these machines when shoreline devices are examined later in this chapter.

16.2.2.1.1 Heaving Buoy Converters

Although the particular arrangement for the conversion of wave energy into electricity described here does not appear to be practical, it serves to illustrate the manner in which more complicated devices of this class operate. The vertical oscillation of the water can act on floats that drive levers and pistons, but such an arrangement may be too complex and too vulnerable to stormy conditions. Simpler schemes may work better. Consider, for instance, a wave-operated pump consisting of a buoy attached to a long vertical pipe equipped with a check-valve at the bottom (see Figure 16.2).

When a wave lifts the device, water in the pipe is accelerated upward because it cannot escape through the check valve. As the wave recedes, the pipe accelerates downward against the movement of the internal water creating a considerable hydraulic pressure. The water can be allowed to escape through the top to drive a turbine. Simultaneously, more water enters through the check valve.

Figure 16.2   Wave-operated pump.

[†] An extensive review of offshore devices can be found in E21 EPRI Assessment.

Let A be the cross-sectional area of the pipe, L its length, δ the density of water, and γ the peak acceleration of the pipe.

The mass of water in the pipe is $M = \delta AL$ and the resulting force is $F = \gamma \delta AL$. The corresponding pressure (when the pipe is accelerated vertically) is $p = \gamma \delta L$. If $L = 100$ m and $\gamma = 5$ ms^{-2}, then the pressure is 0.5 MPa (5 atmos).

AquaBuOY proposed by AquaEnergy is an example of a floating heaving buoy device. The vertically oscillating water column drives a piston, which in turn pumps water at a higher pressure into an accumulator acting as a low pass filter. The accumulator discharges the water into a turbine for generation of electricity. The prototype being tested uses a 6-m diameter buoy and has a pipe that dips 30 m down. (The system requires water depths of 50 meters or more.) Rated power is 250 kW at reasonable wave heights. The system uses slack mooring.

Seadog (Independent Natural Resources, Inc) is a bottom dwelling heaving buoy system. The 5.7-m diameter float pumps water into an elevated basin on shore. Discharge from the basin drives a conventional hydroturbine. The system requires a depth of 20 m.

16.2.2.1.2 Hinged Contour Converters

Both the AquaBuOY and the Seadog mentioned above are **point absorbers** because they interact with only a small ocean area. To generate large amounts of energy, a multitude of these devices must be deployed, each with its own piston and power take off equipment. The solution developed by Ocean Power Delivery, Ltd, although also using a system of buoys, is capable of interacting with a much larger ocean area.

The 750-kW prototype called "Pelamis" installed at the European Marine Energy Centre in Orkney[†] consists of four tubular steel floats measuring 4.63 m in diameter attached to one another by hinges. The length is 150 meters.

The force the waves exert in moving each segment relative to its neighbors is captured by hydraulic rams that press a biodegradable fluid into accumulators, which, in turn, power a number of 125 kW generators.

Figure 16.3   The Pelamis hinged contour converter.

[†] The Orkney Islands form an archipelago some 50 km north of the northernmost tip of Scotland.

The system is moored by a 3-point configuration that allows Pelamis to orient itself normal to the incoming wave front. The natural resonance of the system is automatically altered to match the frequency of the waves. However, when exposed to storm conditions, the system is detuned to minimize the stress on the mooring. Simulations and actual tests show that Pelamis appears to exhibit excellent storm survivability.

16.2.2.1.3 Overtopping Converters

Wave Dragon ApS (Denmark) is developing a fundamentally simple wave energy converter. It consist of a large floating basin or reservoir a few meters above sea level. Waves (concentrated by a pair of reflector arms) hit a ramp rising up and spilling into the reservoir. The water flows out back to the ocean driving a number of Kaplan turbines and, thus, generating electricity. One advantage of this design is that the only moving parts (other than the flowing water) are turbines and generators.

The manufacturer feels confident that the Wave Dragon will withstand intense winds (owing to its low-in-the-water profile) and large waves that simply flow over the installation. A 57-m wide prototype rated at 20 kW has been in operation since March 2003.

Observers are optimistic about the Wave Dragon possibilities. The one obvious disadvantage is the unfavorable mass to power ratio: the proposed 4 MW 300 meter wide machine masses over 30,000 tons!

The consensus, in the end of 2004, was that of the various machines described here, the Pelamis was closest to paying off, while the Wave Dragon needed a bit more development. All other projects were still in the R&D stage.

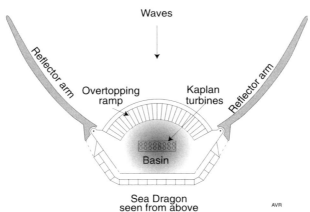

Figure 16.4   The Wave Dragon.

16.2.2.2 Shoreline Wave Energy Converters

Shoreline wave energy converters usually belong to one of two categories: the first is the **tapered channel** type also known as **tapchan**, and the second is the **oscillating water column** (OWC). OWCs can be installed either on shore or in deepwater **offshore** sites. The tapchan by its very nature can only be used as a shoreline device. As discussed in Subsubsection 16.2.1.1, underwater structures can be built to focus a relatively large wave front into a much smaller aperture of a wave energy converter.

16.2.2.2.1 Tapered Channel System [†]

An ingenious wave energy system was developed by a Norwegian company appropriately called Norwave. The system was demonstrated in a fjord at Toftestallen in the vicinity of Bergen, Norway. The general arrangement is sketched out in Figure 16.5. A narrow tapered concrete canal whose walls rise well above sea level connects the ocean with a bay or lake. Waves entering the tapered canal move along it, increasing progressively in height (because of the gradual narrowing) until water spills over the canal walls. Thus, the waves pump water into the lake causing its level to become substantially higher than that of the ocean. This height differential was used to drive a hydraulic turbine that generated electricity.

The only artifact that is directly exposed to the waves is the concrete canal which can be designed to resist any reasonable storm, although in the Norwave prototype it apparently suffered severe damage that caused the discontinuation of the project in the early 1990s after being in intermittent operation since 1986. The actual generating equipment is a normal low-head hydroelectric power plant except that it must be built to resist the corrosive action of the salt water. The Norwave prototype was modest in size. It was equipped with a 350 kW generator fed by a reservoir with 5500 m^2 surface area and a water level 3 m above see level. The opening to the sea was 60 m wide, but the concrete canal started with only a 3-m width and narrowed progressively to 20 cm at its far end. Its length was 80 mm and its depth 7 meters. The system was to operate with a 14 m^3/s pumping capacity. This corresponds to an available water power of slightly more than 400 kW, enough to drive the 350 kW generator.

A problem with the Tapchan is that it will work only in selected sites where the wave regimen is steady and where tides are less than, say, 1 m.

[†] In 1877, Giovanni Schiaparelli, using a favorable opposition of Mars, drew a detailed map of the planet and described what he perceived as "canali." Unfortunately, this was translated into English as "canals" instead of "channels" and supported the view that Mars was inhabited. The expression "tapered channel" suffers from the opposite error. Almost invariably we are referring to artificially built canals, not to natural channels.

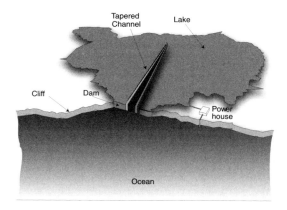

Figure 16.5   NORWAVE wave power plant.

16.2.2.2.2 Wavegen System (OWC)

Another promising scheme for using ocean wave energy was constructed by the Scottish company WAVEGEN and was commission November 2000. It bears the whimsical name **LIMPET** (Land Installed Marine Powered Energy Transformer) and is not a commercial unit, but rather a further research and development unit. It is a scale-up of a 75 kW prototype that operated from 1991 to 1999. LIMPET is a 500 kW plant installed on island of Islay.[†] It is powered by the oscillating water column generated by wave action. A slanted concrete tubelike structure is open on one side to the ocean. Waves cause alternate compression and suction in the air column inside the tube. The resulting forward and backward air flow drives a Wells turbine that has the unusual property of rotating in the same direction regardless of the direction of the air flow. See Problem 15.39.

Water inlet consists of three 6 by 6-meter rectangular ducts whose air columns merge into a common turbine inlet. The system is depicted (in an idealized fashion) in Figure 16.6. The two Wells turbines measure 2.6 m in diameter and operate at between 700 and 1500 rpm. To accommodate this large speed variation and to facilitate interfacing with the local power grid, the output of the two generators is rectified and then inverted into alternating current of the proper frequency.

Among other advantages, the LIMPET represents a low visual intrusion and causes a minimal impact on fauna and flora. According to WAVEGEN, "ongoing projects include development of a floating device and a tunnelled cliff shoreline array in the Faroe Islands (ultimately up to 100 MW)."

[†] Islay is one of the larger islands of a complex archipelago, known as the Hebrides, west of Glasgow and north of Northern Ireland.

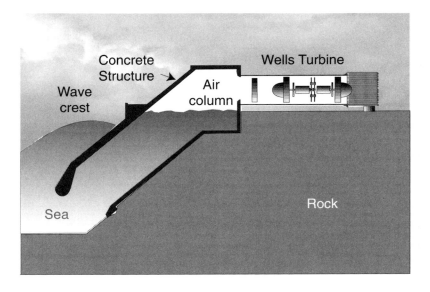

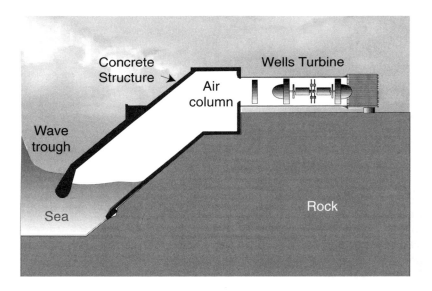

Figure 16.6   The crest of the wave (top picture) corresponds to maximum compression into the Wells turbine, while the trough (bottom picture) corresponds to maximum suction. Owing to the resonance of the water column in the concrete structure, the amplitude of the oscillations in the structure exceeds the amplitude of the waves. The turbine will generate electricity regardless of the direction of the air flow.

16.3   Tidal Energy

Tides can be utilized as energy sources through the currents they cause or through the associated variations in ocean level. The most effective way of taking advantage of these vertical water displacements is through impoundment and this is the only approach being considered seriously. Average tidal fluctuations in the open sea are small. It might be interesting to know that, owing to its relatively small size, there are essentially no tides in the Mediterranean Sea with the result that Greek philosophers were, in general, unacquainted with the phenomenon.

Local resonances can magnify the tidal effect considerably, as demonstrated dramatically by the high tides in Fundy Bay, especially at Cobequid Bay[†] where they reach a 16 m swing.

There is a problem of impounding tides without "detuning" the system as has happened to some extent in the only major tidal power plant in existence—the one at the mouth of the Rance River, in France. It appears that the installation reduced by 20% the power flow of the estuary. Nevertheless, the system delivers an average of 160 MW (peak of 240 MW). It consists of 24 reversible turbines and a dam built across part of the estuary. At high tide, the water flows through the turbines into the estuary. At low tide, the water flows out, again through the turbines, generating power at both ebb and flood. Mean tides are 8.5 m, a large value for tides, but only a modest head for hydraulic turbines. These are of the Kaplan type and have a large diameter: 5.35 m. Near high and low tides, the turbines are used as pumps to accentuate the working head. The Rance River plant, near Saint Malo, has been in operation since the 1970s. One of the difficulties it has experienced is corrosion owing to the use of salt water.

16.4   Energy from Currents

A recommended reading for those interested in this subject is the article by Fraenkel (see "Further Reading" at the end of this chapter).

Currents are an example of the enormous amount of energy stored in the oceans under unfavorably low power densities. As with wind and direct solar, there is a problem with the intermittence of the available power. But, whereas wind and sunlight can vary at random, ocean currents are more predictable, especially when caused by tides. The report "Marine Currents Energy Extraction" estimates that European waters can support at least 12 GW of ocean turbines able to generate an average of 5.5 GW. This is a load factor of 45%, which may be a bit optimistic. Other experts estimate a factor of between 35 and 40%. At any rate, it is much better than the

---

[†] The Bay of Fundy is a substantial body of water between Nova Scotia and New Brunswick on the Canadian mainland. It has a general SSW to NNE orientation and ends in two "horns," the southernmost of which is Cobequid Bay.

20%, or so, for land-based wind turbines. Offshore wind turbines may reach 30%. A 1.2 GW US-operated nuclear power plan delivers an average of 1.05 GW—a load factor 87.6%. European ocean turbines could replace about 5 large nuclear plants. Caveat: We are comparing an estimated performance of ocean turbines, with actual performances of wind turbines and nuclear plants. There is always some degree of optimism in estimates.

Ocean currents are driven by numerous forces: winds, salinity gradients, thermal gradients, the rotation of Earth, and, mainly, by tides. Undisturbed tides are synchronized with the phases of the moon. If the moon had no phases—that is, if, for instance, one had an eternal full moon, then a high tide would come every 12 hours as a given part of the world rotated under the moon. There are two daily tides, one near the sublunar point and one on the opposite side of Earth. However, the moon orbits the Earth in 27.32 days and goes from full moon to new and back to full. On average, two successive full moons recur every 29.53 days. This is called a lunar month or a **lunation**. To understand the distinction between the lunar month and the length of the lunar orbital period, do Problem 12.28.

Meteorological and topographic influences make it impossible for simple theory to predict the exact amplitude or the exact frequency of the tides. In addition, tidal flow is often turbulent so that velocities cannot easily be translated into recoverable energy. Nonetheless, using the expressions developed in Chapter 15, we can estimate available power densities,

$$P_A = \frac{16}{27}\frac{1}{2}\rho v^3 A_v. \tag{8}$$

In a good site, water flow velocities are around 3 m/s. The worlds first underwater turbine servicing consumers on a regular basis uses a 2.5 m/s flow. It is operated by Hammerfest Stroem, in northern Norway, way beyond the Arctic Circle. According do Professor Fraenkel, there are places with velocities as high as 5 m/s. Compared with wind, these are modest velocities, however, the high density of water (1000 kg/m^3,[†] almost 800 times larger than that of air) causes these small velocities to lead to attractive power densities. In good locations (where the local topography accelerates the usually sluggish water flow), power densities can reach 10 kW/m^2, one order of magnitude more than that at a good wind turbine site.

Some of the early proposals for extracting energy from currents seemed impractical. One example is the idea of employing a parachute-like device dragged along by the waters, unwinding a long cable that drives the shaft of a generator. When all the cable has been unwound (perhaps several kilometers), the parachute is collapsed and retrieved by running the generator as a motor, thus rewinding the cable. More practical are schemes that employ what are essentially underwater "wind" turbines.

[†] Ocean water, being salty, has a somewhat higher density—1027 kg/m^3 being typical for surface water where, presumably, most turbines will operate.

Figure 16.7   Prototype of a tidal turbine installed off the Coast of Devon, England. The rotor/turbine assembly is shown in the raised position for maintenance. Source: Marine Current Turbine, Ltd.

16.4.1 Marine Current Turbines System

Marine Current Turbine, Ltd. (MCT) has a program for harnessing ocean currents by using windmill-like underwater turbines. MCT demonstrated a prototype (called "Seaflow") rated at 300 kW and capable of generating an average power of 100 kW. The 11-m diameter rotor can have its pitch adjusted to accommodate tides flowing from opposite directions. The turbine-driven induction generator has its output rectified and inverted to 50-Hz ac which is dissipated in a resistive test load. The company's next project will be a 1-MW system connected to the grid.

Although similar to wind-driven turbines, there are major differences.

16.4.1.1 Horizontal Forces

The power generated by a wind turbine is

$$P_{wind} = \frac{16}{27}\frac{1}{2}\rho_{air}v_{air}^3 A_{wind}\eta_{wind}, \tag{9}$$

and that generated by a water turbine is

$$P_{water} = \frac{16}{27}\frac{1}{2}\rho_{water}v_{water}^3 A_{water}\eta_{water}. \tag{10}$$

Let us assume that both type of turbines have the same efficiency, η. If the two turbines generate the same amount of power, then

$$\rho_{air} v_{air}^3 A_{wind} = \rho_{water} v_{water}^3 A_{water} \tag{11}$$

$$\frac{A_{wind}}{A_{water}} = \frac{\rho_{water}}{\rho_{air}} \left(\frac{v_{water}}{v_{air}} \right)^3. \tag{12}$$

The horizontal forces on the turbines are proportional to $\rho v^2 A$. Hence,

$$\frac{F_{water}}{F_{air}} = \frac{\rho_{water} v_{water}^2 A_{water}}{\rho_{air} v_{air}^2 A_{wind}} = \frac{v_{air}}{v_{water}}. \tag{13}$$

If, for example, the wind velocity is 15 m/s and that of the water is 3 m/s, the horizontal forces on the water turbine will be five times larger than that on a wind turbine with the same power output. The density, ρ, of the fluids plays no role in the above formula because, for a fixed power output and selected flow velocities, the ρA product must be constant. If the density of the fluid is increased the area needed to produce the same power will be correspondingly decreased.

From the above, one should expect substantial horizontal forces in ocean devices. This requires heavy structure (Seaflow masses 130 tons) and strong anchoring systems. Megawatt sized ocean turbines may be subjected to over 100 tons of horizontal forces.

16.4.1.2 Anchoring Systems

The anchoring system of ocean turbines must withstand the extremely large horizontal forces mentioned above. In moderate depth, the turbine might simply be weighed down by an adequate ballast, but it appears that a single pile driven into the ocean floor is being preferred. Great depth would probably require floating platforms tethered to a sunken pile. One difficulty with this arrangement is that as the direction of the current reverses (as happens with tide-driven currents), the floating platform will change position causing all sort of difficulties, especially with the transmission line that carries the electric energy to the consumer.

Fortunately, builders of ocean oil rigs have accumulated a vast experience in constructing undersea structures and in anchoring them.

16.4.1.3 Corrosion and Biological Fouling

The ocean is a hostile environment requiring careful choice of materials and of passivation procedures as well precautions against biological fouling.

16.4.1.4 Cavitation

As we saw in Chapter 15, lift type turbines operate by generating a pressure differential between opposite sides of an air foil (or of a water foil, as it were). The pressure on the suction side may become low enough to cause water to boil generating vapor bubbles which, when reaching regions of higher pressure, will implode releasing energy and reaching high temperatures notwithstanding the environment acting as an excellent heat sink.[†] Such implosions are surprisingly damaging to the rotating blades of hydraulic machines and propellers. Severe pitting and vibration can result and, in case of the submarines, noise is generated reducing the stealthiness of the boat. This phenomenon is called **cavitation**. Clearly, increased depth (because of increased pressure) will retard cavitation, which, therefore, tends to occur at the upper part of the rotor sweep, where the static pressures are at a minimum. The tip speed of the rotor of an ocean current turbine must be kept low enough to avoid cavitation. The safe top speed varies with, among other factors, the depth as explained above. In shallow sites, cavitation may develop when the linear speed exceeds some 15 m/s. One can get a rough estimate of the rotor tip velocity, v_{tip}, that will cause cavitation in typical ocean conditions as a function of the depth, d,

$$V_{tip} \approx 7 + 0.31d - 0.0022d^2. \tag{14}$$

Wavegen reports having relatively minor difficulties with cavitation. The main problem they have experienced from cavitation is loss of efficiency.

16.4.1.5 Large Torque

Because of the large density of the medium, ocean current turbines operate at low rpm. For a given power, large torques are developed. Costly gearing is required to match the characteristics of most electric generators, usually designed for high speed/low torque conditions. In wind turbines, there is a modern trend toward the development of low speed generators. These, if they prove practical, will benefit ocean turbines as well.

When the tip speed of the rotor is limited to a fixed value (owing to cavitation) then, the torque delivered by the turbine is proportional to the generated power raised to 3/2—that is, the torque grows faster than the power, aggravating the larger torque problem.

16.4.1.6 Maintenance

An important factor in the cost of operation is maintenance. Turbine design should be such as to minimize maintenance frequency. Seaflow has the capability of raising the rotor and gear box to make it easily accessible when repairs become necessary. Unattended operation and remote control and metering will probably be an unavoidable requirement for economic operation of the system.

[†] Taleyarkhan et al. report the extremely high temperature of 10 million kelvins of imploding bubbles in deuterated acetone. The temperature is high enough to provoke the nuclear fusion of the deuterium. These results are being disputed by some scientists.

16.4.1.7 Power Transmission

Ocean current turbines must be located at some distance from shore. This calls for expensive undersea transmission lines.

16.4.1.8 Turbine Farms

Ocean turbines of a given power are more compact than equivalent wind turbines, hence the former can be more densely packed than the latter. In addition, in most wind turbine farms, the location of individual units must take into account that the wind direction may be quite variable, and one turbine should not shade the next. The direction of ocean currents is reasonably constant, so that turbines can be set up in close proximity. Packing density of these devices can lead to up to 100 MW/km², one order of magnitude larger than their wind-driven cousins. Close packing has a number of advantages including reduced cost in the power transmission system.

16.4.1.9 Ecology

Arguments can be made that sunken structures in the ocean can act as artificial reefs benefiting the local flora and fauna.

16.4.1.10 Modularity

Modularity of ocean current turbines can contribute significantly to the reduction of operation costs.

16.5 Salination Energy

The highest power density available in the ocean is that associated with the salination energy of fresh river waters mixing with the sea. Indeed, the osmotic pressure of fresh water with respect to sea water is over 2 MPa. Thus, a flow of 1 m³ s⁻¹ will result in the release of energy at a rate of

$$P = p\dot{V} = 2 \times 10^6 \times 1 = 2 \text{ MW}. \tag{15}$$

To produce the same power per unit flow, a hydroelectric plant would require a head of 200 m.

Most proposals for salination engines are based on the use of semipermeable membranes. These can be employed in engines driven by the osmotic pressure between fresh and salt water, or in electrodialysis engines based on the electric potential difference established across a membrane that is permeable to cations and not to anions or vice-versa.

Membranes are expensive and tend to have short lives. The Norwegian company, Statkraft, in Sunndalsøra is developing (2003) special membranes for salination power plants.

Olsson, Wick, and Isaacs (1979) proposed an ingenious salination engine that does not use membranes. It works best between fresh water and brine (defined as a saturated sodium chloride solution) but, presumably, can be made to operate also with ocean water.

Salt water boils at a higher temperature than fresh water. In other words, the former has a lower vapor pressure than the latter. Figure 16.8 shows how the vapor pressure of fresh water depends on temperature. It also shows the difference between the vapor pressure of fresh water and that of brine.

The proposed engine is sketched out schematically in Figure 16.9. Compartment A contains brine, while compartment B contains fresh water. If the liquids are at the same temperature, the vapor pressure in A is lower than that in B, hence water vapor flows from B to A driving the turbine in the interconnecting pipe. Evaporation of the fresh water cools it down lowering the pressure while dilution of the brine causes the temperature (and the pressure) of compartment A to rise. Soon the system will reach equilibrium and the turbine will stop.

This can be avoided by feeding the heat of condensation back from B into A. If all the heat is returned, there will be no temperature change and, eventually, all the water will be transferred to the brine compartment. The heat transfer between the compartments can be accomplished by placing them side by side separated by a thin heat-conducting wall.

In practice, the brine would be replaced by ocean water and rivers would supply the fresh water. Let us determine how much energy can be extracted when 1 kilomole of fresh water is transferred to the brine. Let p_{FR} be the pressure of the water vapor over the fresh water container; p_{BR}, the pressure in the brine container. Let V_{FR} be the volume of the water vapor that flows into the turbine from the fresh water side and V_{BR} be the volume that exits the turbine on the brine side, all in a given time period.

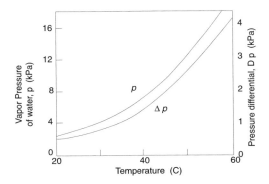

Figure 16.8   Vapor pressure of water and the difference between the pressure over fresh water and that over concentrated brine.

16.17

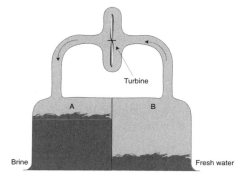

Figure 16.9   A membraneless salination engine.

The water vapor expands through the turbine doing work:

$$W = \int_{p_{FR}}^{p_{BR}} p\,dV, \tag{16}$$

and, since the expansion can be taken as adiabatic,

$$p_{FR} V_{FR}^{\gamma} = p_{BR} V_{BR}^{\gamma}. \tag{17}$$

The integral becomes

$$W = \frac{p_{FR} V_{FR}}{\gamma - 1} \left[1 - \left(\frac{p_{FR}}{p_{BR}} \right)^{\frac{1-\gamma}{\gamma}} \right]. \tag{18}$$

For 1 kilomole of any perfect gas, $pV = RT$. Assume the system operates at a constant temperature of 25 C (298 K), then the $p_{FR} V_{FR}$ product is $8314 \times 298 = 2.49 \times 10^6$ J.

For water, $\gamma = 1.29$.

The work done by 1 kilomole is then

$$W = 8.54 \times 10^6 \left[1 - \left(\frac{p_{FR}}{p_{BR}} \right)^{-0.225} \right]. \tag{19}$$

At 25 C, the fresh water vapor pressure is $p_{FR} = 3.1$ kPa and the pressure difference between the fresh water and brine compartments is 0.59 kPa (see Figure 16.8). This means that $p_{BR} = 3.1 - 0.59 = 2.51$ kPa. Introducing these values into Equation 19, we find that each kilomole of water vapor that flows through the turbine generates 396 kJ of mechanical energy.

Let us compare the above energy with that generated by an OTEC operating between 25 C and 5 C, a ΔT of 20 K. Assume that half of this ΔT is applied across the turbine. In the Lockheed OTEC, the power extracted from the turbine is 90% of the Carnot power. So, let as take the turbine efficiency as equal to the Carnot efficiency which in this case would be 0.033. If 1/4 of the ΔT is the warm water temperature drop, then the input thermal energy (per kmole of warm water) is 10 MJt and the mechanical output from the turbine is 330 J. This is in the same order as the 396 J/kmole calculated for the salination engine. The salination engine could theoretically achieve 100% efficiency. In a laboratory model, Olsson and coworkers demonstrated 40% efficiency.

One problem with this type of engine is the outgassing of the water necessary to assure that the pressures in the compartment are due only to the water vapor, not to the presence on incondensable gases. If instead of concentrated brine more realistically ocean water is used, then the power output of the machine falls substantially.

16.6   The Osmotic Engine

Salination engines are being seriously proposed as a possible energy converter, while the osmotic engine described in this section will, probably, never be more than an academic curiosity. Its study is, nevertheless, an interesting intellectual exercise.

Osmosis is a (quantitatively) surprising phenomenon. If two solutions of different concentrations are separated by an **osmotic membrane** permeable to the solvent but not to the solute, then there is a net flow of the solvent from the more dilute to the more concentrated side. Such flow will persist until the pressure on the concentrated side is sufficiently large.

Osmotic pressures can be measured by means of a U tube with an osmotic membrane at the bottom as suggested in Figure 16.10.   When equilibrium is reached—that is, when the pressures on the two sides of the membrane are the same, the column on the concentrated side is higher than that on the dilute side. The hydrostatic pressure of the brine must equal the sum of the hydrostatic pressure of the fresh water plus the osmotic pressure. What is surprising is the magnitude of the osmotic pressure.

For concentrated NaCl solution at room temperature, the height difference between the two columns would be 4000 meters! The osmotic pressure is 400 atmospheres or 40 MPa.

Osmotic pressure depends on concentration and on temperature. Figure 16.11 shows the osmotic pressure of salt water as a function of salinity at two different temperatures.

To understand the operation of an osmotic engine, consider a cylindrical column of water with base area, A, and a depth, d. If the density of the water column is δ, then the mass is

$$M = A\delta d \tag{20}$$

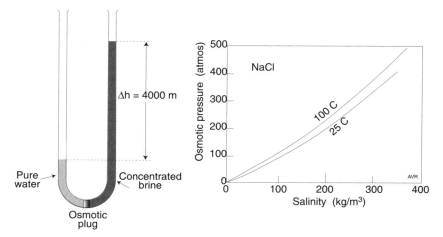

Figure 16.10 Apparatus for
measuring osmotic pressure.

Figure 16.11 The osmotic pressure of
saltwater at two temperatures.

and the weight is

$$W = gA\delta d \qquad (21)$$

where g is the acceleration of gravity. The pressure on the base is

$$p = g\delta d. \qquad (22)$$

Consider now a pipe immersed vertically in the ocean with its top just above the surface and having at its bottom an osmotic plug. Let the pipe be filled with fresh water up to a height, d_F. Let p_S be the pressure exerted by the saltwater (on the outside) on the osmotic plug and p_F the opposing pressure of the fresh water. In addition to the hydrostatic pressures, there is also an osmotic pressure, p_O, tending to force the fresh water into the ocean.

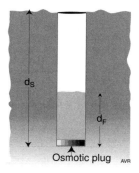

Figure 16.12 A pipe with an osmotic plug in the ocean.

Thus, p_F and p_O act in the same direction and oppose p_S. In equilibrium,

$$p_S = p_F + p_O. \tag{23}$$

The density of fresh water is 1000 kg/m³ while that of ocean water is 1025 kg/m³. Thus

$$1025gd_s = 1000gd_F - p_O, \tag{24}$$

from which

$$d_F = 1.025d_s - \frac{p_O}{1000g}. \tag{25}$$

The osmotic pressure of fresh water with respect to ocean water is 2.4 MPa. Taking $g = 10$ m s^{-2},

$$d_F = 1.025d_S - 240. \tag{26}$$

Down to a depth, d_S, of about 240 m, $d_F < 0$—that is, the osmotic pressure is sufficient to keep the saltwater from entering the pipe. If the pipe goes deeper, reverse osmosis takes place and fresh water from the salty ocean is forced into the pipe. When, for instance, the pipe goes down to 1000 m, the fresh water column will rise 785 m coming within 215 m from the surface.

At what depth will the water rise just to the ocean level? When that happens,

$$d_F = d_S \equiv d = 1.025d - 240, \tag{27}$$

or

$$d = 9600 \text{ m.} \tag{28}$$

If the pipe goes even deeper, fresh water will fountain out of its top.

A possible osmotic power plant could be constructed as shown in Figure 16.13.

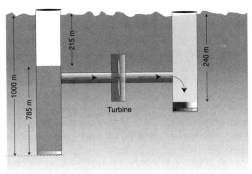

Figure 16.13   A possible configuration for an osmotic engine.

References

Fraenkel, Peter L., Power from marine currents, *Proc. Inst. of Mechanical Engineers*, **216** Part A: Journal of Power and Energy **2002**.

Isaacs, John D., and W. R. Schmitt, Ocean energy: forms and prospects, *Science, 207*, **265**, (**1980**).

Olsson, M., G. L. Wick, and J. D. Isaacs, Salinity gradient power: utilizing vapor pressure difference, *Science, 206*, **452**, **1979**.

Taleyarkhan, R. P., C. D. West, J. S. Cho, R. T. Lahey, Jr. R. Nigmatulin, and R. C. Block, Evidence for Nuclear Emissions During Acoustic Cavitation, *Science* **295**, 1868 (**2002**).

Further Reading

Duckers, L.J. (Coventry Univ, Coventry, UK), Wave energy; Crests and troughs, *Renewable Energy* v 5 n 5–8 Aug. **1994**. p 1444–1452.

E21 EPRI WP – 004 – US, Offshore Wave Energy Conversion Devices, June 16, **2004**

Krock, Hans-Jurgen, editor, Ocean energy recovery, proceedings of the first international conference, ICOER **1989**, Honolulu, Hawaii, November 28–30, 1989, held under the auspices of the Ocean Energy Committee of the Waterway, Port, Coastal, and Ocean Division of the American Society of Civil Engineers, cosponsored by the Pacific International Center for High Technology Research, School of Ocean and Earth Science and Technology of the University of Hawaii, New York, NY, c**1990**. 367 p.

McCormick, Michael E. and Young C. Kim, editors, Utilization of ocean waves—wave to energy conversion, *proceedings of the international symposium*, Scripps Institution of Oceanography, La Jolla, California, June 16–17, 1986, sponsored by the Waterway, Port, Coastal, and Ocean Division of the American Society of Civil Engineers and the National Science Foundation, Symposium on Utilization of Ocean Waves (**1986**: Scripps Institution of Oceanography), New York, NY: c**1987**. vi, 205 p.

McCormick, Michael E., Compendium of international ocean energy activities, prepared under the auspices of the Committee on Ocean Energy of the Waterway, Port, Coastal, and Ocean Division of the American Society of Civil Engineers. New York, NY, c**1989** 86 p.

Michael E. Mccormick, Ocean wave energy conversion, Wiley, New York, c**1981**. xxiv, 233 p.

Practical Ocean Energy Management Systems, Inc.
<http://www.poemsinc.org/home.html>

Richard J. Seymour, editor, Ocean energy recovery: the state of the art. American Society of Civil Engineers, New York, N.Y.: c1992, 309 p.

The exploitation of tidal marine currents (Non-nuclear energy Joule II project results), Report EUR 16683EN, DG Science, Research and Development, The European Commission Office for Official Publications, Luxenburg L-2920, (**1996**).

PROBLEMS

16.1 Reverse osmosis is a technique sometimes employed to extract fresh water from the sea.

Knowing that the osmotic pressure of fresh water with respect to sea water is 2.4 MPa, what is the minimum energy necessary to produce 1 cubic meter of fresh water?

The price of the equipment is \$200/kW and that of electricity is \$50/MWh. What is the cost of a cubic meter of fresh water if there is no operating cost and if the cost of the capital is 20% per year?

How realistic are your results? Point out reasons for having underestimated the cost.

16.2 Refer to the figure. Initially, the shut-off valve is closed and the piston is at a height of 50 m. Pressure of the water in the mains is 3×10^5 Pa. How high will the piston rise when the valve is opened? Assume that the osmotic pressure of the brine-water system is 2 MPa independently of dilution.

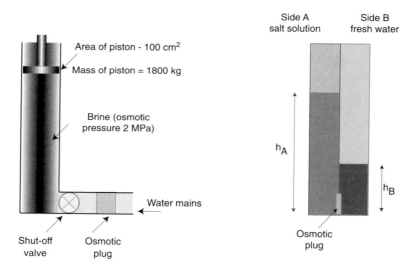

Problem 16.2   Problem 16.3

16.3 At 25 C, the osmotic pressure of a NaCl (versus fresh water) is given approximately by the empirical expression,

$$p = 0.485 + 0.673\sigma + 1.407 \times 10^{-3}\sigma^2,$$

where p is the pressure in atmospheres and σ is the salinity in kg/m^3.

16.24

Two tubes with a square internal cross-section, A, of 8.86×8.86 mm are interconnected via an osmotic wall. In one side (Side A) of the assembly, a NaCl solution containing 2 g of salt and measuring 10 milliliters is poured in. Initially no liquid is poured into Side B. For simplicity assume that the volume of the solution is equal to the volume of the solvent (water).

How high is the saltwater column in Side A? How high is the fresh water column in Side B?

How much distilled water has to be poured into Side B so that, after equilibrium has been reached, the fresh water column is 5 cm high? Again, for simplicity, assume the brine of any concentration has the same density as distilled water.

16.4   Imagine a very tall underwater tower on top of which there is a platform (which, of course, does not change its height above the sea floor). This arrangement, although impractically expensive, permits easy observation of the ocean waves. On a given day, the average height between the trough and the crest of a wave is measured as exactly 2.6 m, and the waves follow one another at a 8.2 seconds interval. How far apart in space are the waves? Assume that the depth of the water is much larger than the wave length of the waves.

16.5   Both the sun and the moon cause ocean tides on Earth.

1. Which tide is bigger? The solar tide or the lunar tide?
2. Calculate the gravitational accelerations at Earth caused by the sun and the moon.
3. If you did the calculations correctly, then you will have found that the gravitational field of the sun on Earth is vastly larger than that of the moon. We must conclude that the tides are not proportional to the gravitational field.

What is the nature of the influence that causes the tides? Calculate the ratio of the lunar to the solar influences.

16.6   Assume that the orbits of Earth around the sun and that of the moon around the Earth are both circular. The former is completed in 3.1558157×10^7 seconds and the latter in 2.36055×10^6 seconds or 27.32 days.

Remember that a full moon occurs when the sun, the Earth, and the moon are in the same plane—that is, their projection on the ecliptic lie in a straight line.

Calculate the number of days between consecutive full moons. Express this with four significant figures.

Prob 16.7   Owing to the possibility of cavitation, the tip speed, v_{tip}, of the rotor of an ocean current turbine must not exceed a certain speed, $v_{tip_{max}}$, which depends on, among other factors, the depth. Show that for a constant tip speed, the torque, Υ, is proportional to $P_g^{3/2}$, where P_g is the generated power.

Subject Index

Abandonment time, market   **1**.10
Abrupt adiabatic process   **2**.6...10
Absorption   **11**.11
Absorptivity   **6**.18
Accelerating potential   **6**.6
Accelerator   **6**.11
Acceptance angle, solar collectors   **12**.15
Acceptor   **14**.20
Acetaldehyde   **13**.12
Acid electrolytes, fuel cell reactions   **9**.9
Acid, strong   **9**.1,2
Activated intermediate species   **10**.34
Activation voltage drop   **9**.59
Activation voltage, fuel cell   **9**.59
Activation, hydride   **11**.12, 15
Adams, A. M.   **9**.63
Adiabatic   **2**.4
Adiabatic processes   **2**.6
Adiabatic processes (abrupt)   **2**.6...10
Adiabatic processes (gradual)   **2**.10...12
Adiabatic turbine   **4**.4
Adsorption storage, hydrogen   **11**.7...8
Adsorption, cesium   **6**.24...26
AFC, alkaline fuel cells   **9**.6
Aiming angle   **12**.17
Air circulation   **12**.12
Airfoil reference plane   **15**.16
Airfoil, asymmetric   **15**.16
Airfoil, symmetric   **15**.16
Airfoils   **15**.16...18
Alcohol   **3**.2, 15...16,   **13**.13
Alcohol thermal decomposition   **10**.5
Alcohol, anhydrous   **13**.10
Alcohol, hydrated   **13**.10
Alcohols, dissociated   **13**.13
Aldehyde   **13**.3
Alga for hydrogen production   **10**.36
Alkali metal thermal electric converter   **7**.1
Alkaline electrolytes, fuel cell   **9**.6
Alkaline electrolytes, fuel cell reactions   **9**.9
Alkaline fuel cells   **9**.6
Alkane hydrocarbons   **13**.2
Alkene hydrocarbons   **13**.2
Alkyne hydrocarbons   **13**.2
Allowable loading rate   **13**.17
Alpha-phase, hydrides   **11**.13
alumina, β' **7**.1

Alvarez   **1**.31
Amalgamation   **9**.2
Ammonia tolerance, fuel cell   **9**.15, 30
Amorphous thin film
  photovoltaic materials   **14**.6
AMTEC efficiency   **7**.9
AMTEC specific resistivity   **7**.9
AMTEC thermodynamics   **7**.12...14
AMTEC, conduction losses   **7**.10...11
AMTEC, electrode efficiency   **7**.10...11
AMTEC, parasitic losses   **7**.10
AMTEC, radiation losses   **7**.10...11
Anaerobic digestion   **13**.14
Anemometers   **15**.8
Angle of attack   **15**.16
Anion-exchange membranes   **9**.28
Anode   **9**.3
Anode reaction   **9**.3
Anomalistic year   **12**.31
Anomaly, mean   **12**.35
Anomaly, true   **12**.28
Anti-knock   **3**.13
Apastron   **12**.28
Aphelion   **12**.28
Apoapsis   **12**.28
Apocenter   **12**.28
Apofocus   **12**.28
Apogee   **12**.28
AquaBuOY   **16**.6
AquaEnergy   **16**.6
Arable land   **1**.8
Archea   **13**.15
Architecture, batteries of thermocouples   **5**.33
Argument of perihelion   **12**.31
Aromatic hydrocarbons   **13**.2
Ascending node   **12**.31
Aspect ratio, wind turbine   **15**.30
Aswan   **10**.16
ATR, Autothermal reaction   **10**.4
Atriplex rosea   **13**.24
Atriplex triangularis   **13**.26
Attitude and orbital control, SPS   **14**.50
Augustus, (Octavian)   **12**.24
Automotive power plants, fuel cell   **9**.38...40
Autothermal reaction, ATR   **10**.4
Autumnal equinox   **12**.31
Available (radio) noise   **8**.1, 2
Available power density, wind   **15**.12
Available power density, wind turbine   **15**.15
Available power, wind turbine   **15**.14...15
Available wind power   **15**.8, 26
Avogadro's number   **2**.3, **9**.43

Axial compressor   **3**.7
Azimuth, solar   **12**.6
Back emission   **6**.26, 32
Back reaction   **10**.34
Bacteria, facultative anaerobes   **13**.15
Bacteria, obligate anaerobes   **13**.15
Bagasse   **13**.9
Ballard   **9**.25
Bandwidth   **8**.1, 2, 5
Bare work function   **6**.25
BASE   **7**.1. . .3
Batch digester   **13**.17
Battery architecture, thermocouples   **5**.33
Beam from space, SPS   **14**.46
Beam splitting   **14**.14
Bergen   **16**.8
β' alumina   **7**.1
Beta prime-phase   **11**.13
Beta-phase, hydrides   **11**.13
Binding energy, nuclear   **1**.24
Biochemistry   **13**.2. . .5
Biogas   **13**.1, 15
Biological filter   **13**.18
Biomass composition   **13**.1
Bipolar   **9**.13
Bipolar electrolyzer   **10**.17
Bipolar fuel cells   **9**.13
Black body   **12**.3
Black body spectrum efficiency   **14**.12
Blending octane value   **3**.15
Boeing   **15**.11
Boltzmann line   **6**.15
Boltzmann's constant   **2**.2
Boltzmann's law   **9**.66
Borden, Peter   **14**.43
Brayton cycle   **3**.6 , 7, 8
Breeder reactor   **1**.26
Brown Bovery   **10**.16
Brush, Charles   **15**.1
Building-integrated photovoltaics   **14**.2
Burned & cropped sugar cane   **13**.9
C3 & C4 plants   **13**.27
Calendar   **12**.23
Calendar, proleptic   **12**.25
Calvin-Benson cycle   **13**.25
Cape Canaveral   **11**.5
CARB   **3**.17
Carbohydrates   **13**.4
Carbon nanotubes   **11**.8
Carbon/nitrogen ratio   **13**.19
Carbon/phosphorus   **13**.19
Carbon-steam reaction   **10**.5

Carboxylic acid   **13**.3
Carnot efficiency   **3**.4
Carrier   **5**.14
Carrier (thermal) conductivity   **5**.16. . .17
Carrier multiplication, semiconductor   **14**.13
Cascade cells   **14**.15
Casson, Lionel   **1**.5
Catalysts for fuel cells   **9**.29. . .30
Catenary   **15**.7
Cathode   **9**.3
Cathode reaction   **9**.4
Cavitation   **16**.14
Celanese   **9**.7
Celestial equator   **12**.31
Cells, electrochemical   **9**.3
Cells, primary   **9**.4
Cells, secondary   **9**.4
Cellulose   **13**.2
Ceramic electrolytes, fuel cell reactions   **9**.10
Ceramic, proton conducting   **9**.17
Cesium adsorption   **6**.24. . .26
Cesium pressure   **6**.23. . .24
Chain reaction   **1**.25
Characteristics, thermionic
 V–J **6**.11, 14, 30. . .32
Chemical dissociation of water   **10**.31. . .32
Chemical nature of electrolyte, fuel cells   **9**.8
Chemical production of hydrogen   **10**.3. . .15
Child-Langmuir law   **6**.7. . .12
Chlamydomonas reihardtii   **10**.36
Chlorophyll   **13**.25
Chloroplast   **13**.21
Chord length   **15**.19
Chromium   **9**.19
Circuit model, fuel cell   **9**.61
Circulation, air   **12**.12
Clarke, Arthur C.   **14**.44
Clathrates   **1**.15. . .16
Close-cycle engines   **3**.5
Closed systems   **2**.17
CO removal   **10**.9. . .11
CO tolerance, fuel cell   **9**.14. . .15, 30
CO_2 disposal   **1**.20. . .22
CO_2 disposal, biological   **1**.20
CO_2 disposal, mineral   **1**.20. . .21
CO_2 disposal, subterranean   **1**.21. . .22
CO_2 disposal, undersea   **1**.22
CO_2 removal   **10**.8. . .9
Coal hydrogenation   **10**.8
Coanda effect   **15**.18
Coanda, Henri-Marie   **15**.18
Cobequid Bay   **16**.11

Coefficient of performance **5**.25
Coefficient of performance,
   hydride heat pump **11**.30. . .31
Coefficient of viscosity **15**.19
Coefficients, conversion **1**.2
Cold fusion **1**.31. . .36
Collector (thermionic) **6**.2, 6
Collectors, solar, flat **12**.14
Colliding beam fusion reactor **1**.30
Collision ionization **6**.23
Combustion, stratified **3**.11
Cominco **10**.16
Compound parabolic concentrators **12**.17
Compression ratio **3**.9
Compression ratio, critical (CRC) **3**.14
Compressor, axial **3**.6
Compressor, hydrogen (hydride) **11**.25
Compressor, radial **3**.7
Concentration **2**.2
Concentration cell voltage **7**.1
Concentration, intrinsic carrier **14**.22
Concentration, solar collectors **12**.15
Concentration-differential cell **7**.1
Concentration-differ.l electrol. **10**.22. . .23
Concentrator, conical **12**.17
Concentrators, compound parabolic **12**.17
Concentrators, solar **12**.15
Condensation, heat of **9**.42. . .43, 71
Conductance, heat **5**.2
Conductance, stomatal **13**.22
Conduction losses, AMTEC **7**.10. . .11
Conduction, electric **5**.10. . .12
Conduction, heat **6**.22
Conductivity, electric **5**.14
Conductivity, thermal **5**.13
Conductivity, thermal, semicond. **5**.16. . .17
Configurations, fuel cells **9**.12. . .37
Conical concentrator **12**.17
Contact ionization **6**.23
Contact potential **9**.67 , **14**.21
Continuous digester **13**.17
Conversion coefficients **1**.2
Conversion, hydrocarbon-to-alcohol **3**.3
Coverage, degree of **6**.24. . .25
Cryogenic engines **3**.25. . .28
Cryogenic hydrogen **11**.5
Crystalline photovoltaic materials **14**.6
Culmination (of the sun) **12**.26
Current from a fuel cell **9**.57. . .58
Cutoff frequency, lower **8**.5
Cutoff frequency, upper **8**.1, 5
Cutoff ratio **3**.10

Cuzco **10**.16
Darrieus wind turbine **15**.7
De Rosier, J. P. **10**.3
Declination **12**.31
Declination, solar **12**.6
Degrees of freedom **2**.1
Demag **10**.16
Demonstration (KOH) fuel cell **9**.12
Departure from ideal,
   mech. heat engines **3**.11. . .13
Depleted region **11**.13
Depolarizer **9**.5
Desorption **11**.11
Desulfurization **10**.8
Detonation **3**.13. . .14
Diaphragms, electrolyzer **10**.17
Diffusers, gas compressors **3**.7
Diffusion constant **14**.37
Digester, batch **13**.17
Digester, continuous **13**.17
Digester, power density **13**.17
Digestion, anaerobic **13**.14
Dillon, William P. **1**.15
Diode, high-pressure, plasma **6**.23
Diode, low-pressure, plasma **6**.23
Diodes, vacuum **6**.22
Diodes, vapor **6**.23. . .36
Direct hydrogenation of coal **10**.8
Direct methanol fuel cell **9**.6
Direct reduction loan **1**.37. . .38
Directions and signs, thermocouples **5**.43
Dispersive system **16**.2
Dispersive, ocean waves **16**.2
Displacer, Stirling **3**.19
Dissociated alcohols **13**.13
District heating **1**.18
Donor **14**.20
Dopants **14**.21
Dormancy **11**.2
Dow ion-exchange membrane **9**.28
Drag **15**.16
Drag coefficient **15**.14, 16
Drag, induced **15**.22
Drag, parasitic **15**.22
Drag, pressure **15**.22
Drag, skin-friction **15**.22
Drag-type wind turbines **15**.4. . .6
Drift velocity **5**.14
Drive train **3**.12. . .13
Duration, wind **16**.1
Dynamic braking **9**.39
Dynamic equilibrium **9**.67

Dynamic forces   **15**.19
Dynamic pressure   **15**.13
Eccentricity, orbital   **12**.36
Ecliptic   **12**.31
Ecliptic longitude   **12**.31
Economy of scales, wind turbines   **15**.3
Effective emissivity   **6**.18
Effective octane number   **3**.15
Efficiency of practical fuel cells   **9**.58
Efficiency, AMTEC   **7**.9
Efficiency, ideal, photodiode   **14**.32
Efficiency,
    maximum of a thermocouple   **5**.10
Efficiency, photocell   **14**.7
Efficiency, photodiode,
    light power dependence   **14**.31
Efficiency, photodiode,
    load match dependence   **14**.39
Efficiency, photodiode,
    rev. sat. curr. dependence   **14**.33
Efficiency, photodiode,
    temperature dependence   **14**.36
Efficiency, practical, photodiode   **14**.42
Efficiency, reversible fuel cells   **9**.46
Efficiency, wind turbine   **9**.8, 15
Egg beater wind turbine   **15**.7
Eichhornia crassipes   **13**.20
Einstein's relation   **7**.2
Electric conductivity   **5**.14
Electrochemical cells   **9**.3, 4
Electrode efficiency, AMTEC   **7**.10. . .11
Electrodes, electrolyzers   **10**.18
Electrolyte   **9**.3
Electrolytes, electrolyzers   **10**.17
Electrolytic hydrogen   **10**.16. . .25
Electrolyzer   **9**.45
Electrolyzer diaphragms   **10**.17
Electrolyzer efficiency   **10**.19. . .20
Electrolyzer, reversible   **9**.45
Electron-rich regimen   **6**.30
Electrons, excess energy   **6**.21
Emission constant, theoretical   **6**.4
Emission limited thermionic device   **6**.7
Emission, back   **6**.26, 32
Emission, ion   **6**.28. . .29
Emission, thermionic   **6**.3
Emissivity   **6**.17
Emissivity, effective   **6**.18
Emissivity, spectral   **6**.17
Emissivity, total   **6**.17
Emitter (thermionic)   **6**.2, 6
Emitter, virtual   **6**.10

End use load   **3**.12, 13
Energy balance, planetary   **1**.4
Energy converters, wave   **16**.4. . .10
Energy converters, wave, offshore   **16**.5. . .7
Energy density, ocean waves   **16**.4
Energy flux, solar   **12**.1
Energy resources   **1**.13. . .15
Energy resources, planetary   **1**.13
Energy utilization   **1**.16. . .18
Energy utilization rate, history   **1**.5
Energy utilization rate, world   **1**.8
Energy, ocean waves   **16**.4
Engine types   **3**.5. . .8
Englehard   **9**.13. . .14
Enrichment of syngas   **10**.6
Enthalpy   **2**.16. . .17, **9**.42
Enthalpy dependence on pressure   **9**.49
Enthalpy dependence on
    temperature   **9**.47. . .49
Enthalpy of formation   **9**.42
Entropy   **1**.4, **3**.19. . .21, **9**.44
Enzymes as catalyst for fuel cells   **9**.30
Eolergometry   **15**.8
Ephemeris time   **12**.22
Epidermis (on a plant leaf)   **13**.21
Equation of center   **12**.36
Equation of time   **12**.5, 33
Equilibrium constant   **10**.28
Equinoctial line   **12**.31
Equipartition of energy   **2**.1
Equivalent weight,
    ion-exchange membrane   **9**.28
Esters   **13**.4
Ethanol   **13**.8. . .13
Ethanol production   **13**.8. . .13
European Marine Centre   **16**.6
Eutectic   **12**.19
Eutrophication   **13**.19
EV   **3**.17
Evacuated tubes solar collectors   **12**.15
Excess energy of electrons   **6**.21
Exchange currents   **9**.68
Exiton   **14**.13
Exothermic reaction   **9**.42
External combustion engines   **3**.6
Extinguished mode   **6**.36
Extra loss, fuel cell   **9**.71
Facultative anaerobes   **13**.15
Faraday   **9**.43
Fast nuclear reactor   **3**.5
Fermentation   **13**.11. . .13
Fermi level   **6**.12

Fermi-Dirac heat capacity   **5**.39
Fertile materials   **1**.25
Fertilizers, nitrogen   **1**.9
Fetch (ocean waves)   **16**.1
Figure of merit   **5**.9
Figure of merit of a material   **5**.11
Fill factor   **14**.30
Filter press electrolyzer   **10**.17
Filtered cells   **14**.16
Filters, nonporous   **10**.9
Financing   **1**.36. . .38
First law of thermodynamics   **2**.4
Fischer-Tropsch   **10**.7
Fisher-Pry   **1**.9. . .11
Fission, thorium   **1**.27
Fission, uranium   **1**.25
Fissionable elements   **1**.25
Fissionables, reserve   **1**.14
Fixation of CO2   **13**.23
Fixed solids   **13**.17
Flat solar collectors   **12**.14
Fleishmann   **1**.31. . .32
Flux   **15**.12
Forces, dynamic   **15**.19
Forces, viscous   **15**.19
Fossil fuel reserves   **1**.14
Fourier, Jean   **5**.2
Franker, P. L.   **16**.11
Free energy   **9**.43
Free energy (reaction)
   dependence on pressure   **9**.54. . .55
Free energy (reaction)
   dependence on T   **9**.50. . .51
Free piston Stirling   **3**.18, 19
Freedom, degrees of   **2**.1
Fructose   **13**.5
Fuel cell (KOH), demonstration   **9**.12
Fuel cell advantages   **9**.38
Fuel cell applications   **9**.37. . .40
Fuel cell configurations   **9**.12
Fuel cell construction (SPFC)   **9**.25. . .31
Fuel cell current   **9**.57
Fuel cell reactions   **9**.8
Fuel cell reactions, acid electrolyte   **9**.9
Fuel cell reactions, alkaline electrolyte   **9**.9
Fuel cell reactions, ceramic electrolyte   **9**.10
Fuel cell reactions, methanol   **9**.10
Fuel cell reactions,
   molten carbonate electrolyte   **9**.10
Fuel cell, *V–I*
   characteristics   **9**.59. . .62, 63. . .70
Fuel cell, activation voltage   **9**.63. . .66

Fuel cell, alkaline electrolyte   **9**.6, 9, 63
Fuel cell, ammonia tolerance   **9**.15, 30
Fuel cell, chemical nature of electrolyte   **9**.8
Fuel cell, circuit model   **9**.61
Fuel cell, CO tolerance   **9**.14, 15, 30
Fuel cell, direct methanol   **9**.31. . .33
Fuel cell, effective surface area   **9**.60
Fuel cell, extra loss   **9**.71
Fuel cell, heat dissipation   **9**.71. . .73
Fuel cell, heat of condens.   **9**.42. . .43, 71
Fuel cell, heat rejection   **9**.71
Fuel cell, internal resistance   **9**.63. . .64, 71
Fuel cell, Joule loss   **9**.71
Fuel cell, kinetics   **9**.6, 62
Fuel cell, load voltage   **9**.63
Fuel cell, lossy   **9**.46
Fuel cell,
   open-circuit voltage   **9**.59, 60, 62, 63
Fuel cell, practical, efficiency   **9**.58, 59
Fuel cell, state of electrolyte   **9**.7
Fuel cell, stationary power plants   **9**.38
Fuel cell, sulfur tolerance   **9**.15
Fuel cell, temperature of operation   **9**.6
Fuel cell, thermodynamic heat power   **9**.71
Fuel cell, type of fuel   **9**.7
Fuel cells for automotive use   **9**.38. . .40
Fuel cells, alkaline   **9**.6
Fuel cells, bipolar   **9**.13
Fuel cells, efficiency (reversible)   **9**.46. . .47
Fuel cells, molten carbonate   **9**.6, 15. . .16
Fuel cells, phosphoric acid   **9**.6, 13. . .15
Fuel cells, solid oxide   **9**.6, 16. . .25
Fuel cells, solid polymer   **9**.6, 25. . .31
Fuel processor   **9**.29, 31
Fuel processor efficiency   **10**.4
Fuel processor, compact   **10**.11, 13. . .15
Full-wave rectifier   **8**.3
Functions of State   **2**.15. . .16
Fundy Bay   **16**.11
Fusion, cold   **1**.31. . .36
Fusion, hot   **1**.27. . .30
Fusion, piezonuclear   **1**.32
Göttingen   **15**.16
Gammas for hydrogen,
   oxygen and water vapor   **9**.53
Garbage dumps   **13**.19
Gas constant   **2**.3
Gas flow plate   **9**.27
Gasifiers, wood   **13**.7. . .8
Gasohol   **3**.15. . .16
Gasoline   **3**.13. . .15
Gasoline, leaded   **3**.15

General Electric   **10**.21, 22, 24, 25
Generator, thermoelectric   **5**.8
Geocentric system   **12**.5
Glazer, Peter   **14**.45
Glenfjord   **10**.16
Glucose   **13**.5
Glyceraldehyde   **13**.11
Glyceric acid   **13**.11
GMT, Greenwich mean time   **12**.23
Gravimetric concentration   **11**.2
Gravimetric storage capacity   **11**.9
Gray body   **6**.17
Green pricing   **15**.2
Greenwich mean time, GMT   **12**.23
Gregorian calendar   **12**.24
Group velocity, ocean waves   **16**.1. . .2
Groves, Sir William   **9**.2
Growian   **15**.4. . .8
Gruman   **15**.8
Gulf Stream   **4**.1
Gyromill   **15**.7
Hahn, Otto   **1**.25
Half-cell reaction   **9**.3
Half-reaction   **9**.3
Half-wave rectifier   **8**.3
Hammerfest Stroem   **16**.12
Harber-Bosch   **10**.16
Head, hydroelectric   **4**.6. . .7
Heat capacity of hydrides   **11**.20, 24
Heat capacity, Fermi-Dirac   **5**.39
Heat conductance   **5**.2
Heat conduction   **6**.22
Heat content   **2**.17
Heat dissipation, fuel cell   **9**.71. . .73
Heat emissivity   **8**.3, 4
Heat engine, solar   **12**.18
Heat exchangers, OTEC   **4**.9. . .10
Heat flow measurement   **5**.5
Heat of combustion   **3**.1, 11
Heat of combustion, higher   **3**.1,   **9**.42
Heat of combustion, lower   **3**.1,   **9**.42
Heat of condensation, fuel cells   **9**.42. . .43
Heat of vaporization   **9**.43
Heat pump, hydride   **11**.29
Heat pumps, thermoelectric   **5**.25. . .32
Heat radiation   **6**.17. . .19
Heat rejection, fuel cell   **9**.71
Heat storing wall   **12**.12
Heaving buoy converters   **16**.5
Heavy-metal nuclear reactor   **1**.23,   **3**.5
Hebrides   **16**.9
Height, ocean waves   **16**.1, 3

Heliocentric system   **12**.5
Heliostat   **12**.18
Hemicellulose   **13**.2
Hero of Alexandria   **1**.5
Hexose   **13**.4
Higher heat of combustion   **3**.1,   **9**.41
High-pressure diodes   **6**.34. . .36
Hindenburg   **10**.2
Holographic concentrators   **14**.17
Holographic solar collectors   **12**.16
Horizontal forces on marine turbines   **16**.13
Horizontal surface, insolation   **12**.10
Hot fusion   **1**.27. . .30
Hour angle, solar   **12**.6
Hybrid automotive engines   **3**.17. . .18
Hydrated methane   **1**.15. . .16
Hydraulic turbines, OTEC   **4**.2. . .3
Hydride heat pump   **11**.29
Hydride hydrogen compressor   **11**.25
Hydride syst. thermodynamics   **11**.21. . .24
Hydrides, hydrogen storage   **11**.11. . .31
Hydrocarbons   **13**.2
Hydrocarbons, alkane   **13**.2
Hydrocarbons, alkene   **13**.2
Hydrocarbons, alkyne   **13**.2
Hydrocarbons, aromatic   **13**.2
Hydrocarbons, oxidation stages   **13**.3
Hydroelectric head   **4**.6. . .7
Hydrogen adsorption   **11**.2
Hydrogen adsorption storage   **11**.7. . .8
Hydrogen advantages   **10**.1
Hydrogen carriers   **11**.2, 10
Hydrogen compression   **11**.2, 3
Hydrogen compressor, hydride   **11**.25
Hydrogen extraction   **10**.9. . .11
Hydrogen gas stations   **10**.17,   **11**.25
Hydrogen liquefaction   **11**.2
Hydrogen only systems   **11**.2
Hydrogen permeation rate   **10**.10
Hydrogen production   **10**.1
Hydrogen production history   **10**.3
Hydrogen production processes   **10**.2
Hydrogen production sequence   **10**.11
Hydrogen purification   **10**.8. . .11
Hydrogen safety   **10**.1, 2
Hydrogen transfer rate   **10**.10
Hydrogen underground storage   **11**.3, 4
Hydrogen uses   **10**.1
Hydrogen, chemical prod. of   **10**.3. . .15
Hydrogen, cryogenic   **11**.5
Hydrogen, electrolytic   **10**.16. . .25
Hydrogen/metal ratio   **11**.13

Hydrogenase   **9**.30,   **10**.36
Hydrogenation of coal   **10**.8
Hydrolysis   **13**.4
Hydronium   **9**.3, 30
Hyperthermophiles   **13**.15
Hysteresis, sorption   **11**.13
ICV   **3**.17
Idatech   **10**.10, 14
Ideal fuel cell   **9**.43, 44, 45, 56, 59
Ideal-diode efficiency   **14**.32
Ignited mode   **6**.36
Impact ionization, semiconductor   **14**.13
Induced drag   **15**.22
Induced velocity   **15**.24
Insolation   **12**.4
Insolation, horizontal surface   **12**.10
Insolation, stationary surface   **12**.7
Insolation, sun tracking surface   **12**.7
Insulation, building   **12**.13
Interelectrode potential   **6**.12. . .13
Internal combustion engines   **3**.6
Internal energy   **2**.3, 4
Internal resistance, fuel cell   **9**.63. . .64, 71
Intrinsic carrier concentration   **14**.22, 37
Inverter, utility-intertie   **14**.2
Ion current magnitude   **6**.31
Ion emission   **6**.28. . .29
Ion exchange membranes   **10**.18
Ionization, collision   **6**.23
Ionization, contact   **6**.23. . .27
Ion-rich regimen   **6**.30
Ion-richness parameter   **6**.30
Iron/water reaction   **10**.3
Isentropic efficiency   **4**.6
Isentropic processes   **2**.11,   **4**.5. . .6
Isentropic turbine   **4**.5. . .6
Isooctane   **3**.14. . .15
Isothermal compression   **9**.55
Isothermal processes   **2**.13. . .15
Isotopes, uranium   **1**.26
Jaques Charles   **10**.3
Jones   **1**.31. . .32
Joule effect   **5**.2
Joule loss, fuel cell   **9**.71
Joule, James   **5**.2
Julian calendar   **12**.25
Julian day number   **12**.25
Julius Caesar   **12**.24
Kelvin, Lord   **5**.39
Kelvin's relations   **5**.39. . .42
Ketone   **13**.3
Kinematic Stirling   **3**.18, 19

Kinetics, fuel cell   **9**.6, 62
Knocking   **3**.11, 12. . .14
Knudsen condition   **7**.5
Lactic acid   **13**.12
Langmuir assumption   **7**.7
Latent heat of condensation   **9**.43
Latent heat of vaporization   **9**.43
Lattice (thermal) conductivity   **5**.16. . .17
Lavoisier, A. L.   **9**.2
Leaded gasoline   **3**.15
Leclanché cells   **9**.5
Lifetime of minority carriers   **14**.25, 37
Lift   **15**.16
Lift coefficient   **15**.16
Lift-type wind turbines   **15**.6. . .7
Light absorption, semiconductors   **14**.7
Lignin   **13**.2
LIMPET   **16**.9
Liquid-fed AMTEC   **7**.3
Load voltage, fuel cell   **9**.63
Loading rates, digesters   **13**.17
Loan, direct reduction   **1**.37. . .38
London, A. L.   **4**.7
Longitude of perihelion   **12**.31
Longitude, ecliptic   **12**.31
Lorenz number   **5**.14,15
Lorenz, Ludwig   **5**.12
Losses in vacuum-diodes   **6**.16. . .22
Louis XVI   **10**.3
Lower cutoff frequency   **8**.5
Lower heat of combustion   **3**.1,   **9**.42
Lowest temperature, thermoelectric   **5**.26
Low-pressure diodes   **6**.23. . .34
Lunation   **16**.12
Lurgi   **10**.16, 18, 24
Marchetti, C.   **1**.11. . .12
Magnus effect   **15**.35. . .36
Magnus effect wind turbine   **15**.7. . .8
Manatee   **13**.20
Manzanares   **12**.18
Marine Current Turbine   **16**.13
Market penetration function   **1**.9. . .13
Marquis d'Arlandes   **10**.3
Maximum heat pumping, thermoelectric   **5**.26
MCFC, molten carbonate
   fuel cells   **9**.6, 15. . .16
MEA, membrane/electrode assembly   **9**.27
Mean anomaly   **12**.35
Mean free path of electron   **8**.2, 4
Mean free path of sodium ions   **7**.6
Mean solar hour   **12**.26
Mean sun   **12**.5

Melis, Anastasios   **10**.36
Membrane, proton conducting   **9**.28
Membrane/electrode assembly (MEA)   **9**.27
Mensis intercalaris   **12**.24
Mercedonius   **12**.24
Mesophilic   **13**.18
Methanation   **10**.7
Methane   **13**.24
Methane yield   **13**.17
Methane, hydrated   **1**.15. . .16
Methanol   **10**.7
Microwave generation, SPS   **14**.47
Mixture (Otto cycle)   **3**.11
Moderators, nuclear   **1**.25
Molten carbonate electrolytes,
    fuel cell reactions   **9**.10
Molten carbonate fuel cells   **9**.6, 15. . .16
Monosaccharides   **13**.4
Montgolfier   **10**.3
Morveau, G.   **9**.2
Multiple streamtubes   **15**.29
n-0ctane   **3**.15
NACA   **15**.16
Nafion   **9**.28
NASA   **15**.16
Neutralization, space charge   **6**.29. . .30
Nickel Metal Hydride Battery
    (NiMH)   **9**.34. . .36
Niter   **12**.19
Nitrogen fertilizers   **1**.9
Non-imaging solar concentrators   **12**.17
Nonporous filters   **10**.9
Norsk Hydro   **10**.16
Norwave   **16**.8
n-region   **14**.20
Nuclear binding energy   **1**.24
Nuclear fission   **1**.22, 25. . .27
Nuclear fusion   **1**.22, 27. . .30
Nuclear masses   **1**.24
Nuclear reactor, fast   **3**.5
Nuclear reactor, heavy metal   **1**.23,   **3**.5
Obligate anaerobes   **13**.15
Obliquity   **12**.31
Obliquity, orbital   **12**.37
Ocean current energy   **16**.11. . .16
Ocean Power Delivery   **16**.6
Ocean temperature profile   **4**.1
Ocean wave energy   **16**.4
Ocean wave velocity   **16**.2
Ocean waves, height   **16**.3
Octane   **3**.14. . .16
Octane rating   **3**.14

Octane value, blending   **3**.15
Octavian (Augustus)   **12**.24
Off-grid electric systems   **14**.2
Offshore wave-energy converters   **16**.5. . .7
Oleofin hydrocarbons   **13**.2
OPEC   **1**.6
Open systems   **2**.17
Open-circuit voltage, AMTEC   **7**.2
Open-circuit voltage, fuel cell   **9**.59
Open-circuit voltage, thermionic   **6**.14. . .15
Open-cycle engines   **3**.5
Operating temperatures, thermionic   **6**.5
Orbital eccentricity   **12**.36
Orbital obliquity   **12**.37
Organic polymers
    photovoltaic materials   **14**.6
Ortho-hydrogen   **11**.5
Oscillating water column converters   **16**.8
Osmotic engine   **16**.19. . .21
Osmotic pressure   **16**.20
OTEC configurations   **4**.2. . .4
OTEC design   **4**.7. . .9
OTEC efficiency   **4**.6. . .7
OTEC siting   **4**.10. . .11
OTEC, dependence on ΔT **4**.7
OTEC, heat exchangers   **4**.9. . .10
OTEC, hydraulic turbines   **4**.2. . .3
OTEC, vapor turbines   **4**.3. . .4
OTEC, water volume   **4**.7
Otto cycle   **3**.9. . .13
Overall reaction   **9**.4
Overtopping converters, wave   **16**.7
Over-voltage   **10**.20
Oxidation   **9**.2
Ozone   **12**.1
P–V Diagrams   **2**.11
PAFC, phosphoric acid fuel cells   **9**.6, 13. . .15
Palladium filter   **10**.10. . .11
Palladium hydrogen filter   **9**.29
Paraffin   **13**.2
Para-hydrogen   **11**.5
Parallel hybrid vehicles   **3**.17
Parasitic drag   **15**.22
Parasitic losses, AMTEC   **7**.10
Partial oxidation, POX   **10**.4. . .5
Pelamis   **16**.6
Peltier coefficient   **5**.4
Peltier effect   **5**.4, 37
Peltier, Jean   **5**.4
Pentose   **13**.4
Perez Alfonso   **1**.6
Perfect-gas law   **2**.2

Performance factor   **11**.3
Periapsis   **12**.28
Periastron   **12**.28
Pericenter   **12**.28
Perifocus   **12**.28
Perigee   **12**.28
Perihelion   **12**.28
Period, ocean waves   **16**.1
Permeation rate, hydrogen   **10**.10
Perovskite   **9**.20
Persian wind turbine   **15**.5
Perveance   **6**.10
Phase velocity, ocean waves   **16**.1. . .2
Phonon-electron interaction   **5**.36
Phonons   **5**.36
Phospho-enol-pyruvate (PEP)   **13**.26
Phosphoglyceric acid (PGA)   **13**.25
Phosphoric acid
    fuel cells (PAFC)   **9**.6, 13. . .15
Photobiological
    hydrogen production   **10**.35. . .36
Photocell efficiency   **14**.7
Photocells, theoretical eficiency   **14**.6
Photodiode   **14**.7, 20. . .40
Photodiode model   **14**.26, 28
Photolytic hydrogen   **10**.33. . .35
Photorespiration   **10**.35,   **13**.27
Photosynthesis   **13**.21. . .28
Photosynthetic efficiency   **13**.24
Photovoltaic diode   **14**.7
Photovoltaic materials,
    amorphous thin film   **14**.6
Photovoltaic materials,
    crystalline photovoltaic materials   **14**.6
Photovoltaic materials,
    organic polymers   **14**.6
Photovoltaic plant, utility-scale   **14**.4. . .5
Photovoltaic systems,
    building-integrated   **14**.2
Physics of thermoelectricity   **5**.33
Piezonuclear fusion   **1**.32
Planetary energy balance   **1**.4
Planetary energy resources   **1**.13
Plank's equation   **14**.7
Plasma diode, high-pressure   **6**.23
Plasma diode, low-pressure   **6**.23
Plasma drop   **6**.35
Plateau   **11**.13
Plutonium   **1**.26
Poisson's equation   **6**.9
Polybenzimidazole   **9**.7
Polytropic law   **2**.11, 12. . .13

Pons, S.   **1**.31. . .32
Population growth   **1**.8
Power delivered, wind turbine   **15**.12
Power density in the wind   **15**.12. . .13
Power density, digester   **13**.17
Power density, solar   **12**.1
Power density, wind   **15**.8
Power output, thermionic   **6**.15. . .16
Power piston, Stirling   **3**.19
Power, ocean waves   **16**.4
POX (partial oxidation)   **10**.4. . .5
Practical fuel cell   **9**.58
Practical fuel cells, efficiency   **9**.58
p-region   **14**.20
Pressure drag   **15**.22
Pressure drop in sodium column   **7**.4
Pressure, cesium   **6**.23. . .24
Pressure-swing adsorption, PSA   **10**.9
Pressure-volume work   **2**.5
Primary cells   **9**.1
Primer, semiconductor   **14**.53. . .59
Processes in heat engines   **3**.5. . .6
Proleptic calendar   **12**.25
Propanotriol   **13**.3
Proton-conducting ceramic   **9**.17
Proton-conducting membrane   **9**.28
PSA, pressure-swing adsorption   **10**.9
Psychrophillic   **13**.18
Pumping friction   **3**.12
Purification of hydrogen   **10**.8. . .11
Pyrolysis   **10**.8
Pyruvic acid   **13**.12
Radial compressor   **3**.7
Radiation losses   **6**.17. . .21
Radiation losses, AMTEC   **7**.10. . .11
Radiation of Heat   **6**.17. . .19
Radiation system, SPS   **14**.48
Radioactive decay   **1**.22
Radioisotope
    thermal generator, RTP   **1**.22,   **5**.20. . .21
Radionuclides in thermoelectrics   **5**.20
Radon   **12**.12
Rance River   **16**.11
Rankine cycle   **3**.5, 6, 7
Reaction rate   **13**.22
Reaction, anode   **9**.3
Reaction, cathode   **9**.4
Reaction, overall   **9**.4
Reactions, fuel cell   **9**.8
Reactions, fuel cell, acid electrolyte   **9**.9
Reactions, fuel cell,
    alkaline electrolyte   **9**.9

Reactions, fuel cell, ceramic electrol.   9.10
Reactions, fuel cell,
    molten carbonate electrolyte   9.10
Reactor, thermal nuclear   3.5
Receiving array, SPS   14.49
Rechargeable cells   9.1
Rectification of alcohol   13.10
Rectifier, full-wave   8.3
Rectifier, half-wave   8.3
Recuperator   12.13
Redox reaction   9.2
Reduction-oxidation reaction   9.2
Reference enthalpy of formation   9.1
Reference plane, airfoils   15.16
Reformate   9.23
Refrigerators, thermoelectric   5.25. . .32
Refuelable cells   9.1
Regenerator, Stirling   3.22
Reserve growth, fuel   1.14
Reserve, fissionables   1.14
Reserve, proved   1.13
Reserve, undiscovered   1.14
Reserves, fossil fuels   1.14
Retarding potential   6.13
Reverse saturation current   14.27, 40. . .42
Reversible electrolyzer   9.45
Reversible fuel cell   9.45
Reversible fuel cells, efficiency   9.46. . .47
Reversible voltage (fuel cell)   9.44
Reynolds number   15.18. . .20
Ribolose (RuDP)   13.25
Richardson's equation   6.4
Right ascension   12.31
Ringbom Stirling   3.18, 19
Robert brothers   10.3
Rotor loading   15.10
RTG, radioisotope
    thermal generator   1.22,   5.20. . .21
RTP   9.42
Rubbing friction   3.12
Sabatier, P.   10.7
Saccharomyces cerevisae   13.12
Safety, hydrogen   10.1, 2
Saha-Langmuir equation   6.29
Saint Malo   16.11
Salination energy   16.16. . .19
Saltpeter   12.19
Samaria-doped ceria, SDC   9.17, 23
Saturated hydrocarbons   13.2
Saturated region   11.13
Saturated thermionic device   6.7
Savonius wind turbine   15.5. . .6

Schiaparelli, G.   16.8
SDC, samaria-doped ceria   9.17, 23
Seadog   16.6
Seaflow   16.13
Secondary cells   9.1
Seebeck coefficient   5.4
Seebeck coefficient of metals   5.18
Seebeck coefficient of semiconductors   5.18
Seebeck coefficient, mean   5.5
Seebeck effect   5.34. . .36
Seebeck, Thomas   5.4
Series hybrid vehicles   3.17
Sewage treatment   13.19
Shift reaction   9.8,   10.6. . .7
Shoreline wave converters   16.8
SI   1.1
Sidereal time   12.26
Single stream tube   15.29
Skin-friction drag   15.22
Slope parameter   11.18
Smith-Putnam wind turbine   15.1
Sodium column, pressure drop   7.4
Sodium ions, mean free path   7.6
Sodium, vapor pressure   7.3
SOFC, solid oxide fuel cells   9.6, 16. . .25
Solar architecture   12.11
Solar azimuth   12.6
Solar chimney   12.18
Solar collectors   12.11
Solar collectors, evacuated tubes   12.15
Solar collectors, flat   12.14
Solar collectors, holographic   12.16
Solar concentrators   12.15
Solar concentrators, nonimaging   12.17
Solar constant   12.1
Solar constant, surface   12.4
Solar day   12.26
Solar declination   12.6
Solar energy conversion (space), SPS   14.46
Solar energy flux   12.1
Solar heat engine   12.18
Solar hour (mean)   12.22
Solar hour angle   12.6
Solar One   12.19
Solar photolysis   10.34. . .35
Solar ponds   4.10. . .11,   12.18
Solar power density   12.1
Solar power satellite (SPS)   14.44
Solar radiation   1.4
Solar spectral power density   12.1
Solar spectral power dens. spectrum   12.3
Solar spectrum   10.35

Solar Two   **12**.19
Solid oxide fuel cells   **9**.6, 16. . .25
Solid polymer
    electrolyzer (SPE)   **10**.18, 21, 22, 24, 25
Solid polymer fuel cells   **9**.6, 25. . .31
Solidity   **15**.26
Solvent extraction, syncrude   **10**.8
Sorption   **11**.11
Sorption hysteresis   **11**.13, 18
Sources of energy   **1**.13
Space charge limited   **6**.8, 29. . .30
Space charge neutralization   **6**.12
Space construction, SPS   **14**.50
Space plane   **11**.5
Space transportation, SPS   **14**.50
Specific heat
    at constant pressure   **2**.5,   **9**.52. . .53
Specific heat at constant volume   **2**.3
Specific resistance of fuel cells   **9**.62
Specific resistivity, AMTEC   **7**.9
Spectral emissivity   **6**.17
Spectral power density, solar   **12**.1 . . .3
Spectrum efficiency, ideal   **14**.19
SPFC, solid polymer fuel cells   **9**.6, 25. . .31
SPS, attitude and orbital control   **14**.50
SPS, beam from space   **14**.46
SPS, microwave generation   **14**.47
SPS, radiation system   **14**.48
SPS, receiving array   **14**.49
SPS, solar energy conversion   **14**.46
SPS, solar power satellite   **14**.44
SPS, space construction   **14**.50
SPS, space transportation   **14**.50
Stagnation point   **11**.7
Stagnation temperature   **11**.6. . .7
Stall, airfoil   **15**.18
Standard enthalpy of formation   **9**.42
Standard time   **12**.23
State (of a gas)   **2**.15
State of electrolyte, fuel cells   **9**.7
Stationary power plants, fuel cell   **9**.38
Stationary surface, insolation   **12**.7
Steam reforming   **9**.7,   **10**.5
Stefan-Boltzmann law   **6**.17,   **14**.8
Stefan-Boltzmann
    radiation constant   **8**.3, 6, 17
Stillage   **13**.11
Stirling engine   **3**.18. . .25
Stirling engine, implementation   **3**.23. . .25
Stoichiometric index   **11**.13
Stoma   **13**.22
Stomatal conductance   **13**.22

Storm waves   **16**.3
STP   **9**.42,   **15**.13
Stratified combustion   **3**.11
Strong acid   **9**.5
Sugar cane yield   **13**.9
Sulfur tolerance, fuel cell   **9**.15
Suntracking surface, insolation   **12**.7
Surface area, fuel cell   **9**.62
Surface solar constant   **12**.4
Swept area   **15**.8
Symbology, in thermodynamics   **9**.41
Syncrude   **10**.8
Syncrude, solvent extraction   **10**.8
Syngas   **9**.7,   **10**.4. . .6
Système International, SI   **1**.1
Tafel equation   **9**.64
Takahashi, Akito   **1**.35
Take over time interval   **1**.10
Tank electrolyzer   **10**.17
Tapchan   **16**.8
Tapered chanel converters   **16**.8
Teledyne   **10**.24
Temperature   **2**.1
Temperature dependence
    of thermocouples   **5**.32. . .33
Temperature of operation, fuel cells   **9**.6. . .7
Temperature profile, ocean   **4**.1
Temperature-swing adsorption, TSA   **10**.9
Tetraethyl lead   **3**.15
Theoretical eficiency, photocells   **14**.6
Thermal conductivity   **5**.13
Thermal conductivity (carrier)   **5**.16. . .17
Thermal conductivity (lattice)   **5**.16. . .17
Thermal conductivity
    in semiconductors   **5**.16. . .17
Thermal conductivity in solids   **5**.16. . .17
Thermal decomposition of alcohols   **10**.5
Thermal velocity of carriers   **5**.14
Thermalization of exitons   **14**.13
Thermionic V–J characteristics
    6.11, 14, 30. . .32
Thermionic emission   **6**.3
Thermionic materials   **6**.5
Thermionic operating temperatures   **6**.5
Thermocouple   **5**.3
Thermocouple, maximum efficiency   **5**.10
Thermocouple, optimizing   **5**.10
Thermocouples, database   **5**.8
Thermocouples, directions and signs   **5**.43
Thermocouples, standardized   **5**.6
Thermodynamic heat power, fuel cell   **9**.71
Thermodynamic of fuel cells   **9**.41. . .45

Thermodynamics
  of hydride systems   **11**.21. . .24
Thermodynamics, AMTEC   **7**.12. . .14
Thermoelectric generator   **5**.8
Thermoelectric generator design   **5**.22. . .24
Thermoelectric gen., applications   **5**.20
Thermoelectric heat pumping, max.   **5**.26
Thermoelectric, lowest temperature   **5**.26
Thermoelectricity physics   **5**.33
Thermolytic hydrogen   **10**.25. . .32
Thermometers, thermoelectric   **5**.6
Thermophilic   **13**.18
Thermophotovoltaic cell   **14**.17
Thomson coefficient   **5**.38. . .39
Thomson effect   **5**.5, 38. . .39
Thomson, William   **5**.39
Thorium fission   **1**.27
Threshold energy, impact ionization   **14**.13
Tidal energy   **16**.10
Tilt angle   **12**.31
Time offset   **12**.5
Time zones   **12**.22
Toftestallen   **16**.8
Topocentric system   **12**.5
Topping cycle   **5**.22
Total emissivity   **6**.17
Trail, BC   **10**.16
Triple alpha reaction   **1**.30. . .31
Triple contact   **9**.17
Tritium   **1**.27. . .29
Tropical year   **12**.26
Troposkein   **15**.7
True anomaly   **12**.28
TSA, temperature-swing adsorption   **10**.9
Tsunamis   **16**.3
Tunneling   **6**.27
Turbines   **4**.4
Turnaround efficiency   **11**.2, 6
Two-bucket wind turbine   **15**.5
Type of fuel, fuel cells   **9**.7
Underground storage, hydrogen   **11**.3, 4
Unignited mode   **6**.36
Unipolar electrolyzer   **10**.16, 17
Units   **1**.1
Universal time, UT   **12**.23
Unsaturated, emission limited   **6**.7
Upper cutoff frequency   **8**.1, 5
Uptake, carbon dioxide   **13**.22
Uranium fission   **1**.25
Uranium isotopes   **1**.26
Utility   **1**.2, 3
Utility-scale photovoltaic plants   **14**.5

Utility-scale photovoltaic systems   **14**.4
Utility-tie electric systems   **14**.2
Utilization factor   **14**.1
Utilization of energy   **1**.16. . .18
V–I characteristics, fuel cell   **9**.59. . .60
V–J characteristics, thermionic   **6**.11, 14
V–I characteristic,
  semiconductor diode   **14**.27, 29
V–I characteristics, AMTEC   **7**.7
V–I characteristics, fuel cell   **9**.59. . .60
Vacuum diodes   **6**.22
Vacuum state devices   **6**.2
van der Waal's forces   **2**.2
Vapor diodes   **6**.23. . .36
Vapor pressure, sodium   **7**.3
Vaporization, heat of   **9**.42. . .43
Variable-density wind tunnel   **15**.20
Variable-speed wind turbine   **15**.12
Vascular aquatic plants   **13**.20
Velocity, induced   **15**.24
Velocity, ocean waves   **16**.2
Virtual emitter   **6**.10
Viscosity, coefficient   **15**.19
Viscous forces   **15**.19
Vitruvius [Pollio], M.   **1**.5
Volatile solids   **13**.17
Volta cell   **9**.5
Volta, Alessandro   **9**.5
Voltage, open-circuit, thermionic diodes   **6**.14.
Volumetric concentration   **11**.2, 15
Volumetric storage capacity   **11**.9
Vortex-type wind machines   **15**.8
Walukiewicz, W.   **14**.15
Water dissociation   **10**.25. . .32
Water hyacinth   **13**.20
Water management, SPFC   **9**.30. . .31
Wave Dragon   **16**.7
Wave energy (ocean)   **16**.1
Wave energy converters   **16**.4. . .10
Wave energy converters,
  offshore   **16**.5. . .7
Wave energy converters,
  shoreline   **16**.8
Wave length, ocean wave   **16**.1, 2
Wavegen system   **16**.9. . .10
Wave-operated pump   **16**.5
Waves, ocean, height   **16**.3
Waves, ocean, velocity   **16**.2
Wells turbine   **16**.9. . .10
Wiedemann, Gustave   **5**.12
Wiedemann-Franz-Lorenz law   **5**.12. . .14
Wien's displacement law   **14**.9

Wind duration   **16**.1
Wind energy availability   **15**.9. . .10
Wind farms   **15**.2
Wind fetch   **16**.1
Wind machines, vortex-type   **15**.8
Wind power density   **15**.8
Wind power, available   **15**.8
Wind pressure   **15**.13
Wind tunnel, variable density   **15**.20
Wind turbine (eggbeater)   **15**.7
Wind turbine characteristics   **15**.10. . .12
Wind turbine configurations   **15**.4. . .8
Wind turbine efficiency   **15**.15
Wind turbine,
    available power   **15**.14. . .15
Wind turbine, available power density   **15**.15
Wind turbine, Darrieus   **15**.7
Wind turbine, efficiency   **15**.8
Wind turbine, Magnus effect   **15**.7
Wind turbine, Persian   **15**.5
Wind turbine, Savonius   **15**.5. . .6
Wind turbine, two-bucket   **15**.5
Wind turbines, drag-type   **15**.4. . .6
Wind turbines, lift-type   **15**.6. . .7
Wood composition   **13**.2
Wood gasifiers   **13**.7. . .8
Work function   **6**.2
Working fluids, heat engines   **3**.6. . .7
Yater, Joseph C.   **8**.1
YSZ, yttria stabilized zirconia   **9**.17
Yttria stabilized zirconia, YSZ   **9**.17
Zenith angle   **12**.5
ZEV   **3**.17
Zinc-acid cells   **9**.36. . .37
Zinc-air cells   **9**.36. . .37
Ziock, H. J.   **1**.21
ZT product   **5**.19
Zymomona mobilis   **13**.10